CONTROL AND
DYNAMIC SYSTEMS

Advances in Theory
and Applications

Volume 55

CONTRIBUTORS TO THIS VOLUME

UBAID M. AL-SAGGAF
CRAIG BARRATT
MAAMAR BETTAYEB
STEPHEN BOYD
REN-JUNG CHANG
BEN M. CHEN
BOR-SEN CHEN
HEPING DAI
ALAN J. LAUB
ROMEO ORTEGA
PRADEEP PANDEY
ALI SABERI
PEDDAPULLAIAH SANNUTI
YACOV SHAMASH
NARESH K. SINHA
MARIO SZNAIER
SPYROS G. TZAFESTAS
WEN-JUNE WANG
KEIGO WATANABE

CONTROL AND DYNAMIC SYSTEMS

ADVANCES IN THEORY AND APPLICATIONS

Edited by

C. T. LEONDES

School of Engineering and Applied Science
University of California, Los Angeles
Los Angeles, California
and
Department of Electrical Engineering
and Computer Science
University of California, San Diego
La Jolla, California

VOLUME 55: DIGITAL AND NUMERIC TECHNIQUES AND THEIR APPLICATIONS IN CONTROL SYSTEMS

Part 1 of 2

ACADEMIC PRESS, INC.
Harcourt Brace Jovanovich, Publishers
San Diego New York Boston
London Sydney Tokyo Toronto

ACADEMIC PRESS RAPID MANUSCRIPT REPRODUCTION

Academic Press, Inc.
1250 Sixth Avenue, San Diego, California 92101-4311

United Kingdom Edition published by
Academic Press Limited
24–28 Oval Road, London NW1 7DX

Library of Congress Catalog Number: 64-8027

International Standard Book Number: 0-12-012755-5

PRINTED IN THE UNITED STATES OF AMERICA

93 94 95 96 97 98 BB 9 8 7 6 5 4 3 2 1

CONTENTS

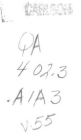

DIGITAL AND NUMERIC TECHNIQUES AND THEIR APPLICATIONS IN CONTROL SYSTEMS

EXTENDED CONTENTS

CONTRIBUTORS

Numbers in parentheses indicate the pages on which the authors' contributions begin.

Ubaid M. Al-Saggaf (51), *Electrical Engineering Department, King Fahd University of Petroleum and Minerals, Dhahran 31261, Saudi Arabia*

Craig Barratt (1), *Information Systems Laboratory, Department of Electrical Engineering, Stanford University, Stanford, California 94305*

Maamar Bettayeb (51), *Electrical Engineering Department, King Fahd University of Petroleum and Minerals, Dhahran 31261, Saudi Arabia*

Stephen Boyd (1), *Information Systems Laboratory, Department of Electrical Engineering, Stanford University, Stanford, California 94305*

Ren-Jung Chang (429), *National Cheng Kung University, Department of Mechanical Engineering, Tainan, Taiwan 701, Republic of China*

Ben M. Chen (195, 263), *Department of Electrical Engineering, State University of New York at Stony Brook, Stony Brook, New York 11794*

Bor-Sen Chen (355), *Department of Electrical Engineering, National Tsing-Hua University, Hsin-Chu, Taiwan, Republic of China*

Heping Dai (149), *Department of Electrical and Computer Engineering, McMaster University, Hamilton, Ontario Canada L8S 4L7*

Alan J. Laub (25), *Department of Electrical and Computer Engineering, University of California, Santa Barbara, California 93106*

Romeo Ortega (471), *Sophia University, Department of Mechanical Engineering, Chiyoda-ku, 102 Tokyo, Japan*

Pradeep Pandey (25), *Department of Mechanical Engineering, University of California, Berkeley, California 94720*

Ali Saberi (195, 263), *School of Electrical Engineering and Computer Science, Washington State University, Pullman, Washington 99164*

Peddapullaiah Sannuti (195, 263), *Department of Electrical and Computer Engineering, Rutgers University, Piscataway, New Jersey 08855*

Yacov Shamash (195, 263), *College of Engineering and Applied Sciences, State University of New York at Stony Brook, Stony Brook, New York 11794*

Naresh K. Sinha (149), *Department of Electrical and Computer Engineering, McMaster University, Hamilton, Ontario Canada L8S 4L7*

Mario Sznaier (305), *Department of Electrical Engineering, University of Central Florida, Orlando, Florida 32816*

Spyros G. Tzafestas (111), *Division of Computer Science, Department of Electrical Engineering, National Technical University of Athens, Zografou, Athens 15773, Greece*

Wen-June Wang (355), *Department of Electrical Engineering, National Central University, Chung-Li, Taiwan, Republic of China*

Keigo Watanabe (111), *Department of Mechanical Engineering, Faculty of Science and Engineering, Saga University, Honjomachi-1, Saga 840, Japan*

PREFACE

Effective control concepts and applications go back over millenia. One very familiar example of this is the windmill, which was designed to derive maximum benefit from windflow, a simple but highly effective optimization technique. Harold Hazen's paper of 1932 in the *Journal of the Franklin Institute* was one of the earliest reference points from which an analytical framework for modern control theory began to be established. There were many other notable items along the way, including the MIT Radiation Laboratory series volume on servomechanisms, the Brown and Campbell book, Bode's book, and Wiener's "yellow peril"—all published shortly after mid-1945. However, it remained for Kalman's papers of the late 1950s (wherein a foundation for modern state space techniques was established) and the tremendous evolution of digital computer technology (which was underpinned by the continuous giant advances in integrated electronics) for truly powerful control systems techniques for increasingly complex systems to be developed. Today we can look forward to a future that is rich in possibilities in a wide variety of areas of major significance, including manufacturing systems, electric power systems, robotics, aerospace systems, and many others with significant economic, safety, cost effectiveness, reliability, and many other implications.

In the 1940s and 1950s the primary techniques for the analysis and synthesis, or design of control systems were Nyquist plots, Bode diagrams, Nichol's charts, root locus techniques, describing function techniques, and phase plane techniques, among others. Basically, these techniques were confined, in their application, to relatively simple single-input–single-output (SISO) systems. Nevertheless, these techniques have continued and will continue to be highly effective, where they may be appropriately utilized. Furthermore, when these techniques can be effectively utilized, they can provide marvelous insights into a system's performance characteristics. In any event, with the trend toward the essential requirement for the analysis and design of increasingly complex multi-input–multi-output (MIMO) systems beginning, most strongly, with the emergence of many diverse advances in state space techniques as well as advances in integrated electronics and computer technology, the stage was set for many diverse and significant developments and advances in digital and numerical techniques for the analysis and design of modern complex control systems of a very wide variety. Thus, this volume is the first volume of a two-part sequence of volumes devoted to the most timely theme of "Digital and Numerical Techniques and Their Application in Control Systems."

The first contribution to this volume is "Closed-Loop Convex Formulation of Classical and Singular Value Loop Shaping," by Craig Barratt and Stephen Boyd. In this contribution significant numerical technique extensions and insights for the "classical" techniques, which were utilized in the 1940s and 1950s, are presented, and so this is a most appropriate contribution with which to begin this volume.

The next contribution is "Numerical Issues in Robust Control Design Techniques," by Pradeep Pandey and Alan J. Laub. This contribution describes numerical and computational techniques for H^∞ control, and it presents an in-depth treatment of a state-space approach to computing the optimal system gain.

The next contribution is "Techniques in Optimized Model Reduction for High Dimensional Systems," by Ubaid M. Al-Saggaf and Maamar Bettayed. The trend toward the necessity for effective techniques for the analysis and synthesis of large scale, complex systems of high dimension, even infinite dimension, i.e., distributed parameter systems, implies the essential requirement for effective modeling techniques which, at the same time, can be dealt with from a computational point of view. This contribution is an extensive treatment of this most fundamentally important issue, and it presents the most highly effective techniques along with a rather comprehensive review and analysis of the literature on this subject.

The next contribution is "Techniques for Adaptive Estimation and Control of Discrete-Time Stochastic Systems with Abruptly Changing Parameters," by Spyros G. Tzafestas and Keigo Watanabe. In the first paragraph of this preface, reference was made to Norbert Wiener's "yellow peril." This was a book published in the mid-1940s on Wiener filtering techniques, and it was colloquially referred to as the "yellow peril" because engineers who might use Wiener filtering techniques were unfamiliar with the abstract mathematics utilized in this book with a yellow cover. In any event, the Kalman filter techniques, which were introduced in the late 1950s and early 1960s, were a "giant step" forward. Since then there have been many important advances in the broad area of stochastic-system techniques. This contribution is a comprehensive treatment of techniques for adaptive state estimation and control problems with abruptly changing parameters, an issue that arises in many diverse applied circumstances.

The next contribution is "Robust Off-line Methods for Parameter Estimation," by Heping Dai and Naresh K. Sinha. The companion article to this contribution by the same authors appears in Volume 53 of this series. It presents six robust recursive identification methods, among other techniques. The contribution appearing here in Volume 55 is a detailed treatment of the essentially important issue of the development of techniques for robust off-line system parameter identification.

The next two contributions are "Loop Transfer Recovery for General Nonminimum Phase Discrete Time Systems, Part 1: Analysis" and "Loop Transfer Recovery for General Nonminimum Phase Discrete Time Systems, Part 2: Design." Both of these contributions are by Ben M. Chen, Ali Saberi, Peddapullaiah Sannuti, and Yacov Shamash. These two companion contributions offer an in-depth treatment of the analysis of and the design of target loop transfer functions — the so-called Loop Transfer Recovery (LTR) problem — for general not necessarily minimum discrete time systems, a broad issue of major importance in many applied instances in modern control system design.

The next contribution is "Set-Induced Norm Based Robust Control Techniques," by Mario Sznaier. Most realistic control problems involve both some type of time-domain constraints and a certain degree of model uncertainty. However, the majority of control design methods currently available focus only on one aspect of this problem. This contribution discusses effective techniques for dealing with both problems, and, as such, represents a contribution of substantial importance for this design problem of major applied significance. A number of important open research areas are also presented.

The next contribution is "Techniques for Robust Nonlinear Large Scale Systems," by Bor-Sen Chen and Wen-June Wang. As implied in the first paragraph of this preface, the demands of today's technology in the planning, design, and realization of sophisticated systems have become increasingly large in scope and complex in structure. This contribution presents a thorough treatment of many of the issues that must be dealt with effectively, and it presents important results for achieving robustness in the design of such systems.

The next contribution is "Extensions in Techniques for Stochastic Dynamic Systems," by Ren-Jung Chang. As noted earlier in this preface, the Kalman filter has led to many major extensions which are important from the point of view of the design and control of modern complex systems. This contribution is a rather comprehensive treatment of many of the extensions that have been achieved and their applied significance.

The concluding contribution for this volume is "Adaptive Control of Discrete-Time Systems: A Performance-Oriented Approach," by Romeo Ortega. Adaptive control theory has grown and expanded continuously since it was first introduced in the early 1960s. The reason for this is that many modern complex systems could benefit from "anthropomorphic-like" characteristics because of the necessity for dealing effectively with widely differing circumstances. This contribution presents a global theory of this broad subject, and it clearly displays the importance of key issues. This contribution is a most appropriate one with which to conclude this volume.

This first volume of a two-volume sequence of companion volumes rather clearly manifests the significance and the power of the digital and numerical techniques that are available and that are under continuing development for control systems. The coauthors are all to be commended for their splendid contributions to this volume that will provide a significant reference source for students, research workers, practicing engineers, and others on the international scene for years to come.

Closed-Loop Convex Formulation of Classical and Singular Value Loop Shaping

Craig Barratt
Stephen Boyd

Information Systems Laboratory
Department of Electrical Engineering
Stanford University
Stanford, CA 94305
U. S. A.

I. INTRODUCTION

In this chapter we show that control system design via *classical loop shaping* and *singular value loop shaping* can be formulated as a *closed-loop convex* problem [1, 2, 3]. Consequently, loop shaping problems can be solved by efficient numerical methods. In particular, these numerical methods can always determine whether or not there exists a compensator that satisfies a given set of loop shaping specifications. Problems such as maximizing bandwidth subject to given margin and cutoff specifications can be directly solved. Moreover, any other closed-loop convex specifications, such as limits on step-response overshoot, tracking errors, and disturbance rejection, can be simultaneously considered.

These observations have two practical ramifications. First, closed-loop convex design methods can be used to synthesize compensators in a framework that is familiar to many control engineers. Second, closed-loop convex

CONTROL AND DYNAMIC SYSTEMS, VOL. 55

design methods can be used to aid the designer using classical loop shaping by computing absolute performance limits against which a classical design can be compared.

We begin by giving a brief overview of classical and singular value loop shaping, which also serves to describe our notation.

A. CLASSICAL LOOP SHAPING

We first consider the standard classical one degree-of-freedom (DOF) single-actuator, single-sensor (SASS) control system shown in figure 1. Here u is the actuator signal, y is the output signal, e is the (tracking) error signal, r is the reference or command signal, and d_{sensor} is a sensor noise. The plant and compensator are linear and time-invariant (LTI), with transfer functions given by P and C, respectively. The plant is given and the compensator is to be designed.

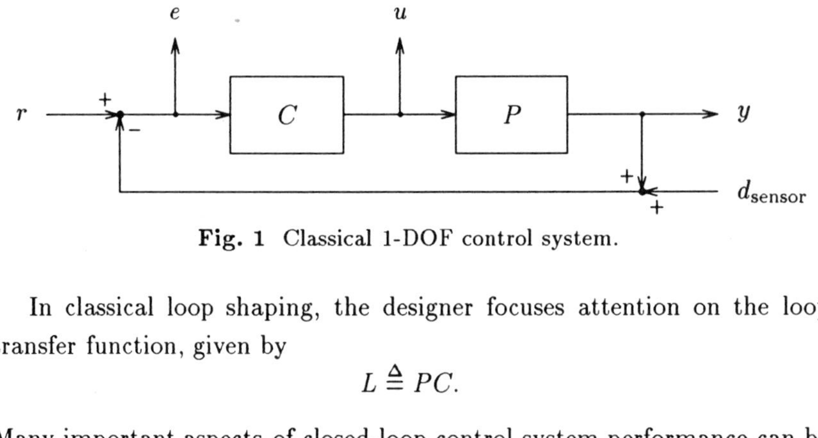

Fig. 1 Classical 1-DOF control system.

In classical loop shaping, the designer focuses attention on the loop transfer function, given by

$$L \triangleq PC.$$

Many important aspects of closed-loop control system performance can be expressed in terms of L. For example, stability of the closed-loop system can be determined from L (provided there are no unstable pole-zero cancellations between P and C). Several important closed-loop transfer functions can be expressed in terms of L.

The *transmission* or input/output (I/O) transfer function

$$T \triangleq L/(1+L)$$

is the closed-loop transfer function from the reference input r to the output

y. Its negative, $-T$, is the closed-loop transfer function from the sensor noise d_{sensor} to the output y.

The *sensitivity* transfer function is given by

$$S \triangleq 1/(1+L).$$

S is the transfer function from the reference input r to the tracking error e. The sensitivity derives its name from the important fact, observed by Bode [4], that to first order, the relative change in T is S times the relative change in P:

$$\frac{\delta T(s)}{T(s)} \simeq S(s)\frac{\delta P_0(s)}{P_0(s)},$$

or, equivalently, S is the first order percentage change in the I/O transfer function divided by the percentage change in the plant transfer function.

Classical loop shaping design is based on two important observations:

- the loop transfer function L has a very simple dependence on the compensator transfer function C, especially in a logarithmic (gain and phase) representation.

- many important requirements for the closed-loop system can be approximately reflected as requirements on the loop gain L.

Loop-shaping specifications constrain the magnitude and possibly the phase of the loop transfer function at each frequency. There are three basic types of loop-shaping specifications, which are imposed in different frequency bands:

- *In-band specifications.* At these frequencies we require $|L|$ to be large, so that S is small and $T \approx 1$. This ensures good command tracking, and low sensitivity to plant variations, two of the most important benefits of feedback.

- *Cutoff specifications.* At these frequencies we require $|L|$ to be small, so that T is small. This ensures that the output y will be relatively insensitive to the sensor noise d_{sensor}, and that the system will remain closed-loop stable in the face of plant variations at these frequencies, for example, excess phase from small delays and unpredictable (or unmodeled) resonances.

- *Crossover (margin) specifications.* Crossover or transition band specifications are imposed between the control bands (where L is large) and cutoff bands (where L is small). At these frequencies the main concern is to keep L a safe distance away from the critical point -1 (closed-loop stability depends on the winding number of L with respect to -1). Classical specifications include gain margin and phase margin. More natural "modern" specifications exclude L from some circle about -1. These modern specifications directly correspond to limiting the peaking of some closed-loop transfer function such as S or T.

The Nyquist criterion (which constrains the winding number of L about -1) is also included as an implicit specification that ensures closed-loop stability.

In many systems the in-band region is at low frequencies, from $\omega = 0$ to $\omega = \omega_B$, the cutoff region is at high frequencies, $\omega > \omega_C$, and the crossover region lies in between, from $\omega = \omega_B$ to $\omega = \omega_C$. In some designs, however, there may be more than one crossover region and one or more in-band and cutoff regions.

A typical set of loop shaping specifications is:

$$\begin{array}{ll} |L(j\omega)| \geq l(\omega) & \text{for } \; 0 \leq \omega \leq \omega_B = 2, \\ |L(j\omega)| \leq u(\omega) & \text{for } \; \omega \geq \omega_C = 5, \\ -150^\circ \leq \angle L(j\omega) \leq 30^\circ & \text{for } \; \omega_B = 2 \leq \omega \leq \omega_C = 5 \end{array}$$

where l and u are the frequency dependent constraint functions shown in figure 2. The in-band and cutoff constraints, which consist of frequency dependent restrictions on the magnitude of L, are conveniently shown on a Bode magnitude plot, while the margin constraint, which is often independent of frequency, is conveniently shown on a Nyquist plot (see figure 3).

In this example, the in-band region is $\omega \leq \omega_B$. Over this region, the large loop gain will ensure good command tracking ($T \approx 1$), and low sensitivity ($|S| \ll 1$). In the cutoff region, $\omega \geq \omega_C$, the small loop gain ensures that sensor noise will not affect the output, and small time-delays and variations in P will not destabilize the closed-loop system. In the in-band region, L cannot be close to the critical point -1 since $|L|$ exceeds +10dB there; similarly, in the cutoff region, $|L|$ is less than -10dB and so cannot

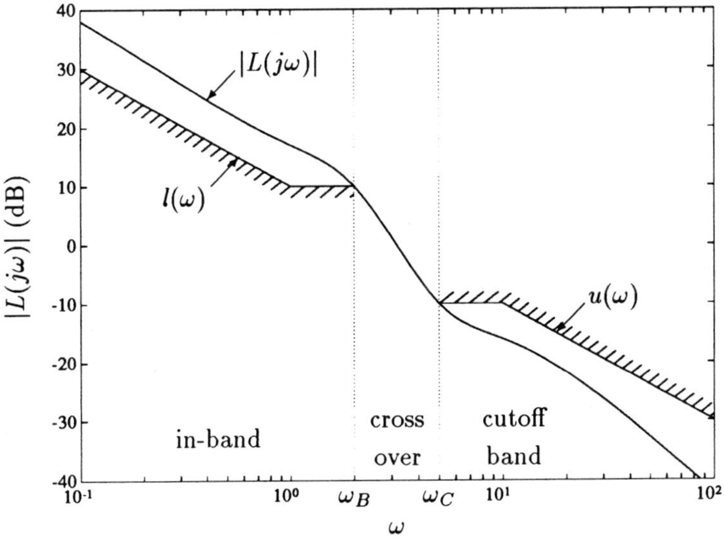

Fig. 2 A typical set of in-band and cutoff specifications. In the in-band region, $\omega \leq \omega_B$, the loop gain magnitude $|L|$ is required to exceed the frequency dependent lower bound $l(\omega)$. In the cutoff region, $\omega \geq \omega_C$, the loop gain magnitude $|L|$ is required to be below the upper bound $u(\omega)$. In the crossover region, $\omega_B < \omega < \omega_C$, the loop gain crosses $|L| = 0\text{dB}$.

be close to -1. The margin specification ensures that L cannot be too close to -1 in the transition region $\omega_B \leq \omega \leq \omega_C$ by constraining $\angle L$. Of course, the phase bounds in the margin constraint can be frequency dependent.

While many important closed-loop properties can be specified via L, some cannot. For example, loop shaping does not explicitly include specifications on $C/(1+PC)$ (actuator effort) and $P/(1+PC)$ (effect of input-referred process noise on y). A design will clearly be unsatisfactory if either of these transfer functions is too large. The specification that these transfer functions should not be too large is usually included as implicit "side information" in a classical loop shape design. Specifications that limit the size of these transfer functions are closed-loop convex, however, and so are readily incorporated in a closed-loop convex formulation.

Given a desired set of loop shaping specifications, the compensator C is typically synthesized by adding dynamics until the various requirements on the loop transfer function L are satisfied (or until the designer suspects that the loop shaping specifications cannot be met).

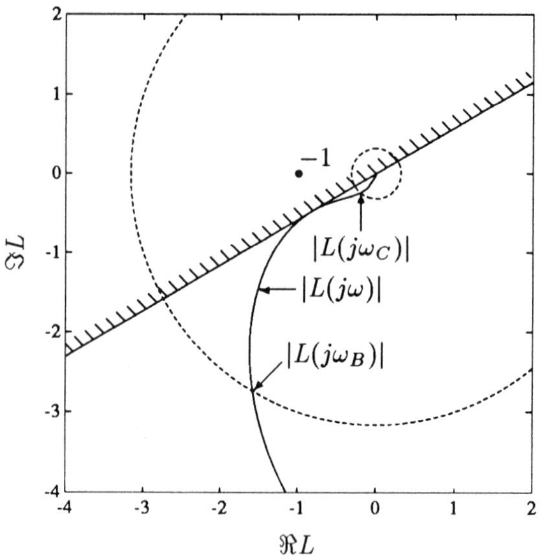

Fig. 3 A typical margin specification requires the phase of the loop transfer function to lie between $-150°$ and $+30°$ over the frequency band $\omega_B \le \omega \le \omega_C$. The dotted circles shown correspond to the magnitude constraints $|L(j\omega)| > +10\text{dB}$ and $|L(j\omega)| < -10\text{dB}$ which must be satisfied for $\omega \le \omega_B$ and $\omega \ge \omega_C$, respectively.

Classical loop shaping is described in many texts; see, for example, [4, 5, 6, 7, 8, 9]. The discussions found in these references emphasize techniques that help the engineer "do" loop-shaping design. With the exception of Bode's work on optimal cutoff characteristics and integral constraints, these references do not consider questions such as:

- Is there a compensator that meets a given set of loop-shaping specifications?
- For a given set of in-band and margin specifications and shape of the cutoff specification, what is the smallest cutoff frequency that can be achieved?
- For a given set of cutoff and margin specifications, how large can the loop gain be made in the in-band region?

The main point of this chapter is that such questions are readily answered.

B. SINGULAR VALUE LOOP SHAPING

We now consider the case in which there are multiple actuators and multiple sensors (MAMS) in the control system shown in figure 1. The plant P and compensator C are given by transfer matrices: P is n_{sens} by n_{act} and C is n_{act} by n_{sens}, where n_{sens} is the number of sensors, and n_{act} is the number of actuators.

Unlike the SASS case, there is no longer a unique choice for the "loop transfer function." A common choice is the loop transfer matrix cut at the sensors:

$$L \triangleq PC.$$

The transmission or input/output (I/O) transfer matrix is

$$T \triangleq (I + L)^{-1}L,$$

and the sensitivity transfer matrix is given by

$$S \triangleq (I + L)^{-1}.$$

These transfer matrices have interpretations that are are similar to those in SASS case. For example, if the plant transfer matrix P changes to $(I+\Delta)P$, then the I/O transfer matrix T, to first order, changes to $(I + S\Delta)T$. (Note that Δ can be interpreted as the output-referred the output-referred fractional change in the I/O transfer matrix T is then given b $S\Delta$ [10, 1].)

In contrast, the loop transfer matrix cut at the actuators is denoted $\tilde{L}$, the complementary loop transfer matrix:

$$\tilde{L} \triangleq CP.$$

Note that the loop transfer matrix and the complementary loop transfer matrix may have different dimensions: L is n_{sens} by n_{sens}, while $\tilde{L}$ is n_{act} by n_{act}. Moreover, loop specifications on L and $\tilde{L}$ are in general different and inequivalent. For example, it is possible for L to be "large" (in the sense to be described below), while $\tilde{L}$ is not "large."

A second difficulty with the extension of SASS loop shaping is choosing a measure for the "size" of the loop transfer matrix. Provided the individual sensor signals are scaled appropriately, a natural (and widely used) measure of the size is based on the singular values of the loop transfer matrix. (The singular values of a matrix M are the square roots of the eigenvalues of the Hermitian matrix M^*M.) Specifically, if all the singular values of the loop transfer matrix are large, then the loop transfer matrix is "large in all directions," and it follows that the sensitivity transfer matrix S is small and $T \approx I$. Similarly, if all the singular values of the loop transfer matrix are small, then the loop transfer matrix is "small in all directions," and it follows that T is small and $S \approx I$. These important ideas are discussed in, for example, [11, 10, 12, 13, 14, 1].

At in-band frequencies, singular value loop shaping specifications have the form

$$\sigma_{\min}(L(j\omega)) \geq l(\omega) > 1,$$

where l is some frequency dependent bound. For cutoff frequencies, singular value loop shaping specifications have the form

$$\sigma_{\max}(L(j\omega)) \leq u(\omega) < 1,$$

where u is some frequency dependent bound.

These specifications are often depicted on a singular value Bode plot, as in figure 4.

(This discussion assumes that there are at least as many actuators as actuators, *i.e.*, $n_{\text{act}} \geq n_{\text{sens}}$. If not, the in-band specifications above are guaranteed to be infeasible since at all frequencies at least one singular value of L is zero. In this case, similar specifications can be imposed on $\tilde{L}$.)

It is difficult to formulate margin specifications that are directly analogous to the gain or phase margin constraints used in the SASS case. The general idea is to ensure that $L + I$ stays "sufficiently invertible" in the crossover band. One effective method simply limits the minimum singular value of this matrix:

$$\sigma_{\min}(L + I) \geq r > 0,$$

or equivalently,

$$\sigma_{\max}(S) \leq 1/r.$$

II. A CLOSED-LOOP CONVEX FORMULATION

A design specification is *closed-loop convex* if it is equivalent to some closed-loop transfer function or matrix (*e.g.*, the sensitivity S) belonging to a convex set.

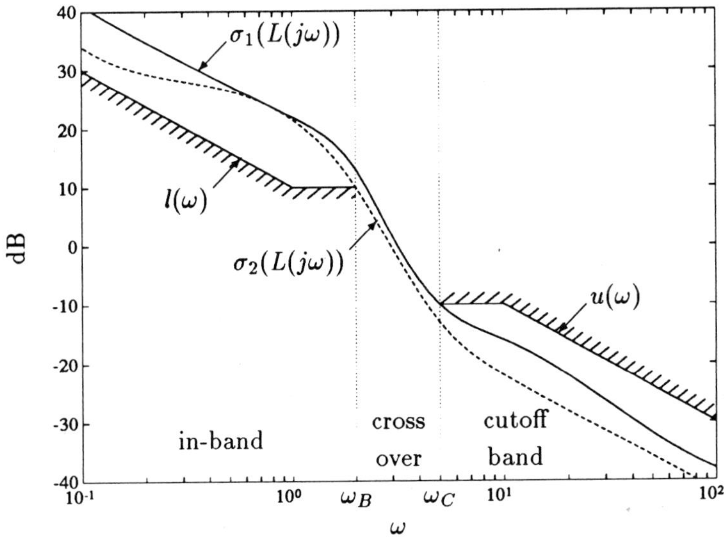

Fig. 4 Examples of in-band and cutoff specifications on the loop gain L in a system with two sensors. In the in-band region, $\omega \le \omega_B$, both singular values of L are required to exceed the lower bound. In the cutoff region, $\omega \ge \omega_C$, both singular values of L are required to be below the upper bound. In the crossover region each singular value of L crosses 0dB.

As a specific example, consider the specification

$$|L(j\omega)| \ge 3 \quad \text{for } 0 \le \omega \le 1. \tag{1}$$

We will see that this is equivalent to

$$|S(j\omega) - 1/8| \le 3/8 \quad \text{for } 0 \le \omega \le 1. \tag{2}$$

Now, the set of transfer functions S that satisfy (2) is convex, since if $S^{(a)}$ and $S^{(b)}$ both satisfy (2), then so does $(S^{(a)} + S^{(b)})/2$. Therefore the specification (1) is closed-loop convex. See [1, 2] for extensive discussions.

The main result of this chapter is that many classical and singular value loop-shaping specifications are closed-loop convex.

A. SASS CASE

1. In-band specifications

We first consider the in-band specification $|L(j\omega)| \geq \alpha$, where $\alpha > 1$. It is closed-loop convex since it is equivalent to the following convex specification on the sensitivity S, (a closed-loop transfer function):

$$|L(j\omega)| \geq \alpha > 1 \iff \left| S(j\omega) - \frac{1}{\alpha^2 - 1} \right| \leq \frac{\alpha}{\alpha^2 - 1}. \tag{3}$$

In other words, requiring $|L| \geq \alpha > 1$ is equivalent to requiring the sensitivity to lie inside a circle centered at $1/(\alpha^2 - 1)$ with radius $\alpha/(\alpha^2 - 1)$. Note that requiring L to be large corresponds to restricting the sensitivity S to lie in a disk that includes the point 0, but is not exactly centered at 0. Figure 5 illustrates this correspondence for $\alpha = 2$.

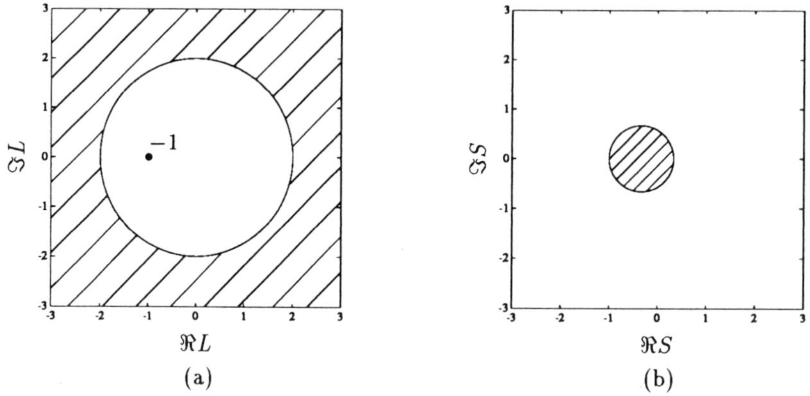

Fig. 5 The region $|L| \geq 2$ in the L-plane is shown in (a). The loop shaping specification $|L| \geq 2$ requires the Nyquist plot of L to lie in the shaded region in (a). The corresponding region in the S-plane is shown in (b), which is a disk that includes but is not centered at 0. This region is convex, and hence the loop gain specification $|L| \geq 2$ is closed-loop convex.

2. Cutoff specifications

We now consider the cutoff specification $|L(j\omega)| \leq \alpha$, where $\alpha < 1$. It is also closed-loop convex since it is equivalent to the following convex specification on the sensitivity:

$$|L(j\omega)| \leq \alpha < 1 \iff \left|S(j\omega) - \frac{1}{1-\alpha^2}\right| \leq \frac{\alpha}{1-\alpha^2}. \tag{4}$$

In other words, requiring $|L| \leq \alpha < 1$ is equivalent to requiring the sensitivity to lie in a disk centered at $1/(1-\alpha^2)$ with radius $\alpha/(1-\alpha^2)$. Note that requiring L to be small corresponds to restricting the sensitivity S to lie in a disk that includes the point 1, but is not exactly centered at 1. Figure 6 illustrates this correspondence for $\alpha = 0.5$.

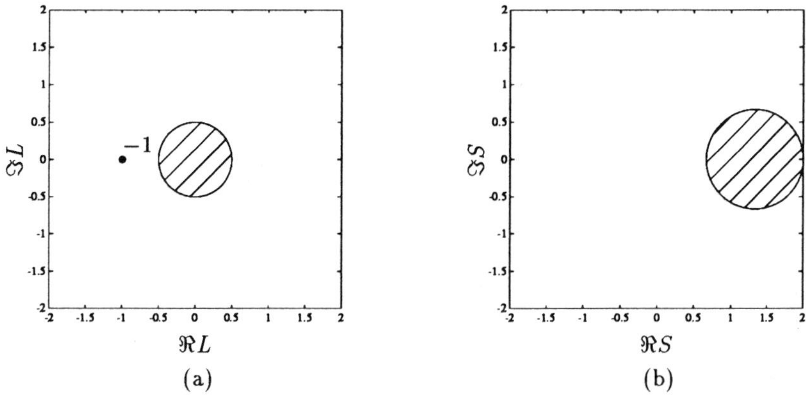

Fig. 6 The region $|L| \leq 0.5$ in the L-plane is shown in (a). The loop shaping specification $|L| \leq 0.5$ requires the Nyquist plot of L to lie in the shaded region in (a). The corresponding region in the S-plane is shown in (b), which is a disk that includes but is not centered at 1. This region is convex, and hence the loop gain specification $|L| \leq 0.5$ is closed-loop convex.

The results (3) and (4) are easily established. Since we give a careful proof for the more general MAMS case, we give a simple discussion here. The in-band loop specification (3) requires L to lie outside a circle of radius α in the complex plane. Since $\alpha > 1$, the critical point -1 lies in the interior of this circle. Since S and L are related by the bilinear transformation $S = 1/(1+L)$, this circle maps to another circle in the S-plane. To find this circle, we note that the points $L = \pm\alpha$ map to $S = 1/(1 \pm \alpha)$, and the circle must be symmetric with respect to the real axis. Moreover since the critical point -1 is mapped to $S = \infty$, the exterior of the $|L| = \alpha$ circle maps to the interior of the circle in the S-plane.

The argument in the case of the cutoff specification (4) is similar, except that the critical point -1 is outside the $|L| = \alpha$ circle and so its interior maps to the interior of the corresponding circle in the S-plane.

We note that the specifications requiring L to be "not too big,"

$$|L(j\omega)| \leq \alpha \quad \text{where } \alpha > 1,$$

and requiring L to be "not too small,"

$$|L(j\omega)| \geq \alpha \quad \text{where } \alpha < 1,$$

are not closed-loop convex, since these specifications are equivalent to $S(j\omega)$ lying outside of the shaded disks in figures 5(b) and 6(b). These specifications, however, are not likely to be used in a practical design. It is interesting that the sensible specifications on $|L|$, given in (3) and (4), turn out to coincide exactly with the specifications on $|L|$ that are closed-loop convex.

3. Phase margin specifications

A common form for a margin specification limits the phase of the loop transfer function in the crossover band:

$$\theta_{\min} \leq \angle L(j\omega) \leq \theta_{\max}.$$

where $-180° < \theta_{\min} < 0°$ and $0° < \theta_{\max} < 180°$. It turns out that such a specification is closed-loop convex if and only if $\theta_{\max} - \theta_{\min} \leq 180°$, in which case S must lie in the intersection of two disks:

$$\theta_{\min} \leq \angle L \leq \theta_{\max} \iff |2S - (1 + j/\tan\theta_{\max})| \leq 1/\sin\theta_{\max} \text{ and} \quad (5)$$

$$|2S - (1 + j/\tan\theta_{\min})| \leq 1/\sin -\theta_{\min}. \quad (6)$$

This is shown in figure 7 for the case $\theta_{\min} = -150°$, $\theta_{\max} = 10°$. The phase margin specification $-150° \leq \angle L \leq 10°$ is equivalent to requiring the sensitivity S to lie in the convex set shown in figure 7(b).

4. General circle specifications

All of the specifications above—in-band, cutoff, and phase margin, are special cases of *general circle specifications.* Consider any generalized circle

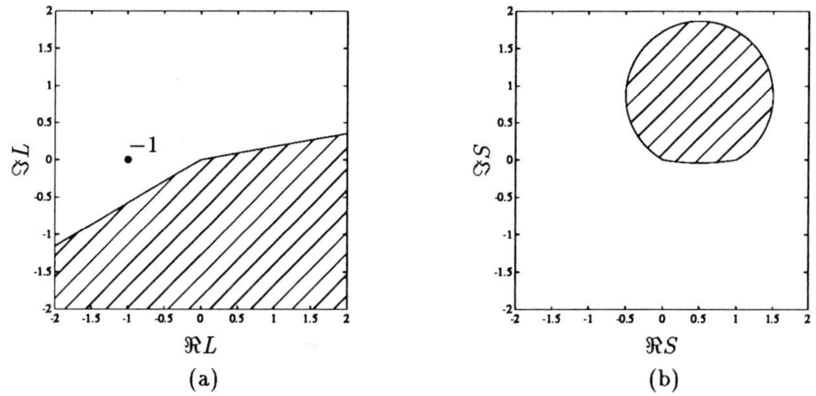

Fig. 7 The region $-150^\circ \leq \angle L \leq 10^\circ$ in the L-plane is shown in (a). The phase margin specification $-150^\circ \leq \angle L \leq 10^\circ$ requires the Nyquist plot of L to lie in the shaded region in (a). The corresponding region in the S-plane is the intersection of two disks, shown in (b), and hence the phase margin specification $-150^\circ \leq \angle L \leq 10^\circ$ is closed-loop convex.

in the complex plane (*i.e.*, a circle or a line, which we consider a "circle" centered at ∞) that does not pass through the critical point -1. Such a circle divides the complex plane into two regions, one of which includes the critical point -1. The specification that the loop transfer function must lie in the region that does not contain -1 is what we call a *generalized circle constraint*, and is readily shown (by a mapping argument) to be closed-loop convex since it is equivalent to S lying inside a circle or half-plane.

The in-band and cutoff specifications are of this form with the circle given by $|L| = \alpha$; in each case the specification requires that the loop transfer function avoid the region that includes the critical point -1. The phase margin constraint can be expressed as the simultaneous satisfaction of the two generalized circle constraints corresponding to the lines that pass through the origin at the angles $\theta_{\min}$ and $\theta_{\max}$, respectively. This explains why the phase margin constraint is equivalent to the sensitivity lying inside the intersection of two disks (see (6)).

We note that generalized circle constraints have appeared in many contexts. Examples include the circle criterion, used in stability and robustness analysis of nonlinear systems (see [15, 16, 17, 18]), the Popov criterion (with a fixed Popov parameter) [19], and many of the specifications in [20, 21, 22].

In the remainder of this section we discuss two particular generalized circle constraints.

One useful generalized circle constraint excludes L from a disk about the critical point -1:

$$|L(j\omega) + 1| \geq \alpha, \tag{7}$$

where $\alpha > 0$. This specification is equivalent to

$$|S(j\omega)| \leq 1/\alpha,$$

which is just a limit on the magnitude of the sensitivity. The case $\alpha = 1$ is shown in figure 8.

Specifications of the form (7) can be used to guarantee a classical phase margin. Since the bounds $l(\omega)$ and $u(\omega)$ are not equal to one in the in-band and cutoff regions, $|L(j\omega)|$ can equal one only in the transition regions. If the specification (7) is imposed at all frequencies in the transition regions, then whenever $|L(j\omega)| = 1$, we have $|L(j\omega)+1| \geq \alpha$, which implies a phase margin of at least $2\arcsin(\alpha/2)$ (and 180° for $\alpha > 2$).

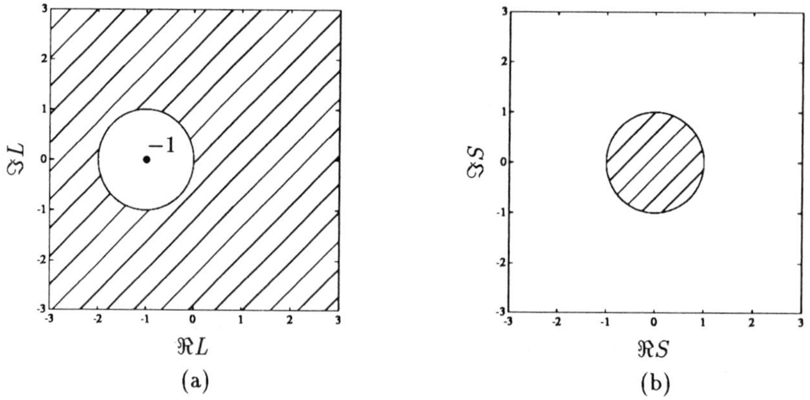

Fig. 8 The region $|L+1| \geq 1$ in the L-plane is shown in (a). The specification $|L+1| \geq 1$ requires the Nyquist plot of L to lie in the shaded region in (a), *i.e.*, to maintain a distance of at least 1 from the critical point -1. The corresponding region in the S-plane is shown in (b), which is a disk around 0. This region is convex, and hence the loop gain specification $|L+1| \geq 1$ is closed-loop convex.

As another example, we consider the specification

$$\Re L(j\omega) \geq -\alpha, \tag{8}$$

where $0 < \alpha < 1$, which can be expressed in terms of the sensitivity as

$$\Re L(j\omega) \geq -\alpha \iff |2S(j\omega) - 1/(1-\alpha)| \leq 1/(1-\alpha).$$

The case $\alpha = 0.5$ is shown in figure 9.

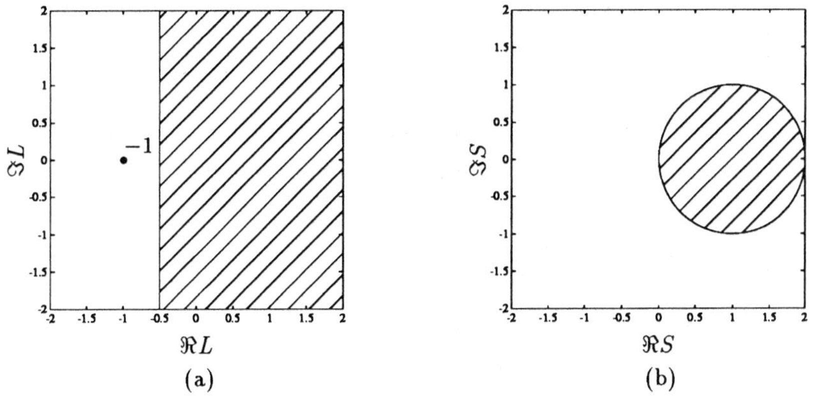

Fig. 9 The region $\Re L \geq -0.5$ in the L-plane is shown in (a). The gain margin specification $\Re L \geq -0.5$ requires the Nyquist plot of L to lie in the shaded region in (a). The corresponding region in the S-plane is the disk shown in (b), and hence the gain margin specification $\Re L \geq -0.5$ is closed-loop convex.

If the specification (8) is imposed at all frequencies, then the closed-loop system will remain stable even if the loop gain is increased by up to $1/\alpha$. Thus it can be interpreted as enforcing a (positive) gain margin of $20\log_{10}(1/\alpha)$dB.

The specification (8), imposed at all frequencies, can also be interpreted as a circle criterion condition that guarantees the system will remain stable if any memoryless nonlinearity in sector $[0,\ 1/\alpha]$ is introduced into the loop. For example, if the specification (8) is imposed at all frequencies (for any $\alpha < 1$) then actuator or sensor saturation cannot destabilize the control system.

B. MAMS CASE

1. In-band and Cutoff Specifications

The analogous results for the MAMS case are:

$$\text{for } \alpha < 1: \quad \sigma_{\max}(L) \le \alpha \iff \sigma_{\max}((1-\alpha^2)S - I) \le \alpha, \tag{9}$$

$$\text{for } \alpha > 1: \quad \sigma_{\min}(L) \ge \alpha \iff \sigma_{\max}((1-\alpha^2)S - I) \le \alpha \tag{10}$$

(we have suppressed the frequency arguments for simplicity). Note that the right-hand sides of (9) and (10) are the same. Thus, the inequality on the right-hand side expresses in one formula all reasonable in-band and cutoff loop shaping specifications:

$$\sigma_{\max}((1-\alpha^2)S - I) \le \alpha \quad \text{both in-band } (\alpha > 1) \text{ and cutoff } (\alpha < 1). \tag{11}$$

(The same correspondences hold with $\tilde{L}$ and $\tilde{S}$.)

We now establish (9). Since $\alpha < 1$, S is nonsingular, and we have

$$\begin{aligned} \sigma_{\max}(L) \le \alpha &\iff \sigma_{\max}(S^{-1} - I) \le \alpha \\ &\iff (S^{-1} - I)^*(S^{-1} - I) \le \alpha^2 I. \end{aligned}$$

Multiplying the last inequality by S^* on the left and S on the right, and multiplying by $1 - \alpha^2 > 0$ (since $\alpha < 1$), gives

$$\begin{aligned} \sigma_{\max}(L) \le \alpha &\iff (1-\alpha^2)^2 S^* S - (1-\alpha^2)S^* - (1-\alpha^2)S + I \le \alpha^2 I \\ &\iff \sigma_{\max}((1-\alpha^2)S - I) \le \alpha, \end{aligned}$$

which is (9).

The in-band result (10) is established in a similar manner. Since $\alpha > 1$, S is nonsingular, and so

$$\begin{aligned} \sigma_{\min}(L) \ge \alpha &\iff \sigma_{\min}(S^{-1} - I) \ge \alpha \\ &\iff (S^{-1} - I)^*(S^{-1} - I) \ge \alpha^2 I. \end{aligned}$$

We proceed as before, except that $1-\alpha^2 < 0$ (since $\alpha > 1$), so the inequality is reversed:

$$\begin{aligned} \sigma_{\min}(L) \ge \alpha &\iff (1-\alpha^2)^2 S^* S - (1-\alpha^2)S^* - (1-\alpha^2)S + I \le \alpha^2 I \\ &\iff \sigma_{\max}((1-\alpha^2)S - I) \le \alpha, \end{aligned}$$

which is (10).

2. General sector specifications

The in-band and cutoff specifications (9) and (10) are special forms of *general sector specifications*, which we now describe. Given complex matrices C and R such that $(I+C)^*(I+C) > R^*R$, the specification

$$(L-C)^*(L-C) \leq R^*R \tag{12}$$

is closed-loop convex. This specification can be interpreted as requiring L to be in a neighborhood of "radius" R about the "center" C that excludes $-I$. The specification (12) reduces to (9) when $C=0$ and $R=\alpha I$.

Similarly, given complex matrices C and R such that $(I+C)^*(I+C) < R^*R$, then

$$(L-C)^*(L-C) \geq R^*R \tag{13}$$

is closed-loop convex. This specification can be interpreted as requiring L to be outside a neighborhood of "radius" R about C that includes $-I$. The specification (13) reduces to (10) when $C=0$ and $R=\alpha I$.

The specifications (12) and (13) are closely connected to the conic sector conditions developed by Zames [23] and Safonov [18]. For example, if C, R, R^{-1}, and L are stable transfer matrices and (12) is imposed at all frequencies, then, using the terminology of Safonov, (12) implies that Graph(L) is inside Cone(C, R). These sector conditions form the basis of various MAMS generalizations of classical frequency domain stability and robustness criteria.

With $C=-I$ and $R=\alpha I$, (13) excludes L from a neighborhood about the critical point $-I$:

$$\sigma_{\min}(L+I) \geq \alpha.$$

This is equivalent to

$$\sigma_{\max}(S) \leq 1/\alpha, \tag{14}$$

which limits the size of the closed-loop sensitivity transfer matrix. Specifications such as (14) that limit the size of a closed-loop transfer matrix, when imposed at all frequencies, can be interpreted as circle criterion constraints that guarantee robustness in the face of various types and locations of nonlinearities.

C. SOME CONVEX AND QUASICONVEX FUNCTIONALS

We have so far considered loop-shaping *specifications*, which are constraints that a given loop transfer function or matrix either satisfies or not. When these specifications are closed-loop convex, then we can use (numerical) nondifferentiable convex optimization methods to determine whether or not the specifications can be achieved [1]. Practical design problems, however, are more often expressed using a combination of hard constraints (specifications) and soft objectives (performance indices), for example: "maximize the bandwidth subject to a fixed set of crossover and cutoff specifications." Of course such a problem can be solved by repeatedly determining whether fixed sets of specifications are feasible, for example, using a bisection on the objective. Many of the performance indices associated with loop-shaping design are *closed-loop quasiconvex* (when they are to be minimized) or *closed-loop quasiconcave* (when they are to be maximized), which means that these problems can be directly solved. We refer the reader to [1] for a precise definition of these terms and descriptions of numerical methods (*e.g.*, the ellipsoid method) that directly solve compensator design problems that are expressed in terms of closed-loop quasiconvex and quasiconcave performance indices.

Many of these performance indices are constructed in the following general way. We have a family of loop-shaping specifications that is indexed by some number u, in such a way that the specifications always become tighter as the parameter u is decreased. Our performance index is then given by the smallest value of u that the current design satisfies the corresponding specification. Similarly, when smaller values of u correspond to looser specifications, we take the performance index to be the largest value of u such that the corresponding specification is satisfied.

For example, consider the family

$$|L(j\omega)| \geq 10 \quad \text{for} \quad \omega \leq u \tag{15}$$

which is indexed by the number u. As u is decreased, the specification (15) becomes looser, that is, if a given compensator satisfies (15) for a given value of u then it satisfies it for any $\tilde{u} \leq u$. In (15), u can be thought of as the 20dB control bandwidth. The performance index, or 20dB control bandwidth, is the largest value of u such that (15) is satisfied.

Similarly, consider the family

$$|L(j\omega)| \geq u \quad \text{for} \quad \omega \leq 1. \tag{16}$$

The parameter u can be thought of as the minimum loop gain over the fixed bandwidth $0 \leq \omega \leq 1$. The performance index, or in-band disturbance rejection, is the largest value of u such that (16) is satisfied.

If the loop shaping specifications in the family indexed by u are all closed-loop convex, then the performance indices described above are *closed-loop quasiconcave* (when the largest u gives the performance index) and *closed-loop quasiconvex* (when the smallest u gives the performance index). For example, the 20dB control bandwidth, given by the largest u such that (15) holds, is closed-loop quasiconcave; whereas the in-band disturbance rejection, given by the minimum loop gain over the bandwidth $0 \leq \omega \leq 1$, is closed-loop quasiconvex.

Common performance indices associated with loop shaping design that are closed-loop quasiconvex or quasiconcave, as appropriate, include:

- *Bandwidth, i.e.*, the smallest frequency for which $|L(j\omega)|$ is less than 3dB, is quasiconcave.
- *System type, i.e.*, the multiplicity (possibly zero) of the pole in L at $s = 0$, is an integer valued quasiconcave performance index. In terms of loop shaping specifications, system type is constrained by forcing $l(\omega)$ to grow as $20t$dB/decade for small ω, where t is the type. (System type determines the multiplicity of the zero in S at $s = 0$.)
- *Classical error constant*, given by the absolute value of the first non-vanishing derivative of S at $s = 0$, is convex (and hence quasiconvex).
- *Cutoff frequency*, defined as the smallest frequency such that $|L(j\omega)|$ is less than some number $\alpha < 1$ (-3dB is typical), is quasiconvex.
- *Cutoff rolloff rate, i.e.*, the asymptotic rate at which $|L(j\omega)|$ decreases, (which for rational plants is a multiple of -20dB/decade), is quasiconcave.

III. CONCLUSIONS

We have shown that many classical and singular value loop shaping problems are closed-loop convex. Consequently, loop shaping problems can be solved by efficient numerical methods. In particular, it can be determined whether or not a compensator exists that satisfies a given set of loop shaping specifications. Loop shaping design problems that are formulated as classical optimization problems, *e.g.*, maximizing bandwidth subject to given margin and cutoff specifications, can be solved by direct numerical methods for quasiconvex optimization.

A consequence of these observations is that closed-loop convex design methods can be used to do compensator design in a classical loop shaping framework which is familiar to many control engineers. In contrast with classical compensator design methods, in which the designer must decide how to vary *parameters* (such as poles, zeros, and gain) in such a way that the loop transfer function meets the specifications, the designer can directly manipulate the loop shaping *specifications*, since the step of finding a suitable compensator, or determining that none exists, can be automated.

We comment, however, that classical loop shaping is an indirect design technique originally developed before the advent of computers, and is not a particularly good method for compensator design, and especially, computer-aided compensator design. In our opinion, every specification that can be expressed in terms of the loop transfer function or matrix can be more directly expressed in terms of some closed-loop transfer function or transfer matrix, so that compensator design directly from closed-loop (convex) specifications is more direct and natural (see [24, 1]).

A closed-loop convex formulation of a classical design technique such as loop-shaping, even if it is indirect, does have an important practical use: it makes use of the large investment we currently have in teaching, understanding, experience with, and engineering intuition about, classical loop shaping.

We close by noting two important *limitations* of loop shaping via closed-loop convex methods. The first is that closed-loop convex methods generate high order compensators (see, for example, [1, 24]), whereas one of the main advantages of classical loop shaping design is that the designer uses "only as much compensator complexity as is needed to meet the specifications." However, the advent of cheap, high performance, digital signal processors has substantially reduced the relevance of compensator order. Also, even if a compensator designed by closed-loop convex methods is not implemented, knowledge that particular loop shaping specifications can or cannot be achieved is very valuable information to the designer. The designer then *knows* exactly how much performance is given up by using a low order compensator, or some other design method.

The second, and in our opinion more important, drawback of closed-loop convex loop shaping is that margin specifications for MAMS systems that are expressed in terms of singular values or more general sector conditions are often overly conservative when one considers more detail about the types of plant variations that can occur (see [25, 26, 27, 28, 29]). Scaling can greatly reduce the conservatism of these specifications. While a fixed scaling preserves the closed-loop convexity of loop shaping specifications, specifications involving optimal (variable) scaling are not closed-loop convex.

This drawback is not present for SASS control systems. In this case, very general robust performance problems turn out to be closed-loop convex (see [1, p246], [30, §4.3]). In chapters 7 and 8 of [30], these closed-loop convex problems are approximately transformed into classical loop shaping specifications.

IV. REFERENCES

1. S. Boyd and C. Barratt, *Linear Controller Design: Limits of Performance*, Prentice-Hall (1991).

2. S. Boyd, C. Barratt, and S. Norman, "Linear controller design: Limits of performance via convex optimization", *Proc. IEEE* **78**(3), pp. 529–574 (1990).

3. E. Polak and S. Salcudean, "On the design of linear multivariable feedback systems via constrained nondifferentiable optimization in $\mathbf{H}_\infty$ spaces", *IEEE Trans. Aut. Control* **AC-34**(3), pp. 268–276 (1989).

4. H. W. Bode, *Network Analysis and Feedback Amplifier Design*, Van Nostrand New York (1945).

5. I. M. Horowitz, *Synthesis of Feedback Systems*, Academic Press New York (1963).

6. K. Ogata, *Modern Control Engineering*, Prentice-Hall Englewood Cliffs, New Jersey second edition (1990).

7. R. C. Dorf, *Modern Control Systems*, Addison-Wesley 5th edition (1988).

8. B. C. Kuo, *Automatic Control Systems*, Prentice-Hall 4th edition (1982).

9. G. F. Franklin, J. D. Powell, and A. Emami-Naeni, *Feedback Control of Dynamic Systems*, Addison-Wesley (1986).

10. F. M. Callier and C. A. Desoer, *Multivariable Feedback Systems*, Springer-Verlag (1982).

11. J. Doyle and G. Stein, "Multivariable feedback design: Concepts for a classical/modern synthesis", *IEEE Trans. Aut. Control* **AC-26**(1), pp. 4–16 (1981).

12. J. S. Freudenberg and D. P. Looze, *Frequency Domain Properties of Scalar and Multivariable Feedback Systems*, Lecture Notes in Control and Information Sciences. Springer-Verlag (1988).

13. J. M. Maciejowski, *Multivariable Feedback Design*, Addison-Wesley (1989).

14. J. Lunze, *Robust Multivariable Feedback Control*, Prentice-Hall (1989).

15. G. Zames, "On the input-output stability of nonlinear time-varying feedback systems—Part II: Conditions involving circles in the frequency plane and sector nonlinearities", *IEEE Trans. Aut. Control* **AC-11**, pp. 465–476 (1966).

16. K. S. Narendra and R. M. Goldwyn, "A geometrical criterion for the stability of certain nonlinear nonautonomous systems", *IEEE Trans. Circuit Theory* **CT-11**, pp. 406–408 (1964).

17. I. W. Sandberg, "A frequency-domain condition for the stability of feedback systems containing a single time-varying nonlinear element", *Bell Syst. Tech. J.* **43**(3), pp. 1601–1608 (1964).

18. M. G. Safonov, *Stability and Robustness of Multivariable Feedback Systems*, MIT Press Cambridge (1980).

19. V. M. Popov, "Absolute stability of nonlinear systems of automatic control", *Automation and Remote Control* **22**, pp. 857–875 (1962).

20. Y-S. Cho and K. S. Narendra, "An off-axis circle criterion for the stability of feedback systems with a monotonic nonlinearity", *IEEE Trans. Aut. Control* **AC-13**, pp. 413–416 (1968).

21. J. C. Willems and G. L. Blankenship, "Frequency domain stability criteria for stochastic systems", *IEEE Trans. Aut. Control* **AC-16**, pp. 292–299 (1971).

22. F. N. Bailey and C. H. Hui, "Loop gain-phase shaping for single-input-single-output robust controllers", *IEEE Control Syst. Mag.* , pp. 93–101 (1991).

23. G. Zames, "On the input-output stability of nonlinear time-varying feedback systems—Part I: Conditions derived using concepts of loop gain, conicity, and positivity", *IEEE Trans. Aut. Control* **AC-11**, pp. 228–238 (1966).

24. S. Boyd, V. Balakrishnan, C. Barratt, N. Khraishi, X. Li, D. Meyer, and S. Norman, "A new CAD method and associated architectures for linear controllers", *IEEE Trans. Aut. Control* **AC-33**(3), pp. 268–283 (1988).

25. J. Doyle, "Analysis of feedback systems with structured uncertainties", *IEE Proc.* **129-D**(6), pp. 242–250 (1982).

26. M. G. Safonov, "Stability margins of diagonally perturbed multivariable feedback systems", *IEE Proc.* **129-D**, pp. 251–256 (1982).

27. M. G. Safonov and J. Doyle, "Minimizing conservativeness of robust singular values", In S. G. Tzafestas, editor, *Multivariable Control*, pp. 197–207. D. Reidel (1984).

28. G. Stein and J. C. Doyle, "Beyond singular values and loop shapes", *J. Guidance and Control* **14**, pp. 5–16 (1991).

29. M.K.H. Fan, A.L. Tits, and J.C. Doyle, "Robustness in the presence of mixed parametric uncertainty and unmodeled dynamics", *IEEE Trans. Aut. Control* **36**(1), pp. 25–38 (1991).

30. J. Doyle, B. Francis, and A. Tannenbaum, *Feedback Control Theory*, Macmillan (1992).

Numerical Issues in Robust Control Design Techniques*

Pradeep Pandey
Dept. of Mech. Eng.
University of California
Berkeley, CA 94720

Alan J. Laub
Dept. of Elec. & Comp. Eng.
University of California
Santa Barbara, CA 93106

I. INTRODUCTION

In the last decade, $\mathcal{H}_\infty$-control methods have emerged as one of the central components of robust control design. This method, along with structured singular value analysis, has provided a powerful tool to synthesize robust controllers for linear systems [2],[3],[4],[5]. The focus of this chapter is on numerical and computational aspects of $\mathcal{H}_\infty$ control.

Interest in using the $\mathcal{H}_\infty$ norm as an optimality criterion was sparked largely by Zames [25]. While initial attempts at solving this problem led to rather complex operator-theoretic methods [6], connecting the state-feedback $\mathcal{H}_\infty$ problem to an (indefinite) Riccati equation was instrumental in recasting the problem in a state-space setting. A definitive answer to the output-feedback case, along with a parallel development for the $\mathcal{H}_2$ problem was given in [5]. This approach relies on formulating the problem in terms of two Riccati equations which depend on a gain parameter, γ. The optimal gain, γ_{opt}, is the infimum over all

*The research described in this paper was supported in part by the National Science Foundation (and AFOSR) under Grant No. ECS87-18897 and the Air Force Office of Scientific Research under Contract No. AFOSR-91-0240.

suboptimal gains, γ, such that three conditions are satisfied, namely, stabilizing solutions exist for two certain Riccati equations and the spectral radius of the product of the solutions is bounded by γ^2.

The results in [5] immediately lend themselves to a bisection-type algorithm since an interval containing γ_{opt} can be isolated by checking the three conditions mentioned above. Such algorithms are, however, slow even when this interval is small. By examining the behavior of the Riccati solutions as a function of γ, we are able to couple a gradient search with bisection [16],[17]. In our gradient method the equation involving the spectral radius is solved using Newton's method which requires the gradient of the spectral radius. Using standard perturbation theory for eigenvalues, this gradient is approximated via the gradient of the Riccati solutions with respect to the gain parameter. The Riccati gradients are computed by solving associated closed-loop Lyapunov equations. In a related development, we show how to avoid explicitly forming the Riccati solutions by working, instead, with Schur vectors of associated Hamiltonian matrices, and we apply this technique to both the bisection and gradient methods.

II. NOTATION

For consistency we use notation that is fairly standard in the $\mathcal{H}_\infty$ control literature. Note that, except where necessary to avoid confusion, we omit all vector and matrix dimensions. $\mathbb{R}$ will denote the real line, $\mathbb{C}$ the complex plane, and $\mathbb{C}_-$, $\mathbb{C}_0$, and $\mathbb{C}_+$ will denote, respectively, the open left half-plane, imaginary axis, and open right half-plane of the complex plane. $\mathbb{R}^{n\times m}$, $\mathbb{C}^{n\times m}$, and $\mathbb{R}(s)^{n\times m}$ will denote, respectively, the sets of real, complex, and real-rational matrices of dimensions $n \times m$.

For a matrix M we will use the following notation:

M^T, M^H	transpose and complex conjugate transpose
$\lambda(M)$, $\Lambda(M)$	an eigenvalue, and the set of eigenvalues of M
$\rho(M)$	spectral radius
$\Lambda_-(M)$, $\Lambda_0(M)$	set of eigenvalues in $\mathbb{C}_-$ and $\mathbb{C}_0$, respectively
$\mathcal{X}_-(M)$	invariant subspace of M associated with $\Lambda_-(M)$
$\sigma(M)$, $\sigma_{max}(M)$	a singular value, and the maximum singular value of M

A real square matrix Q is orthogonal if $Q^TQ = I$. The set of generalized eigenvalues of a matrix pair (M, N) will be denoted by $\Lambda(M, N)$. Matrix norms, unless stated otherwise, will be any consistent matrix norm [22] (i.e., $\|AB\| \leq \|A\|\|B\|$). A subscripted matrix p-norm implies the norm is subordinate to the corresponding vector p-norm. For a $G(s)$ that is real-rational and stable, the 2-norm and ∞-norm are given by

$$\begin{aligned} \|G\|_2^2 &:= \frac{1}{2\pi}\int_{\infty}^{\infty} \text{Tr}\ [G^H(j\omega)G(j\omega)]\ d\omega, \\ \|G\|_\infty &:= \sup_{\omega\in\mathbb{R}}\ \sigma_{max}[G(j\omega)], \end{aligned}$$

where $\text{Tr}(\cdot)$ denotes the trace. We can compute these norms using state-space methods. For a strictly proper system the 2-norm is

$$\|G\|_2^2 = \text{Tr}(CYC^T) = \text{Tr}(B^TXB),$$

where X and Y are respectively the observability and reachability Grammians, i.e.,

$$A^TX + XA + C^TC = 0, \quad \text{and} \quad AY + YA^T + BB^T = 0.$$

The ∞-norm is harder to compute and involves an iterative method. State-space methods were developed to compute this norm in [1].

Following [5], define *dom(Ric)* as the set of Hamiltonian matrices

$$H = \begin{pmatrix} A & -F \\ -G & -A^T \end{pmatrix}$$

such that A, F, and G are square matrices, F and G are symmetric, $\Lambda_0(H) = \emptyset$, and $\mathcal{X}_-(H)$ does not intersect $\text{Im}\begin{pmatrix} 0 \\ I \end{pmatrix}$. For such H, the associated algebraic Riccati equation

$$A^TX + XA - XFX + G = 0$$

has a unique symmetric solution which we denote by $X := Ric(H)$. Further, if $X \geq 0$ we refer to it as *admissible.*

III. $\mathcal{H}_\infty$ CONTROL PROBLEM

In some situations it is effective to model a control problem in the following way. We are given a linear, time-invariant system which is affected by certain exogenous disturbances and we want to regulate certain outputs. In addition, we are given a description of sensors and actuators that can be used for the purpose of control. Our aim is to design a controller which takes its input from the sensors and generates an output that is applied to the system via the actuators such that the system is internally stable and the regulated outputs meet certain design specifications. The following formulation provides a rather general setting for such problems. Consider a linear system with two sets of multivariable inputs and outputs. The state is $x(t) \in \mathbb{R}^n$; inputs are disturbances $w(t) \in \mathbb{R}^{m_1}$ and control $u(t) \in \mathbb{R}^{m_2}$; outputs are errors $z(t) \in \mathbb{R}^{p_1}$ and measurements $y(t) \in \mathbb{R}^{p_2}$. The system equations are:

$$\begin{aligned} \dot{x} &= Ax + B_1 w + B_2 u, \\ z &= C_1 x + D_{11} w + D_{12} u, \\ y &= C_2 x + D_{21} w + D_{22} u. \end{aligned} \tag{1}$$

The transfer function from the inputs $\begin{pmatrix} w \\ u \end{pmatrix}$ to the outputs $\begin{pmatrix} z \\ y \end{pmatrix}$ is:

$$G(s) := \begin{pmatrix} G_{11}(s) & G_{12}(s) \\ G_{21}(s) & G_{22}(s) \end{pmatrix} \tag{2}$$

$$= \begin{pmatrix} D_{11} & D_{12} \\ D_{21} & D_{22} \end{pmatrix} + \begin{pmatrix} C_1 \\ C_2 \end{pmatrix} (sI - A)^{-1} \begin{pmatrix} B_1 & B_2 \end{pmatrix}. \tag{3}$$

Given the system in Eq. (1), the objective in an $\mathcal{H}_\infty$ control design is to find a controller which is internally stabilizing and minimizes the gain of the closed-loop system. To be more precise, suppose we have a feedback controller $K(s) \in \mathbb{R}(s)^{m_2 \times p_2}$ which measures the output y and generates a control input u, i.e., $u = K(s)y$. Then the closed-loop transfer matrix $T_{zw}(s) \in \mathbb{R}(s)^{p_1 \times m_1}$ from the disturbance w to the regulated output z is

$$T_{zw} := G_{11} + G_{12}K(I - G_{22}K)^{-1}G_{21}. \tag{4}$$

Define

$$\mathcal{K} := \{K(s) \in \mathbb{R}(s)^{m_2 \times p_2} \ : \ T_{zw}(s) \text{ in Eq. (4) is internally stable}\}.$$

Hence, $\mathcal{K}$ is the set of all internally stabilizing feedback controllers for the system in Eq. (1). Suppose $K(s) \in \mathcal{K}$ and define

$$\gamma := \|T_{zw}\|_\infty. \tag{5}$$

Then the objective in an $\mathcal{H}_\infty$ control design is to find a controller $K(s) \in \mathcal{K}$ which minimizes γ, the gain of the closed-loop system. The optimal gain is defined by

$$\gamma_{opt} := \inf_{K \in \mathcal{K}} ||T_{zw}||_\infty.$$

A state-space-based solution to this problem is given in [5], [11],[18]. This approach relies on formulating the problem in terms of two Riccati equations which depend on a gain parameter, γ. The optimal gain, γ_{opt}, is the infimum over all suboptimal gains, γ, such that three conditions are satisfied, namely, stabilizing solutions exist for two certain Riccati equations and the spectral radius of the product of the solutions is bounded by γ^2. In this approach certain assumptions are made about the system in Eq. (1). Some of these assumptions are required to ensure that the problem is well posed while others simplify dealing with the problem without loss of generality. These assumptions are:

A1: (A, B_2) is stabilizable and (C_2, A) is detectable.

A2: (A, B_1) is stabilizable and (C_1, A) is detectable.

A3: $D_{22} = 0$.

A4: $D_{12}^T \begin{pmatrix} D_{12} & C_1 \end{pmatrix} = \begin{pmatrix} I & 0 \end{pmatrix}$ and $D_{21} \begin{pmatrix} D_{21}^T & B_1^T \end{pmatrix} = \begin{pmatrix} I & 0 \end{pmatrix}$.

A5: $D_{11} = 0$.

Assumption A1 is necessary for the existence of a stabilizing controller whereas A2 ensures that input-output stability is equivalent to internal stability. Safonov *et al.* [20] have shown that assumptions A3 – A5 can be made without loss of generality, and any system can be transformed to this form using "loop-shifting" techniques. This equivalence was used by Doyle *et al.* [5] where necessary and sufficient conditions were given for the existence of a controller for a system that satisfied the above assumptions.

However, it is evident from [20] that the transformations required to enforce A5 depend on the value of γ under consideration. This makes it harder to characterize the Riccati solutions as a *function* of γ and complicates the computation of gradients. For this reason we define two forms of the problem — the *general* and the *standard* problem. A general problem satisfies assumptions A1 through A4 whereas the standard problem satisfies assumptions A5 also.

We now define the Hamiltonian matrices associated with the two Riccati equations that we need to solve. The notation is essentially from [8]. Define

$$H_\gamma := \begin{pmatrix} A & 0 \\ -C_1^T C_1 & -A^T \end{pmatrix} +$$

$$\begin{pmatrix} B_1 & B_2 \\ -C_1^T D_{11} & 0 \end{pmatrix} \begin{pmatrix} \gamma^2 I - D_{11}^T D_{11} & -D_{11}^T D_{12} \\ -D_{12}^T D_{11} & -I \end{pmatrix}^{-1} \begin{pmatrix} D_{11}^T C_1 & B_1^T \\ 0 & B_2^T \end{pmatrix} \tag{6}$$

and

$$J_\gamma := \begin{pmatrix} A^T & 0 \\ -B_1 B_1^T & -A \end{pmatrix} +$$

$$\begin{pmatrix} C_1^T & C_2^T \\ -B_1 D_{11}^T & 0 \end{pmatrix} \begin{pmatrix} \gamma^2 I - D_{11} D_{11}^T & -D_{11} D_{21}^T \\ -D_{21} D_{11}^T & I \end{pmatrix}^{-1} \begin{pmatrix} D_{11} B_1^T & C_1 \\ 0 & C_2 \end{pmatrix}. \tag{7}$$

Note that these Hamiltonians are functions of the scalar parameter γ. For a problem in standard form these matrices are considerably simplified since $D_{11} = 0$, whereupon Eqs. (6) – (7) reduce to

$$H_\gamma := \begin{pmatrix} A & \gamma^{-2} B_1 B_1^T - B_2 B_2^T \\ -C_1^T C_1 & -A^T \end{pmatrix}, \tag{8}$$

$$J_\gamma := \begin{pmatrix} A^T & \gamma^{-2} C_1^T C_1 - C_2^T C_2 \\ -B_1 B_1^T & -A \end{pmatrix}. \tag{9}$$

To ensure that H_γ and J_γ are well posed we need to ensure that the matrices involving γ in Eqs. (6) – (7) are invertible. Thus define

$$\sigma_{d_1} := \sigma_{max}\left((I - D_{12} D_{12}^T) D_{11}\right), \qquad \sigma_{d_2} := \sigma_{max}\left(D_{11}(I - D_{21}^T D_{21})\right) \tag{10}$$

and

$$\sigma_d := \max(\sigma_{d_1}, \sigma_{d_2}). \tag{11}$$

The following theorem, due to Doyle *et al.*, states necessary and sufficient conditions for the existence of a controller $K(s)$ which is internally stabilizing and for which $\|T_{zw}\|_\infty < \gamma$.

Theorem 1 *For the linear system in (1)-(3), there exists an admissible controller $K(s)$ such that $\|T_{zw}\|_\infty < \gamma$ if and only if $\gamma > \sigma_d$ and the following three conditions hold:*

(i) *$H_\gamma \in dom(Ric)$ and $X_\gamma := Ric(H_\gamma) \geq 0$.*

(ii) $J_\gamma \in dom(Ric)$ *and* $Y_\gamma := Ric(J_\gamma) \geq 0$.

(iii) $\rho(X_\gamma Y_\gamma) < \gamma^2$.

Proof: See [8] and [5]. ∎

Once a γ is found such that the conditions of the above theorem are satisfied, an admissible controller can be synthesized using X_γ and Y_γ (the formulas involved can be consulted in [8]). Now the "optimal" gain can be defined as:

$$\gamma_{opt} := \inf\{\gamma \geq 0 : \text{the 3 conditions of Theorem 1 satisfied}\}. \tag{12}$$

Evidently, this definition immediately lends itself to a bisection-type algorithm since an interval containing γ_{opt} can be isolated by checking the three conditions in Theorem 1. However, this method can be very slow even when γ approaches the optimal value. In contrast, a gradient method like Newton's method performs well in the neighborhood of the solution. By examining the behavior of Riccati solutions as a function of γ we are able to couple a gradient method with bisection which leads to potential speedup near the solution.

IV. CHARACTERIZING RICCATI SOLUTIONS

From Theorem 1 it is clear that existence of admissible Riccati solutions X_γ and Y_γ is *part* of the set of necessary and sufficient conditions for the existence of an controller that achieves a desired closed-loop gain. This is in contrast to the $\mathcal{H}_2$ case where standard assumptions guarantee the existence of admissible solutions to the associated Riccati equations. Therefore, it is natural to investigate the behavior of the Hamiltonian matrices H_γ and J_γ, defined in Eqs. (6)–(7), and the Riccati solutions $X_\gamma := Ric(H_\gamma)$ and $Y_\gamma := Ric(J_\gamma)$ as a function of γ. This section is devoted to discovering when H_γ and J_γ are in *dom*(Ric), and deriving monotonicity results for X_γ, Y_γ, and $\rho(X_\gamma Y_\gamma)$. In what follows we mainly focus on H_γ and X_γ and the results obtained also apply to J_γ and Y_γ by duality.

As a first step we deduce the existence of $\gamma_f < +\infty$ such that for $\gamma \in (\gamma_f, +\infty)$ both H_γ and J_γ are in *dom*(Ric) and $X_\gamma \geq 0$ and $Y_\gamma \geq 0$ exist. For H_γ, the limit as $\gamma \to +\infty$ exists and is given by

$$H_\infty := \lim_{\gamma \to +\infty} H_\gamma = \begin{pmatrix} A & -B_2 B_2^T \\ -C_1^T C_1 & -A^T \end{pmatrix}. \tag{13}$$

The expression for H_γ in Eq. (8), associated with the standard problem, immediately corroborates the above claim. The fact that the claim is also true for H_γ in Eq. (6) will become clear in the sequel. Now our assumptions, namely (A, B_2) stabilizable and (C_1, A) detectable, imply that $H_\infty \in dom(Ric)$ and that $X_\infty := Ric(H_\infty)$ is nonnegative definite [12]. Similarly, $J_\infty \in dom(Ric)$ and $Y_\infty := Ric(J_\infty)$ is nonnegative definite. However, verifying that an admissible solution exists for some finite γ is not as simple as imposing stabilizability/detectability conditions. Nevertheless, by continuity we can deduce the existence of a finite γ such that $\Lambda_0(H_\gamma) = \emptyset$ since $\Lambda_0(H_\infty) = \emptyset$. Further, existence of $X_\infty \geq 0$ and $Y_\infty \geq 0$ settles the question of the existence of a stabilizing controller which achieves a finite closed-loop gain $\gamma_f < +\infty$. Then, by Theorem 1, we have:

$$H_\gamma \text{ and } J_\gamma \text{ in } dom(Ric),\ X_\gamma \geq 0 \text{ and } Y_\gamma \geq 0 \text{ for } \gamma \in (\gamma_f, +\infty). \tag{14}$$

To further characterize the interval $(\gamma_f, +\infty)$, we examine the dependence of H_γ on γ. First, write H_γ as

$$H_\gamma = \begin{pmatrix} A_\gamma & -F_\gamma \\ -G_\gamma & -A_\gamma^T \end{pmatrix} \tag{15}$$

and the associated Riccati equation

$$A_\gamma^T X_\gamma + X_\gamma A_\gamma - X_\gamma F_\gamma X_\gamma + G_\gamma = 0 \,. \tag{16}$$

To obtain explicit expressions for A_γ, F_γ, and G_γ, we introduce a transformation

$$T := \begin{pmatrix} I & 0 \\ -D_{12}^T D_{11} & I \end{pmatrix}.$$

This transformation is useful since it block-diagonalizes the matrix involving γ in Eq. (6) as follows:

$$T^T \begin{pmatrix} \gamma^2 I - D_{11}^T D_{11} & -D_{11}^T D_{12} \\ D_{12}^T D_{11} & -I \end{pmatrix} T = \begin{pmatrix} \gamma^2 I - D_{11}^T (I - D_{12} D_{12}^T) D_{11} & 0 \\ 0 & -I \end{pmatrix}.$$

By defining $\hat{D}_{11} := (I - D_{12} D_{12}^T) D_{11}$ we obtain

$$R_\gamma := \gamma^2 I - D_{11}^T (I - D_{12} D_{12}^T) D_{11} = \gamma^2 I - \hat{D}_{11}^T \hat{D}_{11}. \tag{17}$$

Finally, using a modified version of B_1, defined by $\hat{B}_1 := B_1 - B_2 D_{12}^T D_{11}$, we obtain

$$H_\gamma = \begin{pmatrix} A & -B_2 B_2^T \\ -C_1^T C_1 & -A^T \end{pmatrix} + \begin{pmatrix} \hat{B}_1 \\ -C_1^T D_{11} \end{pmatrix} R_\gamma^{-1} \begin{pmatrix} D_{11}^T C_1 & \hat{B}_1^T \end{pmatrix}. \quad (18)$$

Thus, explicit expressions for A_γ, F_γ, and G_γ in Eq. (16) are:

$$\begin{aligned} A_\gamma &= A + \hat{B}_1 R_\gamma^{-1} D_{11}^T C_1, \\ F_\gamma &= B_2 B_2^T - \hat{B}_1 R_\gamma^{-1} \hat{B}_1^T, \\ G_\gamma &= C_1^T (I + D_{11} R_\gamma^{-1} D_{11}^T) C_1. \end{aligned} \quad (19)$$

This may not appear very enlightening, but is useful for two reasons. First, it is now clear that $R_\gamma \to 0$ as $\gamma \to +\infty$ thus justifying the earlier claim in Eq. (13). Also, for $\gamma > \sigma_d$ we have $R_\gamma > 0$ which implies that $G_\gamma \geq 0$.

Another simplification is obtained by examining the null space of X_γ. It turns out that the singular part of X_γ is independent of γ and is determined by the (C_1, A)-unobservable subspace. Moreover, the Riccati equation associated with H_γ can be reduced to one of smaller order where (C_1, A) is observable.

Lemma 2 *Consider a pair (C, A) and $F = F^T$. Let $\mathcal{V}$ denote the largest A-invariant subspace in* $\ker(C)$ *and let $U := (U_1\ U_2)$ be an orthogonal matrix such that* $\mathrm{Im} U_2 = \mathcal{V}$ *allowing the transformation*

$$U^T A U = \begin{pmatrix} \tilde{A} & 0 \\ * & \bar{A} \end{pmatrix}, \quad CU = \begin{pmatrix} \tilde{C} & 0 \end{pmatrix}, \quad U^T F U = \begin{pmatrix} \tilde{F} & * \\ * & * \end{pmatrix}, \quad (20)$$

where $\bar{A}$ is stable and $(\tilde{C}, \tilde{A})$ is observable. Let

$$H := \begin{pmatrix} A & -F \\ -C^T C & -A^T \end{pmatrix}, \quad \tilde{H} := \begin{pmatrix} \tilde{A} & -\tilde{F} \\ -\tilde{C}^T \tilde{C} & -\tilde{A}^T \end{pmatrix}. \quad (21)$$

Then the following hold:

(i) *$H \in dom(Ric)$ if and only if $\tilde{H} \in dom(Ric)$.*

(ii) *Suppose $\tilde{H} \in dom(Ric)$ and $\tilde{X} = Ric(\tilde{H})$. Then $\tilde{X}$ is nonsingular and $X = Ric(H)$ can be written as*

$$X = U \begin{pmatrix} \tilde{X} & 0 \\ 0 & 0 \end{pmatrix} U^T. \quad (22)$$

Proof: With the transformations in Eq. (20), it is easy to verify that

$$\mathcal{U}^T H \mathcal{U} = \begin{pmatrix} \tilde{A} & 0 & -\tilde{F} & * \\ * & \bar{A} & * & * \\ -\tilde{C}_1^T\tilde{C}_1 & 0 & -\tilde{A} & * \\ 0 & 0 & 0 & -\bar{A} \end{pmatrix}, \quad \text{where } \mathcal{U} := \begin{pmatrix} U & 0 \\ 0 & U \end{pmatrix}.$$

By permuting the second and third block columns and rows of $\mathcal{U}^T H \mathcal{U}$, it is clear that

$$\Lambda(H) = \Lambda(\tilde{H}) \cup \Lambda(\bar{A}) \cup \Lambda(-\bar{A})$$

which provides $\Lambda_0(H) = \emptyset$ if and only if $\Lambda_0(\tilde{H}) = \emptyset$ since $\bar{A}$ is stable. Now, suppose $\tilde{H} \in dom(Ric)$ and let $\mathcal{X}_-(\tilde{H}) = \text{Im}\begin{pmatrix} Z_1 \\ Z_2 \end{pmatrix}$, i.e.,

$$\tilde{H}\begin{pmatrix} Z_1 \\ Z_2 \end{pmatrix} = \begin{pmatrix} Z_1 \\ Z_2 \end{pmatrix} S \tag{23}$$

for some stable S and nonsingular Z_1. With

$$\mathcal{Z}_1 := \begin{pmatrix} Z_1 & 0 \\ 0 & I \end{pmatrix} \quad \text{and} \quad \mathcal{Z}_2 := \begin{pmatrix} Z_2 & 0 \\ 0 & 0 \end{pmatrix}$$

it is easily verified that

$$\mathcal{U}^T H \mathcal{U}\begin{pmatrix} \mathcal{Z}_1 \\ \mathcal{Z}_2 \end{pmatrix} = \begin{pmatrix} \mathcal{Z}_1 \\ \mathcal{Z}_2 \end{pmatrix}\begin{pmatrix} S & 0 \\ * & \bar{A} \end{pmatrix}$$

showing that

$$\mathcal{X}_-(H) = \text{Im}\begin{pmatrix} U\mathcal{Z}_1 \\ U\mathcal{Z}_2 \end{pmatrix}. \tag{24}$$

Now invertibility of Z_1 implies that of $\mathcal{Z}_1$ proving (i).

If $\tilde{H} \in dom(Ric)$, Eq. (23) implies that $\tilde{X} = Z_2 Z_1^{-1}$ and from Eq. (24) it follows that

$$X = \mathcal{Z}_2\mathcal{Z}_1^{-1} = U\begin{pmatrix} Z_2 & 0 \\ 0 & 0 \end{pmatrix}\begin{pmatrix} Z_1 & 0 \\ 0 & I \end{pmatrix}^{-1} U^T$$

has the structure defined in Eq. (22). Now suppose $\tilde{X}$ is singular and let $\text{Im}V = \ker(\tilde{X})$. Since $\tilde{X} = Ric(\tilde{H})$, it is the stablizing solution for

$$\tilde{A}^T\tilde{X} + \tilde{X}\tilde{A} - \tilde{X}\tilde{F}\tilde{X} + \tilde{C}^T\tilde{C} = 0. \tag{25}$$

Pre- and post-multiplying Eq. (25) by V gives $\tilde{C}V = 0$. Post-multiplying the same equation by V gives $\tilde{X}\tilde{A}V = 0$. This implies that V spans an $\tilde{A}$-invariant subspace in $\ker(\tilde{C})$ contradicting observability of $(\tilde{C}, \tilde{A})$. ∎

To apply Lemma 2 to Eq. (16), let

$$\mathcal{V}_x := \text{ the largest A-invariant subspace in } \ker(C_1), \tag{26}$$

and let U be an orthogonal matrix

$$U := (U_1 \; U_2), \quad \text{where} \quad \mathrm{Im} U_2 = \mathcal{V}_x. \tag{27}$$

Then $C_1 U_2 = 0$ and $AU_2 = U_2 \bar{A}$ with $\bar{A}$ stable since (C_1, A) is detectable. Following the development in Lemma 2, define the following transformed matrices:

$$\begin{aligned} U^T A U &= \begin{pmatrix} \tilde{A} & 0 \\ * & \bar{A} \end{pmatrix}, \quad C_1 U = \begin{pmatrix} \tilde{C}_1 & 0 \end{pmatrix}, \\ U^T \hat{B}_1 &= \begin{pmatrix} \tilde{B}_1 \\ * \end{pmatrix}, \quad U^T B_2 = \begin{pmatrix} \tilde{B}_2 \\ * \end{pmatrix} \end{aligned} \tag{28}$$

where $(\tilde{C}_1, \tilde{A})$ is observable. Finally, with these transformations define a reduced version of H_γ:

$$\tilde{H}_\gamma := \begin{pmatrix} \tilde{A} & -B_2 B_2^T \\ -\tilde{C}_1^T \tilde{C}_1 & -\tilde{A}^T \end{pmatrix} + \begin{pmatrix} \tilde{B}_1 \\ -\tilde{C}_1^T \hat{D}_{11} \end{pmatrix} R_\gamma^{-1} \begin{pmatrix} \hat{D}_{11}^T \tilde{C}_1 & \tilde{B}_1^T \end{pmatrix} \tag{29}$$

$$=: \begin{pmatrix} \tilde{A}_\gamma & -\tilde{F}_\gamma \\ -\tilde{G}_\gamma & -\tilde{A}_\gamma^T \end{pmatrix}. \tag{30}$$

Note that stabilizability of (A, B_1) and (A, B_2) imply that of $(\tilde{A}, \tilde{B}_1)$ and $(\tilde{A}, \tilde{B}_2)$, respectively.

Theorem 3 *For $\gamma > \sigma_d$ consider H_γ, $\tilde{H}_\gamma$, and $\mathcal{V}_x$ defined in Eqs. (18), (30), and (26), respectively. Then the following hold:*

(i) *Both H_γ and $\tilde{H}_\gamma$ are well defined and $\mathcal{V}_x$ is also the largest A_γ-invariant subspace in* $\ker(G_\gamma)$.

(ii) *$H_\gamma \in dom(Ric)$ if and only if $\tilde{H}_\gamma \in dom(Ric)$.*

(iii) *Suppose $\tilde{H}_\gamma \in dom(Ric)$ and $\tilde{X}_\gamma = Ric(\tilde{H}_\gamma)$. Then $\tilde{X}_\gamma$ is nonsingular and $X_\gamma = Ric(H_\gamma)$ can be written*

$$X_\gamma = U \begin{pmatrix} \tilde{X}_\gamma & 0 \\ 0 & 0 \end{pmatrix} U^T. \tag{31}$$

Proof: Since $\gamma > \sigma_d \geq \sigma_{d_1}$, from Eq. (10) we see that $R_\gamma > 0$ which implies that both H_γ and $\tilde{H}_\gamma$ are well defined and further that $G_\gamma \geq 0$. Thus $G_\gamma U_2 = 0$ and $A_\gamma U_2 = U_2 \bar{A}$, implying that $\mathcal{V}_x$ is the largest A_γ-invariant subspace in $\ker(G_\gamma)$. This proves (i). With (i) in hand, (ii) and (iii) are immediate from Lemma 2. ∎

As in Lemma 2, $\tilde{X}_\gamma$ is the stabilizing solution of the following reduced Riccati equation associated with $\tilde{H}_\gamma$ in Eq. (30):

$$\tilde{A}_\gamma^T \tilde{X}_\gamma + \tilde{X}_\gamma \tilde{A}_\gamma - \tilde{X}_\gamma \tilde{F}_\gamma \tilde{X}_\gamma + \tilde{G}_\gamma = 0. \tag{32}$$

Since $\tilde{X}_\gamma$ is nonsingular, we can define $Z_\gamma := \tilde{X}_\gamma^{-1}$. By pre- and post-multiplying Eq. (32) with Z_γ we get

$$-\tilde{A}_\gamma Z_\gamma - Z_\gamma \tilde{A}_\gamma^T - Z_\gamma \tilde{G}_\gamma Z_\gamma + \tilde{F}_\gamma = 0 \tag{33}$$

and that $-\tilde{A}_\gamma^T - \tilde{G}_\gamma Z_\gamma$ is stable. In the limit as $\gamma \to +\infty$, this equation becomes

$$-\tilde{A}Z - Z\tilde{A}^T - Z\tilde{C}_1^T \tilde{C}_1 Z + \tilde{B}_2 \tilde{B}_2^T = 0 \tag{34}$$

and $Z \geq 0$ exists since $(\tilde{C}_1, \tilde{A})$ is observable and $(\tilde{A}, \tilde{B}_2)$ is stabilizable and $\tilde{A}_Z := -\tilde{A} - Z\tilde{C}_1^T \tilde{C}_1$ is the corresponding (stable) closed-loop matrix.

Recall that $\Lambda_0(H_\infty) = \emptyset$. By continuity there is a γ_H such that $\Lambda_0(H_\gamma) = \emptyset$ for $\gamma \in (\gamma_H, +\infty)$. The reduction of H_γ to $\tilde{H}_\gamma$ is instrumental in discovering an explicit formula for γ_H as shown in the following theorem.

Theorem 4 *For $\gamma > \sigma_d$ consider H_γ and $\tilde{H}_\gamma$ defined in Eqs. (18) and (30), respectively. Then there is a $\gamma_H \geq \sigma_d$ such that $\Lambda_0(H_\gamma) = \emptyset$ if and only if $\gamma > \gamma_H$ and γ_H can be computed as:*

$$\gamma_H = \|\hat{D}_{11} + \tilde{C}_1 (sI - \tilde{A}_Z)^{-1} B_Z\|_\infty, \tag{35}$$

where

$$\tilde{A}_Z := -\tilde{A} - Z\tilde{C}_1^T \tilde{C}_1, \quad B_Z := \tilde{B}_1 + Z\tilde{C}_1^T \hat{D}_{11},$$

and $Z \geq 0$ is the stabilizing solution to Eq. (34).

Proof: First recall that $\Lambda_0(H_\gamma) = \emptyset$ if and only if $\Lambda_0(\tilde{H}_\gamma) = \emptyset$ by (ii) of Theorem 3. Hence, it suffices to check whether $\tilde{H}_\gamma$ has an eigenvalue on the imaginary axis. Now Eq. (34) has a stabilizing solution $Z \geq 0$ since $(\tilde{C}_1, \tilde{A})$ is observable and $(\tilde{A}, \tilde{B}_2)$ is stabilizable by construction. Thus, the corresponding closed-loop matrix $\tilde{A}_Z$ is stable. With $S_\gamma := \gamma^2 I - \hat{D}_{11}\hat{D}_{11}^T$ let

$$H_Z(\gamma) := \begin{pmatrix} \tilde{A}_Z - B_Z R_\gamma^{-1} \hat{D}_{11}^T \tilde{C}_1 & -B_Z R_\gamma^{-1} B_Z^T \\ \gamma^2 \tilde{C}_1^T S_\gamma^{-1} \tilde{C}_1 & -\tilde{A}_Z^T + \tilde{C}_1^T \hat{D}_{11} R_\gamma^{-1} B_Z^T \end{pmatrix}$$

and

$$\gamma_H := \|\hat{D}_{11} + \tilde{C}_1(sI - \tilde{A}_Z)^{-1} B_Z\|_\infty.$$

Note that $\gamma_H < +\infty$ since $\tilde{A}_Z$ is stable. The Bounded Real Lemma implies that $\Lambda(H_Z(\gamma)) = \emptyset$ if and only if $\gamma > \gamma_H$ (*cf.* [1]). Now, it is easily verified that

$$-\begin{pmatrix} I & -Z \\ 0 & I \end{pmatrix} \tilde{H}_\gamma \begin{pmatrix} I & Z \\ 0 & I \end{pmatrix} = H_Z(\gamma)$$

showing that $\Lambda(-\tilde{H}_\gamma) = \Lambda(H_Z(\gamma))$ and thus $\tilde{H}_\gamma$, or equivalently H_γ, has no eigenvalue on $\mathbb{C}_0$ if and only if $\gamma > \gamma_H$. ∎

Theorem 4 implies that as γ decreases, part of the spectrum of H_γ migrates towards the imaginary axis. The first contact occurs for $\gamma = \gamma_H$ and those eigenvalues which then reach the imaginary axis remain on $\mathbb{C}_0$ for all $\sigma_d < \gamma < \gamma_H$. Let γ_J denote the counterpart of γ_H for J_γ. Then, $\gamma^* := \max(\gamma_H, \gamma_J)$ provides a lower bound for γ_f in Eq. (14). However, algorithms to compute γ^* directly are themselves computationally intensive. Thus, this result is mainly of theoretical interest.

We now examine monotonicity and convexity properties of the Riccati solution X_γ for $\gamma \in (\gamma_f, +\infty)$. Consider an analytic function $\mathcal{F} : \mathbb{R}^{n\times n} \to \mathbb{R}^{n\times n}$. The Fréchet derivative [19] of $\mathcal{F}$ at X is the linear map Ω defined by

$$\lim_{\delta\to 0} \|\frac{\mathcal{F}(X + \delta W) - \mathcal{F}(X)}{\delta} - \Omega(X, W)\| = 0.$$

For example, consider the function $\mathcal{F}(X) := X^2$. Then

$$\mathcal{F}(X + \delta W) = X^2 + \delta XW + \delta WX + \delta^2 W^2,$$

and it is easy to see that $\Omega(X, W) := XW + WX$. Our interest in the Fréchet derivative stems from the fact we can write Eq. (16) as $\mathcal{R}(X, \gamma) = 0$ by defining a function $\mathcal{R} : \mathbb{R}^{n\times n} \times \mathbb{R} \to \mathbb{R}^{n\times n}$ by

$$\mathcal{R}(X, \gamma) := A_\gamma^T X_\gamma + X_\gamma A_\gamma - X_\gamma F_\gamma X_\gamma + G_\gamma.$$

The partial Fréchet derivative of $\mathcal{R}$ is the linear operator defined by

$$\Omega_\gamma(X,W) := A_c^T W + W A_c, \quad \text{where} \quad A_c := A_\gamma - F_\gamma X_\gamma.$$

Since $H_\gamma \in dom(Ric)$ for $\gamma \in (\gamma_f, +\infty)$, then $\Lambda(A_c) \subset \mathbb{C}_-$ which implies that Ω_γ is invertible. Now, the implicit function theorem [21] provides the existence of an analytic function $X : \mathbb{R} \to \mathbb{R}^{n\times n}$ satisfying Eq. (16). Thus, to obtain monotonicity and convexity results we investigate sign definiteness of

$$\frac{d}{d\gamma}X_\gamma \quad \text{and} \quad \frac{d^2}{d\gamma^2}X_\gamma.$$

To obtain such results it will be easier to work with $\tilde{Z}_\gamma := \tilde{X}_\gamma^{-1}$ rather than with $\tilde{X}_\gamma$. Further, the results are rather straightforward for the standard problem, i.e., for $D_{11} = 0$. In this case Eq. (33) becomes

$$-\tilde{A}Z_\gamma - Z_\gamma\tilde{A}^T - Z_\gamma\tilde{C}_1^T\tilde{C}_1 Z_\gamma + \tilde{B}_2\tilde{B}_2^T - \gamma^{-2}\tilde{B}_1\tilde{B}_1^T = 0. \tag{36}$$

Suppose Eq. (36) has a stabilizing solution and the corresponding closed-loop matrix $A_c := -\tilde{A} - Z_\gamma\tilde{C}_1^T\tilde{C}_1$ is stable. Let $Z'_\gamma = \frac{d}{d\gamma}Z_\gamma$. Taking the derivative of Eq. (36) with respect to γ we get

$$A_c Z'_\gamma + Z'_\gamma A_c^T + 2\gamma^{-3}\tilde{B}_1\tilde{B}_1^T = 0$$

which implies that $Z'_\gamma \geq 0$ since A_c is stable. Taking a derivative again we get

$$A_c Z''_\gamma + Z''_\gamma A_c^T - 2Z'_\gamma\tilde{C}_1^T\tilde{C}_1 Z'_\gamma - 6\gamma^{-4}\tilde{B}_1\tilde{B}_1^T = 0$$

implying that $Z''_\gamma \leq 0$. Using the identity $(A^{-1})' = -A^{-1}A'A^{-1}$ we find:

$$\begin{aligned} \frac{d}{d\gamma}\tilde{X}_\gamma &= -Z_\gamma^{-1}Z'_\gamma Z_\gamma^{-1} \leq 0, \\ \frac{d^2}{d\gamma^2}\tilde{X}_\gamma &= Z_\gamma^{-1}\left(2Z'_\gamma Z_\gamma^{-1}Z'_\gamma - Z''_\gamma\right)Z_\gamma^{-1} \geq 0, \end{aligned}$$

implying that $\tilde{X}_\gamma$ and consequently X_γ is nonincreasing and convex. The same results follow for the general problem. Details can be consulted in [7].

Since X_γ is nonincreasing, we can immediately conclude that $\rho(X_\gamma)$ is also nonincreasing. However, since X_γ is symmetric and nonnegative definite, we can show that all the eigenvalues are nonincreasing.

Lemma 5 *Suppose γ_1 and γ_0 are in $(\gamma_f, +\infty)$ with $\gamma_1 < \gamma_0$, and the eigenvalues of the Riccati solutions $\lambda_k(X_{\gamma_1})$ and $\lambda_k(X_{\gamma_0})$ are arranged in decreasing order. Then $\lambda_k(X_{\gamma_1}) \geq \lambda_k(X_{\gamma_0})$ for $k = 1, \ldots, n$.*

Proof: For any positive semi-definite matrices A and E the following identity holds [9]

$$\lambda_k(A) + \lambda_1(E) \geq \lambda_k(A+E) \geq \lambda_k(A) + \lambda_n(E).$$

Now choose $A = X_{\gamma_0}$ and $E = X_{\gamma_1} - X_{\gamma_0}$ and use the second inequality to get

$$\lambda_k(X_{\gamma_0} + X_{\gamma_1} - X_{\gamma_0}) \geq \lambda_k(X_{\gamma_0}) + \lambda_n(X_{\gamma_1} - X_{\gamma_0}).$$

The above implies that $\lambda_k(X_{\gamma_1}) \geq \lambda_k(X_{\gamma_0})$ since $\lambda_n(X_{\gamma_1} - X_{\gamma_0}) \geq 0$. ∎

So far we have shown that the individual Riccati solutions X_γ and Y_γ are nonincreasing functions of γ. However, because of Theorem 1, we are primarily interested in $\rho(X_\gamma Y_\gamma)$. Note that the product is not symmetric but is similar to a nonnegative matrix.

Lemma 6 *Suppose γ_1 and γ_0 are in $(\gamma_f, +\infty)$ with $\gamma_1 < \gamma_0$, Then, $\rho(X_{\gamma_1} Y_{\gamma_1}) \geq \rho(X_{\gamma_0} X_{\gamma_0})$.*

Proof: Suppose A, B, C, and D are symmetric positive semi-definite matrices, and satisfy $A \geq B$ and $C \geq D$. Since these matrices are positive semi-definite they have a positive semi-definite square root. Now, $A \geq B$ implies that $C^{\frac{1}{2}} A C^{\frac{1}{2}} \geq C^{\frac{1}{2}} B C^{\frac{1}{2}}$ from which we conclude that

$$\rho(AC) = \rho(C^{\frac{1}{2}} A C^{\frac{1}{2}}) \geq \rho(C^{\frac{1}{2}} B C^{\frac{1}{2}}) = \rho(BC). \tag{37}$$

Similarly, $C \geq D$ implies that $B^{\frac{1}{2}} C B^{\frac{1}{2}} \geq B^{\frac{1}{2}} D B^{\frac{1}{2}}$, and

$$\rho(BC) = \rho(B^{\frac{1}{2}} C B^{\frac{1}{2}}) \geq \rho(B^{\frac{1}{2}} D B^{\frac{1}{2}}) = \rho(BD). \tag{38}$$

Putting the two inequalitites together we have $\rho(AC) \geq \rho(BD)$. The desired result follows since $X_{\gamma_1} \geq X_{\gamma_0}$ and $Y_{\gamma_1} \geq Y_{\gamma_0}$. ∎

V. GRADIENT METHOD

As noted earlier, Theorem 1 immediately lends itself to a bisection algorithm since an interval containing γ_{opt} can be isolated by checking the three conditions of Theorem 1. We have seen in the previous section that both H_∞ and J_∞ are in *dom*(*Ric*), $X_\infty \geq 0$, and $Y_\infty \geq 0$ implying that all three conditions are satisfied at $\gamma = +\infty$. Further, we deduced the existence of $\gamma_f < +\infty$ such that for $\gamma \in (\gamma_f, +\infty)$ both H_γ and J_γ are in *dom*(*Ric*) and $X_\gamma \geq 0$ and

$Y_\gamma \geq 0$ exist. As γ decreases from $+\infty$, it is of interest to determine which of the three conditions will fail first. It can be argued that condition (iii), namely $\rho(X_\gamma Y_\gamma) < \gamma^2$, is more likely to fail first. Consider condition (i) (the argument for condition (ii) follows by duality). Suppose condition (i) fails first because H_γ has an eigenvalue on $\mathbb{C}_0$. Then there is no controller, stabilizing or not, which makes $\|T_{zw}\|_\infty < \gamma_H$, an unlikely scenario. On the other hand, if $H_\gamma \in dom(Ric)$ but X_γ is indefinite then X_γ must have became unbounded for an even larger value of γ. Hence, it is natural to expect the spectral radius condition to fail first which implies the existence of a nontrivial interval where $X_\gamma \geq 0$, $Y_\gamma \geq 0$, and $\rho(X_\gamma Y_\gamma) > \gamma^2$.

For notational convenience in what follows we introduce a scalar parameter $\alpha := \gamma^{-2}$. Thus, all previously defined scalar- and matrix-valued functions of γ also depend on α in a natural way. Now, our assumption that condition (iii) fails first implies that $\rho(X_\alpha Y_\alpha) = \frac{1}{\alpha}$ at the optimal value. In other words, α_{opt} is a root of the equation

$$h(\alpha) := \alpha\rho_\alpha - 1 = 0, \tag{39}$$

where $\rho_\alpha := \rho(X_\alpha Y_\alpha)$. We can find the root using the Halley-secant ([23], p. 213) (or Newton's) method. To do so we need to form the first and second derivatives of $h(\alpha)$. First, define $h_{\alpha_0} := h(\alpha_0)$ and, for $\delta > 0$, define $h_{\alpha\pm} := h(\alpha_0 \pm \delta)$. Suppose we can compute h_{α_0} and $h_{\alpha\pm}$. Then we can approximate the first and second derivatives as follows:

$$h' := h'(\alpha_0) \approx \frac{h_{\alpha+} - h_{\alpha-}}{2\delta}, \quad h'' := h''(\alpha_0) \approx \frac{h_{\alpha+} - 2h_{\alpha_0} + h_{\alpha-}}{\delta^2}. \tag{40}$$

Using these definitions the Halley-secant method yields

$$\alpha_1 = \alpha_0 + \Delta\alpha, \qquad \text{where} \quad \Delta\alpha = h_{\alpha_0}\left(h' - \frac{h''h_{\alpha_0}}{2h'}\right)^{-1}. \tag{41}$$

We can compute h' and h'' based on our knowledge of how X_α and Y_α vary with α. Since X_α is an analytic function we can write its Taylor series expansion in terms of the scalar parameter α. For $\delta > 0$, let

$$X_{\alpha+\delta} = X_\alpha + \delta X'_\alpha + \frac{\delta^2}{2}X''_\alpha + O(\delta^3). \tag{42}$$

To compute X'_α we first write Eq. (16) in terms of the scalar parameter α,

$$A_\alpha^T X_\alpha + X_\alpha A_\alpha - X_\alpha F_\alpha X_\alpha + G_\alpha = 0, \tag{43}$$

where A_α, F_α, and G_α are counterparts of the expressions defined in Eq. (19). By taking the derivative of Eq. (43) with respect to α we obtain the following Lyapunov equation for X'_α:

$$A_X^T X'_\alpha + X'_\alpha A_X - \begin{pmatrix} I & X_\alpha \end{pmatrix} \mathcal{J} H'_\alpha \begin{pmatrix} I \\ X_\alpha \end{pmatrix} = 0 \tag{44}$$

where

$$\mathcal{J} := \begin{pmatrix} 0 & I \\ -I & 0 \end{pmatrix} \quad \text{and} \quad A_X := A_\alpha - F_\alpha X_\alpha.$$

In a similar manner the following Lyapunov equation can be derived for Y'_α (we omit the details):

$$A_Y Y'_\alpha + Y'_\alpha A_Y^T - \begin{pmatrix} I & Y_\alpha \end{pmatrix} \mathcal{J} J'_\alpha \begin{pmatrix} I \\ Y_\alpha \end{pmatrix} = 0. \tag{45}$$

For reference, the counterpart of Eq. (44) for the standard problem is the following simpler Lyapunov equation:

$$A_X^T X'_\alpha + X'_\alpha A_X + X_\alpha B_1 B_1^T X_\alpha = 0.$$

An analogous equation to (45) can also be written.

The next step is to obtain the gradient of the spectral radius function ρ_α. One way to compute $\rho_{\alpha+\delta}$ is by forming the product $X_{\alpha+\delta} Y_{\alpha+\delta}$ and using a power iteration to compute the spectral radius. Note that since both $X_{\alpha+\delta} \geq 0$ and $Y_{\alpha+\delta} \geq 0$, the product $X_{\alpha+\delta} Y_{\alpha+\delta}$ is similar to a nonnegative definite matrix. However, using first-order approximations for $X_{\alpha+\delta}$ and $Y_{\alpha+\delta}$ results in simple formulas for h' and h''. Forming the products of the first-order approximations leads to

$$\rho_{\alpha+\delta} = \rho(X_\alpha Y_\alpha + \delta(X'_\alpha Y_\alpha + Y'_\alpha X_\alpha) + O(\delta^2)).$$

If we neglect the $O(\delta^2)$ term we can find $\rho_{\alpha+\delta}$ using standard perturbation theory for eigenvalues [22]. Let u and v be the right and left eigenvectors of the product $X_\alpha Y_\alpha$ corresponding to the largest eigenvalue, which we assume to be simple, and define

$$\beta := \frac{u^T E v}{u^T v}, \quad \text{and} \quad E := X'_\alpha Y_\alpha + Y'_\alpha X_\alpha. \tag{46}$$

Then $\rho_{\alpha+\delta} \approx \rho_\alpha + \delta\beta$, and using these values in (40) for h' and h'' leads to

$$h' \approx \rho_\alpha + \alpha\beta \quad \text{and} \quad h'' \approx 2\beta. \tag{47}$$

Now we have all the information needed for the Halley-secant iteration and we can use Eq. (41) to compute α_1. This sets up the basic iteration to find α_{opt}. However, it is not guaranteed that conditions (i) and (ii) of Theorem 1 are met at $\alpha = \alpha_1$, i.e., $H_{\alpha_1} \in dom(Ric)$ and $J_{\alpha_1} \in dom(Ric)$, and stabilizing solutions $X_{\alpha_1} \geq 0$ and $Y_{\alpha_1} \geq 0$ exist. Of course, if α_0 is feasible and $\alpha_1 \leq \alpha_0$ then α_1 is also feasible. The following lemma provides some guidance on how large a step can be taken in this α-iteration.

Lemma 7 *Suppose the Riccati equation*

$$A^T X + XA - XFX + G = 0$$

has an admissible solution $X \geq 0$. *Let* $A_c := A - FX$ *be the closed-loop matrix and* $\Omega(Z) := A_c^T Z + Z A_c$ *be the closed-loop Lyapunov operator. Given a perturbation* ΔA, *the matrix* $A_c + \Delta A$ *is stable if* $\|\Delta A\| \leq (2\|\Omega^{-1}\|)^{-1}$.

Proof: See [10]. ∎

We can use the above lemma to provide bounds for the α step-size. The following shows how to compute the bound for X_α associated with the standard problem. Define $\delta_x := \alpha_1 - \alpha_0$, and the first-order approximation $X_{\alpha_1} \approx X_{\alpha_0} + \delta_x X'_\alpha$, using X'_α determined from Eq. (44). Then the closed-loop matrix at $\alpha = \alpha_1$ is

$$\begin{aligned} A - F_{\alpha_1} X_{\alpha_1} &= A - (F_{\alpha_0} + \delta_x B_1 B_1^T)(X_{\alpha_0} + \delta_x X'_\alpha) \\ &= A - F_{\alpha_0} X_{\alpha_0} - \delta_x \, (B_1 B_1^T X_{\alpha_0} + F_{\alpha_0} X'_\alpha) - \delta_x^2 \, B_1 B_1^T X'_\alpha. \end{aligned}$$

which is in the form $A - F_{\alpha_0} X_{\alpha_0} + \Delta A$. Hence, Lemma 7 yields the following condition for the stability of $A - F_{\alpha_1} X_{\alpha_1}$:

$$\delta_x^2 \, \|B_1 B_1^T X'_\alpha\| + \delta_x \, \|B_1 B_1^T X_{\alpha_0} + F_{\alpha_0} X'_\alpha\| \leq \frac{1}{2\|\Omega_0^{-1}\|}. \tag{48}$$

where $\Omega_0(Z) := A_c^T Z + Z A_c$ is the closed-loop Lyapunov operator. The quadratic equation in δ_x associated with the inequality (48) has two real roots of opposite sign. We solve for and use the positive root. A similar procedure will yield δ_y. Hence, we can use

$$\alpha_b := \min(\delta_x, \delta_y) \tag{49}$$

as an upper bound for the step size computed in (41).

A gradient method, such as the one described above, is guaranteed to converge to a root of a given function if the function is smooth and convex. Since we cannot assume that $h(\alpha)$ is convex we will use a hybrid bisection-gradient method. If the computed value of α_1 in (41) is not in the interval (lb, ub) then we use a bisection step. The bounds are also updated during the iteration. If the conditions of Theorem 1 are met at a given value of α, then the lower bound is updated; otherwise the upper bound is updated. The α-iteration can be summarized as follows:

Gradient Algorithm:

1. Solve for $X_{\alpha_0} = Ric(H_{\alpha_0})$ and $Y_{\alpha_0} = Ric(J_{\alpha_0})$
2. Check conditions of Theorem 1 and update bounds
3. If conditions are not met decrease step size and go to step 1
4. Solve Lyapunov equations (44)–(45) for X'_α and Y'_α, respectively
5. Compute $\Delta\alpha$ using (41)
6. Check stability using Eq. (49)
7. If $\alpha_1 \notin (lb, ub)$ set $\alpha_1 = \frac{1}{2}(ub + lb)$
8. Stop if $\frac{|\alpha_1 - \alpha_0|}{\alpha_0} < tol$. Otherwise, set $\alpha_0 = \alpha_1$ and go to step 1

As we can see from above, the additional workload for the gradient method is the solution of two Lyapunov equations. However, the cost of solving an nth order Riccati equation is approximately $50 - 100n^3$ whereas the cost of solving an nth order Lyapunov equation is about $10 - 12n^3$. Hence, the increase in cost per iteration is small. In the next section we will show how to obtain X'_α essentially for free in case we use the Schur method [14] to solve for X_α.

We now compare the performance of the gradient algorithm described above to a bisection algorithm. In each case we computed an upper and lower bound for γ_{opt} using state-space methods from [4] and used $tol = 10^{-5}$ as the stopping criterion.

Example 1 Consider a linear system which has the following open-loop transfer matrix (as defined in (3)):

$$G(s) = \begin{pmatrix} 0 & 0 & 0 \\ 0 & 0 & 1 \\ 0 & 1 & 0 \end{pmatrix} + \begin{pmatrix} 1 \\ 0 \\ 1 \end{pmatrix} (s+1)^{-1} \begin{pmatrix} 1 & 0 & 1 \end{pmatrix}$$

with $p_1 = 2$ and $m_1 = 2$. The system is open-loop stable and it is easy to verify that the open-loop gain from disturbance to error is $\|T_{open}\|_\infty = 1$. The Hamiltonian $H_\alpha \in \mathbb{R}^{2\times 2}$ defined in Eq. (8) is

$$H_\alpha = \begin{pmatrix} -1 & -(1-\alpha) \\ -1 & 1 \end{pmatrix}.$$

The associated Riccati equation is the scalar equation

$$(1-\alpha)X_\alpha^2 + 2X_\alpha - 1 = 0,$$

which has the positive solution

$$X_\alpha = \frac{1}{1+\sqrt{2-\alpha}}, \quad \text{for} \quad 0 \le \alpha \le 2.$$

In this example $Y_\alpha = X_\alpha$ and we can explicitly compute h' since $\rho_\alpha = X_\alpha^2$ and $\rho'_\alpha = 2X_\alpha X'_\alpha$. We get

$$X'_\alpha = \frac{X_\alpha^2}{2\sqrt{2-\alpha}}, \quad \rho'_\alpha = \frac{X_\alpha^3}{\sqrt{2-\alpha}}, \quad \text{and} \quad h'_\alpha = X_\alpha^2\left(1 + \frac{\alpha X_\alpha}{\sqrt{2-\alpha}}\right).$$

From the gradient algorithm point of view we would compute the closed-loop matrix $A_c = -\sqrt{2-\alpha}$ and solve the Lyapunov equation in Eq. (44) to get

$$X'_\alpha = \frac{X_\alpha^2}{2\sqrt{2-\alpha}},$$

which is the same result as above.

Using methods from Doyle [4] we obtained an upper bound $ub = 0.9215$ and a lower bound $lb = 0.7071$. The optimal norm is $\gamma_{opt} = 0.7320$. The bisection method took 16 iterations whereas the gradient method took 5 iterations.

Example 2 Consider the following example which is a general $\mathcal{H}_\infty$ problem:

$$A = \begin{pmatrix} 0.0904 & 0.8465 \\ 0.6888 & 0.7152 \end{pmatrix}, \qquad B = \begin{pmatrix} 0.7092 & 0.3017 & 0.7001 \\ 0.1814 & 0.9525 & 0.1593 \end{pmatrix},$$

$$C = \begin{pmatrix} 0.3088 & 0.7350 \\ 0.5735 & 0.9820 \\ 0.6644 & 0.5627 \end{pmatrix}, \qquad D = \begin{pmatrix} 0.7357 & 0.4156 & 0.0 \\ 0.2588 & 0.1544 & 1.0 \\ 0.0 & 1.0 & 0.0 \end{pmatrix}$$

where $n = 2$, $m_1 = 2$, $m_2 = 1$, $p_1 = 2$, and $p_2 = 1$. Then $ub = 10.125$, $lb = 1.9525$, and the optimal norm $\gamma_{opt} = 3.1251$. The bisection method took 17 iterations, whereas the gradient method took 6 iterations.

The above examples suggest that the gradient method is much faster than a bisection method. Hence, the gradient method is certainly a step in the right direction in developing algorithms to compute γ_{opt}.

VI. INVARIANT SUBSPACES

In Section IV we proved that the Riccati solutions X_α and Y_α are nondecreasing functions of the scalar parameter α. As the iteration to compute α_{opt} approaches the optimal value, X_α and Y_α can get large and checking the three conditions of Theorem 1 can be a delicate matter. It turns out that we can recast these conditions in terms of certain invariant subspaces of H_α and J_α and thus avoid explicitly computing the Riccati solutions. In the sequel this invariant subspace approach is also applied to the gradient algorithm described in the previous section which leads to further substantial computational savings.

In this section we will be concerned with a fixed value of α and we drop the explicit dependence of all matrices on this parameter. We assume that for this value of α both H_α and J_α are in *dom*(*Ric*). Hence, we can obtain bases for $\mathcal{X}_-(H)$ and $\mathcal{X}_-(J)$ via an ordered Schur decomposition. In other words, we can compute U_1, V_1, U_2, and $V_2 \in \mathbb{R}^{n \times n}$ such that

$$H \begin{pmatrix} V_1 \\ U_1 \end{pmatrix} = \begin{pmatrix} V_1 \\ U_1 \end{pmatrix} S_H \quad \text{and} \quad J \begin{pmatrix} V_2 \\ U_2 \end{pmatrix} = \begin{pmatrix} V_2 \\ U_2 \end{pmatrix} S_J, \tag{50}$$

where $S_H \in \mathbb{R}^{n \times n}$ with $\Lambda(S_H) \subset \mathbb{C}_-$, and $S_J \in \mathbb{R}^{n \times n}$ with $\Lambda(S_J) \subset \mathbb{C}_-$. Hence, the admissible Riccati solutions are $X = U_1 V_1^{-1}$ and $Y = U_2 V_2^{-1}$. Define $W_1 := U_1^T U_2$ and $W_2 := V_1^T V_2$. The following lemma shows that the three conditions of Theorem 1 can be recast in terms of the above invariant subspaces $\mathcal{X}_-(H)$ and $\mathcal{X}_-(J)$.

Lemma 8 *The three conditions of Theorem 1 can be written in terms of the stable invariant subspaces of the Hamiltonians H and J as follows:*

(i) *All* $\lambda_i \in \Lambda(U_1, V_1)$ *are nonnegative and finite,*

(ii) *All* $\lambda_i \in \Lambda(U_2, V_2)$ *are nonnegative and finite,*

(iii) *All* $\lambda_i \in \Lambda(W_1, W_2)$ *are less than* $\frac{1}{\alpha}$.

Proof: Suppose $Xx = \lambda x$, and $w := V_1^{-1}x$. Then

$$Xx = \lambda x \quad \Rightarrow \quad U_1 V_1^{-1} x = \lambda x \quad \Rightarrow \quad U_1 w = \lambda V_1 w.$$

Hence, checking the eigenvalues of X is equivalent to checking the generalized eigenvalues of (U_1, V_1). This proves (i), and (ii) follows similarly. Further, if $XYx = \lambda x$, and $w := V_2^{-1}x$, then

$$XYx = \lambda x \quad \Rightarrow \quad U_1 V_1^{-1} U_2 V_2^{-1} x = \lambda x \quad \Rightarrow \quad W_1 w = \lambda W_2 w,$$

where we have used the fact that $(U_1 V_1^{-1})^T = U_1 V_1^{-1}$. ∎

Therefore, we can check the conditions without explicitly forming the Riccati solutions. However, for the gradient method we need to solve the closed-loop Lyapunov equations (44)–(45). Although it appears that we need the Riccati solutions explicitly, we show below how the entire gradient procedure can be performed without doing so. Suppose $\beta := \frac{\beta_1}{\beta_2}$ is the gradient of the spectral radius $\rho(XY)$, as defined in (46), and let

$$\beta_1 := y^T(X'Y + XY')x, \quad \text{and} \quad \beta_2 := y^T x. \tag{51}$$

Theorem 9 *Let* $\bar{X}' := V_1^T X' V_1$ *and* $\bar{Y}' := V_2^T Y' V_2$. *Also, let* w *and* z *be the respective right and left eigenvectors of* (W_1, W_2) *corresponding to the largest generalized eigenvalue, which we assume to be simple. Then* β_1 *and* β_2 *can be computed as*

$$\beta_1 = z^T(\bar{X}' V_1^{-1} U_2 + U_1^T V_2^{-T} \bar{Y}')w \quad \text{and} \quad \beta_2 = z^T V_1^T V_2 w. \tag{52}$$

Proof: Consider the closed-loop Lyapunov equation in Eq. (44):

$$(A - FX)^T X' + X'(A - FX) - \begin{pmatrix} I & X \end{pmatrix} \mathcal{J} H' \begin{pmatrix} I \\ X \end{pmatrix} = 0.$$

In the above equation substitute $X = U_1 V_1^{-1}$ and pre- and post-multiply by V_1^T and V_1 to get

$$(AV_1 - FU_1)^T X' V_1 + V_1^T X'(AV_1 - FU_1) - \begin{pmatrix} V_1^T & U_1^T \end{pmatrix} \mathcal{J} H' \begin{pmatrix} V_1 \\ U_1 \end{pmatrix} = 0. \tag{53}$$

Now from Eq. (50) we have $AV_1 - FU_1 = V_1 S_H$ whence Eq. (53) becomes

$$S_H^T \bar{X}' + \bar{X}' S_H - \begin{pmatrix} V_1^T & U_1^T \end{pmatrix} \mathcal{J} H' \begin{pmatrix} V_1 \\ U_1 \end{pmatrix} = 0. \tag{54}$$

The above Lyapunov equation can be solved quite cheaply since S_H is already in Schur form. Further, it has a unique solution since S_H is stable [13]. Hence we can solve for $\bar{X}'$ without forming the Riccati solution X. The same remarks apply to $\bar{Y}'$.

Let x and y be the respective right and left eigenvectors of XY corresponding to the largest eigenvalue. Define $w := V_2^{-1}x$. From Lemma 8, w is a right eigenvector of (W_1, W_2). Similarly, $z := V_1{}^{-1}y$ is a left eigenvector of (W_1, W_2). Now the expression for β_1 from Eq. (51) becomes

$$\begin{aligned} \beta_1 &= y^T(X'Y + XY')x \\ &= y^T(V_1^{-T}\bar{X}'V_1^{-1}U_2V_2^{-1} + V_1^{-T}U_1^TV_2^{-T}\bar{Y}'V_2^{-1})x \\ &= z^T(\bar{X}'V_1^{-1}U_2 + U_1^TV_2^{-T}\bar{Y}')w. \end{aligned}$$

Finally, using the definitions of w and z we get

$$\beta_2 = y^T x = z^T V_1^T V_2 w.$$

Thus, we can compute β without forming the Riccati solutions explicitly. ∎

The above result shows that we can compute the gradient of the spectral radius function by working with $\mathcal{X}_-(H)$ and $\mathcal{X}_-(J)$. In fact, this result can be generalized as follows. A significant advance in the numerical solution of Riccati equations was the introduction of the idea of using generalized eigenvalue problems and deflating subspaces rather than eigenvalue problems and invariant subspaces (see [15] for an extensive survey). This is crucial for Riccati equations arising from discrete-time models. In this case, we compute an ordered generalized real Schur form rather than an ordered Schur form, and the Riccati solution is obtained from vectors that span the stable deflating subspace. Fortunately, the technique developed in Theorem 9 can be applied to the deflating vectors as well. Thus, methods developed in this section extend naturally to a much larger class of Riccati equations.

VII. CONCLUSION

We have considered numerical and computational aspects of $\mathcal{H}_\infty$ control and presented an in-depth discussion of a state-space approach to computing the optimal gain. This approach relies on formulating the problem in terms of two Riccati equations that depend on a gain parameter γ. As a first step we characterized the behavior of the solutions to these Riccati equations. This enabled us to couple a gradient method with bisection to compute γ_{opt}, the optimal gain. As expected, numerical experiments showed that the gradient method is much faster than a bisection method. Since we can compute upper and lower bounds for the optimal gain a bisection method is guaranteed to converge. We have derived some important results for the gradient method by proving analyticity, monotonicity, and convexity of the Riccati solutions. The spectral radius function does inherit continuity and monotonicity of the Riccati solutions but the question of convexity is under continued investigation.

We have noted that the Riccati equations can become ill-conditioned as the iteration to compute γ_{opt} approaches the solution. Hence, rather than explicitly forming the Riccati solutions it would be better to work directly with the invariant subspaces of the associated Hamiltonians. We have presented a modified version of the gradient method that avoids explicitly forming the Riccati solutions. It appears that working with the invariant subspace of the Hamiltonians will also lead naturally to an effective algorithm for discrete-time problems.

VIII. REFERENCES

[1] Boyd, S., V. Balakrishnan, and P. Kabamba, "A Bisection Method for Computing the H_∞ Norm of a Transfer Matrix and Related Problems," *Mathematics of Control, Signals, and Systems*, 2(1989), pp. 207–219.

[2] Doyle, J.C. and G. Stein, "Multivariable Feedback: Concepts for a Classical/Modern Synthesis," *IEEE Trans. Auto. Control*, AC-26(1981), pp. 4–16.

[3] Doyle, J., "Analysis of Feedback Systems with Structured Uncertainty," *IEE Proceedings*, Part D, vol. 129(1982), pp. 242–250.

[4] Doyle, J.C., "Lecture Notes on Advances in Multivariable Control," Technical Report, Honeywell, Minneapolis, MN, 1984.

[5] Doyle, J., K. Glover, P.P. Khargonekar, and B.A. Francis, "State-Space Solutions to Standard H_2 and H_∞ Control Problems," *IEEE Trans. Auto. Control*, AC-34 (1989), pp. 831–847.

[6] Francis, B.A., *A Course in H_∞ Control Theory*, Springer-Verlag, New York, 1987.

[7] Gahinet, P. and P. Pandey, "Fast and Numerically Robust Algorithm for Computing the H_∞ Optimum," *Proc. 30th IEEE Conf. Decis. Control*, 1991, to appear.

[8] Glover, K. and J. Doyle, "State-Space Formulas for all Stabilizing Controllers that Satisfy an H_∞ Norm Bound and Relations to Risk Sensitivity," *Sys. and Cont. Letters*, 11 (1988), pp. 167–172.

[9] Golub, G.H. and C.F. Van Loan, *Matrix Computations*, Johns Hopkins University Press, Baltimore, 1983, (second edition: 1989).

[10] Kenney, C. and G. Hewer, "The Sensitivity of the Algebraic and Differential Riccati Equations," *SIAM J. Control Opt.*, 28 (1990), pp. 50–69.

[11] Khargonekar, P., I.R. Petersen, and M.A. Rotea, "H_∞ Optimal Control with State Feedback," *IEEE Trans. Auto. Control*, AC-33 (1988), pp. 786–788.

[12] Kučera, V., "A Contribution to Matrix Quadratic Equations," *IEEE Trans. Auto. Control*, AC-17 (1972), pp. 344–347.

[13] Lancaster, P., *Theory of Matrices*, Academic Press, New York, NY, 1969.

[14] Laub, A.J., "A Schur Method for Solving Algebraic Riccati Equations," *IEEE Trans. Auto. Control*, AC-24 (1979), pp. 913–921, (see also *Proc. 1978 CDC (Jan. 1979)*, pp. 60-65).

[15] Laub, A.J., "Invariant Subspace Methods for the Numerical Solution of Riccati Equations," in *The Riccati Equation*, Bittanti, S., A.J. Laub, and J.C. Willems eds., Springer-Verlag, Berlin, 1991.

[16] Pandey, P., A. Packard, and A.J. Laub, "Characteristics of Riccati Solutions in a General H_∞ Control Problem," *Proc. 28th Annual Allerton Conf. on Communication, Control, and Computing*, Urbana, IL, October 1990.

[17] Pandey, P., C. Kenney, A. Packard, and A.J. Laub, "A Gradient Method for Computing the Optimal H_∞ Norm," *IEEE Trans. Auto. Control*, AC-36 (1991), pp. 887–890.

[18] Petersen, I.R., B.D.O. Anderson, and E.A. Jonckheere, "A First Principles Solution to the Non-singular H_∞ Control Problem," preprint (submitted to *Intl. J. Robust and Nonlinear Cont.*, 1990).

[19] Rudin, W., *Functional Analysis*, McGraw-Hill, New York, 1973.

[20] Safonov, M.G. and D.N.J. Limebeer, "Simplifying the H_∞ Theory via Loop Shifting," *Proc. 27th IEEE Conf. Decis. Control*, pp. 1399–1404, Austin, TX, December 1988.

[21] Sagan, H., *Advanced Calculus*, Houghton Mifflin Company, Boston, MA, 1974.

[22] Stewart, G.W., *Introduction to Matrix Computations*, Academic Press, New York, 1973.

[23] Traub, J.F., *Iterative Methods for the Solution of Equations*, Prentice-Hall, Englewood Cliffs, NJ, 1964.

[24] Wonham, W.M., *Linear Multivariable Control: A Geometric Approach*, Springer-Verlag, New York, 2nd edition, 1979.

[25] Zames, G., "Feedback and Optimal Sensitivity: Model Reference Transformations, Multiplicative Semi-Norms, and Approximate Inverses," *IEEE Trans. Auto. Control*, AC-26 (1981), pp. 301–320.

Techniques in Optimized Model Reduction for High Dimensional Systems

Ubaid M. Al-Saggaf
Maamar Bettayeb

Electrical Engineering Department
King Fahd University of Petroleum and Minerals
Dhahran 31261, Saudi Arabia

I. INTRODUCTION

Mathematical Modeling of most physical systems gives birth to infinite dimensional models. Some examples are network systems, telecommunications, transmission lines, wave propagation, economic systems, chemical reactors and distillation columns. These include systems governed by partial differential equations (distributed systems), delay equations, or integro-differential equations. Concrete analysis and design of general infinite dimensional models are well beyond the existing control theory. Very often, based on engineering judgements or approximations one then derives a finite dimensional approximate model of the physical system. This model is then a "rough" representation of the actual system. However, a large scale complex system may still require a very high dimensional representation which complicates the analysis, simulation and compensator

CONTROL AND DYNAMIC SYSTEMS, VOL. 55

design for such systems.

Compensator synthesis techniques such as linear quadratic Gaussian (LQG) [1-3] and H_∞ optimization techniques [4] require a considerable amount of computation which increases rapidly with the order of the model. These techniques result in high order compensators whose order may even exceed the order of the plant and may be more complex than it would be reasonable to implement. There are two approaches for reduced order controller design. In the first approach, the order of the plant is reduced and then a controller is designed for the reduced order plant. In the second approach, a controller is designed for the full order plant and then reduced. For details on the merits of both approaches and the performance of various reduction algorithms for reduced order controller design, see [65].

In addition, although the advancement of computer technology has enabled handling a large amount of data, such usage of computer time and storage may not be affordable. However, if a system is reasonably approximated by a lower order model, then this can result in efficient and reliable data manipulation, which will result in faster processing, and easier and more accurate system analysis.

The model reduction problem has been a major attraction in system theory literature and considerable attention has been devoted to it in the last few decades and one has only to examine the comprehensive list of references compiled by Genesio et al. [5] to appreciate this fact. Various reduction methodologies have been proposed and algorithms of diverse computational complexity have been presented. The methods are quite diverse, but they can be divided approximately into two classes:

(i) Frequency domain methods

(ii) Time domain methods.

One of the first approaches using state space techniques was the modal analysis of Davison [6]. He basically retains the dominant system eigenvalues and their corresponding eigenvectors. Various modifications of this approach have been subsequently offered [10-11]. In other approaches, Chen et al. [12] use frequency domain expansions, Gibilaro et al. [13] match the moments of the impulse response and Hutton et al. [14] use the Routh approach for high frequency approximation which is later modified by Langholz et al. [15]. Pinguet [16] showed that all of these methods have state space reformulations. What is most striking about all these methods of system approximation is that optimality with respect to some criterion either over all time or all frequencies is rarely considered and never proven. However, optimization techniques in model reduction were considered by some authors. In the time domain, the method of Meier et al. [40] was one of the first attempts. Wilson [17-18] used an optimization approach which is the minimization of the integral squared impulse response error between the full and reduced order models. His technique was generalized and extended by many authors [19-21], [26-31], [37]. In another approach, Obinata et al. [22-23] and Eitelberg [24] minimized the equation error which leads to closed form solutions. Extensions and generalizations of this method were given in [34-36]. Another time domain optimization approach is the minimization of the l_1-norm of the error impulse response [25].

In the frequency domain, optimization techniques for model reduction were considered by Reddy [41], Luus [42], Bistritz et al. [43] and El-Attar et al. [25]. In [44] the author optimized cost functions which are combinations of time domain and frequency domain criteria.

One of the severe drawbacks of most of the above optimization techniques is that they result in non-linear equations in the parameters of the reduced order model. The solution to these non-linear equations is computationally demanding requiring iterative minimization algorithms which suffer from many difficulties such as the choice of the starting guesses, convergence, and multiple local minima.

Recently two developments have dramatically changed the status of model reduction. These are the theories of balanced realizations [52] and optimal Hankel-norm approximations [46-51]. With these techniques it is possible to predict the error between the frequency responses of the full and reduced order models [48], [51], [57-62]. Moreover, the Hankel-norm reduction is optimal, in Hankel-norm sense, has a closed form solution and is computationally simple employing standard matrix software. A preliminary survey of the optimal model reduction methods mentioned above is given in [131].

Consider the state space description of a time-invariant linear dynamical system in continuous time

$$\dot{x}(t) = Ax(t) + Bu(t) \tag{1a}$$

$$y(t) = Cx(t) \tag{1b}$$

where $x(t)$, $u(t)$ and $y(t)$ are, respectively, the state, input and output vectors. Discrete-time systems are represented as

$$x(k+1) = Ax(k) + Bu(k) \tag{2a}$$

$$y(k) = Cx(k) \tag{2b}$$

The dimension of the state space is called the order of the system. The input-output representation of the above systems can also be characterized in frequency domain by their transfer functions

$$G(s) = C(sI - A)^{-1} B$$

$$G(z) = C(zI - A)^{-1} B$$

The state space representation of the reduced order model is assumed to be in the form

$$\dot{x}_r(t) = A_r x_r(t) + B_r u(t)$$

$$y_r(t) = C_r x_r(t)$$

and its transfer function is

$$G_r(s) = C_r (sI - A_r)^{-1} B_r$$

for continuous time systems. For discrete time systems, the state space representation of the reduced order model is given by

$$x_r(k+1) = A_r x_r(k) + B_r u(k)$$

$$y_r(k) = C_r x_r(k)$$

and its transfer function

$$G_r(z) = C_r (zI - A_r)^{-1} B_r$$

In time domain, the objective of model reduction is to find the set of reduced order model matrices A_r, B_r, and C_r such that y_r approximates y as close as possible for all admissible inputs.

In frequency domain, the objective is to find the transfer function of the reduced order model G_r such that G_r is close to G in some specified sense.

The aim of this chapter is mainly to survey optimal model reduction techniques and related methods such as balanced-truncation. Sections II and III of this chapter discuss classical optimal reduction techniques in time and frequency domains. Balanced representations and reductions are treated in section IV as they are used to characterize the optimal Hankel-norm solutions. In section V, optimal Hankel-norm model reduction is reviewed. Finally, in section VI, the different techniques are illustrated with simple examples.

II. OPTIMAL CLASSICAL TIME DOMAIN MODEL REDUCTION TECHNIQUES

Different optimal time domain reduction techniques were proposed. Basically the methods were concerned with the minimization of a function of the output error between the full and reduced order models, the equation error, or an induced norm on the error impulse response. For the minimization of the output error criterion, the optimal approximation has been studied most often in the case when the inputs are delta functions. The most important results were obtained by Wilson [17-18], rederived by Galiana [19], Riggs et al. [20] and Hirzinger et al. [21]. For the equation error approach, the most important contributions were obtained by Obinata et al. [22-23] and Eitelberg [24]. The l_1 -norm minimization approach is due to El-Attar et al. [25]. First, the minimization of the output error criterion is described. Following Wilson [17], the cost function to be minimized assuming a high order stable model is given by:

$$J_1 = \int_0^\infty e^T(t)\, Q\, e(t)\, dt \tag{3}$$

where Q is a positive-definite symmetric matrix, and $e(t)$ is the output error and is given by:

$$e(t) = y(t) - y_r(t) = C\, x(t) - C_r\, x_r(t) \tag{4}$$

Substitution of Eq. (4) in Eq. (3) results in:

$$J_1 = \int_0^\infty z^T(t)\, M\, z(t)\, dt \tag{5}$$

where

$$z(t) = \begin{bmatrix} x(t) \\ x_r(t) \end{bmatrix} \quad \text{and} \quad M = \begin{pmatrix} C^T Q C & -C^T Q C_r \\ -C_r^T Q C & C_r^T Q C_r \end{pmatrix}$$

Minimization of Eq. (5) leads to the following two Lyapunov equations:

$$FR + RF^T + S = 0 \tag{6}$$

$$F^T P + PF + S = 0 \tag{7}$$

where

$$F = \begin{pmatrix} A & 0 \\ 0 & A_r \end{pmatrix} \quad \text{and} \quad S = \begin{pmatrix} BB^T & BB_r^T \\ B_rB^T & B_rB_r^T \end{pmatrix}$$

If P and R are partitioned compatibly with F as:

$$P = \begin{pmatrix} P_{11} & P_{12} \\ P_{12}^T & P_{22} \end{pmatrix} \quad \text{and} \quad R = \begin{pmatrix} R_{11} & R_{12} \\ R_{12}^T & R_{22} \end{pmatrix}$$

Then the necessary conditions for optimality give:

$$A_r = -P_{22}^{-1} P_{12}^T A \, R_{12} R_{22}^{-1}$$

$$B_r = -P_{22}^{-1} P_{12}^T B$$

$$C_r = C \, R_{12} R_{22}^{-1}$$

Clearly Eqs. (6)-(7) are non-linear in the unknown reduced order model matrices A_r, B_r, C_r. This non-linearity is a severe drawback of the method. The method is computationally demanding and requires iterative minimization algorithms which suffer from many difficulties such as the choice of the starting guesses, convergence, and multiple local minima. However, it has been shown, in the particular case where the eigenvalues of A_r are pre-specified, that the problem becomes linear and Eqs. (6)-(7) reduce to that of solving a linear system of equations. In deriving Eqs. (3)-(7), it was assumed that the input is a Dirac delta function. The above approach was extended to multivariable systems [18], and to polynomial input functions [26-27]. Other related results are the geometrical approach [28], and the W-matrix approach [29-30]. The method is also extended to discrete-time systems by Aplevich [31] where now the cost function is defined as:

$$J_1 = \sum_{k=0}^{\infty} e_k^T \, Q \, e_k$$

Another important and closely related approach is the minimization of the equation error [22-24]. This approach was motivated by the work of Bierman [32] and Nosrati et al. [33] in the problem of modeling linear time-varying systems by linear time-invariant systems. Here the results in [23] are summarized. The reduced state vector $x_r(t)$ is taken to approximate a linear combination of $r < n$ state variables of the original full order system:

$$x_r(t) \approx \bar{x}(t) = R \, x(t) \tag{8}$$

where R is a suitable reduction matrix. If the state error is defined as:

$$e_x(t) = R\,x(t) - x_r(t) \tag{9}$$

Then this is a special case of the output error *e(t)* when the output matrices C, and C_r are chosen to be equal to the identity matrix. As seen before, the parameters of the reduced order model which minimize the cost function J_1 of Eq. (3) are not easily obtained because of the non-linear manner in which the parameters appear in $e_x(t)$. Now differentiating and rearranging gives:

$$\dot{e}_x(t) = A_r\, e_x(t) + d(t) \tag{10}$$

where
$$d(t) = (RA - A_r R)\,x(t) + (RB - B_r)\,u(t) \tag{11}$$

d(t) is called the equation error. The reason is that Eq. (11) can be expressed as:

$$d(t) = \dot{\bar{x}}(t) - (A_r \bar{x}(t) + B_r u(t)) \tag{12}$$

i.e. the equation error is the error which arises as a result of substituting $\bar{x}(t)$ of Eq. (8) into the state equations of the reduced order model. From Eq. (10), it is seen that *d(t)* is the forcing term. Thus the error $e_x(t)$ is expected to be small if *d(t)* is small enough. This gives the relation between the state error and the equation error. As will be seen shortly, minimization of a performance index based on *d(t)* is very fruitful in that a closed form solution is derived.

Define the following linearly independent inputs:

$$u^i(t) = [0\,0 \ldots 0\,f(t)\,0 \ldots 0]^T; \quad i = 1, \ldots, m$$

where $u^i(t)$ has zero elements except for the ith element, and *f(t)* is taken to be either a unit delta function or a unit step function. Assume that the full order system is asymptotically stable and define the cost function:

$$J_2 = tr\left\{\int_0^\infty D^T(t)\,P\,D(t)\,dt\right\} \tag{13}$$

where

$$D(t) = [d^1(t)\,d^2(t) \ldots d^r(t)] = (RA - A_r R)X(t) + (RB - B_r)U(t)$$

$$d^i(t) = (RA - A_r R)\,x^i(t) + (RB - B_r)\,u^i(t)$$

$$X(t) = [x^1(t)\;\; x^2(t)\;\; \ldots\;\; x^m(t)]$$

$$U(t) = [u^1(t)\;\; u^2(t)\;\; \ldots\;\; u^m(t)]$$

P is a positive definite weighting matrix, $d^i(t)$ is the equation error, and $x^i(t)$ is the state vector of the full order model when the input $u^i(t)$ is used. Define the matrices:

$$W_x = \int_0^\infty X(t) P X^T(t)\, dt$$

$$W_{xu} = \int_0^\infty X(t) P U^T(t)\, dt$$

$$W_u = \int_0^\infty U(t) P U^T(t)\, dt$$

If W_u and RWR^T are non singular where

$$W = W_x - W_{xu} W_u^{-1} W_{xu}^T$$

then the A_r and B_r that minimize J_2 are given by:

$$A_r = RAWR^T (RWR^T)^{-1} \tag{14}$$

$$B_r = RA\,(I - WR^T (RWR^T)^{-1} R) W_{xu} W_u^{-1} + RB \tag{15}$$

When *f(t)* tends to a unit delta function, Eqs. (14)-(15) reduce to:

$$A_r = RAW_xR^T (RW_xR^T)^{-1}$$

$$B_r = RB$$

and W_x will be given by the following Lyapunov equation:

$$AW_x + W_xA^T + BPB^T = 0$$

When *f(t)* is equal to a unit step function, Eqs. (14)-(15) reduce to:

$$A_r = RW_xA^{-T}R^T RA^{-1} W_x A^{-T} R^T$$

$$B_r = A_r RA^{-1} B$$

Extensions to unstable systems and modification and specialization to single-input-single output systems are given in [34] and [35-36] respectively.

Another method related to output error minimization is given in [37]. In this method, impulse inputs are used. The method is based on using a stability equation and Pade's approximation to get a reduced order model. Two free parameters are incorporated in the reduced order model and the impulse response error is minimized with respect to those two parameters. Of course the resulting reduced order model is only suboptimal with respect to the performance measure J_1 of Eq. (3). However, the computational complexity involved in the minimization is greatly reduced since the technique results in only two non-linear equations to be solved. This method is summarized below.

Given the full order model:

$$g(s)=\frac{n(s)}{d(s)}=\frac{b_0+b_1 s+\ldots+b_{n-1}s^{n-1}}{1+a_1 s+\ldots+a_n s^n}=\sum_{i=0}^{\infty} h_i s^i \tag{16}$$

An rth reduced order model which is a function of two parameters k_1 and k_2 is given by:

$$g_r(s)=\frac{n_r(s,k_1,k_2)}{d_r(s,k_1,k_2)}=\frac{\overline{b}_0+\overline{b}_1 s+\ldots+\overline{b}_{r-1}s^{r-1}}{1+\overline{a}_1 s+\ldots+\overline{a}_r s^r} \tag{17}$$

This reduced order model is obtained by the following algorithm:

1. Express *d(s)* as $d(s)=d_e(s)+d_o(s)$ where

$$d_e(s)=1+a_2 s^2+a_4 s^4+\ldots=\prod_{i=1}^{m_1}\left(1+\frac{s^2}{z_i^2}\right)$$

$$d_o(s)=a_1 s+a_3 s^3+\ldots=a_1 s\prod_{i=1}^{m_2}\left(1+\frac{s^2}{p_i^2}\right);\quad a_1>0$$

and m_1 and m_2 are the integer parts of *n/2* and *(n-1/2)* or *(n+1/2)* respectively.

2. Using Michailov's Stability Criterion and Pade's approximation, a denominator polynomial $d_r(s)$ is defined by discarding the factors with the larger magnitudes of z_i^2 and p_i^2 as done by Chen et al. [38]:

$$d_{re}(s)=\prod_{i=1}^{\overline{m}_1}\left(1+\frac{s^2}{z_i^2}\right)$$

$$d_{ro}(s)=a_1 s\prod_{i=1}^{\overline{m}_2}\left(1+\frac{s^2}{p_i^2}\right);$$

$$d_r(s)=d_{re}(s)+d_{ro}(s)$$

where $\overline{m}_1$ and $\overline{m}_2$ are the integer parts of *r/2* and *(r-1/2)* or *(r+1/2)* respectively.

3. Define $d_r(s,k_1,k_2)$ by

$$d_{re}(s)=\prod_{i=1}^{\overline{m}_1}\left(1+\frac{k_1 s^2}{z_i^2}\right)$$

$$d_{ro}(s)=a_1 s\prod_{i=1}^{\overline{m}_2}\left(1+\frac{k_1 s^2}{p_i^2}\right)$$

$$d_r(s,k_1,k_2)=d_{re}(s,k_1)+k_2 d_{ro}(s,k_1)$$

4. Match the r initial moments of the systems of Eqs. (16)-(17):

$$\bar{b}_i = h_i + \sum_{j=1}^{i} \bar{a}_j h_{i-j}; \quad i = 0, 1, \ldots, r-1$$

and define $n_r(s, k_1, k_2)$ by:

$$n_r(s, k_1, k_2) = \sum_{i=1}^{r-1} \bar{b}_i s^i$$

Note that with this construction, the coefficients of $n_r(s, k_1, k_2)$ are explicitly known functions of k_1 and k_2.

Among the class of the rth reduced order models given by Eq. (17), one is selected that minimizes J_1 of Eq. (3). Since the cost function is only a function of k_1 and k_2, only two non-linear equations have to be solved. Also as discussed in [37], the optimal values for k_1 and k_2 usually lie between 0 and 1 which simplify the solution of the optimization problem.

The last important time domain model reduction method with an optimization criterion is due to El-Attar et al. [25]. The cost function is defined as:

$$J_3 = |h_{e0}| + \|\bar{h}_e(t)\|_1 \tag{18}$$

where $$\|\bar{h}_e(t)\|_1 = \int_o^{\infty} |\bar{h}_e(t)| \, dt$$

and $h_e(t) = h_{e0}\, \delta(t) + \bar{h}_e(t)$ is the error impulse response and $\delta(t)$ is the Dirac delta function.

An important feature of the above cost function is that it is the L_1-norm of the impulse response of the error. Thus minimization of Eq. (18) will give a uniformly good approximation over all L_1 or L_∞ -inputs. This is in contrast to the cost functions J_1 and J_2 which depend on the input used and changing the input will change the resulting reduced order model that minimizes J_1 or J_2. The minimization of J_3, which is equivalent to determining a best L_1 approximation, is a very complicated task. Therefore in [25] the discretized problem is solved using the algorithm in [39]. That is, the cost function which is minimized is given by

$$\bar{J}_3 = \sum_{i=1}^{N} |h_e(t_i)|$$

where $t_l \leq t_1 < t_2 < \ldots . < t_N \leq t_u$ is a set of N points and $[t_l, t_u]$ is the time interval of interest.

III. OPTIMAL CLASSICAL FREQUENCY-DOMAIN MODEL REDUCTION TECHNIQUES

Different methods were proposed to find a reduced order model that optimize a frequency-domain criterion. The most important results were obtained by Reddy [41], Luus [42], El-Attar et al. [25], and Bistritz et al. [43].

Reddy [41] minimizes the integral of the square of the error between corresponding real and imaginary parts of original and assumed transfer functions numerator and denominator to evaluate the coefficients of the approximate model. He assumes that the transfer function of the full and reduced order models are expressed for, $s = j\omega$, by

$$g(j\omega) = \frac{n(j\omega)}{d(j\omega)} = \frac{F_1(\omega) + j\omega F_2(\omega)}{F_3(\omega) + j\omega F_4(\omega)}$$

$$g_r(j\omega) = \frac{n_r(j\omega)}{d_r(j\omega)} = \frac{R_1(\omega) + j\omega R_2(\omega)}{R_3(\omega) + j\omega R_4(\omega)}$$

The parameters of the reduced order model $g_r(s)$ are found by minimizing the following four cost functions:

$$E_i = \int_{\omega_l}^{\omega_u} (F_i - R_i)^2 \, d\omega , \quad i = 1 , .. , 4 \tag{19}$$

where $\omega_l \leq \omega \leq \omega_u$ represents the frequency range of interest. Equation (19) is then translated into a set of linear equations that can be uniquely solved for the reduced order model parameters.

One advantage of Reddy's approach is that it is computationally simple. However, the reduced model was found to be considerably worse than that obtained by a non-optimal approach such as the continued fraction method of Chen et al. [12]. The problem lies in considering the numerator and denominator polynomials separately in the minimization. To overcome this, Luus [42] proposed the following cost function:

$$J_4 = \sum_{i=1}^{N} q_i^2 \, | g(j\omega_i) - g_r(j\omega_i) |^2 \tag{20}$$

where q_i is a weighting sequence and $\omega_l \leq \omega_1 < \omega_2 < \ldots . < \omega_N \leq \omega_u$ is a selected set of frequencies in the range of interest.

This method resulted in reduced order models that approximate the full order model more "closely" than Reddy's method. However, the minimization of the

cost function J_4 results in non-linear equations in the reduced order model parameters. The same severe drawbacks of J_1 are also applicable here. Moreover, the reduced order model is a function of the discrete frequencies ω_i and changing them may give a different reduced order model.
El-Attar et al. [25] proposed to minimize the L_∞- norm to get the reduced order model, i.e. they proposed to minimize

$$J_5 = \| g(s) - g_r(s) \|_\infty = \sup_{\omega \geq 0} | g(j\omega) - g_r(j\omega) | \tag{21}$$

The minimization of J_5 is equivalent to determining a best L_∞ approximant. The solution to this problem is a very complicated task. Thus they proposed to use the algorithm in [45] to solve the discretized problem, i.e. the cost function to be minimized is now

$$\bar{J}_5 = \max_{i \in [1,N]} | g(j\omega_i) - g_r(j\omega_i) |$$

where $\omega_l \leq \omega_1 < \omega_2 < \dots < \omega_N \leq \omega_u$ is a selected set of frequencies in the range of interest. A similar criterion was given by Bistritz et al. [43] where they proposed an algorithm for deriving local best Chebyshev rational approximations based on weighted least squares approximations at discrete frequencies.

IV. BALANCED REPRESENTATIONS AND MODEL REDUCTION

A. MODEL REDUCTION VIA BALANCING

Model reduction based on balancing does not generally lead to optimal solutions. However, balancing tools are essential to the optimal approximation problems to be treated in the next section. Actually, optimal Hankel norm solutions are given in closed form in terms of the balanced representation of the system. Besides, reduced models via balancing proved to be near optimal in many circumstances.

Balanced representations for discrete time systems were first introduced by Mullis et al. [7-8] in the synthesis of minimum roundoff noise fixed point digital filters. They considered the problem of finding optimal word length to compromise between storage and quantization efficiencies. In control systems terminology, storage and quantization effects correspond to controllability and observability properties, respectively. The best tradeoff between high controllability with low observability and low controllability with high observability is provided by internally balanced realizations. The states of such realizations are balanced between controllability and observability. Thus they represent a convenient structure for model reduction.

The results of [7-8] were extended by Moore [52] for continuous time systems who gave a geometrical interpretation of internally balanced representation and used it for model reduction.

Discrete time balanced realizations have also been considered by Kung [9]. He proposed a minimal realization algorithm based on singular value analysis of the Hankel matrix. His realization is balanced and is identical to the one obtained using balancing [46].

As will be seen later, this realization is essentially unique. Thus in the sequel the term balanced realization will be used to describe it.

A geometrical interpretation for the discrete time balanced representation can be given. Let S_r be the set of points in the x-state space that are reachable from zero initial conditions with some input sequence $\mathbf{U}$ such that $\|\mathbf{U}\| \leq 1$ where

$$\mathbf{U} = [\, \ldots\ldots \quad u_1^T \quad u_0^T \,]^T$$

i.e.

$$S_r = \{\, x : x = \mathbf{C}\,\mathbf{U}\,, \quad \|\mathbf{U}\| \leq 1 \,\} \tag{22}$$

where $\mathbf{C} = [B \quad AB \quad A^2B \ldots\ldots]$ is the infinite controllability matrix. So if $x \in S_r$, then

$$\mathbf{U} = \mathbf{C}^T (\mathbf{C}\,\mathbf{C}^T)^{-1} x = \mathbf{C}^T W_c^{-1} x \tag{23}$$

where $W_c = \mathbf{C}\,\mathbf{C}^T$ is the controllability gramian. Thus $\mathbf{U}$ in Eq. (23) is the minimum norm input sequence that will take the system from zero initial conditions to state x. Now

$$\|\mathbf{U}\| \leq 1 \quad \Rightarrow \quad x^T W_c^{-1}\, \mathbf{C}\,\mathbf{C}^T W_c^{-1} x = x^T W_c^{-1} x \leq 1$$

thus

$$S_r = \{\, x : x^T W_c^{-1} x \leq 1 \}$$

If $\quad W_c = V\,Z\,V^T\,, \quad V = [v_1 \quad v_2 \ldots v_n]\,, \quad Z = diag\ \{z_1\,, z_2 \ldots z_n\}\,, \quad V\,V^T = I$

then S_r is an ellipsoid with principal axes $v_1\,, v_2\,, \ldots\,, v_n$ whose lengths are $\sqrt{z_1}\,, \sqrt{z_2}\,, \ldots, \sqrt{z_n}$, respectively. Now let $x = v_k$ and let $\mathbf{U}^k$ be the minimum norm control sequence that will take the system from the origin to x. Then $\|\mathbf{U}^k\|^2 = \frac{1}{a_k}$.

Thus if $z_i < z_j$, then $\| \mathbf{U}^i \| > \| \mathbf{U}^j \|$, i.e. the direction v_i is less controllable than the direction v_j.

Similarly let S_0 be the set of points in the x-state space that when used as initial conditions with zero input will produce an output sequence **Y** such that $\| \mathbf{Y} \| \leq 1$ where

$$\mathbf{Y} = [y_0^T \quad y_1^T \quad \ldots\ldots]^T$$

i.e.

$$S_0 = \{ x : \mathbf{Y} = \mathbf{O}\, x \, , \quad \| \mathbf{Y} \| \leq 1 \} \tag{24}$$

Since $\mathrm{Y} = \mathbf{O}\, x$ where $\mathbf{O} = [C^T \quad A^T C^T \quad (A^2)^T C^T \ldots\ldots]^T$ is the infinite observability matrix, then

$$\| \mathbf{Y} \| \leq 1 \quad \Rightarrow \quad x^T \mathbf{O}^T \mathbf{O}\, x = x^T W_o\, x \leq 1$$

thus

$$S_o = \{ x : x^T W_o\, x \leq 1 \} \tag{25}$$

If $W_o = W\, S\, W^T$, $W = [w_1 \quad w_2 \ldots w_n]$, $B = diag\ \{s_1 , s_2 \ldots s_n\}$, $W\, W^T = I$

then S_0 is an ellipsoid with principal axes $w_1 , w_2 , \ldots , w_n$ whose lengths are $\frac{1}{\sqrt{s_1}}, \frac{1}{\sqrt{s_2}}, \ldots, \frac{1}{\sqrt{s_n}}$ respectively.

As with the controllability gramian, it can be argued that if $s_j > s_i$ then an initial condition in the direction of w_j has more effect in the output than an initial condition in the direction w_i. i.e. the direction w_j is more observable than the direction w_i.

If a similarity transformation T is used such that the axes of the ellipsoids and their lengths are the same for the new state variables, then every state direction will be as controllable as observable. Also the lengths of the ellipsoids principal axes will give a measure of how controllable and observable a certain state direction is.

The controllability gramian W_c , and the observability gramian W_o can be found as the unique solutions of the following Lyapunov equations:

$$A\, W_c\, A^T - W_c + B\, B^T = 0 \tag{26a}$$

$$A^T\, W_o\, A - W_o + C^T\, C = 0 \tag{26b}$$

If the representation of the system of Eq. (2) is transformed to another representation using the nonsingular transformation T, then the new state space representation of the system is $(\overline{A}, \overline{B}, \overline{C})$ where

$$\overline{A} = T^{-1} A\, T, \qquad \overline{B} = T^{-1} B, \qquad \overline{C} = C\, T$$

and the gramians will be transformed to

$$\overline{W}_c = T^{-1} W_c T^{-T}, \quad \overline{W}_o = T^T W_o T, \quad \overline{W}_c \overline{W}_o = T^{-1} W_c W_o T$$

Thus the gramians W_c and W_o depend on the state-space coordinates. However, the eigenvalues of their product $W_c W_o$ are invariant under state space transformations and thus are input/output invariants. The square root of the eigenvalues of $W_c W_o$ are called the Hankel singular values of the system $G(z) = C(zI - A)^{-1}B$.

A realization $(\overline{A}, \overline{B}, \overline{C})$ of $G(z)$ is said to be balanced, if $\overline{W}_c = \overline{W}_o = \Sigma$, where $\Sigma = diag\{\sigma_1, \sigma_2, \dots, \sigma_n\}$, $\sigma_1 \geq \sigma_2 \geq \dots \geq \sigma_n > 0$. If

$$W_c = R^T R, \quad U^T R W_o R^T U = \Sigma^2, \quad U^T U = I$$

then the state space transformation:$T_b = R^T U \Sigma^{-1/2}$ will transform the system (A, B, C) to a balanced representation.

The states of a balanced representation are balanced between controllability and observability. That is, the input-to-state coupling and the state-to-output coupling are weighted equally so that those state components which are weakly coupled to both the input and the output are discarded. The Hankel singular values provide a means of determining those states. Thus the states corresponding to the smallest Hankel singular values can be neglected. If the realization $(\overline{A}, \overline{B}, \overline{C})$ of $G(z)$ is balanced and $\overline{A}, \overline{B}, \overline{C}$, and Σ are partitioned compatibly as

$$\overline{A} = \begin{pmatrix} A_{11} & A_{12} \\ A_{21} & A_{22} \end{pmatrix}, \quad \overline{B} = \begin{pmatrix} B_1 \\ B_2 \end{pmatrix}, \quad \overline{C} = [C_1\ C_2] \quad \Sigma = \begin{pmatrix} \Sigma_1 & 0 \\ 0 & \Sigma_2 \end{pmatrix} \tag{27}$$

where the dimensions of A_{11} and Σ_1 are $r \times r$, then eliminating the least controllable-observable states gives the following balanced-truncation reduced order model: $G_{br} = C_1 (zI - A_{11})^{-1} B_1$.

The balanced representation has several interesting properties as do the reduced order models obtained from the balanced representation. For example, a balanced representation of an asymptotically stable, controllable, and observable system is unique up to an arbitrary transformation T such that $T\Sigma = \Sigma T$ and $T^T T = I$ and if the Hankel singular values are distinct, then T is just a sign matrix [51]. Pernebo et al. [64] showed that $\|\overline{A}\|_s \leq 1$, where $\|.\|_s$ denotes the spectral norm, and strict inequality holds if the Hankel singular values are distinct. Also, they showed that the balanced-truncation reduced order model is asymptotically stable and that if $\sigma_r > \sigma_{r+1}$ it is also controllable and observable. In [66] it was shown that the balanced system of Eq. (27) satisfies the following inequalities:

$$\|\overline{B}\|_s \leq \sigma_1^{1/2}, \qquad \|B_2\|_s \leq \sigma_{r+1}^{1/2}, \qquad \|A_{21}\|_s \leq \left(\frac{\sigma_{r+1}}{\sigma_r}\right)^{1/2}$$

$$\|\overline{C}\|_s \leq \sigma_1^{1/2}, \qquad \|C_2\|_s \leq \sigma_{r+1}^{1/2}, \qquad \|A_{12}\|_s \leq \left(\frac{\sigma_{r+1}}{\sigma_r}\right)^{1/2}$$

Moreover, it was shown that the balanced-truncation reduced order model enjoys the following L_∞-norm error bound [62]

$$\overline{\sigma}\,(G(e^{j\theta}) - G_{br}(e^{j\theta})) \leq 2 \sum_{i=r+1}^{n} \sigma_i \quad \text{for all} \quad \theta$$

where $\overline{\sigma}$ denotes the largest singular value.

Reliable algorithms for computing the balancing transformation are described in [53]. However, for the purpose of getting a reduced order model, a balancing transformation T_b need not be computed. This transformation is intrinsically badly conditioned for systems with some nearly uncontrollable and/or unobservable modes. A reliable algorithm is described in [67] for computing the reduced order model without the need for computing the balancing transformation T_b. Moreover, this algorithm does not require the system (A, B, C) to be controllable and observable (i.e. minimal).

The discrete balanced-truncation reduced order model is not balanced. In [59] use is made of the least controllable and observable states of the discrete balanced representation to get a balanced reduced order model. Assuming $(\overline{A}, \overline{B}, \overline{C})$ is balanced and partitioned as in Eq. (27), the reduced order model is defined as

$$A_r = A_{11} - A_{12}\,(\alpha I + A_{22})^{-1}\,A_{21}$$

$$B_r = B_1 - A_{12}\,(\alpha I + A_{22})^{-1}\,B_1$$

$$C_r = C_1 - C_2\,(\alpha I + A_{22})^{-1}\,A_{21}$$

$$D_r = -\,C_2\,(\alpha I + A_{22})^{-1}\,B_2$$

where α is any complex number such that $|\alpha| = 1$. This reduced order model is asymptotically stable, and the reduction error defined by

$$E(z) = \overline{C}\,(zI - \overline{A})^{-1}\,\overline{B} - (D_r + C_r\,(zI - A_r)^{-1}\,B_r)$$

has a transmission zero at $z = -\alpha$ and its Chebyshev norm is bounded by twice the trace of Σ_2. The choice of α shapes the reduction error and thus the modeling error can be forced to zero at $z = \pm 1$. Other properties of this reduced order model can be found in [59].

For continuous time systems, balanced representations are defined in a similar way. However, the controllability gramian W_c, and the observability gramian W_o are given by

$$W_c = \int_0^\infty e^{At} B B^T e^{A^T t} dt$$

$$W_o = \int_0^\infty e^{A^T t} C^T C e^{At} dt$$

and can be found as the unique solution of the following Lyapunov equations:

$$A W_c + W_c A^T + B B^T = 0 \qquad (28a)$$

$$A^T W_o + W_o A + C^T C = 0 \qquad (28b)$$

If the realization $(\overline{A}, \overline{B}, \overline{C})$ of $G(s)$ is balanced and $\overline{A}, \overline{B}, \overline{C}$, and Σ are partitioned compatibly as in Eq. (27) then it is shown in [64] that $\| e^{A t} \|_s < 1$, $t > 0$ and that the reduced order model (A_{11}, B_1, C_1) is balanced and if $\sigma_r > \sigma_{r+1}$, then it is asymptotically stable, controllable and observable. Moreover, it enjoys the following L_∞-norm error bound [51],[58]

$$\overline{\sigma}(G(j\omega) - G_{br}(j\omega)) \le 2 \sum_{i=r+1}^{n} \sigma_i \quad \text{for all} \quad \omega$$

For symmetric systems (i.e. systems such that $G(s) = G^T(s)$), a cross-gramian W_{co} can be defined [70-73]. This matrix is given, for continuous time systems, as the solution of the following matrix equation:

$$A W_{co} + W_{co} A + B C = 0$$

and is similarly defined for discrete time systems. A fundamental property of the cross-gramian is that [71] $W_{co}^2 = W_c W_o$. Moreover, since for a symmetric system there exists a symmetric matrix R satisfying [74]

$$R A = A^T R , \qquad R B = C^T$$

then it can be shown [75] that $W_{co} = R^{-1} W_o = W_c R$. For a balanced system $W_o = W_c = \Sigma$ and R will be a diagonal sign matrix which implies that the balanced representation for a symmetric system will be sign symmetric. Other properties of balancing are given [132-134].

B. FREQUENCY-WEIGHTED BALANCED REPRESENTATIONS AND MODEL REDUCTION

In model reduction, sometimes it is important that the reduction error is small at a certain frequency band. This is especially important when using the reduced order model in feedback control system design [57-58]. There an accurate approximation of the full order system is needed at the crossover region. This idea motivated some authors to use frequency weighting for system approximation [57-59].

Enns [69] developed a method of frequency weighting, but no error bound for his method is known. Consider the asymptotically stable weighting $W_i(s) = D_i + C_i(sI - A_i)^{-1}B_i$ as an input weighting to the asymptotically stable system $G(s)$. Enns idea is to modify the controllability gramian in a way that reflects the presence of the input weighting. Then this "frequency-weighted" gramian is used in place of the system controllability gramian in deriving the balancing transformation. To see this, let the system matrices of the cascade system be defined as

$$\hat{A} = \begin{bmatrix} A & BC_i \\ 0 & A_i \end{bmatrix}, \quad \hat{B} = \begin{bmatrix} BD_i \\ B \end{bmatrix}$$

Let

$$\hat{W}_c = \begin{bmatrix} W_c & W_{c12} \\ W_{c12}^T & W_{c22} \end{bmatrix}, \quad \hat{W}_o$$

be the solution of the following Lyapunov equations

$$\hat{A}\hat{W}_c + \hat{W}_c\hat{A}^T + \hat{B}\hat{B}^T = 0$$

$$A^T W_o + W_o A + C^T C = 0$$

and let T be a state transformation of (A,B,C) such that in the new state space $(\overline{A},\overline{B},\overline{C})$ we have $\overline{W}_c = \overline{W}_o = \Sigma$. Assume the system is partitioned as in Eq. (27). Now a frequency weighted reduced order model is defined by eliminating the states that correspond to the smallest diagonal entries of Σ. This reduced order model is generically asymtotically stable. Moreover, there exists a dual procedure that uses an output frequency weighting. Also one can use both input and output frequency weightings to come up with a two-sided frequency-weighted balanced representation. However, there is no proof for the asymptotic stability in this case.

Al-Saggaf et al. [59] developed frequency weighting techniques for continuous and discrete systems that extends the method of balancing and gave Chebyshev norm error bounds for their method. An important distinct feature of the method in [59] that is different from other frequency weighted model reduction techniques is that the frequency weighting need not be stable and the reduction error has transmission zeros at the poles of the frequency weighting. The frequency weighting method described in [59] assumes a frequency weighting of the form $W_o(s) = C_o\,(sI - A_o)^{-1} B_o$ where A_o, $C_o \in R^{m\times m}$ and $B_o \in R^{m\times q}$. Then a frequency weighted output normal representation is defined as one for which [59]

$$A\,X - X\,A_o + B\,C_o = 0$$

$$A\,\Sigma^2 + \Sigma^2 A^T + X\,B_o\,B_o^T\,X^T = 0$$

$$A + A^T + C^T C = 0$$

Now if (A, B, C) and Σ are partitioned as in Eq. (27) and X is partitioned compatibly as $X^T = [X_1^T \;\; X_2^T]^T$ then a frequency-weighted reduced order model is defined as:

$$G_r(z) = D_r + C_r\,(zI - A_r)^{-1} B_r$$

where

$$D_r = C_2 X_2 C_o^{-1}, \quad C_r = C_1, \quad A_r = A_{11}, \quad \text{and} \quad B_r = B_1 + A_{12} X_2 C_o^{-1}$$

This reduced order model is asypmtotically stable and the reduction error $E_{n-r}(s) = G(s) - G_r(s)$ has zeros at the poles of the frequency weighting $W_o(s)$. Moreover

$$\| E_{n-r}(s)\, W(s) \|_\infty \le 2\, tr(\Sigma_2)$$

C. EXTENSIONS OF BALANCING

Balanced state space representations were an attraction in system theory as can be seen from the large number of publications that extend both the theory and the application of balanced-state space representations. Jonckheere et al. [77] developed closed-loop balancing which is based on simultaneously diagonalizing the solutions to the control and estimation Riccati equations. Youssuff et al. [78] have pointed out some limitation of the method of [77]. Shokoohi et al. [79], [108] have considered balancing linear time-variable systems and Verriest et al. [80] have considered balancing the general class of analytic time-varying linear systems. Liu et al. [82] proposed a model reduction technique based on fractional representation and balancing. Davis et al. [81] suggested yet another modification to the standard balancing approach.

Different techniques were proposed to determine which states are to be retained in the reduced order model. Kabamba [84] considered a criterion based on the contribution of each state to the L_2 magnitude of the impulse response and Skelton et al. [89] showed how this is related to component cost analysis of balanced states. Similarly, Davidson [86] suggested an alternative criterion for the selection of states in terms of their input/output contribution. However, it was shown, by an example, in [92] that straightforward truncation of states in some cases may lead to a better approximation of the impulse response than the methods suggested in [84] and [86]. Other criteria were also proposed in [87-88]. Singular perturbation approximation of balanced systems were also proposed as an alternate way of defining the reduced order model [59],[76],[94-98], [134].

Some algorithms were developed to obtain balanced state space representations utilizing structural properties or transfer functions. Fairman et al. [101] consider the case of scalar continuous time systems having simple poles, Seinsson et al. [100] employ a technique based on the Markov parameters, Fairman et al [105] use I/O map decomposition for continuous time systems, and Therapos [99] gives a technique for discrete SISO systems. Other techniques were given by Tsai et al. [107].

Other developments were in the parameterization of balanced realizations [91] and in the generalization using bilinear mapping [90]. Balanced representations were extended to 2D systems [83],[103-104],[111], 3D systems [106], systems involving delays [56] and infinite dimensional systems [55], [110]. There were developments to improve the frequency behavior of balanced-truncation reduced order models [93],[96],[113]. Balancing stochastic systems was also considered [85],[102],[109],[112].

For unstable systems, Kenney et al. [114] and Therapos [115] gave necessary and sufficient conditions such that the product $W_c W_o$ is diagonalizable with all its eigenvalues real and positive and Al-Saggaf [116] developed generalized normal representations for which the results in [114-115] are special cases. The reduction method in [116] results in reduced order models with the same number of unstable poles as the full order model and with an a priori upper bound on the reduction error which is very important when using the reduced order models in feedback control system design.

Some authors considered approximately balanced representations. Jonckheere et al. [117] showed that, under some assumptions, balanced-truncation and optimal Hankel-norm approximation are equivalent to modal truncation in an asymptotic sense as the damping is reduced to zero. Jonckheere et al. [118-119] used a parameterization of balanced SISO systems to show the same result. These results were shown for a more general system by Gregory [120] and then for even more general system by Belloch et al. [121].

V. OPTIMAL HANKEL-NORM APPROXIMATION

In this section, we review relatively recent developments [46-51] for the derivation of optimal system reduction schemes with tractable algorithms in explicit closed forms. Both time domain and frequency domain criteria are considered. First, some background material needed to motivate and define the criteria used in the approximation is given. The optimal problem in the multivariable case is much more involved than the scalar case. However, the solution in the scalar case is unique, has a simple closed form and gives insight to the multivariable case. Therefore for completeness, both cases are treated separately. The order of the presentation of the various subsections also follow the chronological development of the problem solution.

A. OPTIMAL HANKEL-NORM DISCRETE TIME SCALAR PROBLEM

Denoting by u_i the input at instant i, the output of a causal linear discrete system at instant k is given by:

$$y_k = \sum_{i=1}^{\infty} h_{k-i} u_i = \sum_{j=k}^{\infty} h_j u_{k-j}$$

where h_i is the output response to a unit pulse input, i.e., the pulse response of the system. Define the vector of the infinite sequence of inputs

$$u_i, \quad i = 1, 2, \dots \qquad \text{as} \quad u_{[1 \quad \infty]} = [u_1 \quad u_2 \quad \dots \quad u_\infty]^T$$

the corresponding infinite sequence of outputs is concatenated as

$$y_{[1 \quad \infty]} = [y_1 \quad y_2 \quad \dots \quad y_\infty]^T$$

The above linear convolution equation can then be displayed in matrix form as

$$\begin{pmatrix} y_1 \\ y_2 \\ y_3 \\ \cdot \\ \cdot \\ \cdot \end{pmatrix} = \begin{pmatrix} h_1 & h_2 & h_3 & \cdot & \cdot & \cdot \\ h_2 & h_3 & \cdot & \cdot & \cdot & \cdot \\ h_3 & \cdot & \cdot & \cdot & \cdot & \cdot \\ \cdot & \cdot & \cdot & \cdot & \cdot & \cdot \\ \cdot & \cdot & \cdot & \cdot & \cdot & \cdot \\ \cdot & \cdot & \cdot & \cdot & \cdot & \cdot \end{pmatrix} \begin{pmatrix} u_0 \\ u_{-1} \\ u_{-2} \\ \cdot \\ \cdot \\ \cdot \end{pmatrix}$$

or

$$y_{[1 \quad \infty]} = H \, u_{[-\infty \quad 0]} \tag{29}$$

Note, for a stable system, H is bounded in l_2 (square summable sequences). As shown above, the Hankel matrix H is a linear operator which takes past inputs into future outputs of a linear system.

The input-output relations of the system can also be expressed in the frequency domain. The z-transforms of the input sequence u_i , pulse sequence h_i , and output sequence y_i are respectively defined as

$$u(z) = \sum_{i=1}^{\infty} u_i \, z^{-i}, \quad f(z) = \sum_{i=1}^{\infty} h_i \, z^{-i}, \quad y(z) = \sum_{i=1}^{\infty} y_i \, z^{-i} = f(z) \, u(z)$$

Equation (29) then has the following transform representation,

$$y(z) = \Pi_{-}[f(z) \, z^{-1} \, u(z^{-1})] \tag{30}$$

where II_ is the projection operator which takes only negative powers of z. This representation clearly shows that the Hankel operator (associated with $f(z)$) maps anti-causal inputs into strictly causal outputs.

Note that to every causal transfer function there corresponds a unique Hankel matrix whose elements are the positive inverse z-transform coefficients of $f(z)$, i.e.,

$$h_i = \frac{1}{2\pi j} \oint_C f(z) \, z^{i-1} \, dz \,, \quad i = 1, 2, \ldots.$$

where C is any contour of integration. As relevant to system stability theory, C is taken to be the unit circle. Then, the elements h_i of H become simply the positive Fourier coefficients of $f(z)$,

$$h_i = \frac{1}{2\pi} \int_0^{2\pi} f(z) \, z^i \, d\theta \,, \quad z = e^{j\theta}$$

The Hankel matrix associated with $f(z)$, $H\{f(z)\}$ is denoted by H for simplicity, i.e., $H\{f(z)\}=H$.

The Hankel matrix associated with a function is defined here for any bounded function $g(z) \in l_\infty$ as

$$H\{g(z)\} = H\left\{\sum_{i=-\infty}^{\infty} g_i z^{-i}\right\} = H\left\{\sum_{i=1}^{\infty} g_i z^{-i}\right\}$$

so that the Hankel matrix gets no contribution from any analytic function inside the unit circle.

The spectral norm of a linear operator is defined by

$$\|K\|_s = \sup_{\|v\|_2=1} \|Kv\|_2$$

where $\|v\|_2$ denotes the l_2 norm of the vector v. The Chebyshev norm of any function $g(z)$ in l_∞ is also defined as

$$\|g(z)\|_\infty = \operatorname{ess}\sup_{0\le\theta\le 2\pi} |g(e^{j\theta})|$$

For a causal transfer function $f(z)$ and its corresponding Hankel matrix H, the above norms have the following input-output characterizations

$$\|H\|_s = \sup_{u_{[-\infty\ 0]}} \frac{\|y_{[1\ \infty]}\|_2}{\|u_{[-\infty\ 0]}\|_2}$$

and

$$\|f(z)\|_\infty = \sup_{u_{[1\ \infty]}} \frac{\|y_{[1\ \infty]}\|_2}{\|u_{[1\ \infty]}\|_2}$$

So that the Chebyshev norm leads to uniformly good approximations over all l_2-inputs. The optimal approximation problems to be reviewed in this section are now stated [46],[48]

Time Domain Problem:

Given a high order discrete time system model (A, b, c), where A is $n \times n$, b is $n \times 1$ and c is $1 \times n$, or equivalently an infinite Hankel matrix H, find a Hankel matrix Λ of given rank r such that $\|H - \Lambda\|_s$ is minimized.

Interestingly, the above problem is directly related to the following transform domain problem.

Frequency Domain Problem:

Given a high order transfer function $f(z)$ defined on the circle $|z| = 1$, find a function $g(z)$ bounded on $|z| = 1$ such that $\|f(z) - g(z)\|_\infty$ is minimized, $g(z)$ belongs to the class of functions representative in the form $g(z) = g_S(z) + g_A(z)$ where $g_S(z)$ is a proper rational function analytic in $|z| > 1$ having exactly r poles in the stable region $|z| < 1$ and $g_A(z)$ is any (possibly) improper rational function analytic in $|z| \leq 1$.

Consider the infinite Hankel matrix with a corresponding state space realization (A, b, c), i.e., $h_i = cA^{i-1}b$, $i=1, 2, \ldots$

$$H = [c^T \ A^T c^T \ (A^2)^T c^T \ \ldots\ldots\ldots\ldots\ldots]^T \ [b \ Ab \ A^2 b \ \ldots\ldots\ldots\ldots\ldots] = O\ C$$

For a rank n Hankel matrix H, the minimal realization (A, b, c) is of order n. The Hankel singular values σ_i defined in section IV are also the non zero singular values of the Hankel matrix H. The following Hankel factorization obtained from the singular value decomposition of H leads to a balanced realization [46],[9].

$$H = U\,\Sigma\,V^T = (U\,\Sigma^{1/2})\,(\Sigma^{1/2}\,V^T) = O\ C$$

where
$$U = [u_1\ u_2\ \ldots.\ u_n]$$
$$V = [v_1\ v_2\ \ldots.\ v_n]$$

u_i and v_i are infinite length singular vectors. Note that

$\Delta_1 = U_1\,\Sigma\,V_1^T$ where $U_1 = [u_1\ u_2\ \ldots.\ u_r]$, $V_1 = [v_1\ v_2\ \ldots.\ v_r]$, and $\Sigma_1 = diag(\sigma_1, \ldots., \sigma_r)$ is a best spectral norm approximation to H, i.e., $\varepsilon = \min_{S_r} \|H - S_r\|_s = \|H - \Delta_1\|_s$ where S_r is any matrix of rank r. Moreover

$\varepsilon = \sigma_{r+1}$. Unfortunately, Δ_1 is not, in general Hankel and, therefore, this result is not applicable to our problem.

For the optimal Hankel solution, consider the scalar transfer function $f(z) = c(zI - A)^{-1}b$ where A is $n \times n$, and (A,b,c) is minimal. Then, $f(z) = \frac{n(z)}{d(z)}$ where $d(z)=det\ (zI-A)$ is of degree n and the degree of $n(z)$ is less than n. Let

$$\mu(z) = \sigma_{r+1}^{-1/2}\,\bar{c}(zI - \bar{A})^{-1}\,e_{r+1} = \sigma_{r+1}^{-1/2} c(zI - A)^{-1}\,s_{r+1}$$

and
$$\nu(z) = \sigma_{r+1}^{-1/2}\,e_{r+1}^T(I - z\bar{A})^{-1}\,\bar{b} = \sigma_{r+1}^{-1/2} t_{r+1}(I - z\,A)^{-1}\,b$$

where $(\overline{A},\overline{b},\overline{c})$ is the balanced realization of $f(z)$ or H. e_{r+1} denotes the (r+1)st column of the $n \times n$ identity matrix. s_{r+1} is the (r+1)st column of $S = T^{-1}$ and t_{r+1} is the (r+1)st row of the balanced transformation T of (A, b, c). Then [46],

$$\mu(z) = \frac{m(z)}{d(z)}$$

where $m(z)$ is a polynomial of degree less than n.
From the sign symmetry property of balancing,

$$\nu(z) = \pm\frac{m^*(z)}{d^*(z)}$$

where $m^*(z)$ is the reverse polynomial of $m(z)$, that is, $m^*(z) = z^k\, m\left(\frac{1}{z}\right)$ if $m(z)$ has degree k. Also, define the Fourier coefficients of $f(z)$ to be $c_k(f) = \frac{1}{2\pi}\int_0^{2\pi} z^k f(z)\, dz\,,\quad z = e^{j\theta}$. The results of [46] are summarized below:

Time Domain Solution

Let $\sigma_{r+1} < \sigma_r$ denote the (r+1) st singular value of a bounded Hankel matrix H. Then there exists a unique Hankel matrix Λ^r of rank r which minimizes $\| H - \Lambda \|_s$ over all bounded Hankel matrices Λ of rank r and $\| H - \Lambda^r \|_s = \sigma_{r+1}$. Moreover, $\Lambda^r = H - H(\phi^r(z))$ where

$$\phi^r(z) = \sigma_{r+1}\frac{\mu(z)}{\nu(z)}$$

Frequency Domain Solution

Let $f \in l_\infty$ and let $\sigma_{r+1} < \sigma_r$ denote the (r+1)st singular value of $H(f)$. Then there exists a unique function g^r which minimizes $\| f - g \|_\infty$ over the class of functions $g = g_S + g_A$ where g_S is a strictly proper rational function of degree r with poles in $|z| < 1$ and g_A is any function bounded on $|z| = 1$ with $c_k(g_A) = 0$, $k > 0$. Moreover, $g^r(z) = f(z) - \phi^r(z)$, and $\| f - g^r \|_\infty = \sigma_{r+1}$.

Note that from a system theoretic point of view the frequency domain solution is not as interesting as the time domain solution, since the minimizing function need not be causal. However, since its proper causal part, $g_S^r(z)$, is precisely of degree r, $H(g_S^r(z)) = H(g^r) = \Lambda^r$ is the minimizing Hankel matrix of the time domain solution. Note also that $\frac{\mu(z)}{v(z)} = \pm\frac{m(z)d^*(z)}{m^*(z)d(z)}$ is an all pass function. That is,

$\left|\frac{\mu(z)}{v(z)}\right| = 1$ for $z = e^{j\theta}$. Let $\lambda = \pm\sigma_{r+1}$, the (r+1)st eigenvalue of H, then the frequency domain result can be further refined to give [46]:

$$g^r(z) = \frac{p(z)}{m^*(z)}$$

where
$$p(z) = \frac{n(z)m^*(z) - \lambda d^*(z)m(z)}{d(z)}$$

is a polynomial of degree less than n and $m^*(z)$ is the numerator polynomial of $\sigma_{r+1}^{-1} e_{r+1}^T (I - z\overline{A})^{-1}\overline{b}$ having degree less than n and precisely r zeros in $|z| < 1$. It should also be noted, that in general $g_S^r(z)$ is not optimal in the l_∞ norm unless $r = n - 1$ (one step approximation). Nevertheless, in [51], a direct term d is added to the causal part of the optimal approximation and an upper bound is derived and is given by

$$\|f - g_s^r - d\|_\infty \le (\sigma_{r+1} + \sigma_{r+2} + \ldots.. + \sigma_n)$$

Note that the Hankel norm is independent of d. Also, the lower bound is given by

$$\|g - g_s^r\|_\infty \ge \sigma_{r+1}$$

and the equality is achieved only with the additional anticausal part. However, in many practical situations the singular values decrease fairly quickly and the upper bound gets quite close to the theoretical lower bound. These error bounds motivated the authors in [54] to apply optimal Hankel norm approximations to reduced closed-loop control system design and good performances were obtained.

The basic steps in computing g_s^r can be summarized as follows [46]:

1) Solve the Lyapunov Equations for W_o and W_c
2) Compute the eigenvalues for W_oW_c.
3) Find t_{r+1}, the (r+1)st row of the balancing transformation T.
4) Find $m^*(z)$
5) Partial fraction expand $\lambda\frac{md^*}{m^*d} = \frac{n}{d} - \frac{p}{m^*}$ to obtain the causal part $g_S^r(z)$

It should be noted that step 4 can be converted to an eigenvalue problem since if $t_{r+1}\, b \neq 0$ (which it will be generically), then the zeros of $m^*(z)$ are the eigenvalues (modulo an eigenvalue at $z = 0$) of the matrix

$$A - \frac{1}{t_{r+1}\, b}\, b\, t_{r+1} A$$

Other algorithms, along these lines, exploiting further the system structure were derived in [46] and [48].

A different proof of the Hankel-norm approximation, avoiding the complex functional analytic method presented above, has also been derived recently by Michalestzky [122] using state space stochastic realizations of scalar-valued stationary processes.

Finite Data Approximation:

In a real input-output experiment (with pulse input), one cannot indefinitely run the experiment to display an infinite sequence of pulse response data. In practical situations only a finite number n of Markov parameters is given. The corresponding (generically) rank n Hankel matrix generates a minimal realization of order n. Such a preliminary realization also results when the high order Markov parameters are assumed to be approximately zero. This is justified in the case of a very stable system. However, the order n can be very large and a lower order approximate realization is needed. In this special case of interest, the pulse response transfer function to be approximated is simply given by

$$f(z) = h_1 z^{-1} + h_2 z^{-2} + \dots + h_n z^{-n} = \frac{h_1 z^{n-1} + h_2 z^{n-2} + \dots + h_n z}{z^n}$$

which corresponds to $d(z) = z^n$.

Let

$$H_n = \begin{pmatrix} h_1 & h_2 & h_3 & \cdot & \cdot & h_n \\ h_2 & h_3 & & & h_n & \\ h_3 & \cdot & \cdot & h_n & & \\ \cdot & \cdot & \cdot & & & \\ \cdot & \cdot & & & & \\ h_n & & & & & \end{pmatrix}$$

Then, the optimal approximation of lower order r is explicitly given by [46]:

$$g^r(z) = \frac{h_1 m_{n-1} z^{n-2} + (h_1 m_{n-2} + h_2 m_{n-1}) z^{n-3} + \dots + (h_1 m_1 + h_2 m_2 + \dots + h_{n-1} m_{n-1})}{m_{n-1} z^{n-1} + m_{n-2} z^{n-2} + \dots + m_0}$$

where $m = [m_0\ m_1 \dots m_{n-1}]^T$ is the eigenvector corresponding to the (r+1)st eigenvalue of H_n.

In frequency domain approximation, a meaningful system theoretic problem is to approximate $f(z)$ by a stable rational function $g_R(z)$ such that $\|f(z) - g_R(z)\|_\infty$ is minimized. However, a best Chebyshev rational approximation of a desired order does not necessarily exist. Even if a solution exists, it is not necessarily unique [63]. Consequently, solutions are not constructive and the minimal error is not satisfactorily characterized. One can, at best, derive algorithms to find *local* best Chebyshev approximations as done in [43]. Closed form solutions were reviewed in this section by extending the approximation problem to allow non analytic functions in the solution. As a consequence, the optimal solution of order r is unique and the minimal error is simply given by σ_{r+1}. Note that in the stable rational case, if a solution exists, the optimal error is bounded below by σ_{r+1}.

B. SCALAR CONTINUOUS TIME CASE

All discrete time results derived in this section are generalizable to continuous time systems. Bettayeb [48] and Bettayeb et al. [47] derived the continuous time results which turn out to be quite similar to the discrete time results. Technical differences in the results will be discussed.

Consider the following controllability and observability maps F_1 and F_2 associated with Eqs. (1a)-(1b), respectively

$$x_0 = F_1 u = \int_{-\infty}^{0} e^{-A\tau} b\, u(\tau)\, d\tau$$

$$y(t) = F_2 x_0 = c\, e^{At} x_0, \quad 0 \le t < \infty$$

with their adjoints defined by

$$F_1^* x_0 = b^T e^{-A^T t} x_0$$

$$F_2^* y = \int_0^\infty e^{A^T t} c^T y(t) dt$$

in terms of these maps the controllability and observability gramians are then $W_c = F_1 F_1^*$ and $W_0 = F_2^* F_2$. The Hankel operator is defined as

$$(\Gamma u)(t) = \int_{-\infty}^{0} c e^{A(t-\tau)} b \, u(\tau) \, d\tau, \quad 0 \leq t < \infty$$

$$= \int_{0}^{\infty} c \, e^{A(t+\lambda)} \, b \, u(-\lambda) \, d\lambda = \int_{0}^{\infty} h(t+\lambda) \, u(-\lambda) \, d\lambda \quad (31)$$

where the function $h(t) = ce^{At}b$ is the impulse response of the system (A,b,c) with transfer function

$$f(s) = \mathcal{L}[h(t)] \underline{\Delta} \int_{0}^{\infty} h(t) e^{-st} dt = c(sI - A)^{-1} b \quad (32)$$

The Hankel operator Γ is the continuous time analog of the Hankel matrix in discrete time. The Schmidt pair y_i and u_i, corresponding to the singular value σ_i verify the following relations for the operator Γ and its adjoint Γ^*:

$$(\Gamma u_i)(t) = \sigma_i y_i(t)$$

$$(\Gamma^* y_i)(t) = \sigma_i u_i(t)$$

or, in transform domain:

$\Pi_-[f(s) u_i(s)] = \sigma_i y_i(s)$, $u_i(s) = \mathcal{L}[u_i(t)]$, $y_i(s) = \mathcal{L}[y_i(t)]$ and $\Pi_+[f(-s) y_i(s)] = \sigma_i u_i(s)$

where the projection $\Pi_-(.)$ takes only the stable terms in the partial fraction expansion of (.) and $\Pi_+ = 1 - \Pi_-$.

In general, to each function $f(s)$ bounded on the $j\omega$ axis ($f \in L_\infty$) a Hankel operator Γ is defined as

$$(\Gamma u)(t) = \int_{0}^{\infty} h(t+\lambda) \, u(-\lambda) \, d\lambda, \quad t \geq 0$$

where $h(t) = \mathcal{L}^{-1}[\Pi_-(f(s))]$ and $u(t) \in L_2$. The induced norm of Γ is given by

$$\|f(s)\|_\infty \underline{\Delta} \sup_{\|u(t)\|_2 = 1, \; 0 \leq t < \infty} \|(\Gamma u)(t)\|_2 = \sup_{s = j\omega} |f(s)|$$

Also the spectral norm of the Hankel operator Γ can be defined as,

$$\|\Gamma\{f(s)\}\|_s \underline{\Delta} \|\Gamma\|_s \underline{\Delta} \sup_{0 \leq t < \infty, \; -\infty < \lambda \leq 0} \frac{\|(\Gamma u)(t)\|_2}{\|u(\lambda)\|_2} = \sigma_{\max}^{1/2}(\Gamma \Gamma^*)$$

where the L_2 norm is defined by $\|u(t)\|_2 = \int_0^\infty |u(t)|^2 dt$

The continuous time analogue of the discrete time results developed earlier are summarized below [47,48].

Time domain optimality solution:

Let $\sigma_{r+1} < \sigma_r$ denote the (r+1)st singular value of a bounded Hankel operator Γ of rank n $(r<n)$. Then there exists a unique Hankel operator Λ^r of rank r which minimizes $\|\Gamma-\Lambda\|_s$ over all bounded Hankel operators Λ of rank r and $\|\Gamma-\Lambda^r\|_s=\sigma_{r+1}$. Moreover, $\Lambda^r=\Gamma-\sigma_{r+1}\Gamma\{\Phi_{r+1}(s)\}$ where

$$\Phi_{r+1}(s) \underline{\Delta} \frac{y_{r+1}(s)}{u_{r+1}(s)} = \frac{c(sI-A)^{-1}e_{r+1}}{e_{r+1}^T(-sI-A)^{-1}b}$$

$\Phi_{r+1}(s)$ is all pass and *(A,b,c)* is balanced.

Frequency domain optimality solution:

Let $f(s)=c(sI-A)^{-1}b=\mathcal{L}\{ce^{At}b\}$ be a stable transfer function of order n and let $\sigma_{r+1}<\sigma_r$ denote the (r+1)st singular value of Γ. Then there exists a unique function $g^r(s)$ which minimizes $\|f-g\|_\infty$ over the class of functions $g=g_R+g_I$ where $g_R(s)$ is a strictly proper causal rational function of degree r $< n$ and g_I is any function which is uniformly bounded on the $j\omega$-axis and analytic in the left half plane. Moreover, $g^r(s)=f(s)-\sigma_{r+1}\frac{y_{r+1}(s)}{u_{r+1}(s)}$ and $\|f-g^r\|_\infty=\sigma_{r+1}$.

Let $f(s) \underline{\Delta} \dfrac{n(s)}{d(s)}$, $y_{r+1}(s)=\dfrac{m(s)}{d(s)}$, $u_{r+1}(s)=\pm y_{r+1}(-s)=\pm\dfrac{m(-s)}{d(-s)}$

then, as in the discrete time case, we have,

$$g^r(s)=\frac{q(s)}{m(-s)d(s)}=\frac{p(s)}{m(-s)} \quad \text{where} \quad q(s)\underline{\Delta}\, n(s)m(-s)\pm\sigma_{r+1}m(s)d(-s)$$

and $p(s)$ is a polynomial of degree $<n$.

Discussion

(1) In continuous time systems, the optimal approximation is proper since $g^r(s)$ is the sum of two proper rational functions. However, in discrete time, the optimal approximation is not necessarily proper [46].

(2) Here the approximation of finite dimensional systems (operators) by reduced systems of lower order is considered. However, these results are also true

for infinite dimensional systems, i.e. irrational functions. This also allows us to approximate infinite dimensional models by reduced systems of finite order, a very desirable step in the analysis and design of complex systems.

(3) In terms of physical interpretation of the approximation problem, there exists an important difference between discrete time and continuous time input-output approximations. In the discrete case, the Markov parameters that constitute the Hankel matrix to be approximated are also the pulse responses of the system at discrete instants. Therefore, in discrete time, the Hankel norm approximation is directly related to the input-output characteristics of the system. This is no longer true for continuous time systems. There, the Markov parameters are related to the impulse response in a less direct way. Let $f(s) = c(sI - A)^{-1} b = \sum_{i=1}^{\infty} cA^{i-1}b\, s^{-i} = \sum_{i=1}^{\infty} h_i s^{-i}$. If $h(t)$ is the impulse response, then

$$c\, A^i\, b = h_{i+1} = \frac{d^i h(t)}{dt^i}\Big|_{t=0}; \quad i = 0, 1, \ldots$$

C. MULTIVARIABLE DISCRETE TIME TOLERANCE PROBLEM

Given the multivariable discrete time system (A,B,C) with a square $p \times p$ transfer function $F(z)$ and the corresponding Hankel matrix H, the tolerance problem is defined as follows:

Find an approximate rational function $F^r(z)$ of minimal order, with a corresponding Hankel matrix H^r, such that $\|H - H^r\|_s \leq \rho$

If ρ is such that $\sigma_{r+1} < \rho < \sigma_r$, the tolerance problem has a solution $H^r = H\{F^r(z)\}$ where H^r has exactly r stable poles. The solution to the tolerance problem is given in terms of the delayed transfer matrix of $F(z)$

$$\overline{F}(z) = H_0 z^{-1} + F(z) z^{-1}$$

with its associated extended Hankel matrix,

$$\overline{H} \underline{\Delta} H\{\overline{F}(z)\} = \begin{pmatrix} H_0 & CB & CAB & . & . & . \\ CB & CAB & . & . & . & . \\ CAB & . & . & . & . & . \\ . & . & . & . & . & . \\ . & . & . & . & . & . \\ . & . & . & . & . & . \\ . & . & . & . & . & . \end{pmatrix} \tag{33}$$

The idea is to find (if it exists) a $p \times p$ real constant matrix H_0 so that the tolerance ρ is the (r+1)[st] singular value of $\overline{H}$ with multiplicity p, i.e.

$$\overline{\sigma}_r > \rho = \overline{\sigma}_{r+1} = \ldots = \overline{\sigma}_{r+p} > \overline{\sigma}_{r+p+1}$$

where $\overline{\sigma}_i$, $i = 1, 2\ldots$ are the singular values of $\overline{H}$.

A state space realization of $\overline{F}(z)$ is given by

$$\overline{A} = \begin{pmatrix} A & 0 \\ C & 0 \end{pmatrix}, \quad \overline{B} = \begin{pmatrix} B \\ H_0 \end{pmatrix} \quad \text{and} \quad \overline{C} = [0 \quad I_p]$$

It was shown by Bettayeb [48] that the required H_0 is solution to the following algebraic matrix Riccati equation:

$$R_\rho(H_0) = CW_cC^T + H_0H_0^T - \rho^2 I_p$$

$$-(AW_cC^T + BH_0^T)^T (W_oW_c - \rho^2 I_n)^{-1} W_o (AW_cC^T + BH_0^T) = 0$$

Note that $W_oW_c - \rho^2 I_n$ is invertible since $\rho^2 \neq \lambda_i(W_oW_C)$ by assumption $(\sigma_{r+1} < \rho < \sigma_r)$. Existence of a real solution to $R_\rho(H_0) = 0$ was also established and explicitly given in [48].

Let $(\overline{A}, \overline{B}, \overline{C})$ be a balanced realization of $\overline{F}(z) = H_0 z^{-1} + F(z)z^{-1}$ where H_0 is a solution to the above Riccati equation for a given ρ and define

$$\overline{X}(z) = \overline{B}^T(I - z\overline{A}^T)^{-1}E_r, \quad Y(z) = \overline{C}(zI - \overline{A})^{-1}E_r, \quad \text{and} \quad E_r = [e_{r+1}, \quad \ldots \quad, e_{r+p}]$$

The solution to the tolerance problem is summarized as:

Tolerance problem solution

Given (A,B,C) with transfer function matrix $F(z)$ and Hankel matrix H, if ρ is such that $\sigma_{r+1} < \rho < \sigma_r$ then there exists a real matrix H_0 such that

$$\overline{H}' = H - \rho\overline{H}[z\overline{Y}(z)\overline{X}^{-1}(z)]$$

has rank r. Moreover, $\| H - \overline{H}' \|_s \leq \rho$ and $\| z\overline{F}(z) - \overline{F}'(z) \|_\infty = \rho$

where $\overline{F}^r(z) = z\overline{F}(z) - \rho z\overline{Y}(z)\overline{X}^{-1}(z)$ and $\|F(z)\|_\infty \underline{\Delta}\; ess \sup_{0\le\theta<2\pi} \|G(e^{j\theta})\|_s$

As for the scalar case, some useful properties of the solution can be obtained if one exploits advantageously the rational function representation of $\overline{X}(z)$ and $\overline{Y}(z)$. However, these are not included for briefly. Details are given in [48-50].

Remarks

(1) The strictly proper causal part $\overline{F}_s^r(z)$ of $\overline{F}^r(z)$ is optimal with respect to the spectral norm of the Hankel matrix, i.e., $\|H - H\{\overline{F}_s^r(z)\}\|_s \le \rho$ since the possibly nonproper anticausal part of $\overline{F}^r(z)$ is analytic inside the unit circle and therefore has no contribution to the Hankel matrix. However, the performance evaluation of $\overline{F}_s^r(z)$ in the frequency domain is not as trivial. Note that $\|H_0 + F(z) - \overline{F}^r(z)\|_\infty = \rho$, i.e., an "artificial" constant term is added to the original function *F(z)* to compensate for the error. This complicates the frequency domain interpretation of the approximation.

(2) It is interesting to note that for single-input multi-ouput or multi-input single-output systems, the optimal approximation is solved if $\rho = \sigma_{r+1}$ (i.e. no extension is needed). Also when $\rho \ne \sigma_{r+1}$, the tolerance solution is simpler. For example in the multi-input single-output case, the Riccati equation reduces to a scalar second order equation.

(3) The spectral error between the original high order system and the optimal rth order model is simply bounded by the tolerance which is given a priori, i.e. $\sigma_{r+1} < \rho < \sigma_r$. One then can evaluate any rth order reduced order model obtained by any procedure to judge its performance just by locating the tolerance with respect to the singular values σ_r and σ_{r+1}.

(4) The magnitudes of the singular values of the Hankel matrix associated with the high order system dictate the natural order for the underlying reduced system. A good trade off between the accuracy of the reduced system and its complexity (order) could be detected a priori by displaying the singular values of the original system. For example, if $\sigma_r \gg \sigma_{r+1}$, one can choose the tolerance ρ to be close to σ_{r+1} Actually, once σ_{r+1} is known, one can arbitrarily choose ρ to approach σ_{r+1}.

Then, the rth order tolerance solution approaches the optimal approximation in performance. Even then, one still has some flexibility in the solution since the Riccati equation generally admits many solutions. To each solution H_0 to the Riccati equation there is a corresponding solution to the tolerance problem. Even though these solutions all have the same performance error, some other characteristics (i.e. steady state step error, initial error) can differ. It is an interesting problem to study the practical implications of each solution in terms of performance evaluation. Generalization to continuous time systems is also possible.

D. OPTIMAL HANKEL NORM APPROXIMATION: THE MULTIVARIABLE CASE

Characterization of all solutions to the optimal Hankel norm approximation of linear multivariable continuous time systems have been solved completely in [51].

Let Γ_G be the continuous time Hankel operator associated with the stable transfer function $G(s) = C(sI - A)^{-1}B$. Denote the Hankel singular values of $G(s)$ by $\sigma_i(G(s)) \underline{\Delta}\, \sigma_i$. The Hankel norm of $G(s)$ is defined as: $\| G(s) \|_H \underline{\Delta} \| \Gamma_G \|_s = \sigma_1(\Gamma_G)$.

The optimal Hankel-norm approximation problem is then:
Given a stable $G(s)$ of McMillan degree n, find a stable $\hat{G}(s)$ of McMillan degree $k<n$ such that $\| G(s) - \hat{G}(s) \|_H$ is minimized.

Glover [51] gave characterization of all optimal Hankel-norm approximations both in state space and transfer function forms. As in the optimal Hankel scalar case treated earlier, the solution involves a characterization of all rational functions $\hat{G}(s) + F(s)$ that minimizes $\| G(j\omega) - \hat{G}(j\omega) - F(j\omega) \|_\infty$, where $F(s)$ is anticausal. Both, the optimal Hankel error and the optimal L_∞ error turned out to be equal to σ_{k+1}, the (k+1)st singular value of the Hankel operator Γ_G . However, as mentioned earlier, the L_∞ norm of the approximation error is not meaningful in the context of model reduction as it includes an anticausal part $F(s)$. Glover [51] gave a specific optimal Hankel-norm approximation algorithm with good frequency domain bounds on the reduction error $\| G(s) - \hat{G}(s) \|_\infty \underline{\Delta} \| E(s) \|_\infty$. He explicitly derived an algorithm to compute a constant term $\hat{D}$ such that

$\| G(s) - \hat{G}(s) - \hat{D} \|_\infty \leq \sum_{i=k+1}^{n} \sigma_i$. In [51], it is argued that this upper bound will often be close to the lower bound of σ_{k+1} which can be only achieved if an anti-causal term is included. If $\hat{D} = D$ (the direct-term of *G(s)*), then the upper bound on the reduction error above is doubled. Note that the Hankel norm of the error is optimum ($\| G(s) - \hat{G}(s) \|_H = \sigma_{k+1}$ for any $\hat{D}$ since the direct term does not affect the Hankel operator.

The following specific algorithm for the optimal Hankel norm model reduction is described in [51]. It is assumed that a state space realization *(A,B,C,D)* of a stable rational function with McMillan degree *n* and a reduced order $k<n$ (*k* can be dictated by the singular values separation of the Hankel operators) are given.

Optimal Hankel Norm Model Reduction Algorithm

(1) Form a balanced realization of *G(s)*. Let the Hankel singular values be $\sigma_1 \geq \sigma_2 \ldots \geq \sigma_k > \sigma_{k+1} > \sigma_{k+2} \geq \ldots \geq \sigma_n > 0$, and let the balanced realization of state dimension *n* be reordered so that

$$A = \begin{pmatrix} A_{11} & A_{12} \\ A_{21} & A_{22} \end{pmatrix}, \quad B = \begin{pmatrix} B_1 \\ B_2 \end{pmatrix}, \quad C = (C_1 \quad C_2)$$

$$\Sigma = diag(\sigma_1, \sigma_2, \ldots, \sigma_k, \sigma_{k+2}, \ldots, \sigma_n, \sigma_{k+1}) = diag(\Sigma_1, \sigma_{k+1})$$

with *(A,B,C)* partitioned conformally with Σ (i.e. A_{11} is $(n-1) \times (n-1)$).

(2) Form a state space representation for $\hat{G}(s) + F(s)$

$$U = -(C_2 B_2)/(B_2 B_2^T)$$

$$\Gamma = (\Sigma_1^2 - \sigma_{k+1}^2 I)$$

$$\hat{A} = \Gamma^{-1} (\sigma_{k+1}^2 A_{11}^T + \Sigma_1 A_{11} \Sigma_1 - \sigma_{k+1} C_1^T \; U B_1^T)$$

$$\hat{B} = \Gamma^{-1} (\Sigma_1 B_1 + \sigma_{k+1} C_1^T U)$$

$$\hat{C} = C_1 \Sigma_1 + \sigma_{k+1} U B_1^T$$

(3) Block Diagonalize $\hat{A}$

(a) Reduce $\hat{A}$ to real upper Schur form, i.e. find V_1 such that $V_1^T V_1 = I$ and $V_1^T \hat{A} V_1$ is in upper Schur form.

(b) Find an orthogonal matrix V_2 such that

$$V_2^T V_1^T \hat{A} V_1 V_2 = \begin{pmatrix} \hat{A}_{11} & \hat{A}_{12} \\ 0 & \hat{A}_{22} \end{pmatrix}$$

where $Re(\lambda_i(\hat{A}_{11})) < 0, Re(\lambda_i(\hat{A}_{22})) > 0$ and that $\hat{A}_{11} \in R^{k \times k}$

(c) Find $X \in R^{k \times (n-k-1)}$ such that

$$\hat{A}_{11} X - X \hat{A}_{22} + \hat{A}_{12} = 0$$

(d) Let

$$T = V_1 V_2 \begin{pmatrix} I & X \\ 0 & I \end{pmatrix} = (T_1 \quad T_2)$$

$$S = \begin{pmatrix} I & -X \\ 0 & I \end{pmatrix} V_2^T V_1^T = \begin{pmatrix} S_1 \\ S_2 \end{pmatrix}$$

(e) Let

$$\hat{B}_1 = S_1 \hat{B}$$

$$\hat{C}_1 = \hat{C} T_1$$

(4) The reduced order model of order k is given by

$$\hat{G}(s) = \hat{C}_1 (sI - \hat{A}_{11})^{-1} B_1 + \hat{D}$$

where the computations of $\hat{D}$ is given in [51].

The extension of these results to the case of approximation of unstable systems is trivial [51]. Optimal Hankel approximation of infinite dimensional systems has also been approached in [55-56].

In [123], it is shown that the theory of rational approximation in Hankel norm is closely related to a generalized form of the Nevanlinna-Pick interpolation problem. A two-variable polynomial approach is used in [124] to rederive the results of [128]. Jonchheere et al. [90] generalized the standard optimal Hankel-norm model reduction problem to an optimal Hankel-norm model reduction problem over a disc in the complex plane by using a bilinear transformation that maps the right half plane onto the disc. This generalization has a good practical significance as one can tune his approximation in relation to the frequency content of transient speed of the signals in consideration. A generalized eigenvalue algorithm for the computation of the optimal Hankel norm approximation related to those developed in [46-48], is derived in [125] for single input single output continuous time systems. Safonov et al. [126] established the equivalence between the L_∞ sensitivity optimization problem and the zeroeth order optimal Hankel approximation problem and used it to design a feedback compensator minimizing weighted L_∞ norm of the sensitivity function of a Multivariable linear time invariant system. Ball et al. [129] used a linear fractional map to parameterize all solutions to the tolerance problem for discrete time systems. They recovered Glover's [51] results for the tolerance problem using a bilinear transformation. The optimal Hankel norm model reduction problem has been treated in [130] in continuous time using the same approach.

E. FREQUENCY-WEIGHTED OPTIMAL HANKEL REDUCTION

Frequency weighting can be introduced in the Hankel-norm approximation problem to shape the approximation error at various frequency ranges of interest. Latham et al. [60] proposed to minimize the frequency-weighted Hankel-norm approximation error $\| W(s)\,[G(s) - \hat{G}(s)] \|_H$ where $G(s)$ is a given scalar stable transfer function of McMillan degree n, $\hat{G}(s)$ an optimal Hankel-norm stable approximation of some prescribed degree k which minimizes the above error and the frequency weighting rational function $W(s)$ is such that $W(-s)$ is stable and of minimal phase.

The solution to the above problem was developed in [60] for discrete time systems. Frequency weighting in continuous time was performed using the discrete time result after transforming the problem to the discrete time set up via the bilinear transformation. No upper bound was given in [60], in the L_∞ error of the resulting stable approximation. An L_∞ bound of the weighted error is derived in [127], however, its evaluation is not straightforward. Hung et al. [61] solved the continuous time frequency-weighting Hankel-norm approximation for a slightly more general weighting $W(s)$, i.e. by relaxing the minimal phase condition on $W(-s)$. The resulting solution is essentially the continuous time version of the approximation scheme proposed in [60]. Letting now, $W(s)$ to be an unstable biproper rational matrix and G(s) a stable rational transfer function matrix, the approximation scheme developed in [61] is summarized as:

(1) Let $W(s)G(s) = W_1(s) + G_1(s)$
where $W_1(s)$ and $G_1(s)$ are the unstable and stable projections of *W(s) G(s)* respectively.

(2) Find $\hat{G}_1(s)$, the *k*-th order optimal Hankel-norm approximation of $G_1(s)$.

(3) Assuming $\hat{G}_1(s)$ and $W^{-1}(s)$ have no common poles, perform the decomposition

$$W^{-1}(s)\,\hat{G}_1(s) = \hat{G}(s) - \overline{W}(s)$$

where the poles of $\hat{G}_1(s)$ and $\overline{W}(s)$ are contained in the poles of $\hat{G}_1(s)$ and $W^{-1}(s)$ respectively (Note that $\hat{G}(s)$ satisfies $W(s)\hat{G}(s) = \hat{W}_1(s) + \hat{G}_1(s)$ with $\hat{W}_1(s) = W(s)\overline{W}(s)$ and $\hat{W}_1(s)$ is unstable).

(4) $\hat{G}(s)$ of step 3 minimizes the frequency weighted Hankel norm error with

$$\| W(s)\,[G(s) - \hat{G}(s)] \|_H = \sigma_{k+1}(W(s)G(s))$$

No L_∞ norm upper bound of the error for this approximation is available in [61]. Hung et al. [61] generalized the above to stable weightings for the restricted case of first order scalar weight $W(s) = \frac{s-\beta}{s+\alpha}$, $\alpha > 0$. They showed that the solution to this problem for this simple weight is closely related to the solution to the above unstable weighting problem. Furthermore, an L_∞ error bound is provided for the case $\beta > 0$.

VI. EXAMPLES

Three illustrative examples will be considered. In the first example, the optimal Hankel norm approximation is compared to optimal classical time domain approximations. For brevity, the best uniform approximations for all L_1- and $L_\infty-$inputs of [25] is taken as a representative example of the optimal time domain methods. Similarly, in the second example, the optimal Hankel norm approximation is compared to the optimal frequency domain model reduction technique developed in [42]. In example 3, balanced, weighted balanced, optimal Hankel and weighted optimal Hankel reduction schemes are compared.

Example 1:

Consider the 5th order system with transfer function [25]

$$g(s) = \frac{35.8223\,s^3 - 120.9286\,s^2 + 2327.8\,s - 2863}{s^5 + 9.8\,s^4 + 162.9\,s^3 + 872.3\,s^2 + 4284.3\,s + 5751.6}$$

This system is chosen for its highly oscillatory and non-minimium phase behavior. The system singular values are 0.5936, 0.5276, 0.4294, 0.3898 and 0.1433. A 4th order optimal Hankel approximation is computed as suggested by the largest separation in the singular values, and is found to be

$$g_H(s) = \frac{-0.1433\, s^4 + 3.5368\, s^3 - 19.5681\, s^2 + 217.4474\, s - 246.3792}{s^4 + 5.0870\, s^3 + 94.8419\, s^2 + 270.1647\, s + 694.665}$$

Note that $g_H(s)$ is stable as it is a one step approximation. The 4th order uniform approximation for all L_1- and $L_\infty-$inputs [25] is given by

$$g_V(s) = \frac{33.8059\, s^2 - 7.1888\, s + 2330.9}{s^4 + 10.974\, s^3 + 227.7036\, s^2 + 1018.5\, s + 10154.5}$$

The time domain impulse responses and corresponding errors are plotted in figures 1 and 2 respectively. Hankel approximation gave better results as evidenced by the graphs and the following calculated errors on the criteria used by the two reduction schemes:

$$\| g(s) - g_H(s) \|_\infty = 0.1433 \quad \text{and} \quad \| g(s) - g_V(s) \|_\infty = 0.7273$$

$$e_H := |d_0| + \sum_{i=1}^{61} | h(t_i) - h_H(t_i) | = 25.1232 \quad \text{and} \quad e_V := \sum_{i=1}^{61} | h(t_i) - h_V(t_i) | = 25.1932$$

where $h(t)$, $h_H(t)$ and $h_V(t)$ are the impulse responses associated with $g(s)$, $g_H(s)$ and $g_V(s)$ respectively, d_0 is the direct term of $g_H(s)$, and $t_i = \frac{i-1}{20}$, $i = 1, 2, \ldots, 61$. It is interesting to note that the optimal Hankel reduced order model approximated well the non-minimum phase behavior represented by the zeros of the system. The zeros of the 5th order system $(1.2829, 1.0464 \pm j\, 7.8232)$ are approximately retained in the optimal Hankel 4th order model $(1.2423, 1.0004 \pm j\, 7.972)$.

Example 2:

Consider the all pole 5th order system [42] given by

$$g(s) = \frac{432}{2.8\, s^5 + 15.76\, s^4 + 30.94\, s^3 + 26.08\, s^2 + 9.1\, s + 1}$$

The system singular values are 285.4717, 80.0141, 11.4023, 0.8888 and 0.0288. Third order models derived from the optimal Hankel approximation and the

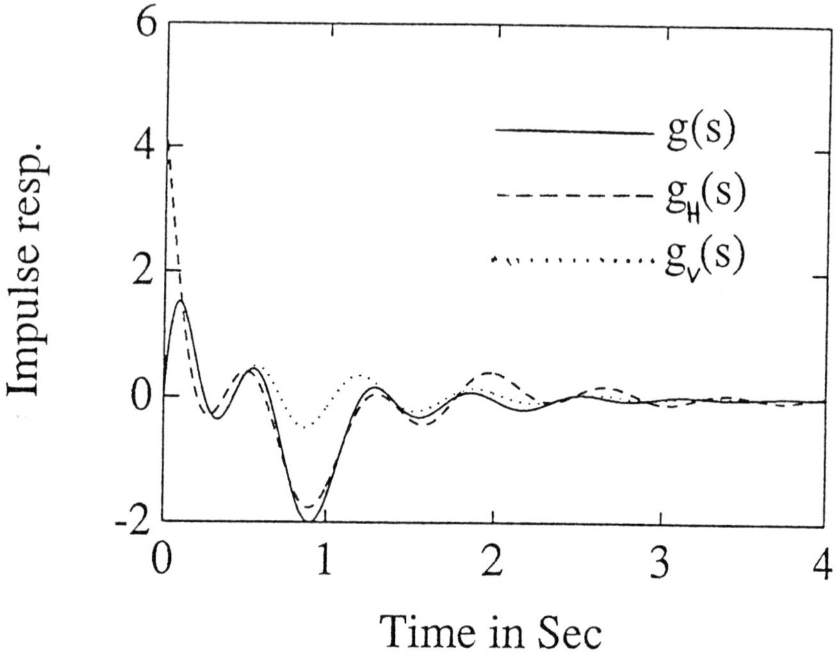

Figure 1. Impulse Response for Example 1.

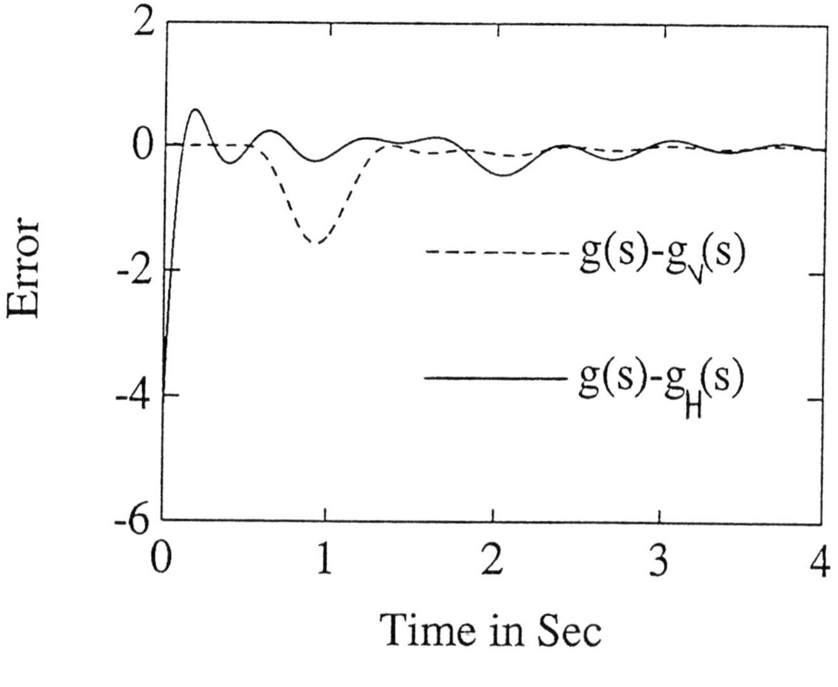

Figure 2. Error Response for Example 1.

optimal frequency domain model reduction technique of Luus [42] are compared for illustration. These reduced order models are respectively given by:

$$g_H(s) = \frac{6.4960\, s^2 - 15.6009\, s + 25.315}{s^3 + 1.1865\, s^2 + 0.4971\, s + 0.0584}$$

$$g_L(s) = \frac{432}{30.27\, s^3 + 24.6\, s^2 + 9.09\, s + 1.0}$$

($g_H(s)$ is the strictly proper causal part of the optimal Hankel norm approximation). The errors for the two criteria of the two methods are computed as :

$$\| g(s) - g_H(s) \|_\infty = 1.7691 \quad \text{and} \quad \| g(s) - g_L(s) \|_\infty = 12.8876$$

$$J_{4H} := \sum_{i=1}^{100} | g(j\omega_i) - g_H(j\omega_i) |^2 = 127.1161 \quad \text{and} \quad J_{4L} := \sum_{i=1}^{100} | g(j\omega_i) - g_L(j\omega_i) |^2 = 1885.4$$

The frequencies ω_i are generated by:

$$\omega_1 = .01\,, \qquad \omega_i = 1.10\, \omega_{i-1}\,, \quad i = 2\,, \ldots\,, 100$$

Hankel approximation is therefore performing much better than the optimal frequency response method of [42] with respect to both criteria. The superiority of Hankel approximation for this example is also seen from the impulse response and the frequency response errors plotted in figures 3 and 4 respectively for both methods. Note that one can still improve further the Hankel approximation by adding, to the strictly stable proper reduced model $g_H(s)$, the direct term derived in [51].

Example 3:

This example consists of a steam unit connected to an infinite bus and is taken from [135]. The system consists of a 5th order winding representation for a synchronous machine, a 4th order automatic voltage regulator and exciter, a 2nd-order shaft and a 2nd-order turbine and governor system.

The state vector of the linearized small-perturbation 13th order model is:

$$x^T(t) = [\Delta i_q \quad \Delta i_d \quad \Delta i_{kq} \quad \Delta i_{kd} \quad \Delta i_f \quad \Delta\delta \quad \Delta\omega \quad \Delta E_{fd} \quad \Delta v_a \quad \Delta v_r \quad \Delta v_s \quad \Delta P_m \quad \Delta P_g]$$

and the outputs are taken to be $\Delta\delta$ (rotor angle), Δi_d (d-axis current), and Δi_q (q-axis current).

The inputs to the system are the change in load power demand (ΔP_L) and the change in the exciter voltage (ΔV_C).

The four methods; balanced, optimal Hankel, frequency weighted balanced

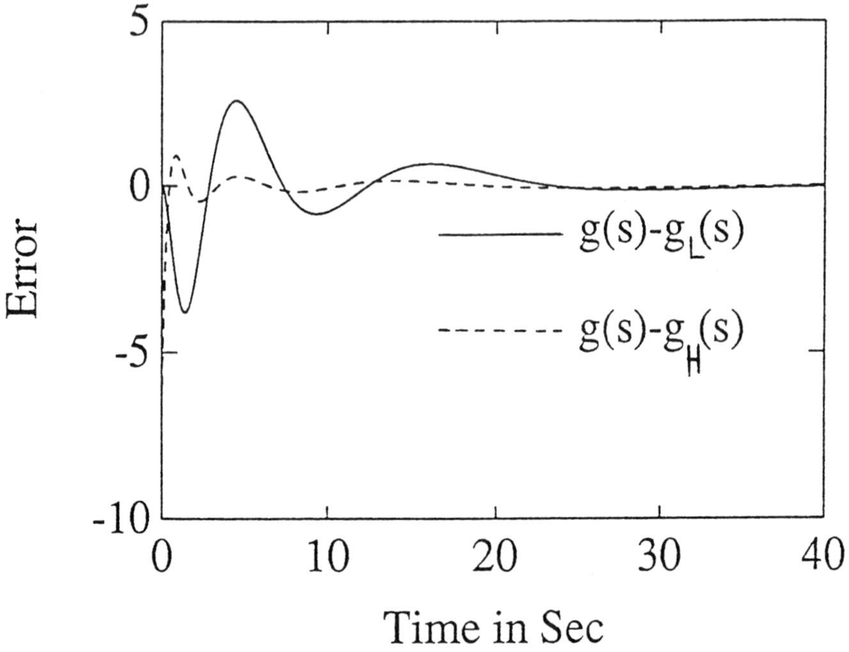

Figure 3. Error Impulse Response for Example 2.

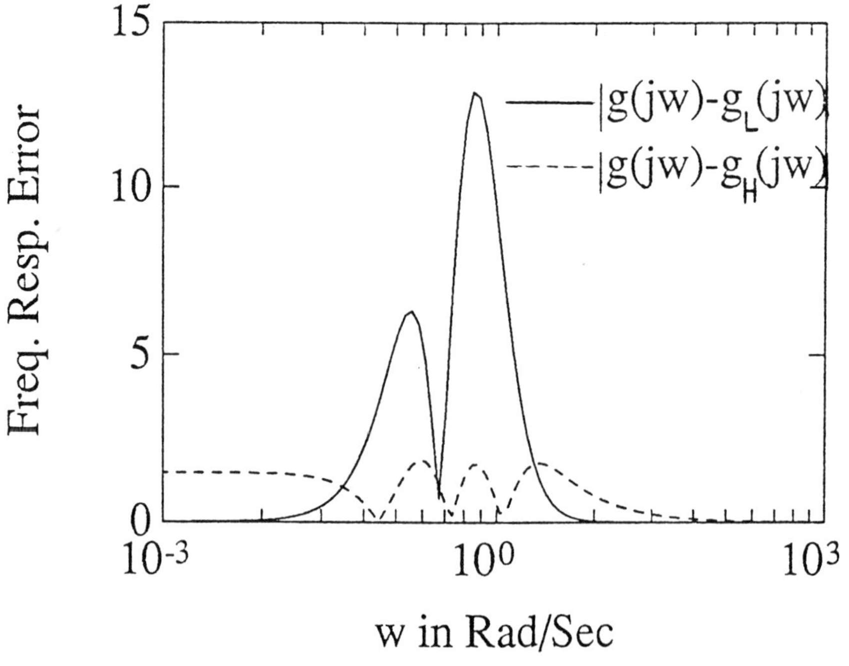

Figure 4. Error Frequency Response for Example 2.

and frequency weighted optimal Hankel model reduction schemes described in the text are applied to this example. Using the balancing algorithm we compute the Hankel singular values of the system:

$\{2.9007,\ 1.7819,\ 1.3289,\ .8787,\ 0.1345,\ 0.8236\times10^{-1},\ 0.1864\times10^{-1},$

$0.166\times10^{-1},\ 0.8472\times10^{-2},\ 0.8033\times10^{-2},\ 0.3845\times10^{-2},\ 0.1759\times10^{-2},\ 0.973\times10^{-8}\}$

The Hankel singular values suggest a fourth order reduced model is appropriate for balanced and optimal Hankel approximations. Therefore, for comparison purposes, a reduced order model of order four will be computed for the four methods. To eliminate the steady state error to step inputs between the original system and the balanced reduced order model, a frequency weighting $W_1(s)=I/s$, where I, is the identity matrix, is used in the weighted balanced approximation. As this frequency weighting can not be used in the optimal weighted Hankel approximation, the weighting $W_2(s)=\frac{s-10000}{s+0.0001}I$ is used in the later method. This weight captures the low frequency behavior of $W_1(s)=I/s$ and meets the constraint imposed by the optimal weighted Hankel approximation on the weighting. For both optimal Hankel and frequency weighted optimal Hankel reduced models, a constant term $\hat{D}$ is included in the approximation . Figure 5 gives the maximum singular value of the reduction error for the four methods. From the figure, the balanced reduction method (bal.) gives poor performance at low frequencies. The Hankel approximation with the direct term (hank.) gives basically the same error at all frequencies. Both the weighted balanced (wbal.) and optimal weighted Hankel (whank.) reduced order models gave major performance improvement at low frequencies. However, the errors at high frequencies are increased. Similarly, the weighted balanced and weighted Hankel approximations gave excellent time domain performance (at transient and steady state).

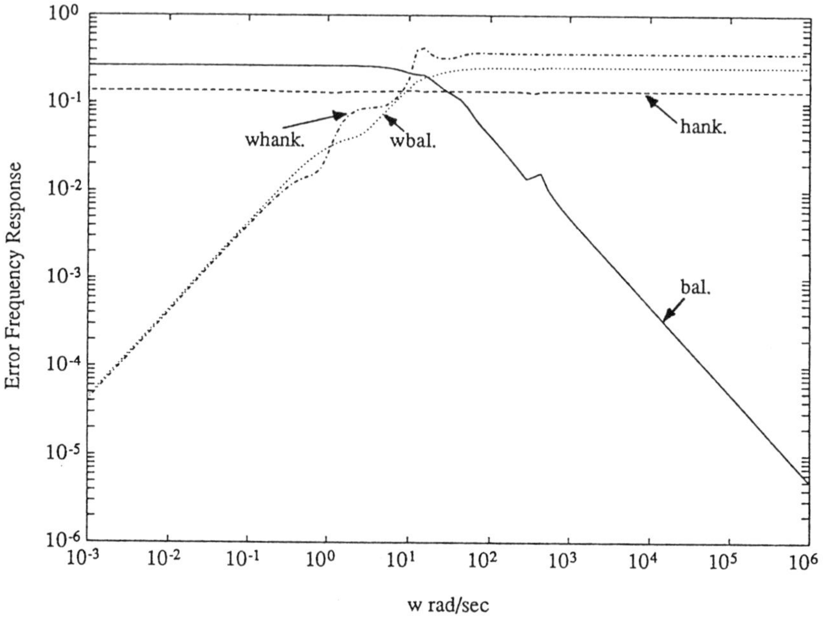

Figure 5. Error Frequency Response for Example 3.

ACKNOWLEDGEMENT

The authors gratefully acknowledge the support of King Fahd University of Petroleum & Minerals.

REFERENCES

1. M. J. Grimble & M. A. Johnson, Optimal Control and Stochastic Estimation theory and applications, Volumes 1 & 2, John Wiley & Sons, 1988.

2. A. E. Bryson, Jr and Y-C. Ho, Applied Optimal Control, Hemisphere Publishing Corporation, New York, 1975.

3. H. Kwakernaak and R. Sivan, Linear Optimal Control Systems, John Wiley and sons, New York, 1972.

4. B.A. Francis & J.C. Doyle, "Linear Control Theory with an H^∞ Optimality Criterion", *SIAM J. CONTROL AND OPTIMIZATION,* Vol. 25, No. 4, July 1987.

5. R. Genesio and M. Milanese,"A note on the derivation and use of reduced order models," *IEEE Trans. Automat. Contr.*, Vol. AC-21, pp. 118-122, 1976.

6. E. J. Davison, "A method for simplifying linear dynamic systems," *IEEE Trans. Automat. Contr.*, Vol. AC-11, pp. 93-101, Jan. 1966.

7. C. T. Mullis and R. A. Roberts, "Synthesis of minimum roundoff noise fixed point digital filters", *IEEE Trans. Circuits Syst.* **CAS-23**, pp. 551-561 (1976).

8. C. T. Mullis and R. A. Roberts, "Roundoff noise in digital filters: Frequency Transformations and invariants", *IEEE Trans. Acoust., Speech, Signal Processing,* **ASSP-24**, pp. 538-550 (1976).

9. S. Kung, "A New identification and model reduction algorithm via singular value decompositions", *In Proc. 12th Annu. Asilomar Conf. Circuits, Syst., Comput.*, pp. 705-714 (1978).

10. M. R. Chidambara, "Further comments by M. R. Chidambara," *IEEE Trans. Automat. Contr.*, Vol. AC-12, pp. 799-800, Dec. 1967. E. J. Davison, "Further reply by E. J. Davison," *IEEE Trans. Automat. Contr.*, Vol. AC-12, p. 800, Dec. 1967.

11. M. R. Chidambara, "Two simple techniques for the simplification of large dynamic systems," *Proc. 1969 JACC*, PP. 669-674, 1969.

12. C. F. Chen and L. S. Shieh,," A novel approach to linear model simplification," *Int. J. Contr.*, vol. 8,pp 561-570,1968

13 L. G. Gibilaro and F. P. Lees," The reduction of complex transfer function models to simple models using the method of moments," *Chem. Engng. Sci.*,, vol. 24, pp 85-93, 1969.

14. M. Hutton and B. Friedland, "Routh approximations for reducing order of linear, time invariant systems," *IEEE Trans. Automat. Contr.* AC-20(3), pp. 329-337, 1975.

15. G. Langholz and D. Feinmesser, "Model reduction by Routh Approximations," *Int. J. Systems Sci.*, 9(5) pp. 493-496, 1978.

16. P.J.M. Pinguet, "State space formulation of a class of model reduction methods," Master Thesis, University of Cambridge, September 1978.

17. D.A. Wilson, "Optimum Solution of model reduction problem," *Proc. IEE*, vol. 117, June 1970.

18. D.A. Wilson, "Model reduction for multivariable systems," *Int. J. Contr.*, vol. 20, pp. 57-64, 1974.

19. F.D. Galiana, "On the approximation of multiple input-multiple output constant linear systems,"*Int. J. Control,* Vol. 17, No. 6, pp. 1313-1324, 1973.

20. J.B. Riggs and T.F. Edgar, "Least squares reduction of linear systems using impulse response," *Int. J. Control*, Vol. 20, No. 2, pp.213-223, 1974.

21. G. Hirzinger and G. Kreisselmeir, "On optimal approximation of high-order linear systems by low-order models," *Int. J. Control,* Vol. 22, No. 3, pp. 399-498, 1975.

22. G. Obinata and H. Inooka, "A Method for Modeling Linear Time-Invariant Systems by Linear Systems of Low order," *IEEE Trans. Automat. Contr.*, vol. AC-21, pp. 602-603, Aug, 1976; see also "Correction to"

A Method for Modeling Linear Time-Invariant Systems by Linear Systems of Low order," *IEEE Trans. Automat. Contr.*, vol. AC-22, p. 286, April 1977.

23. G. Obinata and H. Inooka, "Authors' Reply to "comments on Model Reduction by Minimizing the Equation Error," *IEEE Trans. Automat. Contr.*, vol. AC-28, pp. 124-125, Jan. 1983.

24. E. Eitelberg, "Model Reduction by Minimizing the Weighted Equation Error," *Int. J. Control*, vol. 34, No. 6, pp. 1113-1123, 1981.

25. R.A. El-Attar, and M. Vidyasagar," Order reduction by l_1 and l_∞ - Norm Minimization," *IEEE Trans. Automat. Contr.*, Vol. AC-23, No. 4, pp. 731-734, 1978.

26. D.A. Wilson and R.N. Mishra, "Optimal reduction of Multivariable Systems," *Int. J. Contr.*, Vol. 29, No. 2, pp. 267-278, 1979.

27. J.M. Siret, G. Michailesco, and P. Bertrand, "Optimal approximation of high-order Systems subject to polynomial inputs," *Int. J. Contr.*, Vol. 26, No. 6, pp. 963-971, 1977.

28. J.H. Anderson, "Geometrical approach to reduction of dynamical systems," *Proc. IEE*, Vol. 114, pp. 1014-1018, 1967.

29 D. Mitra, "W-matrix and the geometry of model equivalence and reduction," *Proc. IEE*, Vol. 116, pp. 1101-1106, 1969.

30. D. Mitra, "Analytical results on the use of reduced models in the control of linear dynamical systems," *Proc. IEE*, Vol. 116, pp. 1439-1444, 1969.

31. J.D. Aplevich, "Approximation of discrete linear systems," *Int. J. Control*, Vol. 17, No. 3, pp. 565-575, 1973.

32. G.J. Bierman, "Weighted least squares stationary approximation to linear systems", *IEEE Trans. Automat. Contr.*, Vol. AC-17, pp. 1313-1324, 1973.

33. H. Nosrati and H.E. Meadows, "Modeling of linear time-varying systems by linear time-invariant systems of low order," *IEEE Trans. Automat. Contr.*, Vol. AC-18, pp. 50-52, 1973.

34. E. Elitelberg, "Comments on Model Reduction by Minimizing the Equation Error,"*IEEE Trans.* Automat. Contr., Vol. AC-27, No. 4, pp. 1000-1002, 1982.

35. A. Lepschy and U. Viaro, "Model Reduction using the output equation error," *Int. J. Systems Sci.*, Vol. 15, pp. 1011-1021, 1984.

36. A. Lepschy, G.A. Mian and U. Viaro, "A method for optimal linear model reduction," *Syst. Contr. letters,* Vol. 8, pp. 405-410, 1987.

37. N.N. Puri and D.P. Lan, "Stable Model Reduction by Impulse Response Error Minimization Using Michailov Criterion and Pade's Approximation," *J. Dynamic Syst. Meas. and Contr.*, Vol. 110, pp. 389-394, 1988.

38. T.C. Chen, C.Y. Chang, and K.W. Han, "Model Reduction using the Stability-Equation Method and the Pade's Approximation method," *J. Franklin Inst.*, Vol. 309, pp. 473-490, 1980.

39. R.A. El-Attar, M. Vidyasagar, and S.R.K. Dutta, "Optimality Conditions for l_1-norm minimization", in *Proc. 19th Midwest Symp. Circuits and Systems*, pp. 272-275, 1976.

40. L. Meier, III and D.G. Luenberger, "Approximation of linear Constant Systems," *IEEE Trans. Automat. Contr.*, Vol. AC-12, pp. 585-588, 1967.

41. A.S.S.R. Reddy, "A method for frequency domain simplification of transfer functions", *Int. J. Contr.*, Vol. 23, No. 3, pp. 403-408, 1976.

42. R. Luus, "Optimization in model reduction," *Int. J. Contr.*, Vol. 32, No. 5, pp. 741-747, 1980.

43. Y. Bistritz and G. Langholz, "Model Reduction by best Chebyshev rational approximations in the complex plane," *Int. J. Contr.*, Vol. 30, No. 2, pp. 277-289, 1979.

44. U.M. Al-Saggaf, "Approximation of discrete linear systems based on a Dual Criterion," *Int. J. Syst. Sciences*, Vol. 22, No. 2, pp 383-392, 1991.

45. S. R. Dutta and M.Vidyasagar, "New Algorithms for constrained mini-max optimization," *Mathematical Programming*, vol. 13,pp. 140-155, 1977.

46. L.M. Silverman and M. Bettayeb, "Optimal approximation of linear Systems", in *JACC, Vol. 2, p. FA 8-A,* August 1980.

47. M. Bettayeb, L.M. Silverman and M.G. Safonov, "Optimal approximation of continuous time systems, "Proc. *19th I.E.E.E. CDC,* Albuquerque, New Mexico, December 1980.

48. M. Bettayeb, "Approximation of linear systems: New Approaches based on Singular value decomposition", Ph.D. Thesis, University of Southern California, Los Angeles, August 1981.

49. S.Y. Kung and D. W. Lin, "Optimal Hankel Norm Model Reductions: Multivariable Systems, *IEEE Trans. Auto. Contr.*, AC-26, pp. 832-852, 1981.

50. S.Y. Kung and D.W. Lin, "A state-space formulation for optimal Hankel-norm model approximations," *IEEE Trans. Auto. Contr.*, No. AC-26, No. 4, pp. 942-946, August 1981.

51. K. Glover,"All optimal Hankel-norm approximations of Linear Multivariable systems and their L^∞ -error bounds", *Int. J. Control*, Vol. 39, No. 6, pp. 1115-1193, 1984.

52. B.C. Moore, "Principal components analysis in linear systems: Controllability, observability and model reduction", *IEEE Transactions on Automatic control,* Vol. AC-26, pp. 17-32, 1981.

53. A.J. Laub, M.T. Heath, C.C. Paige and R.C. Ward, "Computation of System balancing Transformations and other Applications of simultaneous Diagonalization Algorithms",*IEEE Transactions on Automatic Control,* Vol. AC-32, No. 2, pp. 115-121, 1987.

54. K. Glover and D.J.N. Limebeer, "Robust Multivariable Control System Design using Optimal Reduced-order Plant Models", *Proc. Amer. Contr. Conf.*, San Francisco, pp. 644-649, 1983.

55. K. Glover, R.F. Curtain, and J.R. Partington, "Realization and approximation of linear infinite dimensional systems with error bounds", *SIAM J. Contr. Optimiz.,* Vol. 26, pp. 863-898, 1988.

56. K. Glover, J. Lam, and J.R. Partington, "Balanced realization and Hankel-norm approximation of systems involving delays", in *Proc. IEEE Conf. Decision Contr.,* Athens, Greece, pp. 1810-1815, Dec. 1986.

57. U.M. Al-Saggaf, "On Model Reduction and Control of discrete-time Systems", Ph.D. dissertation, Information Systems Laboratory, Dept. of Electrical Engineering, Stanford University, June 1986.

58. D. F. Enns, "Model Reduction for Control System Design," Ph.D. dissertation, Dept. of Aeronautics and Astronautics, Stanford University, June 1984.

59. U.M. Al-Saggaf and G.F. Franklin, "Model Reduction via Balanced Realizations, an Extension and Frequency Weighting Techniques", *IEEE Trans. on Automatic Control,* Vol. AC-33, No. 7, pp. 687-692, July 1988.

60. A. Latham and B.D.O. Anderson, "Frequency-weighted optimal Hankel-norm approximation of stable transfer functions", *Systems & Control Letters,* Vol. 5, pp. 229-236, 1985.

61. Y. S. Hung and K. Glover, "Optimal Hankel-norm approximations with first-order stable weighting functions," *Systems & Control Letters,* Vol. 7, pp. 165-172, 1986.

62. U.M. Al-Saggaf and G.F. Franklin, "An Error Bound for a Discrete Reduced Order Model of a Linear Multivariable System", *IEEE Trans. on Automatic Control,* Vol. AC-32, No. 9, pp. 815-819, Sept., 1987.

63. J.L. Walsh, *Interpolation and Approximation by Rational Functions in the Complex Domain, 2nd. Ed.,* American Mathematical Society, 1956.

64. L. Pernebo and L. M. Silverman, "Model Reduction via balanced state space representations," *IEEE Trans. Automat. Contr.,* vol. AC-27, No. 2, Apr. 1982.

65. B. D. O. Anderson and Y. Liu, "Controller Reduction:Concepts and approaches," *IEEE Trans. Automat. Contr.*, vol. Ac-34, No. 8, Aug. 1989.

66. U. M. AL-Saggaf, "Subsystem Interconnection and Near Optimum Control of Discrete Balanced Systems", *IEEE Trans. Automat. Control* **(accepted).**

67. M. G. Safonov and R. Y. Chiang, "A Schur method for balanced-truncation model reduction", *IEEE Trans. Automat. Contr.,* **AC-34**, pp. 729-733 (1989).

68. K. V. Fernando and H. Nicholson, "Minimality of SISO Linear Systems", *Proc. IEEE* **70**, pp. 1241-1242 (1982).

69. D. F. Enns, "Model reduction with balanced realizations: An error bound and a frequency weighted generalization, in *Proc. 23rd Conf. Decision Contr.*, Las Vegas, NV, 1984, PP. 127-132.

70. A. J. Laub, L. M. Silverman, and M. Verma, "A note on cross-Gramians for symmetric realizations", *Proc. IEEE*, 71, pp. 904-905 (1983).

71. K. V. Fernando and H. Nicholson, "On the cross-Gramian for symmetric MIMO systems", *IEEE Trans. Circuits Syst.*, **CAS-32**, pp. 487-489, (1985).

72. J. A. De Abreu-Garcia and F. W. Fairman, "A Note on cross-Gramians for orthogonally symmetric realizations", *IEEE Trans. Automat. Contr.*, **AC-31**, pp. 866-868, (1986).

73. J. A. De Abreu-Garcia and F. W. Fairman, "On using permutation symmetric Jordan realizations to achieve SISO balancing", *Int. J. Syst. Sci.*, 18, pp. 441-448 (1987).

74. R. R. Bitmead and B. D. O. Anderson, "The matrix Cauchy index: Properties and applications", *SIAM J. Appl. Math.*, 33, pp. 655-671 (1977).

75. J. A. De Abreu-Garcia and F. W. Fairman, "Balanced Realizations of Orthogonally Symmetric Transfer Function Matrices", *IEEE Trans. Circuits Syst.*, **CAS-34**, pp. 997-1010 (1987).

76. K. V. Fernando and H. Nicholson, "Singular Perturbational Approximations for Discrete-Time Balanced Systems", *IEEE Transactions on Automatic Control*, **AC-28**, Feb. 1983, pp. 240-242.

77. E. A. Jonckheere and L. M. Silverman, "A New Set of Invariants for Linear Systems-Application to Reduced Order Compensator Design", *IEEE Transactions on Automatic Control,* Vol. **AC-28**, Oct. 1983, pp. 953-964.

78. A. Yousuff and R. E. skelton, "A Note on Balanced Controller Reduction", *IEEE Transactions on Automatic Control,* Vol. **AC-29**, March 1984, PP. 254-257.

79. S. Shokoohi, L. M. Silverman, and P. M. Van Dooren, "Linear Time-Variable Systems: Balancing and Model Reduction", *IEEE Transactions on Automatic Control,* Vol. **AC-28**, Aug. 1983, pp. 810-822.

80. E. I. Verriest and T. Kailath, "On Generalized Balanced Realizations", *IEEE Transactions on Automatic Control,* Vol. **AC-28**, Aug. 1983, PP. 833-844.

81. J. A. Davis and R. E. Skelton, "Another Balanced Controller Reduction Algorithm", *Systems & Control Letters,* Vol. 4, April 1984, pp. 79-83.

82. Y. Liu and B. D. O. Anderson, "Controller Reduction via Stable Factorization and Balancing", *International Journal of Control,* Vol. 44, No. 2, Aug. 1986, pp. 507-531.

83. F. Clara and L. M. Silverman, "Two-dimensional discrete space-varying systems: identification, balancing and model reduction", *in* "Identification and system parameter estimation", (G. A. bekey and G. N. Saridis, ed.), Pergamon, Oxford, England, 1982.

84. P. T. Kabamba, "Balanced gains and their significance for L^2 model reduction", *IEEE Trans. Automat. Contr.* **AC-30(7)**, PP. 921-923 (1985).

85. R. J. Vaccaro, "Deterministic balancing and stochastic model reduction, " *IEEE Trans. Automat. Control* **AC-30(9)**, pp. 921-923 (1985).

86. A. M. Davidson, "Balanced systems and model reduction", *Electron Lett.* **22(10)**, pp. 531-532 (1986).

87. G. J. Lastman and N. K. Sinha, "An error analysis of the balanced matrix method of model reduction, with application to the selection of reduced-order models", *Large Scale Syst. Theory & Appl.* **9(1)**, pp. 63-71 (1985).

88. L. Thiele, "Balanced model reduction in time- and frequency-domain", *Proc. of the 1985 International Symposium on Circuits and Systems"*, pp. 345-348 (1985).

89. R. E. Skelton and P. Kabamba, "Comments on Balanced gains and their significance for L^2 model reduction", *IEEE Trans. Automat. Contr.* **AC-31(8)**, 796-797 (1986).

90. E. Jonckheere and R. Li, "Generalization of optimal Hankel-norm and balanced model reduction by bilinear mapping" *Int. J. Contr.* **45(5)**, pp. 1751-1769 (1987).

91. R. J. Ober, "Balanced realizations: canonical form, parametrizations, model reduction", *Int. J. Contr.* **46(2)**, pp. 643-670 (1987).

92. P. Agathoklis and V. Sreeram, "Truncation criteria for model reduction using balanced realization", *Electron. Lett.* **24(14)**, pp. 837-838 (1988).

93. V. Sreeram and P. Agathoklis, "Model reduction using balanced realizations with improved low frequency "behaviour", *Syst. Control Lett.* **12(1)**, pp. 33-38 (1989).

94. Y. Liu and B. D. O. Anderson, "Singular perturbation approximation of balanced systems, *Int. J. Control* **50(4)**, pp. 1379-1405 (1989).

95. K. V. Fernando and H. Nicholson, "Singular perturbation model reduction of balanced systems", *IEEE Trans. Automat. Contr.* **AC-27**, pp. 466-468 (1982).

96. K. V. Fernando and H. Nicholson, "Singular perturbational model reduction in the frequency domain", *IEEE Trans. Automat. Contr.* **AE-27**, PP. 969-970 (1982).

97. K. V. Fernando and H. Nicholson, "On the structure of balanced and other principal representations of SISO systems, *IEEE Trans. Automat. Contr.*, **AC-28**, PP. 228-231 (1983).

98. K. V. Fernando and H. Nicholson, "Reciproical transformations in balanced model-order reduction *proceedings of the Institution of Electrical Engineers, Pt D,* **130, pp. 359-362 (1983).**

99. C. P. Therapos, "Internally balanced minimal realization of discrete SISO systems", *IEEE Trans. Automat. Contr.* **AC-30(3)**, pp. 297-299 (1985).

100. J. R. Seinsson and F. W. Fairman, "Minimal balanced realization of transfer function matrices using Markov parameters", *IEEE Trans. Automat. Contr.* **AC-30(10)**, pp. 1014-1016 (1985).

101. F. W. Fairman, S. S. Mahil and J. A. De-Abreu, "Balanced Realization Algorithm for scalar continuous-time systems having simple poles", *Int. J. Syst. Sci.* **15(6)**, pp. 685-694 (1984).

102. U. B. Desai and D. Pal, "A Realization approach to stochastic model reduction and balanced stochastic realizations", *Proc. 21st IEEE Conf. Dec. & Contr.*, pp. 1105-1112 (1982).

103. J. R. Seinsson, F. W. Fairman and A. Kumar, "Separately balanced realization of two-dimensional separable-denominator transfer functions", *Int. J. Syst. Sci* **18(3)**, pp. 419-425 (1987).

104. A. Kumar, F. W. Fairman, and J. R. Seinsson, "Separately balanced realization and model reduction of 2-D separable-denominator transfer function from input-output data", *IEEE Trans. Circuits & Syst.* **CAS-34(3)** pp. 233-239 (1987).

105. F. W. Fairman and B. S. Lee, "Direct minimal balanced realization from impulse response matrices using I/O map decomposition", *SIAM J. Control & Optimiz.* **25(1)**, pp. 1-17 (1987).

106. T. Hinamoto; S. Maekawa, J. Schimonishi, and A. N. Venetsanopoulos, "Balanced realization and model reduction of 3-D separable-denominator transfer functions", *J. Franklin Inst.* **325(2)**, pp. 207-219 (1988).

107. Mi-Ching Tsai and Yen-Ping Shih, "Balanced minimal realization via singular value decomposition of a Sarason operator", *Automatica* **24(5)**, pp. 701-705 (1988).

108. S. Shokoohi and L. M. Silverman, "Model Reduction of discrete time-variable systems via balancing", proc. of 20th IEEE Conf. on Dec. & Control, pp. 676-680 (1981).

109. P. Harshavardhana, E. A. Jonckheere, and L. M. Silverman, "Stochastic Balancing and Approximation-Stability and Minimality", *IEEE Trans. Automat. Contr.* **AC-29 (8)**, pp. 744-746 (1984).

110. R. Ober, "A parameterization approach to infinite-dimensional balanced Realizations and their approximation, *IMA J. Math. Control Inf.* **4(4)**, pp. 263-279 (1987).

111. W. S. Lu, E. B. Lee and Q. T. Zhang, "Balanced Approximation of Two-Dimensional and Delay-Differential Systems", *International Journal of Control* **46(6)**, pp. 2199-2218 (1987).

112. M. Green, "Relative error bound for balanced stochastic truncation", *IEEE Trans. Automat. Contr.* **AC-33(10)**, pp. 961-965 (1988).

113. U. M. Al-Saggaf, "Reduced order models for Dynamic Control of Power Plants with an improved Transient and Steady State Behavior", **submitted.**

114. C. Kenney and G. Hewer, "Necessary and Sufficient Conditions for balancing unstable systems", *IEEE Trans. Automat. Contr.* **AC-32**, pp. 157-160 (1987).

115. C. P. Therapos, "Balancing Transformations for unstable non-minimal linear systems", *IEEE Trans. Automat. Contr.* **AC-34**, pp. 455-457., 1989.

116. U. M. Al-Saggaf, "Model reduction for discrete unstable systems based on generalized normal representations", *International Journal of Control,* **(Accepted)**.

117. E. A. Jonckheere and L. M. Silverman, "Singular value analysis of Deformable Systems", *Circuits, Systems Signal Process,* **Vol. 1**, Birkhauser, Boston (1982).

118. E. A. Jonckheere and P. Opendenacker, "Singular value analysis, Balancing and Model Reduction of Large Space Structures", *Proceedings of American Control Conference,* (1984).

119. E. A. Jonckheere, "Principal Component Analysis of Flexible Systems-Open Loop Case", *IEEE Trans. Automat. Contr.* **AC-29**, pp. 1095-1097 (1984).

120. C. Z. Gregory, "Reduction of Large Flexible Spacecraft Models Using Internal Balancing Theory", *AIAA Journal of Guidance and Control* **7**, pp. 725-732 (1984).

121. P. A. Belloch, D. L. Mingori and J. D. Wei, "Perturbation Analysis of Internal Balancing for Lightly Damped Mechanical Systems with Gyroscopic and Circulatory Forcess", *AIAA Journal of Guidance and Control* **10(4)**, pp. 406-410 (1987).

122. Gy. Michaletzky, "Hankel-norm approximation of a rational function using stochastic realizations", Systems Control Letters 13, pp. 211-216, 1989.

123. P. H. Delsante, Y. Genin and Y. Kamp, "On the role of the Nevanlinna-Pick problem in circuit and system theory", Int. J. Circuit Theory Application, Vol. 9, pp. 177-187, 1981.

124. Y. V. Genin, S. Y. Kung, "A two-variable approach to the model reduction problem with Hankel norm criterion", IEEE Trans. Circuits System, Vol. CAS-28, No. 9, pp. 912-924, 1981.

125. P. Hanshavardhana, E. A. Jonckheere and L. M. Silverman, "Eigenvalue and Generalized Eigenvalue formulations for Hankel norm reduction directly from polynomial data", Proc. of 23rd IEEE Conference on Decision and Control, pp. 111-119, Las Vegas, NV, December 1984.

126. M. G. Safonov and H. S. Verma, "L_∞ sensitivity optimization and Hankel approximation", IEEE Trans. Automat. Contr., Vol. AC-30, pp. 279-200, March 1985.

127. B. D. O. Anderson, "Weighted Hankel-norm approximation: Calculation of bounds," *Systems and Control Letters,* No. 7, (1986).

128. V. M. Adamjan, D. Z. Arov and M. G. Kerin, "Analytic propoerties of Schmidt pairs for a Hankel operator and the generalized Schur-Takagi problem" *Math. USSR Sbornik*, Vol. 15, pp. 31-73, (1971).

129. J. A. Ball and A. C. M. Ran, "Optimal Hankel norm model reduction and Wiener-Hopf Factorizations I: The canonical case", *SIAM J. Contr. Optimization*, Vol. 25, No. 2, p. 362, March 1987.

130. J. A. Ball and A. C. M. Ran, "Optimal Hankel norm model reduction and Wiener-Hopf Factorizations I: The non-canonical case", *J. Integral Equations and Operator Theory*, Vol. 10, p. 416, 1987.

131. U. M. Al-Saggaf and M. Bettayeb, "Optimization in model reduction", Arab. J. Sci. and Engg., Vol. 15, No. 4B, pp. 705-719, Oct. 1990.

132. M. Bettayeb and S. Djennoune,"Resultats nouveaux sur l'interconnection des sous-systemes d'une representation equilibree",Accepted for publication in RAIRO,APII,1989.

133. M. Bettayeb,"New interpretaion of balancing state space representations as input-output energy minimization problem", *Int. J. Syst. Sci.*, Vol. 22, No. 2, pp. 325-331, Feb. 1991.

134. M. Bettayeb and S. Djennoune,"Structure of singularly perturbed systems and balancing", *Accepted, IEE Proc. Pt.D*, 1991.

135. M. Bettayeb and U. M. Al-Saggaf,"Practical model reduction techniques for power systems", *Submitted.*

Techniques for Adaptive Estimation and Control of Discrete-Time Stochastic Systems with Abruptly Changing Parameters

Spyros G. Tzafestas
Division of Computer Science,
Department of Electrical Engineering,
National Technical University of Athens,
Zografou, Athens 15773, Greece

Keigo Watanabe
Department of Mechanical Engineering,
Faculty of Science and Engineering,
Saga University,
Honjomachi-1, Saga 840, Japan

I. INTRODUCTION

The estimation problem for the state of a linear stochastic system with abruptly changing parameters can be solved by using a Kalman filter only if the jump parameters are observed. However, the jump parameters are generally unknown, and therefore we can not solve the above problems by using a single Kalman filter. One approach is to use the multiple model adaptive filtering (MMAF) approach (Lainiotis [1,2]) developed for systems having unknown constant parameters, setting an upper (or lower) bound on the a posteriori probabilities (Gustafson and co-workers [3], Willsky [4], Watanabe [5,6]). Another approach is to extend the MMAF technique to stochastic systems with possibly unknown, time-varying parameters, which are modeled as a finite-state Markov (or semi-Markov) chain state with known transition statistics. The latter approach seems to be more natural estimation one, if the transition probabilities are known.

CONTROL AND DYNAMIC SYSTEMS, VOL. 55

The motivation for considering system models with jumps arises from the applicability of such models to a large class of realistic problems, namely:

(i) fault detection for a dynamic system with failures in components or subsystems (see Willsky [4,7], Gustafson and co-workers [3], Watanabe [5,6]),

(ii) target tracking for a moving vehicle with sudden maneuvers (see Ricker and Williams [8], Moose, VanLandingham and McCabe [9], Chang and Tabaczynski [10]) and

(iii) approximation of a nonlinear system by a set of linearized models to cover the entire dynamic range (Moose, VanLandingham and Zwicke [11]).

It is well-known (Ackerson and Fu [12]) that to evaluate the minimum mean-squared error (MMSE) estimate of the system in switching environment, the computational and storage requirements increase exponentially with time, which renders the optimal solution impractical. To circumvent this problem, there are some suboptimal algorithms: e.g.,

(a) random sampling algorithm (RSA) (Akashi and Kumamoto [13]),

(b) detection-estimation algorithm (DEA) (Tugnait and Haddad [14], Tugnait [15], Mathews and Tugnait [16]),

(c) generalized pseudo-Bayes algorithm (GPBA) (Ackerson and Fu [12], Jaffer and Gupta [17], Chang and Athans [18], Sugimoto and Ishizuka [19]), and

(d) interacting multiple model algorithm (IMMA) (Blom [20,21], Blom and Bar-Shalom [22]).

It is worth noting that the IMMA performs nearly as well as the 2nd order GPBA method with notably less computation.

It is also interesting to note that several simulations results presented by Tugnait [15] indicate that, in general, the GPBA is to be preferred, compared with the RSA and DEA methods, though the performance of the various algorithms is very much dependent upon the system model under consideration. The GPBA due to Ackerson and Fu [12] and Chang and Athans [18] are a special case for the GPBA due to Jaffer and Gupta [17]. The result of Sugimoto and Ishizuka [19] is the same as that of Jaffer and Gupta [17].

On the other hand, the control problems for systems in which the jump parameters are included in the plant or observation model have received considerable attention (see, e.g., Griffiths and Loparo [23], Mariton [24]). The results can be mainly classified into two categories:

(i) the parameters are observed or known, and

(ii) the parameters are not observed or unknown.

In the first category, Sworder [25] and Wonham [26] solved the so-called optimal control *jump linear quadratic* (JLQ) regulation problem for a case where both the system state and jump parameters are perfectly measurable. Sworder [27] also studied a similar JLQ regulation problem for a case in which the measurement of the jump parameters is close to the true value. Assuming that the jumps of the model parameters are perfectly observed, Mariton [24] solved the optimal jump linear quadratic Gaussian (JLQG) regulator problem for stochastic systems with state- and control-dependent noises. Note that all these results are derived for continuous-time systems. Similar results for discrete-time systems can be found in Blair and Sworder [27], and Chizeck, Willsky and Castanon [28].

If the jump parameters are unknown, then the control problem becomes a *dual* problem, in the sense that the optimal controller must simultaneously identify the jump parameters and regulate the system states (Bar-Shalom [29]). The JLQG control problem under imperfect state observations has been considered by many authors (e.g., Fujita and Fukao [30], Akashi, Kumamoto and Nose [31], Tugnait [32], Watanabe and Tzafestas [33] or [37]) for the case where the jumps are confined to the observation system. For the general case where the jumps are included in both the state and observation equations, Griffiths and Loparo [23] have proposed a suboptimal control algorithm which reduces the computational complexity of the dual control problem. VanLandingham and Moose [34], and Moose, VanLandingham and Zwicke [9] have also studied a JLQG control problem for nonlinear systems with semi-Markov jump parameters by directly applying a multiple model adaptive control (MMAC) technique developed for the jumpless case (Deshpande, Upadhyay and Lainiotis [35], Lainiotis [36]).

When extending the MMAC concept, developed for the jumpless case, to the general control problem, in which the Markovian jump parameters are included in both the plant and observation models and are not observed, the optimal control in the Bayesian sense needs s^N parallel Kalman filters and s^N sequences of control gains at time k, where s denotes the number of Markov chain states and N is a final time. Therefore, as in the estimation problem, to get a practical controller for large N, one must resort to a suboptimal control strategy.

The goal of this chapter is to provide a view of some GPBAs proposed in the literatures [12,17,38] for the state estimation and system structure detection problem and to apply them to a suboptimal passive-type MMAC problem. The system models under consideration are described by discrete-time stochastic systems with unknown jump parameters, which can be modeled by a finite-state Markov chain with known transition statistics.

In Section II, two GPBAs are introduced which are called "one-step measurement update method" and "n-step measurement update method", respectively. The former is due to Jaffer and Gupta [17], and the latter to Watanabe and Tzafestas [38]. In particular, the latter method is similar to an algorithm for fixed-lag smoothing for a lag of $n-1$ units of time, and consists of reprocessing $n-1$ measurement data with s parallel filtered estimates as starting conditions at time $k-n$. To avoid the exponential growth of the size of elemental controllers, a suboptimal passive-type MMAC [33,37,39] is proposed in Section III. Algorithms for elemental control gains are given by a set of s coupled Riccati-like equations that can be completely computed off-line, while a GPBA based on an n-step measurement update method is applied for the elemental filter mechanism.

II. GENERALIZED PSEUDO-BAYES ESTIMATION AND DETECTION

A. Problem Statement

Let $\alpha(k) \in S = \{1, 2, ..., s\}, k \in \{1, 2, ...\}$, denote a discrete finite-state, Markov chain with completely known time-invariant transition probabilities

$$p_{ij} \triangleq \Pr\{\alpha(k) = j | \alpha(k-1) = i\}, \quad i, j \in S \tag{1}$$

and initial probability distribution $p_i = \Pr\{\alpha(0) = i\}, i \in S$. Let $\pi = [p_{ij}]$, an $s \times s$ matrix, denote the transition probability matrix. Now $\alpha(k)$ governs the structure of a stochastic dynamical system under the normal or failure mode. The system state equation is given by

$$x(k+1) = A(\alpha(k+1))x(k) + G(\alpha(k+1))u(k) + B(\alpha(k+1))w(k) \tag{2}$$

where $x(k) \in \mathcal{R}^{n_1}$ is the state vector, $u(k) \in \mathcal{R}^q$ is the input vector, and $w(k) \in \mathcal{R}^p$ is a zero-mean Gaussian white noise sequence with covariance Q. The observation equation associated with (2) is modeled by

$$z(k) = C(\alpha(k))x(k) + D(\alpha(k))v(k) \tag{3}$$

where $z(k) \in \mathcal{R}^{m_1}$ is the measurement vector, and $v(k) \in \mathcal{R}^{m_1}$ is a zero-mean Gaussian white measurement noise with covariance R such that $D_i R D_i^T > 0$ ($i \in S$) where $D(\alpha(k)) \in \{D_i, i = 1, 2, ..., s\}$. The initial state is assumed to possess the following Gaussian distribution:

$$x(0) \sim N(\bar{x}_0, P_0) \tag{4}$$

Finally, $x(0), w(k), v(k)$ and $\alpha(k)$ are assumed to be mutually independent.

The objective is to find the minimum mean-squared error (MMSE) state estimate $\hat{x}(k|k)$ of $x(k)$ given the observation data $Z_k = \{z(i), 1 \leq i \leq k\}$ and all past inputs $U_{k-1} = \{u(i), 0 \leq i \leq k-1\}$, and to decide on the value of $\alpha(k)$ (system fault detection), given Z_k and U_{k-1}, minimizing the probability of error.

B. Optimal Solution

1. State Estimation

It is well-known that the MMSE filtered state estimate $\hat{x}(k|k)$ is given by the conditional mean

$$\hat{x}(k|k) = E[x(k)|Z_k, U_{k-1}] \tag{5}$$

Define a Markov chain state sequence $I(k)$ as

$$I(k) \triangleq \{\alpha(1), ..., \alpha(k)\} \tag{6}$$

and let $I_j(k)$ denote a specific sequence from the space of all possible sequences $I(k)$ which contains s^k elements. If the state estimate conditioned on a specific sequence is written as

$$\hat{x}_j(k|k) = E[x(k)|I_j(k), Z_k, U_{k-1}] \tag{7}$$

then

$$\hat{x}(k|k) = \sum_{j=1}^{s^k} \hat{x}_j(k|k) p(I_j(k)|Z_k, U_{k-1}) \tag{8}$$

where $p(I_j(k)|Z_k, U_{k-1})$ is the a posteriori probability of $I_j(k)$ given Z_k and U_{k-1}, which is subject to

$$p(I_j(k)|Z_k, U_{k-1}) = \frac{f(z(k)|I_j(k), Z_{k-1}, U_{k-1})p(I_j(k)|Z_{k-1}, U_{k-2})}{\sum_{l=1}^{s^k} f(z(k)|I_l(k), Z_{k-1}, U_{k-1})p(I_l(k)|Z_{k-1}, U_{k-2})} \tag{9}$$

where $f(\cdot|\cdot)$ is the conditional probability density of the observation $z(k)$ given the past observations Z_{k-1}, the past controls U_{k-1}, and the particular state mode sequence $I_j(k)$. Furthermore, it is found that (Mathews and Tugnait [16])

$$p(I_j(k)|Z_{k-1}, U_{k-2}) = p(\alpha(k)|\alpha(k-1) = i)p(I_m(k-1)|Z_{k-1}, U_{k-2}), \quad i \in S \tag{10}$$

because

$$\begin{aligned} p(\alpha(k)|I_m(k-1), Z_{k-1}, U_{k-2}) &= p(\alpha(k)|I_m(k-1)) \\ &= p(\alpha(k)|\alpha(k-1) = i), \quad i \in S \end{aligned} \tag{11}$$

where the first equality in (11) follows from the conditional independence of $\{\alpha(k)\}$ and $\{z(k), u(k)\}$, and the second equality in (11) from the Markovian nature of $\alpha(k)$. Here, $I_m(k-1)$ denotes a specific sequence from the space of all possible sequences $I(k-1)$ as defined in (6). The associated state estimation error covariance matrix

$$P(k|k) = E\{[x(k) - \hat{x}(k|k)][x(k) - \hat{x}(k|k)]^T|Z_k, U_{k-1}\} \tag{12}$$

is given by (Lainiotis [10]):

$$\begin{aligned} P(k|k) = \sum_{j=1}^{s^k}\{&P_j(k|k) + [\hat{x}_j(k|k) - \hat{x}(k|k)] \\ &\times [\hat{x}_j(k|k) - \hat{x}(k|k)]^T\}p(I_j(k)|Z_k, U_{k-1}) \end{aligned} \tag{13}$$

where

$$P_j(k|k) = E\{[x(k) - \hat{x}_j(k|k)][x(k) - \hat{x}_j(k|k)]^T|I_j(k), Z_k, U_{k-1}\} \tag{14}$$

Now, having available the initial information (4) and the sequence $I_j(k)$, one can obtain $\{\hat{x}_j(k|k), P_j(k|k)\}$ recursively through Kalman filters matched to the sequences $I_j(k), j = 1, ..., s^k$. Furthermore, since $f(z(k)|I_j(k), Z_{k-1}, U_{k-1})$ is Gaussian, the weighting probability $p(I_j(k)|Z_k, U_{k-1})$ can also be computed from the information supplied by the same Kalman filters. Thus, the optimal estimator (8), which is a weighted sum of s^k estimates $\hat{x}_j(k|k)$, requires an exponentially increasing memory and computational capability with time. Therefore, one has to resort to suboptimal schemes to circumvent this difficulty.

2. Structure Detection

In a structure (or fault) detection problem, we find $\alpha(k)$ such that

$$\hat{\alpha}(k) = \arg\left\{\max_{\alpha\in S} p(\alpha(k)|Z_k, U_{k-1})\right\} \tag{15}$$

Now we have

$$p(\alpha(k)|Z_k, U_{k-1}) = \sum_{j=1}^{s^{k-1}} p(\alpha(k), I_j(k-1)|Z_k, U_{k-1}) \tag{16}$$

Thus, s^{k-1} Kalman filters in parallel are necessary to compute (16) for given $\alpha(k) \in S$. The approximation used to alleviate this difficulty is the same as for the state estimation problem.

C. Suboptimal Algorithms

The various approximations to state estimation or structure detection available in the literature have been classified into some categories as indicated in the introduction. In this subsection we shall introduce two GPBAs proposed in [17] and [38].

1. Generalized Pseudo-Bayes Algorithm

The essential assumption of GPBA is that the probability density of the system state at time k conditioned on Z_k and U_{k-1}, and the Markov chain state sequence $I_j(k, k-n) \triangleq \{\alpha(i), k-n \leq i \leq k\}, j = 1, ..., s^{n+1}, n \geq 1$ is Gaussian, whereas, in reality, it is a Gaussian sum. That is, it is assumed that

$$f(x(k)|I_j(k, k-n), Z_k, U_{k-1}) \sim N(\hat{x}_j(k|k), P_j(k|k)) \tag{17}$$

Under this assumption, the state estimate $\hat{x}(k|k)$ and the associated state estimation error covariance matrix $P(k|k)$ are approximated by

$$\hat{x}(k|k) = \sum_{j=1}^{s^{n+1}} \hat{x}_j(k|k) p(I_j(k, k-n)|Z_k, U_{k-1}) \tag{18}$$

and

$$P(k|k) = \sum_{j=1}^{s^{n+1}} \{P_j(k|k) + [\hat{x}_j(k|k) - \hat{x}(k|k)] \times [\hat{x}_j(k|k) - \hat{x}(k|k)]^T\} p(I_j(k, k-n)|Z_k, U_{k-1}) \tag{19}$$

It is important to note here that, given an observation $z(k)$, one can consider two approaches for updating the conditional estimates that have been obtained up to time $k-1$. In the first, the s^{n+1} conditional estimates $\hat{x}_j(k|k)$ can be obtained by updating the one-step past s^n conditional estimates at time $k-1$. This method is called the "one-step measurement update method". The second is based on computing the s^{n+1} conditional estimates by updating the n-step past s conditional estimates at time $k-n$, but under the condition that $z(k-1), ..., z(k-n+1)$ and $u(k-2), ..., z(k-n)$ have already been stored. This method is called the "n-step measurement update method".

The approximation used for the state estimation algorithm also leads to a suboptimal detection algorithm. Equation (16) is approximated by

$$p(\alpha(k)|Z_k, U_{k-1}) = \sum_{j=1}^{s^n} p(\alpha(k), I_j(k-1, k-n)|Z_k, U_{k-1}) \tag{20}$$

because $I_j(k, k-n) = \{\alpha(k), I_j(k-1, k-n)\}$.

2. One-Step Measurement Update Method

Almost all approaches appeared in the literature belong to this category. The fundamental form was first proposed by Ackerson and Fu [12] under the assumption that $f(x(k)|Z_k, U_{k-1})$ is Gaussian.

To facilitate the presentation, we introduce the hypothesis:

$$H_{i_k} : \alpha(k) = i_k, \quad i_k \in S \tag{21}$$

Then, in the one-step measurement update approach equations (18) and (19) can be realized as the following algorithm.

One-step measurement update algorithm. It is assumed that the following quantities are available at time $k-1$ for $k > n+1$:

$$\hat{x}_{i_{k-n},\dots,i_{k-1}}(k-1|k-1) \triangleq E[x(k-1)|H_{i_{k-n}}, \dots, H_{i_{k-1}}, Z_{k-1}, U_{k-2}],$$

$$\begin{aligned}
&P_{i_{k-n},\ldots,i_{k-1}}(k-1|k-1) \\
&\quad \triangleq E\{[x(k-1) - \hat{x}_{i_{k-n},\ldots,i_{k-1}}(k-1|k-1)] \\
&\qquad \times [x(k-1) - \hat{x}_{i_{k-n},\ldots,i_{k-1}}(k-1|k-1)]^T | H_{i_{k-n}}, \ldots, H_{i_{k-1}}, Z_{k-1}, U_{k-2}\}
\end{aligned}$$

and $p(H_{i_{k-n}}, \ldots, H_{i_{k-1}} | Z_{k-1}, U_{k-2})$. Then, having available H_{i_k} and $z(k)$, equations (18) and (19) can be rewritten as

$$\hat{x}(k|k) = \sum_{i_{k-n}=1}^{s} \cdots \sum_{i_k=1}^{s} \hat{x}_{i_{k-n},\ldots,i_k}(k|k) p(H_{i_{k-n}}, \ldots, H_{i_k} | Z_k, U_{k-1}) \tag{22}$$

$$\begin{aligned}
P(k|k) = \sum_{i_{k-n}=1}^{s} \cdots \sum_{i_k=1}^{s} &\{P_{i_{k-n},\ldots,i_k}(k|k) + [\hat{x}_{i_{k-n},\ldots,i_k}(k|k) - \hat{x}(k|k)] \\
&\times [\hat{x}_{i_{k-n},\ldots,i_k}(k|k) - \hat{x}(k|k)]^T\} p(H_{i_{k-n}}, \ldots, H_{i_k} | Z_k, U_{k-1})
\end{aligned} \tag{23}$$

where $\hat{x}_{i_{k-n},\ldots,i_k}(k|k)$ and $P_{i_{k-n},\ldots,i_k}(k|k)$ are obtained recursively by the parallel Kalman filters:

$$\begin{aligned}
\hat{x}_{i_{k-n},\ldots,i_k}(k|k-1) = A(\alpha(k) = i_k)&\hat{x}_{i_{k-n},\ldots,i_{k-1}}(k-1|k-1) \\
&+ G(\alpha(k) = i_k) u(k-1)
\end{aligned} \tag{24}$$

$$\begin{aligned}
P_{i_{k-n},\ldots,i_k}(k|k-1) = A(\alpha(k) = i_k)&P_{i_{k-n},\ldots,i_{k-1}}(k-1|k-1) A^T(\alpha(k) = i_k) \\
&+ B(\alpha(k) = i_k) Q B^T(\alpha(k) = i_k)
\end{aligned} \tag{25}$$

$$K_{i_{k-n},\ldots,i_k}(k) = P_{i_{k-n},\ldots,i_k}(k|k-1) C^T(\alpha(k) = i_k) V^{-1}_{i_{k-n},\ldots,i_k}(k|k-1) \tag{26}$$

$$\begin{aligned}
\hat{x}_{i_{k-n},\ldots,i_k}(k|k) = \hat{x}_{i_{k-n},\ldots,i_k}(k|k-1) &+ K_{i_{k-n},\ldots,i_k}(k) \\
&\times [z(k) - C(\alpha(k) = i_k)\hat{x}_{i_{k-n},\ldots,i_k}(k|k-1)]
\end{aligned} \tag{27}$$

$$P_{i_{k-n},\ldots,i_k}(k|k) = [I - K_{i_{k-n},\ldots,i_k}(k) C(\alpha(k) = i_k)] P_{i_{k-n},\ldots,i_k}(k|k-1) \tag{28}$$

where $V_{i_{k-n},\ldots,i_k}(k|k-1)$ is defined by

$$\begin{aligned}
V_{i_{k-n},\ldots,i_k}(k|k-1) = C(\alpha(k) = i_k)&P_{i_{k-n},\ldots,i_k}(k|k-1) C^T(\alpha(k) = i_k) \\
&+ D(\alpha(k) = i_k) R D^T(\alpha(k) = i_k)
\end{aligned} \tag{29}$$

The initial conditions for the above equations are

$$\hat{x}_{i_0}(0|0) = \hat{x}(0), \qquad P_{i_0}(0|0) = P(0) \tag{30}$$

for all $i_0 \in S$. The a posteriori probability $p(H_{i_{k-n}}, ..., H_{i_k}|Z_k, U_{k-1})$ can be computed in the recursive form by:

$$\begin{aligned} &p(H_{i_{k-n}}, ..., H_{i_k}|Z_k, U_{k-1}) \\ &= \frac{p(z(k)|H_{i_{k-n}}, ..., H_{i_k}, Z_{k-1}, U_{k-2})}{\sum_{i_{k-n}=1}^{s} \cdots \sum_{i_k=1}^{s} p(z(k)|H_{i_{k-n}}, ..., H_{i_k}, Z_{k-1}, U_{k-2})} \\ &\quad \times \frac{p(H_{i_{k-n}}, ..., H_{i_k}|Z_{k-1}, U_{k-2})}{p(H_{i_{k-n}}, ..., H_{i_k}|Z_{k-1}, U_{k-2})} \end{aligned} \tag{31}$$

where

$$\begin{aligned} &p(z(k)|H_{i_{k-n}}, ..., H_{i_k}, Z_{k-1}, U_{k-2}) \\ &\sim N[C(\alpha(k) = i_k)\hat{x}_{i_{k-n},\dots,i_k}(k|k-1), V_{i_{k-n},\dots,i_k}(k|k-1)] \end{aligned} \tag{32}$$

and

$$p(H_{i_{k-n}}, ..., H_{i_k}|Z_{k-1}, U_{k-2}) = p_{i_{k-1},i_k} p(H_{i_{k-n}}, ..., H_{i_{k-1}}|Z_{k-1}, U_{k-2}) \tag{33}$$

Before going to the next stage, one has to store the following quantities:

$$p(H_{i_{k-n+1}}, ..., H_{i_k}|Z_k, U_{k-1}) = \sum_{i_{k-n}=1}^{s} p(H_{i_{k-n}}, ..., H_{i_k}|Z_k, U_{k-1}) \tag{34}$$

$$\hat{x}_{i_{k-n+1},\dots,i_k}(k|k) = \sum_{i_{k-n}=1}^{s} \frac{\hat{x}_{i_{k-n},\dots,i_k}(k|k) p(H_{i_{k-n}}, ..., H_{i_k}|Z_k, U_{k-1})}{p(H_{i_{k-n+1}}, ..., H_{i_k}|Z_k, U_{k-1})} \tag{35}$$

$$P_{i_{k-n+1},\dots,i_k}(k|k) = \sum_{i_{k-n}=1}^{s} \frac{P^*_{i_{k-n},\dots,i_k}(k|k) p(H_{i_{k-n}}, ..., H_{i_k}|Z_k, U_{k-1})}{p(H_{i_{k-n+1}}, ..., H_{i_k}|Z_k, U_{k-1})} \tag{36}$$

where

$$\begin{aligned} P^*_{i_{k-n},\dots,i_k}(k|k) &= P_{i_{k-n},\dots,i_k}(k|k) + [\hat{x}_{i_{k-n},\dots,i_k}(k|k) - \hat{x}_{i_{k-n+1},\dots,i_k}(k|k)] \\ &\quad \times [\hat{x}_{i_{k-n},\dots,i_k}(k|k) - \hat{x}_{i_{k-n+1},\dots,i_k}(k|k)]^T \end{aligned} \tag{37}$$

It should be noted that Chang and Athans [18] and Tugnait [15] use a refined equation to evaluate the covariance matrix $P(k|k)$, whereas in Ackerson and Fu [12] the following approximation is used

$$P(k|k) = \sum_{i=1}^{s} P_{i_k}(k|k)p(H_{i_k}|Z_k, U_{k-1}) \tag{38}$$

However, this covariance matrix is not studied in the original paper of Jaffer and Gupta [17]. Although Tugnait [15] claimed that this difference is minor and had not shown any appreciable difference in computer simulations, a similar modification is crucial, as it will be found in the later simulations, in evaluating equations (36) and (37). On the basis of these facts, we sometimes call equations (22)-(37), a modified Jaffer-Gupta algorithm.

The one-step measurement update algorithm requires the storage of s^n estimates (n_1-dimension), s^n covariances ($n_1(n_1+1)/2$-dimension) and s^n a posteriori probabilities (scalar), regardless of the numbers of the storage the system runs. As $n \to k-1$, of course, the method approaches the optimal algorithm. It is remarked that for $k \leq n+1$ the suboptimal method coincides with the optimal one. Figure 1 shows the time evolution of elemental Kalman filters for this method with $s = n = 2$. Notice that if all evolutions in Fig.1 (b) except for two evolutions, $\hat{x}_{11} \to \hat{x}_{111}$ and $\hat{x}_{22} \to \hat{x}_{222}$, are neglected, then one gets the algorithm of Ackerson and Fu [12].

The suboptimal detection algorithm (20) can be rewritten as

$$p(H_{i_k}|Z_k, U_{k-1}) = \sum_{i_{k-n}=1}^{s} \cdots \sum_{i_{k-1}=1}^{s} p(H_{i_{k-n}}, \ldots, H_{i_{k-1}}, H_{i_k}|Z_k, U_{k-1}) \tag{39}$$

or

$$p(H_{i_k}|Z_k, U_{k-1}) = \sum_{i_{k-n+1}=1}^{s} \cdots \sum_{i_{k-1}=1}^{s} p(H_{i_{k-n+1}}, \ldots, H_{i_{k-1}}, H_{i_k}|Z_k, U_{k-1}) \tag{40}$$

3. n-Step Measurement Update Method

As discussed in Section II.C.1, this method starts by reprocessing the past observations $z(k-1), \ldots, z(k-n+1)$ and controls $u(k-2), \ldots, u(k-n)$ with initial conditions at time $k-n$. After that, having available $z(k)$ and $u(k)$ the algorithm

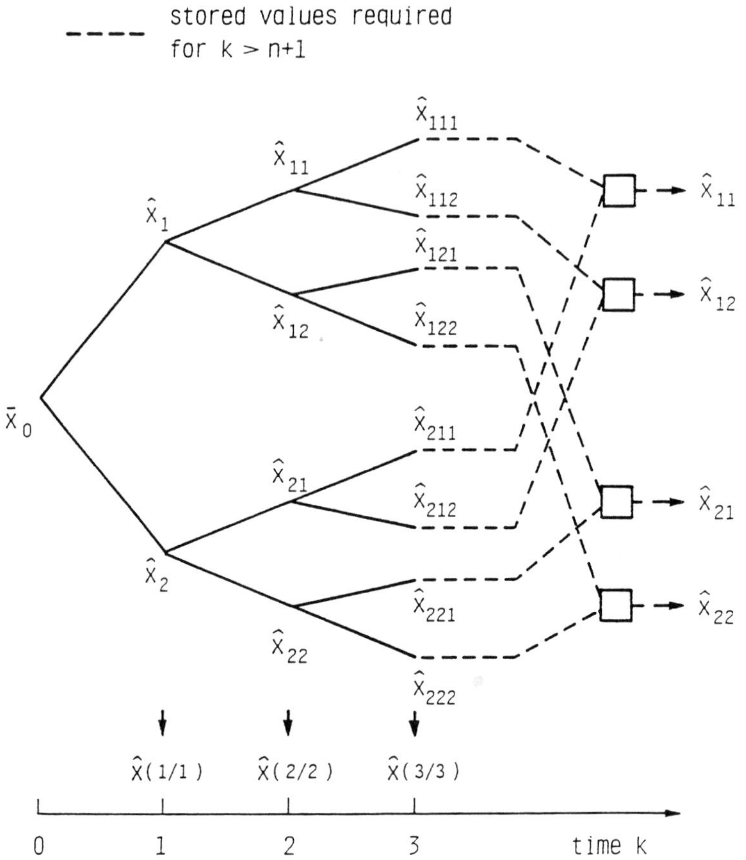

Fig.1(a). Time evolution of elemental Kalman filters in the one-step measurement update method for $k \leq n+1$, where $s = n = 2$.

produces s^{n+1} conditional estimates by applying the most recent estimates (i.e., estimates at time $k-1$), which have already been generated by the reprocessing scheme. The resulting algorithm is summarized as follows.

n-step measurement update algorithm. At time $k-n, k > n+1$, the following quantities are provided:

$$\hat{x}_{i_{k-n}}(k-n|k-n) \triangleq E[x(k-n)|H_{i_{k-n}}, Z_{k-n}, U_{k-n-1}],$$

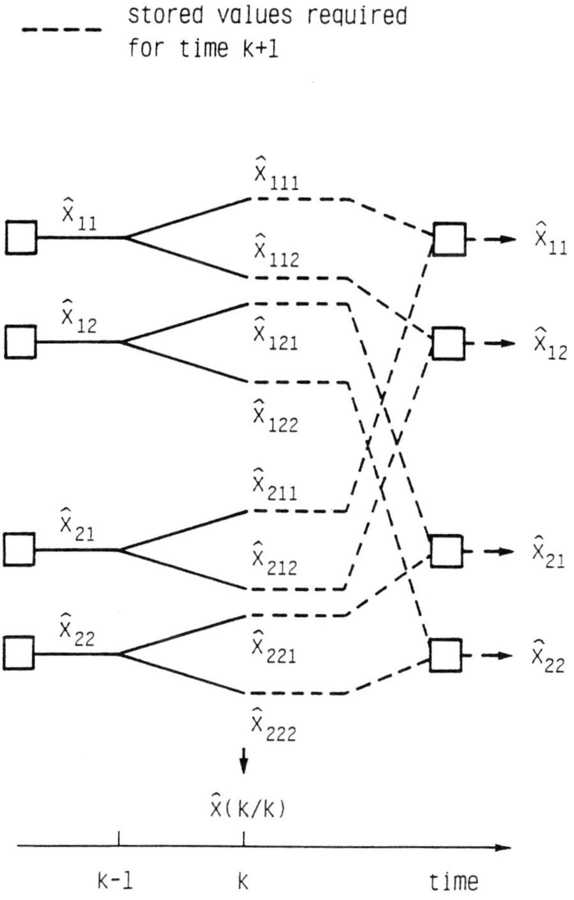

Fig.1(b). Time evolution of elemental Kalman filters in the one-step measurement update method for $k > n+1$, where $s = n = 2$.

$$
\begin{aligned}
P_{i_{k-n}}(k-n|k-n) \\
&\triangleq E\{[x(k-n) - \hat{x}_{i_{k-n}}(k-n|k-n)] \\
&\quad \times [x(k-n) - \hat{x}_{i_{k-n}}(k-n|k-n)]^T | H_{i_{k-n}}, Z_{k-n}, U_{k-n-1}\}
\end{aligned}
$$

and $p(H_{i_{k-n}}|Z_{k-n}, U_{k-n-1})$. Then, given $H_{i_{k-n+1}}$, Z_{k-n+1} and, U_{k-n} at time

$k-n+1$, we compute

$$\hat{x}_{i_{k-n},i_{k-n+1}}(k-n+1|k-n+1) \triangleq E[x(k-n+1)|H_{i_{k-n}}, H_{i_{k-n+1}}, Z_{k-n+1}, U_{k-n}]$$

and

$$P_{i_{k-n},i_{k-n+1}}(k-n+1|k-n+1) \triangleq E\{[x(k-n+1)-\hat{x}_{i_{k-n},i_{k-n+1}}(k-n+1|k-n+1)][x(k-n+1) - \hat{x}_{i_{k-n},i_{k-n+1}}(k-n+1|k-n+1)]^T|H_{i_{k-n}}, H_{i_{k-n+1}}, Z_{k-n+1}, U_{k-n}\}$$

by applying the s^2 parallel Kalman filters.

$$\hat{x}_{i_{k-n},i_{k-n+1}}(k-n+1|k-n) = A(\alpha(k-n+1)=i_{k-n+1})\hat{x}_{i_{k-n}}(k-n|k-n) + G(\alpha(k-n+1)=i_{k-n+1})u(k-n) \tag{41}$$

$$P_{i_{k-n},i_{k-n+1}}(k-n+1|k-n) = A(\alpha(k-n+1)=i_{k-n+1})P_{i_{k-n}}(k-n|k-n) \times A^T(\alpha(k-n+1)=i_{k-n+1}) + B(\alpha(k-n+1)=i_{k-n+1}) \times QB^T(\alpha(k-n+1)=i_{k-n+1}) \tag{42}$$

$$K_{i_{k-n},i_{k-n+1}}(k-n+1) = P_{i_{k-n},i_{k-n+1}}(k-n+1|k-n)C^T(\alpha(k-n+1)=i_{k-n+1}) \times V^{-1}_{i_{k-n},i_{k-n+1}}(k-n+1|k-n) \tag{43}$$

$$\hat{x}_{i_{k-n},i_{k-n+1}}(k-n+1|k-n+1) = \hat{x}_{i_{k-n},i_{k-n+1}}(k-n+1|k-n) + K_{i_{k-n},i_{k-n+1}}(k-n+1)[z(k-n+1) - C(\alpha(k-n+1)=i_{k-n+1}) \times \hat{x}_{i_{k-n},i_{k-n+1}}(k-n+1|k-n)] \tag{44}$$

$$P_{i_{k-n},i_{k-n+1}}(k-n+1|k-n+1) = [I - K_{i_{k-n},i_{k-n+1}}(k-n+1)C(\alpha(k-n+1)=i_{k-n+1})] \times P_{i_{k-n},i_{k-n+1}}(k-n+1|k-n) \tag{45}$$

where $V_{i_{k-n},i_{k-n+1}}(k-n+1|k-n)$ is defined as

$$\begin{aligned}&V_{i_{k-n},i_{k-n+1}}(k-n+1|k-n)\\&\triangleq C(\alpha(k-n+1)=i_{k-n+1})P_{i_{k-n},i_{k-n+1}}(k-n+1|k-n)\\&\quad\times C^T(\alpha(k-n+1)=i_{k-n+1})\\&\quad+D(\alpha(k-n+1)=i_{k-n+1})RD^T(\alpha(k-n+1)=i_{k-n+1})\end{aligned}\tag{46}$$

and the initial conditions are

$$\hat{x}_{i_0}(0|0)=\bar{x}_0,\qquad P_{i_0}(0|0)=P_0\tag{47}$$

for all $i\in S$. In addition, one computes the probability density functions

$$\begin{aligned}&f(z(k-n+1)|H_{i_{k-n}},H_{i_{k-n+1}},Z_{k-n},U_{k-n})\\&\sim N[C(\alpha(k-n+1)=i_{k-n+1})\hat{x}_{i_{k-n},i_{k-n+1}}(k-n+1)|k-n)\\&\quad,V_{i_{k-n},i_{k-n+1}}(k-n+1|k-n)]\end{aligned}\tag{48}$$

Applying the same scheme for the time stages $k-n+2,...,k$, we compute the parallel Kalman filters $s^3,...,s^{n+1}$. Then, equations (18) and (19) can be rewritten as

$$\hat{x}(k|k)=\sum_{i_k=1}^{s}\hat{x}_{i_k}(k|k)p(H_{i_k}|Z_k,U_{k-1})\tag{49}$$

$$\begin{aligned}P(k|k)=&\sum_{i_k=1}^{s}\{P_{i_k}(k|k)+[\hat{x}_{i_k}(k|k)-\hat{x}(k|k)][\hat{x}_{i_k}(k|k)-\hat{x}(k|k)]^T\}\\&\times p(H_{i_k}|Z_k,U_{k-1})\end{aligned}\tag{50}$$

where $\hat{x}_{i_k}(k|k)=E[x(k)|H_{i_k},Z_k,U_{k-1}]$, $P_{i_k}(k|k)=E\{[x(k)-\hat{x}_{i_k}(k|k)][x(k)-\hat{x}_{i_k}(k|k)]^T|H_{i_k},Z_k,U_{k-1}\}$, and $p(H_{i_k}|Z_k,U_{k-1})$ are given by:

$$\hat{x}_{i_k}(k|k)=\sum_{i_{k-n}=1}^{s}\cdots\sum_{i_{k-1}=1}^{s}\hat{x}_{i_{k-n},...,i_k}(k|k)p(H_{i_{k-n}},...,H_{i_{k-1}}|H_{i_k},Z_k,U_{k-1})\tag{51}$$

$$\begin{aligned}&P_{i_k}(k|k)\\&=\sum_{i_{k-n}=1}^{s}\cdots\sum_{i_{k-1}=1}^{s}\{P_{i_{k-n},...,i_k}(k|k)+[\hat{x}_{i_{k-n},...,i_k}(k|k)-\hat{x}_{i_k}(k|k)]\\&\times[\hat{x}_{i_{k-n},...,i_k}(k|k)-\hat{x}_{i_k}(k|k)]^T\}p(H_{i_{k-n}},...,H_{i_{k-1}}|H_{i_k},Z_k,U_{k-1})\end{aligned}\tag{52}$$

$$p(H_{i_k}|Z_k, U_{k-1}) = \frac{\sum_{i_{k-n}=1}^{s} \cdots \sum_{i_{k-1}=1}^{s} [\prod_{j=0}^{n-1} F(i,j,k,n)] p(H_{i_{k-n}}|Z_{k-n}, U_{k-n-1})}{\sum_{i_{k-n}=1}^{s} \cdots \sum_{i_{k-1}=1}^{s} \sum_{i_k=1}^{s} [\prod_{j=0}^{n-1} F(i,j,k,n)] p(H_{i_{k-n}}|Z_{k-n}, U_{k-n-1})} \quad (53)$$

with

$$F(i,j,k,n) \triangleq f(z(k-j)|H_{i_{k-n}}, ..., H_{i_{k-j}}, Z_{k-1-j}, U_{k-1-j}) p(H_{i_{k-j}}|H_{i_{k-1-j}}) \quad (54)$$

The probability $p(H_{i_{k-n}}, ..., H_{i_{k-1}}|H_{i_k}, Z_k, U_{k-1})$ is given by

$$p(H_{i_{k-n}}, ..., H_{i_{k-1}}|H_{i_k}, Z_k, U_{k-1}) = \frac{[\prod_{j=0}^{n-1} F(i,j,k,n)] p(H_{i_{k-n}}|Z_{k-n}, U_{k-n-1})}{\sum_{i_{k-n}=1}^{s} \cdots \sum_{i_{k-1}=1}^{s} [\prod_{j=0}^{n-1} F(i,j,k,n)] p(H_{i_{k-n}}|Z_{k-n}, U_{k-n-1})} \quad (55)$$

Note that $\hat{x}_{i_k}(k|k), p_{i_k}(k|k)$, $p(H_{i_k}|Z_k, U_{k-1})$ and $u(k)$ must be stored as the starting conditions for the $k+n$ stage.

For the structure detection, equation (20) can now be directly combined with (53). Therefore, no additional computations are needed for choosing the normal or fault state corresponding to the largest value of (20).

Figure 2 shows the time evolution of elemental Kalman filters for the proposed method with $s = n = 2$. Note that $\hat{x}_{i_{k-1}}(k-1|k-1), P_{i_{k-1}}(k-1|k-1)$ and $p(H_{i_k}|Z_{k-1}, U_{k-2})$ have already been stored as the starting conditions for the $k+1$ stage. Thus, the present algorithm requires the storage of $s \times n$ estimates (n_1-dimension), $s \times n$ covariances ($n_1(n_1+1)/2$-dimension), $s \times n$ a posteriori probabilities (scalar) and $n-1$ observations (m_1-dimension), if $u(\cdot) \equiv 0$. It is now interesting to note that the present method requires algebraically growing storage with s and n, whereas the one-step measurement update method, described in Section II.C.2, requires exponentially growing storage. Table 1 gives their storage requirements;

$$T_1 = s^n[n_1 + n_1(n_1+1)/2 + 1] \quad (56)$$

for the one-step measurement update method, and

$$T_n = sn[n_1 + n_1(n_1+1)/2 + 1] + (n-1)m_1 \quad (57)$$

for the n-step measurement update method, where $n_1 = 10$ and $m_1 = 5$. One can see from Table I that a remarkable difference between the storage requirements of

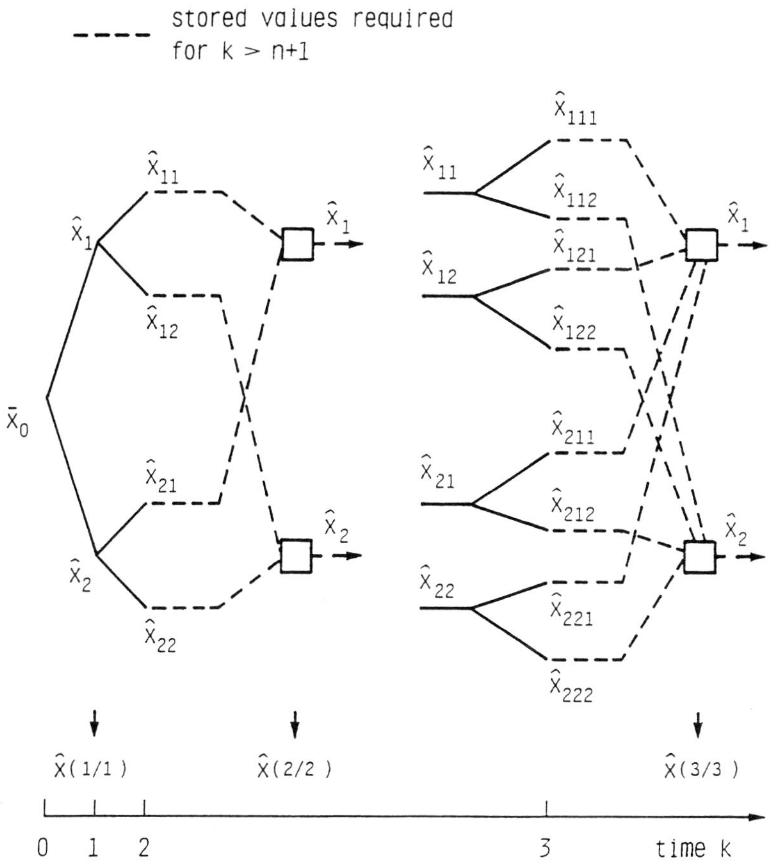

Fig.2(a). Time evolution of elemental Kalman filters in the n-step measurement update method for $k \le n+1$, where $s = n = 2$.

the two methods appears as s and n become large. Note however that the computational speed of the present algorithm is slightly slower than that of the one-step measurement update algorithm. This is, for example, confirmed by comparing the numbers of elemental Kalman filters required for one-step and n-step measurement

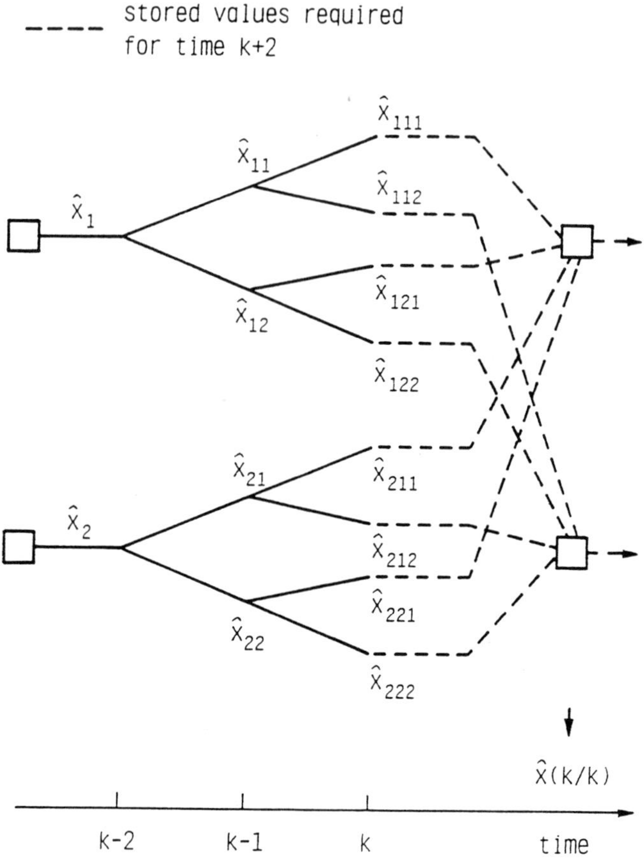

Fig.2(b). Time evolution of elemental Kalman filters in the n-step measurement update method for $k > n+1$, where $s = n = 2$.

updates, i.e.,

$$E_1 = s^{n+1} \quad \text{and } E_n = \sum_{i=1}^{n} s^{i+1} \tag{58}$$

It can be seen from Table II that as s becomes large the difference between the two methods disappears for any n. From these facts, the n-step measurement update method seems to be effective for reducing the storage requirements of

Table I. Storage requirements for GPBAs, where the upper and lower figures refer to the one- and n-step measurement update methods, respectively.

	$s=2$	$s=3$	$s=4$	$s=5$
$n=1$	132	198	264	330
	132	198	264	330
$n=2$	264	594	1056	1650
	269	401	533	665
$n=3$	528	1782	4224	8250
	406	604	802	1000
$n=4$	1056	5346	16896	41250
	543	807	1071	1335
$n=5$	2112	16038	67584	206250
	680	1010	1340	1670

Table II. Numbers of elemental Kalman filters for GPBAs, where the upper and lower figures refer to the one- and n-step measurement update methods, respectively.

	$s=2$	$s=3$	$s=4$	$s=5$
$n=1$	4	9	16	25
	4	9	16	25
$n=2$	8	27	64	125
	12	36	80	150
$n=3$	16	81	256	625
	28	117	336	775
$n=4$	32	243	1024	3125
	60	360	1360	3900
$n=5$	64	729	4096	15625
	124	1089	5456	19525

systems with a Markov chain state of relatively large size.

Note that for $n = 1$ the two methods are completely identical in their implementation form; this is in fact the algorithm of Chang and Athans [18].

D. Examples

Here, two simulation examples are presented for comparing the estimation and detection performances of four suboptimal algorithms:

(a) Ackerson and Fu (A-F) method [12],
(b) Jaffer and Gupta (J-G) method [17],
(c) modified J-G method using (22)-(37),
(d) n-step measurement update method given by (41)-(55).

Example 1

Consider a scalar dynamical system described by the following equations:

$$x(k+1) = 1.04x(k) + w(k)$$

$$z(k) = x(k) + D(\alpha(k))v(k), \quad k = 1, 2, ...$$

$$s = 2, \text{ i.e., } \alpha(k) \in \{1, 2\}$$

The initial conditions are assumed to be $x(0) \sim N(30, 400)$. The sequences $\{w(k)\}$ and $\{v(k)\}$ are mutually independent zero-mean Gaussian white noises with covariances $Q = 0.1$ and $R = 1.0$, respectively. The process $\alpha(k)$ is modeled by a Markov chain with transition probability matrix:

$$\pi = \begin{bmatrix} 0.5 & 0.5 \\ 0.5 & 0.5 \end{bmatrix}$$

We take $p_1 = p_2 = 0.5$. Finally, we have $D(1) = 10, D(2) = 1$. This system models a failure mode due to the abrupt change of measurement noise.

The four suboptimal algorithms were simulated and their state estimation performance is compared in Fig.3 (rms errors in state estimation are compared for various n). The performance was evaluated by averaging over 50 Monte Carlo runs. A lower bound is obtained by running a Kalman filter which knows the true values of the switching parameters. Note also that the modified J-G method with $n = 1$ and the n-step measurement update method with $n = 1$ are the same as the algorithm of Chang and Athans [18]. The information in Fig.3 is also summarized in Table 3 after averaging over 30 time stages, together with the average probability of error in structure detection.

From Fig.3 and Table III, one observes that the performances of the modified J-G and the n-step measurement update methods are better than those of the other methods in both the estimation and detection cases; in particular the n-step measurement update method has the best estimation and detection performances. Note also that an increase in n leads to improvement in estimation performance, but does not necessary lead to improvement in detection performance.

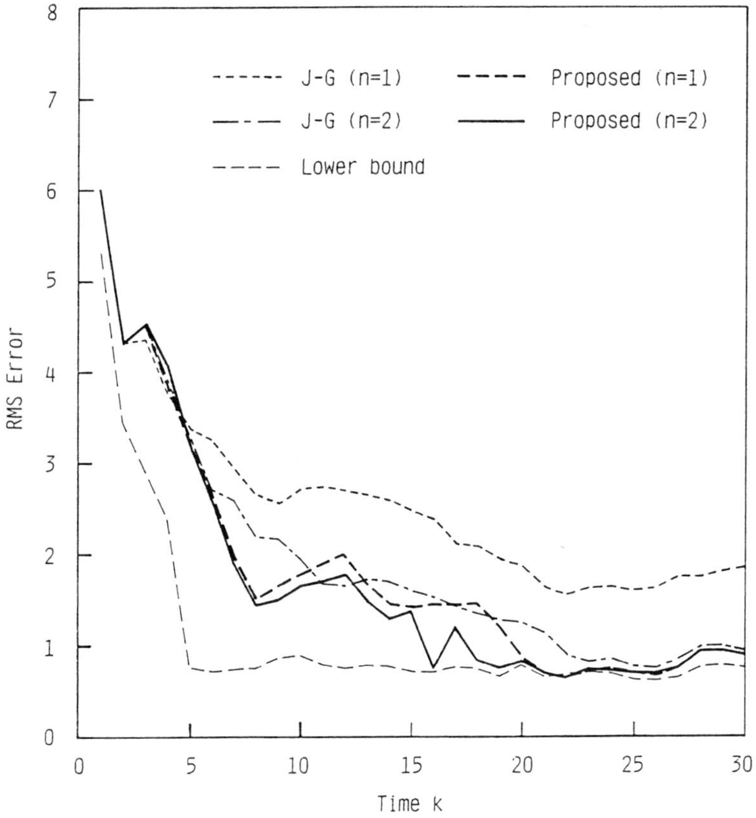

Fig.3(a). Comparison of rms state estimation errors due to suboptimal algorithms, J-G and n-step measurement update methods, for Example 1.

Example 2

Consider a relatively high-dimensional Markov chain, i.e. the system of Example 1, except that $p_i = 0.25$ for $i = 1, ..., 4$,

$$D(1) = 10, \ D(2) = 7, \ D(3) = 4, \ D(4) = 1$$

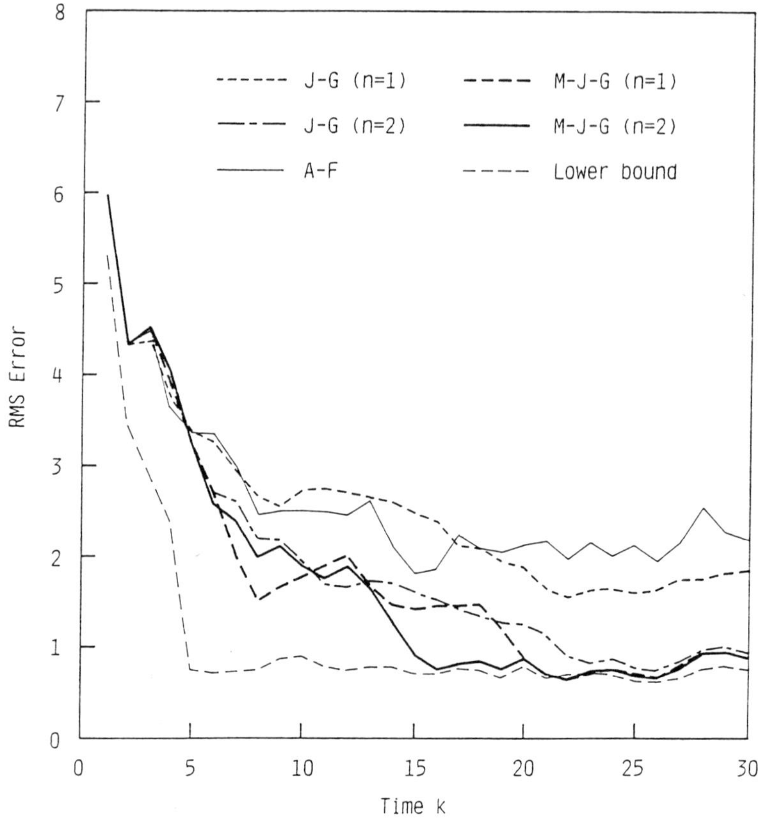

Fig.3(b). Comparison of rms state estimation errors due to suboptimal algorithms, A-F, J-G and modified J-G methods, for Example 1.

and

$$\pi = \begin{bmatrix} 0.15 & 0.15 & 0.2 & 0.5 \\ 0.2 & 0.2 & 0.3 & 0.3 \\ 0.3 & 0.2 & 0.1 & 0.4 \\ 0.5 & 0.15 & 0.2 & 0.15 \end{bmatrix}$$

The results of averaging over 50 Monte Carlo runs are shown in Fig.4 and Table IV. Observe that, in this case, the n-step measurement update method has comparable estimation and detection performances with the modified J-G method. However, from (56) and (57) with $n_1 = m_1 = 1$ and $s = 4$, it follows that the present method

Table III. RMS state estimation error and probability of error in structure detection for Example 1.

Algorithm	Average RMS Error in State Estimation	Average Probability of Error in Structure Detection
A-F method	2.629	0.3133
J-G method with		
n = 1	2.548	0.2133
n = 2	1.926	0.1973
n = 3	1.715	0.1900
Modified J-G method with		
n = 1	1.803	0.1887
n = 2	1.736	0.1867
n = 3	1.712	0.1893
n-step measurement update method with		
n = 1	1.803	0.1887
n = 2	1.694	0.1793
n = 3	1.672	0.1820
Lower bound	1.102	*

requires about 50% less storage than the existing GPBA methods for the $n = 2$ case. It is also found from Table IV that the actual ratio of computational speed between the J-G (or modified J-G) method and the n-step measurement update for the $n = 2$ case is approximately close to the ideal ratio (i.e., 1.25) presented in Table II.

III. CONTROL FOR SYSTEMS WITH JUMP PARAMETERS

A. Problem Statement

Reconsider the system described by (1)-(4), in which the input $u(k)$ is explicitly regarded as control input.

The objective is to find control sequences $\{u(k), 0 \leq k \leq N-1\}$, which

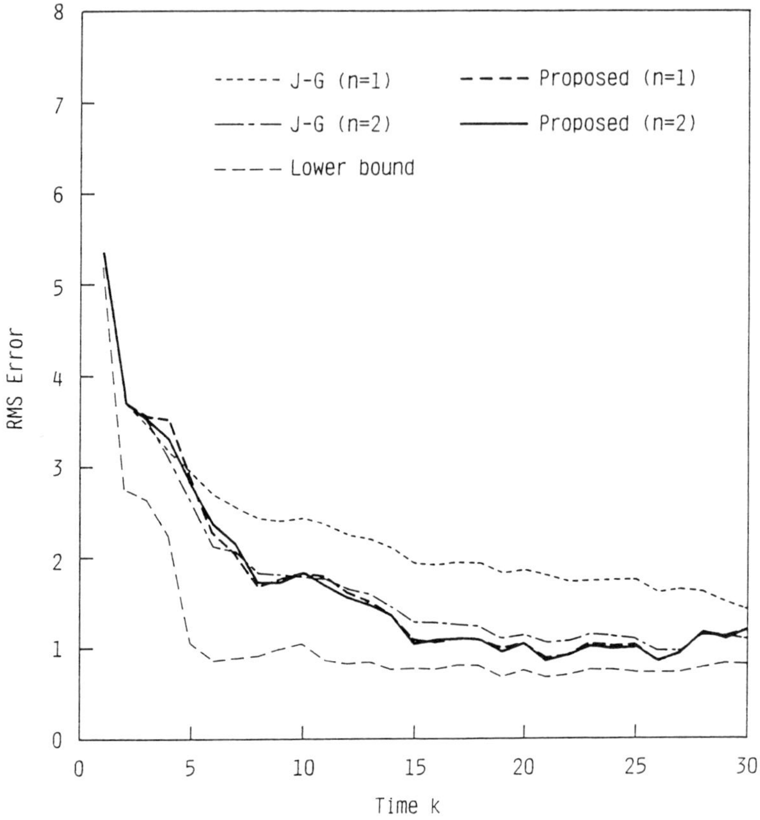

Fig.4(a). Comparison of rms state estimation errors due to suboptimal algorithms, J-G and n-step measurement update methods, for Example 2.

approximately minimize the following quadratic cost functional

$$J = E\left\{ \|x(N)\|^2_{Q_c(N)} + \sum_{k=0}^{N-1} \|x(k)\|^2_{Q_c} + \|u(k)\|^2_{R_c} \right\} \tag{59}$$

where the matrices $Q_c(N)$ and Q_c are positive semidefinite, and R_c is positive definite. We expect that the controller is of the form

$$u(k) = \phi\{k, Z_k, U_{k-1}\} \tag{60}$$

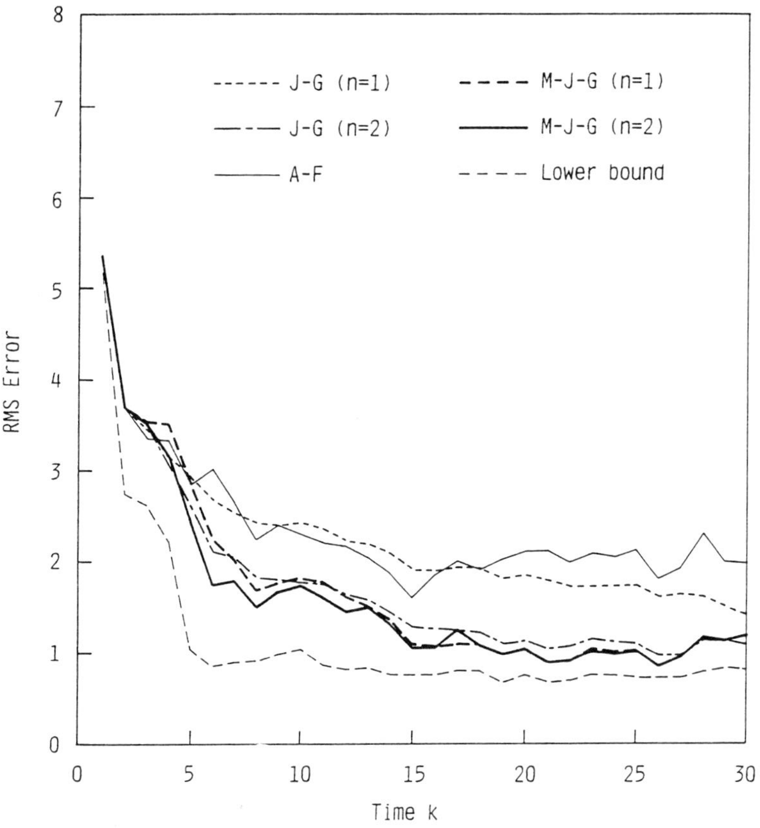

Fig.4(b). Comparison of rms state estimation errors due to suboptimal algorithms, A-F, J-G and modified J-G methods, for Example 2.

where ϕ denotes a measurable function of the observation data Z_k and all past inputs U_{k-1}.

B. Multiple Model Adaptive Control

Extending the MMAC algorithm derived in the jumpless case (Deshpande, Upadhyay and Lainiotis [35], Lainiotis [36] to the above control problem, the

Table IV. RMS state estimation error and probability of error in structure detection for Example 2.

Algorithm	Average RMS Error in State Estimation	Average Probability of Error in Structure Detection	Average Computation Time(sec)
A-F method	2.379	0.5733	1.664
J-G method with			
n = 1	2.258	0.5147	3.322
n = 2	1.739	0.4973	10.442
n = 3	1.727	0.4867	39.480
Modified J-G method with			
n = 1	1.704	0.4860	3.650
n = 2	1.633	0.4840	11.722
n = 3	1.676	0.4827	44.422
n-step measurement update method with			
n = 1	1.704	0.4860	3.770
n = 2	1.693	0.4853	12.886
n = 3	1.688	0.4853	49.100
Lower bound	1.127	*	0.646

control law is defined as

$$u(k) = \sum_{j=1}^{s^N} u_j(k) p(I_j(N)|Z_k, U_{k-1}), \qquad 0 \le k \le N-1 \tag{61}$$

where $p(I_j(N)|Z_k, U_{k-1})$ denotes the *a posteriori* probability of $I_j(N)$ given Z_k and U_{k-1}. The "elemental" control law $u_j(k)$ is of the form

$$u_j(k) = F_j(k)\hat{x}_j(k|k) \tag{62}$$

where $\hat{x}_j(k|k)$ is given by (7) and $F_j(k)$ denotes the control gain matrix for the j-th elemental controller obtained by

$$F_j(k) = -[R_c + G_j^T S_j(k+1) G_j]^{-1} G_j^T S_j(k+1) A_j \tag{63}$$

in which the matrices $S_j(k)$ are given by the Riccati difference equation

$$S_j(k) = A_j^T[S_j(k+1) - S_j(k+1)G_j\{R_c + G_j^T S_j(k+1)G_j\}^{-1} \times G_j^T S_j(k+1)]A_j + Q_c, \qquad S_j(N) = Q_c(N), \quad j = 1, ..., s^N \tag{64}$$

Here, A_j and G_j stand for $A(\alpha(k+1))$ and $G(\alpha(k+1))$ under a specific sequence $I_j(N)$ from all possible sequences $I(N)$.

It should be noted that, unlike the MMAC problem for a system with only Markovian faulty sensors(Tugnait [32], Watanabe and Tzafestas [33] or [37]), we need in general s^N sequences of control gains. For large final time N, this is not practical.

C. Suboptimal Control

Here, some suboptimal control laws will be presented by applying the aforementioned GPBAs for state estimation.

Algorithm C.1

Given the observation data Z_k and past control U_{k-1}, a suboptimal MMAC algorithm for system (1)-(4) is given by

$$u(k) = \sum_{i_k=1}^{s} F_{i_k}(k)\hat{x}_{i_k}(k|k)p(H_{i_k}|Z_k, U_{k-1}) \tag{65}$$

where $\hat{x}_{i_k}(k|k)$ and $p(H_{i_k}|Z_k, U_{k-1})$ are provided by (51) and (52), respectively, and the elemental control gains $F_{i_k}(k)$ are given by

$$F_{i_k}(k) = -N^{-1}(k) \sum_{i_{k+1}=1}^{s} p_{i_k,i_{k+1}} G^T(\alpha(k+1) = i_{k+1}) \times S_{i_{k+1}}(k+1)A(\alpha(k+1) = i_{k+1}) \tag{66}$$

$$N(k) = \left[R_c + \sum_{i_k=1}^{s} \sum_{i_{k+1}=1}^{s} p_{i_k,i_{k+1}} G^T(\alpha(k+1) = i_{k+1}) \times S_{i_{k+1}}(k+1)G(\alpha(k+1) = i_{k+1})p(H_{i_k}|Z_k, U_{k-1})\right] \tag{67}$$

The matrices $S_{i_{k+1}}(k+1)$ are determined by a set of s coupled Riccati-like equations, namely:

$$S_{i_k}(k) = Q_c + \sum_{i_{k+1}=1}^{s} p_{i_k,i_{k+1}} A^T(\alpha(k+1) = i_{k+1}) \times S_{i_{k+1}}(k+1) A(\alpha(k+1) = i_{k+1}) - F_{i_k}^T(k) N(k) F_{i_k}(k), \qquad S_{i_N}(N) = Q_c(N) \tag{68}$$

Proof

The derivation is performed by using the Dynamic Programming technique. The details are given in Watanabe and Tzafestas [39].

Note that the matrices $S_{i_k}(k)$ and $F_{i_k}(k)$ described above cannot be computed off-line, because $N(k)$ contains the *a posteriori* probability $p(H_{i_k}|Z_k, U_{k-1})$. In order to compute the Riccati-like equation (68) off-line, we construct an approximation to Algorithm C.1. This is summarized below without proof.

Algorithm C.2

The elemental control gains $F_{i_k}(k)$ can be determined on-line by (66) and (67), but with the following Riccati-like equations:

$$\begin{aligned} S_{i_k}(k) = Q_c &+ \sum_{i_{k+1}=1}^{s} p_{i_k,i_{k+1}} A^T(\alpha(k+1) = i_{k+1}) S_{i_{k+1}}(k+1) A(\alpha(k+1) = i_{k+1}) \\ &- \left[\sum_{i_{k+1}=1}^{s} p_{i_k,i_{k+1}} A^T(\alpha(k+1) = i_{k+1}) S_{i_{k+1}}(k+1) G(\alpha(k+1) = i_{k+1})\right] \\ &\times \left[R_c + \sum_{i_{k+1}=1}^{s} p_{i_k,i_{k+1}} G^T(\alpha(k+1) = i_{k+1}) S_{i_{k+1}}(k+1) G(\alpha(k+1) = i_{k+1})\right]^{-1} \\ &\times \left[\sum_{i_{k+1}=1}^{s} p_{i_k,i_{k+1}} G^T(\alpha(k+1) = i_{k+1}) S_{i_{k+1}}(k+1) A(\alpha(k+1) = i_{k+1})\right], \end{aligned}$$

$$S_{i_N}(N) = Q_c(N) \tag{69}$$

If the same approximation is applied to (63), then one obtains the practical control gains $F_{i_k}(k)$ as in the following algorithm.

Algorithm C.3

The elemental control gains $F_{i_k}(k)$ can be modified as

$$F_{i_k}(k) = -N_{i_k}^{-1}(k) \sum_{i_{k+1}=1}^{s} p_{i_k,i_{k+1}} G^T(\alpha(k+1) = i_{k+1}) \times S_{i_{k+1}}(k+1)A(\alpha(k+1) = i_{k+1}) \tag{70}$$

where

$$N_{i_k}(k) = \left[R_c + \sum_{i_{k+1}=1}^{s} p_{i_k,i_{k+1}} G^T(\alpha(k+1) = i_{k+1}) \times S_{i_{k+1}}(k+1)G(\alpha(k+1) = i_{k+1}) \right] \tag{71}$$

and $S_{i_{k+1}}(k+1)$ are given by (69).

Note that this gain calculation is completely computed off-line, and $F_{i_k}(k)$, $i_k = 1, ..., s, 0 \leq k \leq N-1$ can be stored. In the case of no system and measurement noises, the MMAC controller (65) can be simply expressed as

$$u(k) = F_{i_k}(k)x(k) \tag{72}$$

where $F_{i_k}(k)$ are given by Algorithm C.2. This is a generalization of the result of Birdwell, Castanon and Athans [40] to problems including dynamic jump parameters.

In the jumpless case ($p_{i_k,i_{k+1}} = \delta_{i_k,i_{k+1}}$), Algorithm C.1 reduces to the result of Lee and Sims [41]. It is easily seen that in this case Algorithms C.1 and C.2 coincide with the results of Upadhyay and Lainiotis [42] and Deshpande, Upadhyay and Lainiotis [35], respectively. When the dynamic system (1) does not contain the jump parameter $\alpha(k)$, (i.e., $A(\alpha(k)) = A, G(\alpha(k)) = G$ and $B(\alpha(k)) = B$), one obtains an MMAC algorithm for a system with Markovian faulty sensors as studied in [33] or [37].

D. Examples

To illustrate the application of the above result, we choose a simple scalar system of the form:

$$x(k+1) = 1.04x(k) + G(\alpha(k+1))u(k) + w(k)$$

Table V. Performance measures for Example 3.

Algorithm	Average Cost	RMS Error
A-F method	2977	0.6099
J-G method		
n=1	2973	0.5506
n=2	2979	0.6300
n=3	2994	0.7602
Present method		
n=1	2973	0.5494
n=2	2973	0.5492
n=3	2973	0.5492
Lower bound	2972	0.5279

$$z(k) = x(k) + D(\alpha(k))v(k)$$

with cost function

$$J = E\left\{x^2(30) + \sum_{k=0}^{29}[x^2(k) + u^2(k)]\right\}$$

where $w(k) \sim N(0, 0.1), v(k) \sim N(0, 1)$. The initial conditions are $x(0) \sim N(30, 400), \alpha(k) \in \{1, 2\}$, and $p_i = 0.5$ for $i = 1, 2$. The Monte Carlo runs were obtained using three MMACs with the following suboptimal filtering techniques:

(a) Ackerson and Fu(A-F) method [12],

(b) Jaffer and Gupta(J-G)method [17],

(c) n-step measurement update method given by (41)-(55)

Note that the form of elemental control gains given by Algorithm C.3 was used for all MMACs. The cost function was averaged over 50 Monte Carlo runs.

Example 3

It is assumed that $G(1) = 1, D(1) = 1, G(2) = 1.5, D(2) = 1$, and

$$\pi = \begin{bmatrix} 0.5 & 0.5 \\ 0.5 & 0.5 \end{bmatrix}$$

The second structure represents an actuator hard-over failure.

In Table V, the average cost function values for various suboptimal MMACs are tabulated together with the rms estimation error averaged over 30 time stages

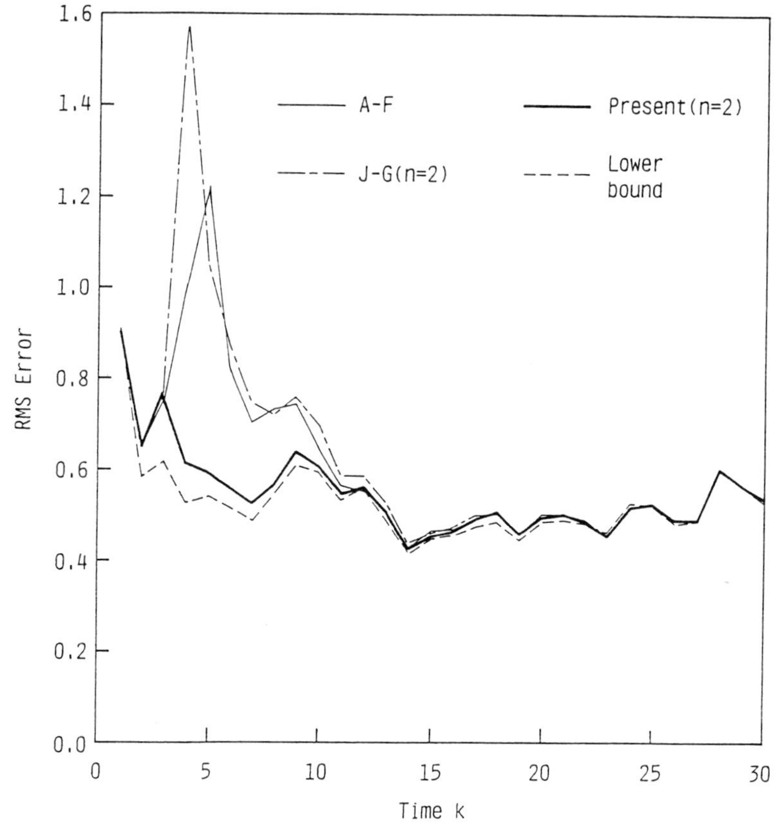

Fig.5. RMS state estimation error for Example 3.

and 50 Monte Carlo runs. A lower bound is provided by realizing an MMAC which uses the Kalman filter with known jump parameters.

It is seen from Table V and Fig.5 that the control performance of all MMACs is almost the same, but the estimation performance of the present approach is better than that of the other methods.

Example 4

Consider the system of Example 3 except that $D(2) = 10$. In this case, the second structure represents a simultaneous actuator and sensor failure. Table VI and Fig.6 summarize the corresponding control and estimation performances.

Table VI. Performance measures for Example 4.

Algorithm	Average Cost	RMS Error
A-F method	3422	3.535
J-G method		
n=1	3087	1.420
n=2	3206	2.405
n=3	3266	2.659
Present method		
n=1	3084	1.385
n=2	3087	1.401
n=3	3087	1.403
Lower bound	3049	0.9396

Observe that the present method outperforms the other methods in both the estimation and control cases.

IV. CONCLUSIONS

In this chapter, the adaptive state estimation and control problems have been considered for linear discrete-time stochastic systems with abruptly changing parameters. The changes were modeled by a Markov chain with known transition statistics. Since the optimal solution of the tate estimation problem is impractical, a suboptimal approach called the "generalized pseudo-Bayes algorithm" was presented. In particular, one-step and n-step measurement update methods were discussed in detail, and their differences were demonstrated through several simulation studies.

In the multiple model adaptive control case, some suboptimal control algorithms were introduced for systems with unknown Markov jump parameters in both the state and observation equations, because the optimal solution is also impractical. The elemental control gains are given by a set of coupled Riccati-like difference equations as proposed under perfect observations by Birdwell, Castanon and Athans [40], while some generalized pseudo-Bayes algorithms were used for the elemental filter mechanisms. The features of the present approaches were illustrated by simulation examples.

Some issues which were not discussed here are:

(i) the feasibility of knowing the transition probabilities for practical prob-

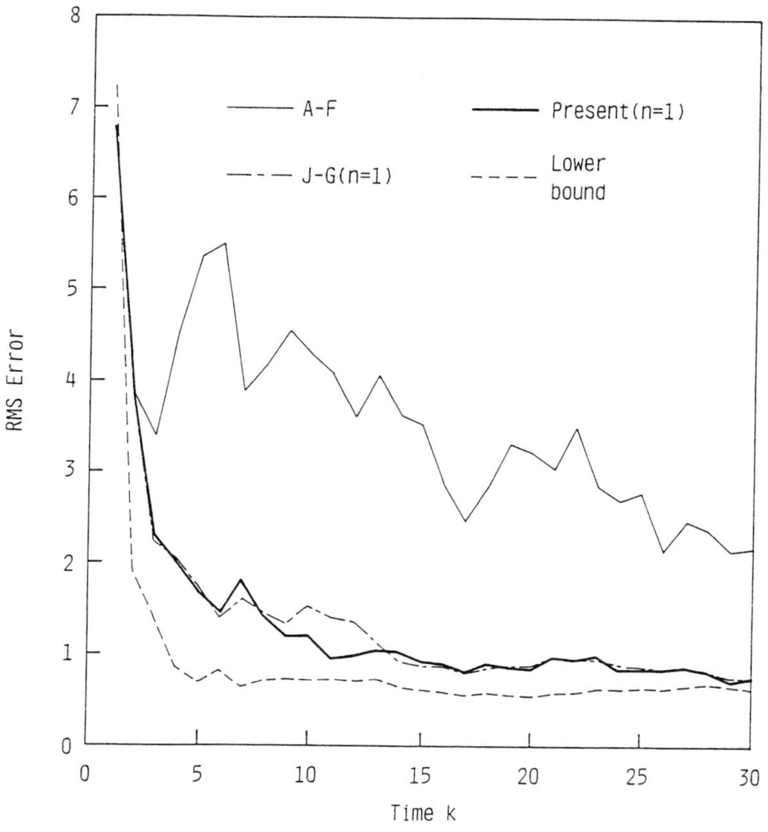

Fig.6. RMS state estimation error for Example 4.

lems,

(ii) the robustness of the Markov approach to errors in the transition probability assignments, and

(iii) the feasibility of estimating the transition probabilities.

We are planning to examine these issues in our future work. For some discussion of item (iii), the reader is referred to Sawaragi, Katayama and Fujishige [47]; Tugnait and Haddad [45,46], and Tugnait [43,44].

The IMMA technique extended to the distributed-sensor networks is reported in Chang and Bar-Shalom [48], and the algorithm is applied for tracking a maneuvering target in a cluttered and low detection environment. A similar hierarchical multiple model adaptive filtering and control problem is studied in Watanabe [49], and Watanabe and Tzafestas [50]. Some other topics concerning the JLQ or the JLQG control problem are addressed in Mariton [51]. General aspects regarding the multiple model adaptive (or partitioning) techniques of estimation and control are studied in detail in Watanabe [52].

V. REFERENCES

1. D.G. Lainiotis, " Optimal Adaptive Estimation: Structure and Parameter Adaptation", *IEEE Trans. Auto. Control* **AC-16**, pp.160-170 (1971).
2. D.G. Lainiotis, "Partitioning: A unifying framework for adaptive systems, I: Estimation", *Proc. IEEE* **64**, pp.1126-1142 (1976).
3. D.E. Gustafson, A.S. Willsky, J.Y. Wang, M.C. Lancaster, and J.H. Triebwasser, "ECG/VCG Rhythm Diagnosis using Statistical Signal Analysis, I. Identification of Persistent Rhythms", *IEEE Trans. Biomedical Engng.* **BME-25**(4), pp.344-353 (1978).
4. A.S. Willsky, "Failure Detection in Dynamic Systems", *AGARD* **109** (1980).
5. K. Watanabe, "A Multiple Model Adaptive Filtering Approach to Fault Diagnosis in Stochastic Systems", in R.J. Patton, P.M. Frank and R.N. Clark (Eds), *Fault Diagnosis in Dynamic Systems*, Prentice Hall, London, U.K., pp.411-438 (1989).
6. K. Watanabe, "A Decentralized Multiple Model Adaptive Filtering for Discrete-Time Stochastic Systems", *ASME Trans. J. Dynamic Syst., Measure., and Control* **111**, pp.371-377 (1989).
7. A.S. Willsky, "A Survey of Design Methods for Failure Detection in Dynamic Systems", *Automatica* **12**, pp.601-611 (1976).
8. G.G. Ricker and J.R. Williams, "Adaptive Tracking Filter for Maneuvering Targets", *IEEE Trans. Aero. Electron. Syst.* **AES-14**(1), pp.185-193 (1978).
9. R.L. Moose, H.F. VanLandingham, and D.H. McCabe, "Modeling and Estimation for Tracking Maneuvering Targets", *IEEE Trans. Aero. Electron. Syst.* **AES-15**(3), pp.448-456 (1979).
10. C.B. Chang and J.A. Tabaczynski,"Application of State Estimation to Target Tracking", *IEEE Trans. Auto. Control* **AC-29**(2), pp.98-109 (1984).

11. R.L. Moose, H.F. VanLandingham, and P.E. Zwicke, "Digital Set Point Control of Nonlinear Stochastic Systems", *IEEE Trans. Industrial Electron. and Contr. Instrumentation* **IECI-25**(1), pp.39-45 (1978).
12. G.A. Ackerson and K.S. Fu, "On the State Estimation in Switching Environments", *IEEE Trans. Auto. Control* **AC-15**(1), pp.10-17 (1970).
13. H. Akashi and H. Kumamoto, "Random Sampling Approach to State Estimation in Switching Environments", *Automatica* **13**, pp.429-434 (1977).
14. J.K. Tugnait and A.H. Haddad, " A Detection-Estimation Scheme for State Estimation in Switching Environments", *Automatica* **15**, pp.477-481 (1979).
15. J.K. Tugnait, "Detection and Estimation for Abruptly Changing Systems" , *Automatica* **18**(5), pp.607-615 (1982).
16. V.J. Mathews and J.K. Tugnait, "Detection and Estimation with Fixed Lag for Abruptly Changing Systems", *IEEE Trans. Aero. Electron. Syst.* **AES-19**(5), pp.730-739 (1988).
17. A.G. Jaffer and S.C. Gupta, "On Estimation of Discrete Processes Under Multiplicative and Additive Noise Conditions", *Inform. Sciences* **3**, pp.267-276 (1971).
18. C.B. Chang and M. Athans, "State Estimation for Discrete Systems with Switching Parameters", *IEEE Trans. Aero. Electron. Syst.* **AES-14**(3), pp.418-425 (1978).
19. S. Sugimoto and I. Ishizuka, "Identification and Estimation Algorithms for a Markov Chain plus AR Process", *Proc. IEEE Intl. Conf. on Acoustics, Speech and Signal Processing, ICASSP 83*, pp.247-250 (1983).
20. H.A.P. Blom, "An Efficient Filter for Abruptly Changing Systems", *Proc. 23rd IEEE Conf. on Decision and Control*, pp.656-658 (1984).
21. H.A.P. Blom, "An Efficient Decision-Making-Free Filter for Processes with Abrupt Changes", *Proc. IFAC Identification and System Parameter Estimation 1985*, pp.631-636 (1985).
22. H.A.P. Blom and Y. Bar-Shalom, "The Interacting Multiple Model Algorithm for Systems with Markovian Switching Coefficients", *IEEE Trans. Auto. Control* **AC-33**(8), pp.780-783 (1988).
23. B.E. Griffiths and K.A. Loparo, "Optimal control of jump-linear gaussian systems", *Int. J. Control* **42**(4), pp.791-819 (1985).
24. M. Mariton, "Joint Estimation and Control of Jump linear Systems With

Multiplicative Noises" , *ASME J. Dynamic Systems, Measure. and Control* **109**(1), pp.24-28 (1987).

25. D.D. Sworder, "Feedback Control of a Class of Linear Systems with Jump Parameters", *IEEE Trans. Auto. Control* **AC-14**(1), pp.9-14 (1969).
26. W.M. Wonham, "Random differential equations in control theory", in A.T. Bharucha-Reid (Ed), *Probabilistic Methods in Applied Mathematics*, **II**, Academic Press, New York, N.Y. (1970).
27. W.P. Blair and D.D. Sworder, "Feedback control of a class of linear discrete systems with jump parameters and quadratic cost criteria", *Int. J. Control* **21**(5), pp.833-841 (1975).
28. H.J. Chizeck, A.S. Willsky, and D. Castanon, "Discrete-time Markovian-jump linear quadratic optimal control", *Int. J. Control* **43**(1), pp.213-231 (1986).
29. Y. Bar-Shalom, "Stochastic Dynamic Programming: Caution and Probing, *IEEE Trans. Auto. Control* **AC-26**(5), pp.1184-1195 (1981).
30. S. Fujita and T. Fukao, "Optimal Stochastic Control for Discrete-time Linear System with Interrupted Observations", *Automatica* **8**, pp.425-432 (1972).
31. H. Akashi, H. Kumamoto and K. Nose, "Application of Monte Carlo method to optimal control for linear systems under measurement noise with Markov dependent statistical property", *Int. J. Control* **22**, pp.821-836 (1975).
32. J.K. Tugnait, "Control of Stochastic Systems with Markov Interrupted Observations", *IEEE Trans. Aero. and Electron. Systems* **AES-19**, (2), pp.232-239 (1983).
33. S.G. Tzafestas and K. Watanabe, "Stochastic Control Algorithms for Systems with Markovian Faulty Sensors", *Proceedings of 12th IMACS World Congress On Scientific Computation*, Paris, **2**, pp.681-684 (1988).
34. H.F. VanLandingham and R.L. Moose, "Digital Control of High Performance Aircraft using Adaptive Estimation Techniques", *IEEE Trans. Aero. and Electron. Systems* **12**, pp.112-119 (1977).
35. J.G. Deshpande, T.N. Upadhyay, and D.G. Lainiotis, "Adaptive Control of Linear Stochastic systems", *Automatica* **9**, pp.107-115 (1973).
36. D.G. Lainiotis, "Partitioning: A Unifying Framework for Adaptive Systems II: Control", *Proc. IEEE* **64**, pp.1182-1198 (1976).
37. K. Watanabe and S.G. Tzafestas, "Stochastic Control for Systems with Faulty Sensors", *ASME J. Dynamic Systems, Measure. and Control* **112**(1), pp.143-

147 (1990).

38. K. Watanabe and S.G. Tzafestas, "A new generalized pseudo-Bayes estimation for discrete-time systems with jump failure parameters, *Proceedings of 12th IMACS World Congress on Scientific Computation*, Paris, **2**, pp.427-430 (1988).
39. K. Watanabe and S.G. Tzafestas, "Multiple model adaptive control for jump-linear stochastic systems", *Int. J. Control* **50**(5), pp.1603-1617 (1989).
40. J.D. Birdwell, D.A. Castanon, and M. Athans, "On Reliable Control System Designs", *IEEE Trans. Systems, Man, and Cyber.* **SMC-16**(5), pp.703-711 (1986).
41. A.Y. Lee and C.S. Sims, "Adaptive estimation and stochastic control for uncertain models", *Int. J. Control* **19**(3), pp.625-639 (1974).
42. T.N. Upadhyay and D.G. Lainiotis, "Joint Adaptive Plant and Measurement Control of Linear Stochastic Systems", *IEEE Trans. Auto. Control* **AC-18**(5), pp.567-571 (1974).
43. J.K. Tugnait, "Adaptive Estimation and Identification for Discrete Systems with Markov Jump Parameters", *IEEE Trans. Auto. Control* **AC-27**(5), pp.1054-1065 (1982).
44. J.K. Tugnait, "On Identification and Adaptive Estimation for Systems with Interrupted Observations", *Automatica* **19**(1), pp.61-73 (1983).
45. J.K. Tugnait and A.H. Haddad, "State Estimation Under Uncertain Observations with Unknown Statistics", *IEEE Trans. Auto. Control* **AC-24**(2), pp.201-210 (1979).
46. J.K. Tugnait and A.H. Haddad, "Adaptive Estimation in Linear Systems with Unknown Markovian Noise Statistics", *IEEE Trans. Inform. Theory* **IT-26** (1), pp.66-78 (1980).
47. Y. Sawaragi, T. Katayama, and S. Fujishige, "Adaptive Estimation for a Linear System with Interrupted Observation", *IEEE Trans. Auto. Control* **AC-18**(2), pp.152-154 (1973).
48. K.C. Chang and Y. Bar-Shalom, "Distributed Adaptive Estimation with Probabilistic Data Association", *Automatica* **25**(3), pp.359-369 (1989).
49. K. Watanabe, "A Decentralized Multiple Model Adaptive Filtering for Discrete-Time Stochastic Systems", *Trans. ASME, J. Dynamic Syst. Meas. Control* **111**, pp.371-377 (1989).

50. K. Watanabe and S.G. Tzafestas, "A Hierarchical Multiple Model Adaptive Control of Discrete-time Stochastic Systems for Sensor and Actuator Uncertainties", *Automatica* **26**(5), pp.875-886 (1990).
51. M. Mariton, *Jump Linear Systems in Automatic Control*, Marcel Dekker, New York, N.Y.(1990).
52. K. Watanabe, *Adaptive Estimation and Control: Partitioning approach*, Prentice Hall, London, U.K. (1991).

Robust Off-line Methods for Parameter Estimation

Heping Dai
Naresh K. Sinha

Department of Electrical and Computer Engineering
McMaster University
Hamilton, Ontario
Canada L8S 4L7

I. INTRODUCTION

It is well known that conventional estimation methods fail to give unbiased estimates when the input-output data are contaminated with some large errors (called outliers). To reduce the effect of outliers on parameter estimation, a special kind of estimation method is needed. This is called "robust algorithm".

Following Huber's minimax principle [1], several robust identification methods have been proposed. Masreliez and Martin [2] have given a paper on the robust Bayesian estimation and the robustified Kalman filter for the linear model. A class of robust stochastic approximation algorithms for system identification has been proposed by Stanković and Kovacević [3]. Other methods have been described by Puthenpura and Sinha. These are modified maximum likelihood method [4], robust identification from impulse and step response [5], and robust recursive identification algorithms [6,7]. In addition, six robust recursive identification methods have been described in the previous chapter and some earlier papers [8,9,10,11] for identification and state estimation of bilinear systems.

In the above methods, the robustness of parameter estimates is obtained by introducing an influence function for the purpose of reducing the effect of outliers on the estimate. Unfortunately, in none of these methods a mention has been made of the procedure for directly generating an estimate of the turning point of influence functions while the parameter is estimated. In other words, it is always necessary to have *a priori* knowledge of some statistical properties of the noise in order that an appropriate influence function can be determined beforehand and both robustness of estimates and minimization of robust criterions can be secured. Obviously, as concluded in the previous chapter, applications of these methods are limited due to the lack of a procedure for determining the proper influence function.

This chapter is especially dedicated to solving the problem of robust system identification. In section II, we shall first give a detailed discussion about robust criterions which have been widely used in previous work. Subsequently, a class of more general robust criterions will be used to replace the previous criterions when developing robust identification algorithms. This will result in an off-line method, the robust iterative least squares method with modified residuals (RILSMMR). The most attractive feature of the RILSMMR is that it can provide robust estimates of parameters and the turning point of influence functions at the same time. In section IV, a counterpart of the method will be utilized for estimating a set of expansion coefficients of orthogonal functions which will approximate very noisy time series. As an application of these robust coefficient estimates, an algorithm will be described for robust identification of both linear and bilinear continuous-time systems. In the final section, several interesting comments and conclusions will be provided.

II. DISCUSSION ABOUT ROBUST CRITERIONS

It is known that robust identification problem can be solved by different approaches. A basic and commonly used approach is to obtain robust estimates of system parameters by introducing robust criterions. In general, these criterions should possess several desirable features as pointed out earlier. In this section, a detailed discussion will be included. It will bring out a class of general cost functions for deriving a more practical and useful robust identification method.

II.A. Comments on Previous Robust Criterions

Most robust methods indicated previously are based on minimizing a sum of less rapidly increasing functions of residuals. That is

$$\frac{1}{N}\sum_{t=1}^{N} \rho\big(r(t)\big) = \min \tag{1}$$

with the residual $r(t)$, at the t-th instant, defined as

$$r(t) = y(t) - \phi^{T}(t)\,\theta. \tag{2}$$

Suppose that $\rho(x)$ is a kind of convex functions. Then, letting the derivative of Eq.(1) with respect to θ equal to zero gives the following equation

$$\frac{1}{N}\sum_{t=1}^{N} \phi(t)\,\psi\big(y(t) - \phi^{T}(t)\,\theta\big) = 0 \tag{3}$$

with

$$\psi(\nu) = \frac{d\rho(\nu)}{d\nu}.$$

Usually, $\psi(\nu)$ is called the influence function. Eq.(3) can be solved for solving robust estimates of parameters.

To reduce the effect of outliers on parameter estimates, it is desirable to somehow introduce an influence function. The influence function should have a turning point properly selected beforehand. Here a problem arises, as influence functions may not be determined directly by minimizing the given robust criterion in Eq.(1).

A number of efforts have been made to solve this problem. One effort uses empirical formulas for calculating the turning point of the influence function [7] if the noise (or residual) variance is known beforehand. The second suggests that the turning point of influence functions may be obtained only by guessing different values of the residual variance when no information about variance is available. Ljung [13] has recommended an approach, based on robust statistics, to determine influence functions for system identification. Instead of guessing the residual variance, an estimate of the residual variance was recommended once residuals are obtained. The approach, however, leads to some practical and theoretical difficulties. First, it is not feasible to calculate an estimate of the residual variance. Second, proof of convergence of the procedure would be more complicated. In addition, the way to select the estimate of the residual variance cannot guarantee minimization of a robust criterion even though the selection of the residual variance estimate is indeed appropriate. Therefore, from the practical point of view, uses of those robust identification methods are restricted.

One of the basic reasons for the problem is that the criterion in (1) does not provide any hint on how to estimate the residual variance if it is not known beforehand. In other words, the turning point of the influence function does not appear in the corresponding criterion. Therefore, it is hard to determine the turning point of the influence function and consequently, parameter estimates, or the robustness of parameter estimation may deteriorate significantly if a poor influence function is used.

Furthermore, even if somehow a reasonably good estimate of the residual variance were obtained, there is no rule to update the estimate as the residual variance changes. Therefore, minimization of the robust criterion is not guaranteed. In other words, the criterion described in Eq.(1) is suitable only for those cases where the residual variance is invariant.

Definitely, there is a great need for improving the commonly used robust criterion in order to obtain the desirable robustness of parameter estimates.

II.B. A Class of General Robust Criterions

To overcome the drawbacks of the cost functions described in Eq.(1), one possible approach is to introduce the residual variance in the corresponding criterion. This leads to a modified robust criterion [1]

$$J(\theta,\sigma) = \frac{1}{N}\sum_{t=1}^{N}\left[\rho_0\left(\frac{r(t)}{\sigma}\right) + \beta\right]\sigma. \tag{4}$$

In the above equation, $\rho_0(x)$ is a convex function with $\rho_0(0)=0$, the symbol σ is variance of residuals, and the quantity β is a positive constant, introduced for the purpose of obtaining the consistent estimate of the residual variance. Clearly, the difference between (1) and (4) is the residual variance σ introduced to compensate the drawback of the previous criterion.

To obtain robust estimates of parameters as well as the residual variance, the following equations should be solved simultaneously. The equations are obtained by setting the partial derivatives of (4) with respect to θ and σ equal zero. They are

$$\frac{1}{N}\sum_{t=1}^{N}\phi(t)\psi_0\left(\frac{r(t)}{\sigma}\right) = 0 \tag{5.a}$$

$$\frac{1}{N}\sum_{t=1}^{N}\chi_0\left(\frac{r(t)}{\sigma}\right) = \beta. \tag{5.b}$$

The functions $\psi_0(.)$ and $\chi_0(.)$ are defined as below:

$$\psi_0(x) = \frac{d\rho_0(x)}{dx} \tag{5.c}$$

$$\chi_0(x) = x\psi_0(x) - \rho_0(x). \tag{5.d}$$

In order to have consistent estimate of the residual variance of the normal model and to recapture the estimate for the classical choice $\rho_0(x)=x^2/2$, the β is assumed as

$$\beta = \frac{N-P}{N}E_{\mathfrak{n}}(\chi_0). \tag{6}$$

Here, the quantity N is the number of data points to be processed and P the number of parameters to be identified. The symbol $E_{\mathfrak{n}}(\chi_0)$ stands for the expectation of χ_0 with the Gaussian probability density function, i.e.

$$E_{\mathfrak{n}}(\chi_0(x)) = \frac{1}{\sqrt{2\pi}}\int_{-\infty}^{+\infty}\chi_0(x)e^{-\frac{x^2}{2}}dx. \tag{7}$$

Thus, the problem of robust identification of system is reduced to a corresponding optimization problem. This should be an easier problem. With any optimization technique, the robust estimates $\hat{\theta}$ and $\hat{\sigma}$ can be calculated by solving Eqs.(5.a) and (5.b) simultaneously.

III. ROBUST IDENTIFICATION OF DISCRETE-TIME SYSTEMS

It is well known that there are many well developed optimization techniques. In particular, the robust iterative least squares method with modified residuals can be derived by applying the ordinary iterative Gauss-Newton approach to the optimization problem. It leads to a robust off-line method for identification of discrete-time systems, called the robust iterative least squares method with modified residuals.

III.A. Problem Formulation

For simplicity, a class of discrete-time, linear, and single-input single-output systems is considered. The algorithm to be derived here, however, can be easily extended to multi-input multi-output linear systems. As usual, *a priori* knowledge of the structure and order of the systems is assumed.

There are different ways to represent linear systems. One commonly used model is the linear difference equation. That is

$$A(z^{-1})y(t) = B(z^{-1})u(t) + v(t) \tag{8}$$

where $y(t)$ and $u(t)$ represent the output and input of linear systems, respectively, at the t-th instant. The polynomials $A(z^{-1})$ and $B(z^{-1})$ are defined as

$$\left.\begin{aligned} A(z^{-1}) &= 1 - a_1z^{-1} - \cdots - a_nz^{-n} \\ B(z^{-1}) &= b_1z^{-1} + \cdots + b_nz^{-n} \end{aligned}\right\} \tag{9}$$

where the constant n is the order of the system. The quantity $\nu(t)$ is the t-th element of a sequence of random noises $\{\nu(t)\}$. Their distributions belong to the so-called "ϵ-contaminated family" $\mathcal{P}_\epsilon$ [1]

$$\wp_\epsilon = \{ F \mid F = (1-\epsilon)G + \epsilon H, \qquad 1 \geq \epsilon \geq 0 \}. \tag{10}$$

Usually, G is assumed to be a normal distribution. Corresponding to the distribution of outliers, H is assumed to be unknown but belonging to some classes of symmetric distributions with zero mean and finite variance. Without losing generality, both G and H are defined as normal density $N_1(.\mid 0,\sigma_1^2)$ and $N_2(.\mid 0,\sigma_2^2)$, respectively. Then, to simulate the behaviour of outliers, we may choose $\sigma_2^2 \gg \sigma_1^2$. The quantity ϵ is the probability of occurrence of outliers in the assumed Gaussian distribution $N_1(.\mid 0,\sigma_1^2)$.

Clearly, Eq.(8) can be put in the equation error form

$$y(t) = \phi^{\mathrm{T}}(t)\,\theta + \nu(t). \tag{11}$$

The vector $\phi(t)$ is the observation vector defined as

$$\phi^{\mathrm{T}}(t) = \begin{bmatrix} y(t-1) & \cdots & y(t-n) & u(t-1) & \cdots & u(t-n) \end{bmatrix} \tag{12.a}$$

and the parameter vector, θ, is given by

$$\theta^{\mathrm{T}} = \begin{bmatrix} a_1 & \cdots & a_n & b_1 & \cdots & b_n \end{bmatrix}. \tag{12.b}$$

Now, we can see that the problem of robust identification is to estimate the parameter vector θ in (11) in a manner that parameter estimates are distorted very little by outliers that are present in the input-output data. In particular, the objective of this chapter is to develop a robust approach which can not only fulfil the task without any noise information but also guarantee the minimization of the robust criterion.

III.B. Robust Iterative Least Squares Method with Modified Residuals

The basic idea of this method is to apply the ordinary iterative Gauss-Newton approach to the particular robust criterion (4) with the difference that the residual $r(t)$ is replaced by its metrically Winsorized version $r_*(t)$. This is described below.

Let

$$r_*(t) = \psi_0\left(\frac{r(t)}{\sigma^{(\mathrm{m})}}\right)\sigma^{(\mathrm{m})} \tag{13}$$

where $\sigma^{(\mathrm{m})}$ is the estimate of the residual variance σ at step m. Suppose further that

$$\frac{d\psi_0(x)}{dx} = 1 \tag{14}$$

when $|x| < c$, $1 < c < 2$ is an adjusted constant. Then, taking the derivative of $r_*(t)$ with respect to the parameter vector θ leads to

$$\frac{\partial r_*(t)}{\partial \theta} = -\phi(t). \tag{15}$$

Suppose that a reasonably good estimate $\theta^{(\mathrm{m})}$ of the parameter vector θ at step m has been obtained. Expanding $r_*(t)$ at $\theta^{(\mathrm{m})}$, using the Taylor series which retains the first two terms gives

$$r_*(t) \approx r_*^{(\mathrm{m})}(t) - \phi^{\mathrm{T}}(t)\left(\theta - \theta^{(\mathrm{m})}\right) \tag{16.a}$$

or,

$$r_*(t) \approx r_*^{(m)}(t) - \phi^T(t)\Delta\theta \tag{16.b}$$

with

$$r_*^{(m)}(t) = \psi_0\left(\frac{r^{(m)}(t)}{\sigma^{(m)}}\right)\sigma^{(m)}. \tag{16.c}$$

Hence, the problem of calculating the parameter estimate $\theta^{(m+1)}$, at step $m+1$, is equivalent to a problem of estimating $\Delta\theta$.

Substituting Eq.(16) into Eq.(5.a) gives

$$\frac{1}{N}\sum_{t=1}^{N}\phi(t)\left(r_*^{(m)}(t) - \phi^T(t)\Delta\theta\right) = 0. \tag{17}$$

It is interesting to notice that, at the step $m+1$, the robust estimation problem of $\Delta\theta$ is reduced to solving the classical least squares problem, that is

$$\theta^{(m+1)} = \theta^{(m)} + q\Delta\theta \tag{18}$$

where $0<q<2$ is an arbitrary relaxation factor, and $\Delta\theta$ is determined by solving Eq.(17). This gives

$$\Phi^T(N)\Phi(N)\Delta\hat{\theta} = \Phi^T(N)r_*^{(m)} \tag{19}$$

where the matrix $\Phi(N)$ and the vectors $\Delta\hat{\theta}$ and $r_*^{(m)}$ are given as

$$\left.\begin{aligned} \Phi^T(N) &= \left[\phi(n+1) \;\cdots\; \phi(N)\right] \\ \Delta\hat{\theta}^T &= \left[\Delta\hat{\theta}_1 \;\cdots\; \Delta\hat{\theta}_{2n}\right] \\ \left(r_*^{(m)}\right)^T &= \left[r_*^{(m)}(n+1) \;\cdots\; r_*^{(m)}(N)\right]. \end{aligned}\right\} \tag{20}$$

The algorithm developed so far is based on the assumption that the residual variance is known already. It simply indicates that the influence function has been appropriately determined before estimating parameters. However, this is not true in most cases. To solve this problem, an estimate of residual variance is needed, which can be obtained by solving Eq.(5.b). To implement the algorithm, however, an iterative formula is derived to generate the estimate of residual variance $\sigma^{(m+1)}$ if the trial value $\theta^{(m)}$ at step m have been obtained. That is

$$\left(\sigma^{(m+1)}\right)^2 = \frac{1}{N\beta}\sum_{t=1}^{N}\chi_0\left(\frac{r(t)}{\sigma^{(m)}}\right)\left(\sigma^{(m)}\right)^2. \tag{21}$$

It will be proved in lemma 2 that the estimate of residual variance given by Eq.(21) minimizes the robust criterion (4).

Combining the above equations together, the entire algorithm of the robust iterative least squares method with modified residuals can be carried out by tracking the following procedure:

1. Let $\theta^{(m)}$ and $\sigma^{(m)}$ be reasonably good trial values of θ and σ;
2. Calculate $r_*(t)$ from Eq.(13);
3. Solve Eq.(19) for $\Delta\hat{\theta}$;
4. From Eq.(18), compute the new estimate of the parameter vector $\theta^{(m+1)}$ at step $m+1$;
5. Using Eq.(21), update the estimate of the residual variance $\sigma^{(m+1)}$, at step $m+1$;

6. Stop iterating if the parameter changes by less than δ times the estimate of residual variance, i.e.

$$|\Delta \hat{\theta}_i| < \delta \sigma^{(m+1)} \tag{22}$$

for all $i=1, ..., p$. Otherwise, let $\theta^{(m)} = \theta^{(m+1)}$ and $\sigma^{(m)} = \sigma^{(m+1)}$ and go back to step 2 until the stopping criterion (22) is satisfied.

Notice that the implementation of the algorithm requires some additional considerations.

The first is how to select the cost function $\rho_0(x)$ described in Eq.(4). According to Huber [1], $\rho_0(x)$ is chosen

$$\rho_0(x) = \begin{cases} \frac{1}{2}x^2 & |x| \le c \\ c|x| - \frac{1}{2}c^2 & |x| > c \end{cases} \tag{23.a}$$

where c is an adjusted constant in the range of 1 to 2. In many cases, the turning point is a product of the residual variance and the constant c. Therefore, the selection of the constant c may slightly affect the robustness of the estimates. Following the definition in Eqs.(5.c) and (5.d), $\psi_0(x)$ and $\chi_0(x)$ can be derived

$$\psi_0(x) = \begin{cases} x & |x| \le c \\ c\, sign\,(x) & |x| > c \end{cases} \tag{23.b}$$

$$\chi_0(x) = \frac{1}{2}\psi_0^2(x) = \begin{cases} \frac{1}{2}x^2 & |x| \le c \\ \frac{1}{2}c^2 & |x| > c. \end{cases} \tag{23.c}$$

In particular, replacing the argument x in Eq.(23.b) by the ratio $r(t)/\sigma$ leads to the conventional expression of the influence function. That is

$$\psi_0(r(t)) = \begin{cases} r(t) & |r(t)| \le c\sigma \\ c\sigma\, sign(r(t)) & |r(t)| > c\sigma. \end{cases} \tag{24}$$

Clearly, for an appropriately selected constant c, the turning point of influence function only depends on the residual variance, a key factor of determining influence functions.

The second consideration is the appropriate choice of the initial values $\theta^{(0)}$ and $\sigma^{(0)}$. Based on robust statistics, the estimate of σ using the median absolute deviation

$$\hat{\sigma}^{(0)} = med(|r(t)|) \tag{25.a}$$

seems to be a good choice. As to the initial values of parameters, the least squares estimate is preferred

$$\theta^{(0)} = \hat{\theta}_{LS}. \tag{25.b}$$

From the algorithm investigated above, the main advantage of the method over the previous methods for robust identification is very evident. It is that the robustness of parameter estimates and optimization of the criterion can be always guaranteed no matter whether variance of noise (or residuals) is known beforehand. In the following part of this section, robustness of the method will be verified by comparison of the derived method with non-robust methods.

III.C. Convergence Analysis

As an important part of developing a new robust identification method, the local convergence analysis of the developed method is given below. It should be mentioned here that the proofs of lemmas 2 to 4 and theorem 1 are similar to those given by Huber [1]. The basic difference, however, lies in the model assumptions. For simplicity, the first and second derivatives of $\rho_0(x)$ with respect to x are denoted as $\dot{\rho}_0(x)$ and $\ddot{\rho}_0(x)$, respectively.

Lemma 1

(i) $\rho_0(x)$ is a differentiable function with $\rho_0(0) = 0$;

(ii) $\dot{\rho}_0(x)$ is a continuous and piecewise continuous differentiable function with $\dot{\rho}_0(0)=0$;

(iii) The second derivative of $\rho_0(x)$ is supposed to be

$$0 \le \ddot{\rho}_0(x) = k_1 \le 1 \qquad for \quad |x| \le x_0$$
$$0 \le \ddot{\rho}_0(x) = k_2 \le k_1 \qquad for \quad |x| > x_0$$

where k_1 and k_2 are constants, and x_0 is called the turning point, then

$$\ddot{\rho}_0(x) - \frac{2}{x^2}\chi_0(x) \le 0.$$

■

Proof: See Appendix VII.A.

Comments: It is easy to show that Eq.(23) satisfies the assumptions in lemma 1. Besides, the conclusion in lemma 1 implies that $\rho_0(x) \ge 0$ is a convex function, and $\rho_0(x)/x$ is convex for $x<0$ and concave for $x>0$. Finally, if $k_1 = k_2 > 0$, the cost function becomes the normal least squares criterion.

Lemma 2

Based on the assumptions in lemma 1, if

$$\left(\sigma^{(m+1)}\right)^2 = \frac{1}{N\beta}\sum_{t=1}^{N}\chi_0\left(\frac{r(t)}{\sigma^{(m)}}\right)\left(\sigma^{(m)}\right)^2$$

then, the following inequality is true

$$J\left(\theta^{(m)},\sigma^{(m)}\right) - J\left(\theta^{(m)},\sigma^{(m+1)}\right) \ge \beta\,\frac{\left(\sigma^{(m+1)} - \sigma^{(m)}\right)^2}{\sigma^{(m)}}.$$

In particular, J is strictly decreasing unless Eq.(5.b) is already satisfied.

■

Proof: See Appendix VII.A.

Lemma 3

Based on the assumptions in lemma 1, if

$$\theta^{(m+1)} = \theta^{(m)} + q\Delta\hat{\theta}$$

where $\Delta\hat{\theta}$ is determined by (19), then

$$J\left(\theta^{(\mathrm{m})},\sigma^{(\mathrm{m})}\right) - J\left(\theta^{(\mathrm{m}+1)},\sigma^{(\mathrm{m})}\right) \geq \frac{q(2-q)}{2\sigma^{(\mathrm{m})}N}\|\Delta\hat{\theta}\|^2$$
$$= \frac{q(2-q)}{2\sigma^{(\mathrm{m})}N}\left(\Phi^{\mathrm{T}}(N)\boldsymbol{r}_*\right)^{\mathrm{T}}\left(\Phi^{\mathrm{T}}(N)\Phi(N)\right)^{-1}\left(\Phi^{\mathrm{T}}(N)\boldsymbol{r}_*\right)$$

where $\boldsymbol{r}_*$ is a column vector of the modified residuals defined as

$$\boldsymbol{r}_*^{\mathrm{T}} = \left[r_*(n+1) \quad \cdots \quad r_*(N)\right]$$

and the norm is defined as

$$\|x\|^2 = x^{\mathrm{T}}\left(\Phi^{\mathrm{T}}(N)\Phi(N)\right)x.$$

In particular, unless Eq.(5.a) is already satisfied, J is strictly decreasing.

■

Proof: See Appendix VII.A.

Lemma 4

Assume that $\Phi(N)$ has full rank and that input $u(t)$ and output $y(t)$ are bounded. Then, the sets of the form

$$\mathfrak{C}_{\mathrm{b}} = \left\{(\theta,\sigma) \mid \sigma \geq 0, \quad J(\theta,\sigma) \leq b < \infty\right\}$$

are compact.

■

Proof: See Appendix VII.A.

Theorem 1

Based on the assumptions of lemmas 1 to 4, the following statements are true:

(1) The sequence $(\theta^{(\mathrm{m})},\sigma^{(\mathrm{m})})$ in $\mathfrak{C}_{\mathrm{b}}$ has at least one accumulation point $(\hat{\theta},\hat{\sigma})$;

(2) Every accumulation point $(\hat{\theta},\hat{\sigma})$ with $\hat{\sigma}>0$ is a solution of Eq.(5) and minimizes Eq.(4).

■

Proof: See Appendix VII.A.

III.D. Results of Simulation

To show advantages of the proposed method, simulation results given by both non-robust and robust methods are presented.

Consider a linear model of order 2, which can be expressed by Eqs.(8) and (9). The true parameters are specified

$$\left.\begin{aligned} A(z^{-1}) &= 1 - 1.75z^{-1} + 0.9z^{-2} \\ B(z^{-1}) &= z^{-1} + 0.7z^{-2}. \end{aligned}\right\} \tag{26}$$

To identify the parameters of the given system, a PRBS of magnitude 2 is chosen as an input series. The noise term $\nu(t)$ is generated in such a way that its distribution belongs to the "ϵ-contaminated family" $\mathfrak{P}_\epsilon$. Both G and H are assumed to be normal distributions with zero means and finite variances. That is

$$G \sim \mathfrak{N}(0,\sigma_1^2); \qquad H \sim \mathfrak{N}(0,\sigma_2^2). \tag{27}$$

To simulate the behaviour of outliers, σ_2^2 is set much larger than σ_1^2. In this example, we have selected $\sigma_1^2=0.07$ and $\sigma_2^2=8.0$. The probability that outliers occur in the observations is taken as 0.1. It indicates that 10 percent of the noise are outliers which might severely affect the parameter estimates obtained by conventional methods.

To implement the simulation, the quantity δ is taken as 0.001, the constant c is chosen as 1.5 and the relaxation factor q is selected 1. The parameter error norm is defined by the following equation:

$$n(t) = \frac{\|\hat{\theta} - \theta_0\|}{\|\theta_0\|} \tag{28}$$

where the Euclidean norm is used.

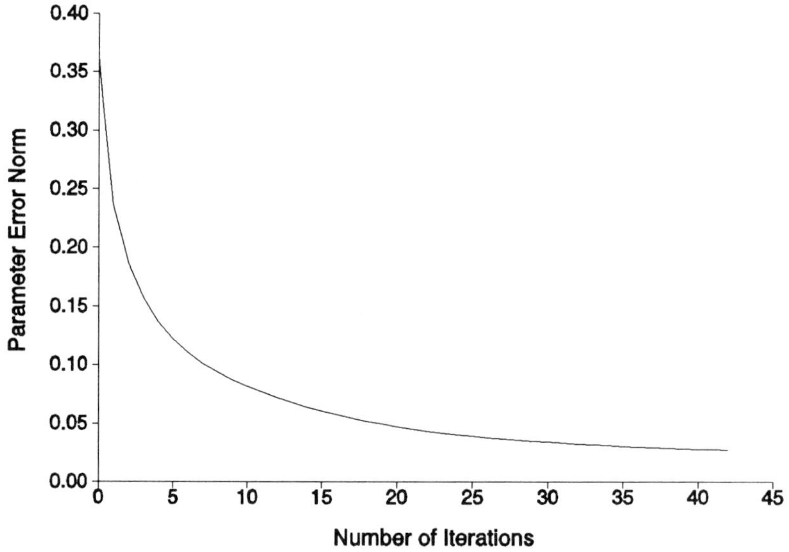

Figure 1. Illustrations of convergence and robustness of parameter estimates of linear systems by the RILSMMR

Table I.
Parameter estimates of linear systems obtained by LS and the RILSMMR

	True parameters	LS estimates	Robust estimates
a_1	1.75	1.389813	1.714177
a_2	-0.90	-0.554520	-0.866419
b_1	1.00	1.070881	1.028008
b_2	0.70	1.372003	0.729916
Parameter norm	0.0	0.362767	0.027618

The simulation results, obtained after 42 iterations, are listed in Table I. From the results, the robustness of the parameter estimates obtained by the developed method is easily demonstrated. To further investigate the convergence of this method, Fig.1 is included. From the figure, it may be observed that the parameter error norm decreases rapidly as the number of iterations increases. It plainly verifies that both robustness and convergence of the estimates using the robust iterative least squares method with modified residuals are very satisfied.

It should be emphasized once more that the advantage of the proposed method over earlier robust methods is obvious. That is, the estimate of residual variance can be obtained directly so that minimization of (4) is achieved by using the influence function determined from the estimate of the residual variance.

III.E. Extension To Discrete-Time Bilinear System Identification

The extension of the proposed method to bilinear system identification is very straightforward because of the linearity of bilinear system parameters. The only difference lies in the selection of observation vectors.

Suppose that the bilinear systems under consideration can be described by their observability canonical form as follows:

$$\left.\begin{aligned} \boldsymbol{x}(t+1) &= \boldsymbol{P}\boldsymbol{x}(t) + \boldsymbol{Q}\boldsymbol{x}(t)u(t) + \boldsymbol{R}u(t) \\ w(t) &= \boldsymbol{S}\boldsymbol{x}(t) \\ y(t) &= w(t) + e(t) \end{aligned}\right\} \tag{29.a}$$

where we have $\boldsymbol{x}(t) \in \Re^{n}$, $u(t)$, $w(t)$, and $y(t) \in \Re$. The matrices and vectors are

$$\boldsymbol{P} = \begin{bmatrix} 0 & & I \\ a_n & \cdots & a_1 \end{bmatrix}; \qquad \boldsymbol{Q} = \begin{bmatrix} & 0 & \\ b_n & \cdots & b_1 \end{bmatrix}; \tag{29.b}$$

$$\boldsymbol{R}^{T} = \begin{bmatrix} c_1' & \cdots & c_n' \end{bmatrix}; \qquad \boldsymbol{S} = \begin{bmatrix} 1 & 0 & \cdots & 0 \end{bmatrix}$$

and the quantity n is the order of the discrete-time bilinear systems.

To fit the problem into the framework of recursive identification, the input-output representation of the state-space model (29) should be derived. That is

$$\begin{aligned} \{A(z^{-1}) + u(t-n)B(z^{-1})\}y(t) &= \{C(z^{-1}) + u(t-n)D(z^{-1})\}u(t) \\ &\quad + \{A(z^{-1}) + u(t-n)B(z^{-1})\}e(t) \end{aligned} \tag{30.a}$$

where

$$\left.\begin{aligned} A(z^{-1}) &= 1 - a_1 z^{-1} - \cdots - a_n z^{-n} \\ B(z^{-1}) &= -b_1 z^{-1} - \cdots - b_n z^{-n} \\ C(z^{-1}) &= c_1 z^{-1} + \cdots + c_n z^{-n} \\ D(z^{-1}) &= d_1 z^{-1} + \cdots + d_n z^{-n} \end{aligned}\right\} \tag{30.b}$$

and z^{-1} is the backward shift operator. The coefficients $\{c_i\ i=1 \ldots n\}$ and $\{d_i\ i=1 \ldots n\}$

are given as

$$\begin{bmatrix} c_1 \\ c_2 \\ \vdots \\ c_n \end{bmatrix} = \begin{bmatrix} 1 & & & \\ -a_1 & 1 & & \\ \vdots & \vdots & \ddots & \\ -a_{n-1} & \cdots & -a_1 & 1 \end{bmatrix} \begin{bmatrix} c_1' \\ c_2' \\ \vdots \\ c_n' \end{bmatrix} \tag{31.a}$$

$$\begin{bmatrix} d_1 \\ d_2 \\ \vdots \\ d_n \end{bmatrix} = \begin{bmatrix} 0 & & & \\ -b_1 & 0 & & \\ \vdots & \vdots & \ddots & \\ -b_{n-1} & \cdots & -b_1 & 0 \end{bmatrix} \begin{bmatrix} c_1' \\ c_2' \\ \vdots \\ c_n' \end{bmatrix} \tag{31.b}$$

For the sake of brevity, Eq.(30) can be rewritten as

$$y(t) = \phi^{\mathrm{T}}(t)\,\theta + \nu(t) \tag{32}$$

where $\phi(t)$ and θ are the observation vector and the true parameter vector, respectively. They are defined as

$$\phi^{\mathrm{T}}(t) = \left[\boldsymbol{Y}^{\mathrm{T}}(t-1), \boldsymbol{Y}^{\mathrm{T}}(t-1)u(t-n), \boldsymbol{U}^{\mathrm{T}}(t-1), \boldsymbol{U}^{\mathrm{T}}(t-1)u(t-n)\right] \tag{33}$$

$$\boldsymbol{U}^{\mathrm{T}}(t-1) = \left[u(t-1)\cdots u(t-n)\right]$$

$$\boldsymbol{Y}^{\mathrm{T}}(t-1) = \left[y(t-1)\cdots y(t-n)\right]$$

and

$$\theta^{\mathrm{T}} = \left[\boldsymbol{a}^{\mathrm{T}}, \boldsymbol{b}^{\mathrm{T}}, \boldsymbol{c}^{\mathrm{T}}, \boldsymbol{d}^{\mathrm{T}}\right] \tag{34}$$

$$\boldsymbol{a}^{\mathrm{T}} = \left[a_1 \cdots a_n\right] \qquad \boldsymbol{b}^{\mathrm{T}} = \left[b_1 \cdots b_n\right]$$

$$\boldsymbol{c}^{\mathrm{T}} = \left[c_1 \cdots c_n\right] \qquad \boldsymbol{d}^{\mathrm{T}} = \left[d_1 \cdots d_n\right].$$

And the residual $\nu(t)$ is given as

$$\nu(t) = e(t) - \boldsymbol{a}^{\mathrm{T}}\boldsymbol{E}(t-1) - \boldsymbol{b}^{\mathrm{T}}\boldsymbol{E}(t-1)u(t-n) \tag{35}$$

with

$$\boldsymbol{E}^{\mathrm{T}}(t-1) = \left[e(t-1) \;\; \cdots \;\; e(t-n)\right].$$

The additive noise term $e(t)$ belongs to the so-called "ϵ-contaminated family" $\wp_\epsilon$, defined in Eq.(10).

It is very straightforward to derive a corresponding algorithm for robust identification of bilinear systems if the vector $\phi(t)$ and θ in (12) are replaced by those specified in (33) and (34). The whole algorithm studied in the preceding section can be directly used for estimating parameters of bilinear systems. The proof of convergence of the developed method should remain valid since it is relatively independent of observation and parameter vectors.

As to robustness of the presented method, simulation results including those for both robust and non-robust methods are provided. The bilinear system considered is given by Eq.(29), and the parameter vector is given as

$$\theta^T = \left[\, 1.75 \quad -0.9 \quad -0.2 \quad 0.1 \quad 1.0 \quad 0.7 \quad 0.0 \quad 0.2 \,\right]. \tag{36}$$

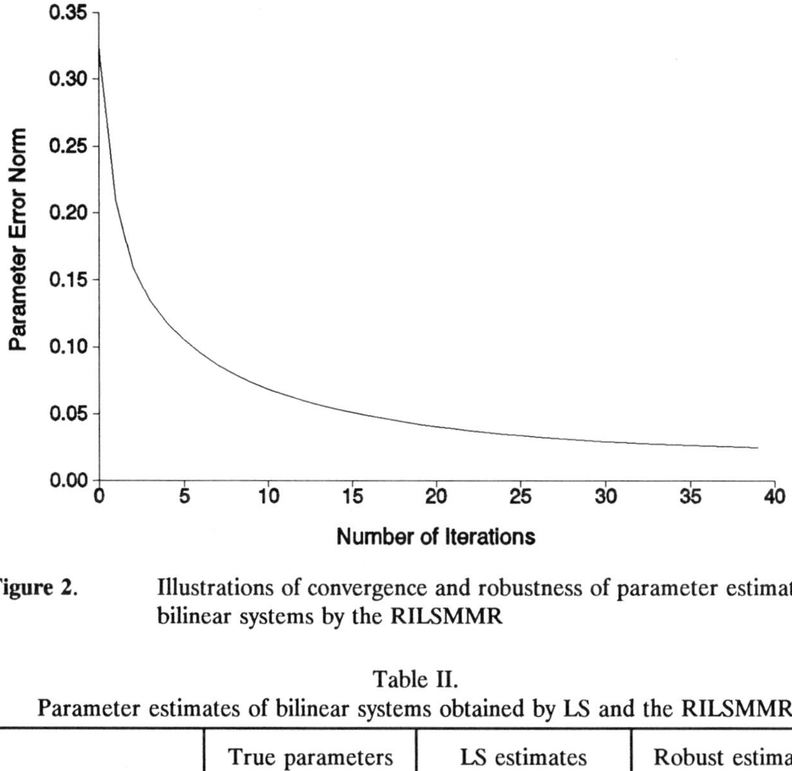

Figure 2. Illustrations of convergence and robustness of parameter estimates of bilinear systems by the RILSMMR

Table II.
Parameter estimates of bilinear systems obtained by LS and the RILSMMR

	True parameters	LS estimates	Robust estimates
a_1	1.75	1.408821	1.713323
a_2	-0.90	-0.577670	-0.867026
b_1	-0.20	-0.174334	-0.193946
b_2	0.10	0.094682	0.097442
c_1	1.00	1.011444	1.017077
c_2	0.70	1.293478	0.719971
d_1	0.00	0.026302	0.016278
d_2	0.20	0.130629	0.197913
Parameter norm	0.0	0.325822	0.025102

The noise term $e(t)$ is generated according to Eq.(10). The variances σ_1^2 and σ_2^2 are taken as 0.07 and 8.0, respectively. The probability of outliers occurring in the normal noise takes the value 0.1. The input signal is chosen as a PRBS of magnitude 1.5. To implement the algorithm, the constants c and δ are taken as 1.5 and 0.001, respectively. The relaxation factor q is selected as 1. Again, the parameter error norm used here is defined by Eq.(28).

The simulation results, obtained after 39 iterations, are shown in Table II, and the corresponding parameter error norm is plotted in Fig.2. From the results, it is clear that both robustness and convergence of the estimates by the robust iterative least squares method with modified residuals are much better than those by previous robust or non-robust methods.

IV. ROBUST ESTIMATION OF COEFFICIENTS OF ORTHOGONAL FUNCTIONS

Representation of a time function by the superposition of members of a set of simple functions, which are easy to generate, is a useful attribute. However, only sets of orthogonal functions can be used to completely synthesis any time function to a required degree of accuracy. In other words, any time function, which is absolutely integrable in an interval $[0,T)$, can be represented by an infinite series of orthogonal functions. There have been many applications of this technique in various areas of engineering, such as system identification and control, communication, signal processing, pattern recognition, image processing, acoustic signal processing, and speech processing.

Unfortunately, representation of a given time series is not a simple and easy problem in practice. The estimation of expansion coefficients of a noisy time series is a particularly challenging topic. The importance of the topic is clear considering its many applications in widely different areas.

In this section, given time series contaminated with noise or even outliers, the method proposed in the last section will be applied to the estimation of expansion coefficients of orthogonal functions. In fact, the procedure to be discussed here has been described in two previous papers [14,15] by the authors[1]. A proof of local convergence and simulation results of the method will be included.

IV.A. Problem Formulation

Consider a time function, $f(t)$, defined over a time interval $[0,T)$ as being represented by an orthogonal series, $g_m(t)$ $m=0, 1, \ldots$. Thus

$$f(t) = \sum_{m=0}^{\infty} a_m g_m(t) \tag{37}$$

where $\{a_m\ m=0, 1, \ldots\}$ is set of coefficients for the orthogonal function series $\{g_m(t)$ $m=0, 1, \ldots\}$ is said to be orthogonal over the interval $T>t\geq 0$ if

$$\int_0^T g_n(t) g_m(t)\, dt = \begin{cases} T & \text{if } n = m \\ 0 & \text{if } n \neq m \end{cases} \tag{38}$$

when n and m have integer values.

From the practical point of view, only a finite number of terms, M, can be used

---

[1] Figures 3 to 7, lemmas 1, 5 to 7 and theorem 2 have been reproduced from reference 15, by kind permission of IEE.

for practical realization of the series given by Eq.(37). Therefore, it is necessary to choose the coefficients $\{a_m\}$ to minimize the mean-square approximation error (*M.S.E.*) as follows

$$M.S.E. = \int_0^T \left[f(t) - \sum_{m=0}^{M-1} a_m g_m(t) \right]^2 dt. \tag{39}$$

Clearly, it can be realised by making

$$a_m = \frac{1}{T} \int_0^T f(t) g_m(t) dt \tag{40}$$

for $m=0, 1, ..., M\text{-}1$.

For complete orthogonal function series, such as Walsh functions, block-pulse functions, etc., the *M.S.E.* monotonically decreases to zero as M tends to infinity.

For engineering applications, it is clear that minimizing the mean-square error, *M.S.E.* in (39) is a fundamental step. None of the applications may be achieved if the *M.S.E.* is not properly minimized, or formula (40) calculating coefficients a_m, $m=0, 1, ..., M\text{-}1$, does not work well. This restriction leads to a common basic assumption for calculating expansion coefficients. That is, the time function to be approximated should contain little or even no noise and measurement errors; otherwise, the approximation may not be good and the results may considerably deteriorate.

Unfortunately, this simplifying assumption is not satisfied in a variety of real-life situations where time series may be contaminated by noise, sometimes, outliers. These common reasons arise from failure of transducers, analog to digital conversion errors, large disturbances, and even data transmission errors. In these cases, the traditional expansion method, minimizing the *M.S.E.*, fails to produce accurate estimates of coefficients. The corresponding approximation solution may, therefore, deviate considerably from the true ones. In the following part of this section, the problem of obtaining robust estimates of coefficients from noise-contaminated time functions will be discussed in detail.

IV.B. Robust Coefficient Estimation of Orthogonal Functions

For physical applications, it is desirable to weaken, or even to remove the noise-free, especially the outliers-free assumption. This goal has long been pursued due to its distinct significance. As far as the authors are aware, no previous attempts have succeeded in solving the problem. In this section, the problem will be solved by replacing the traditional expansion criterion, *M.S.E.*, with a robust criterion discussed in [1]. As a result, it is possible to employ orthogonal functions to solve various physical problems in the presence of noise, measurement errors, and even outliers in the input-output data.

Suppose that the contaminated output $f(t)$ of a physical system can only be described as

$$f(t) = g(t) + e(t). \tag{41}$$

Here, $g(t)$ is the noise-free output of the system and $e(t)$ stands for a random noise, or measurement errors containing outliers. It is customary to represent the noise term through the "ϵ-contaminated family" $\wp_\epsilon$ described in Eq.(10).

Without lose of generality, assume that $g(t)$ can be expanded by a sum of M Walsh functions

$$g(t) = \sum_{m=0}^{M-1} a_m g_m(t) = \boldsymbol{a}^T \boldsymbol{g}_{(M)}(t) \tag{42}$$

with

$$\boldsymbol{a}^T = \begin{bmatrix} a_0 & a_1 & \cdots & a_{M-1} \end{bmatrix} \tag{43.a}$$

$$\boldsymbol{g}_{(M)}^T = \begin{bmatrix} g_0(t) & g_1(t) & \cdots & g_{M-1}(t) \end{bmatrix}. \tag{43.b}$$

Combining (42) with (41) gives

$$f(t) = \boldsymbol{a}^T \boldsymbol{g}_{(M)}(t) + e(t). \tag{44}$$

In the presence of outliers, minimizing $M.S.E.$ usually gives very poor estimates. Intuitively, if we follow the line used in the last section, a robust estimate of the coefficient vector $\boldsymbol{a}$ will be generated even though the data may contain outliers. However, there is a basic difference between the two estimation problems discussed in the last section and this section. In the preceding section, the estimation problem is concerned with identifying parameters of systems in discrete-time models. Here, using any set of orthogonal functions in continuous-time expression, the problem is estimation of the expansion coefficients for approximating a given time function $f(t)$.

In order to derive the approach, a class of general robust criterion should be minimized instead of the traditional cost function, the $M.S.E.$.

Introduce

$$J(\boldsymbol{a},\sigma) = \frac{1}{T}\int_0^T \left\{ \rho_0\left(\frac{f(t) - \boldsymbol{a}^T \boldsymbol{g}_{(M)}(t)}{\sigma}\right) + \beta \right\} \sigma \, dt. \tag{45}$$

Similar to section III, we assume $\rho_0(.)$ is a convex function with $\rho_0(0)=0$, σ is variance of residuals, and β is a positive constant introduced for obtaining a consistent estimate of σ.

The estimates of both coefficients and the residual variance can be obtained by minimizing Eq.(45). Let the derivatives of Eq.(45) with respect to the coefficient vector $\boldsymbol{a}$ and variance σ be zero. That is

$$\int_0^T \psi_0\left(\frac{f(t) - \boldsymbol{a}^T \boldsymbol{g}_{(M)}(t)}{\sigma}\right) \boldsymbol{g}_{(M)}(t) \, dt = 0 \tag{46.a}$$

$$\int_0^T \chi_0\left(\frac{f(t) - \boldsymbol{a}^T \boldsymbol{g}_{(M)}(t)}{\sigma}\right) dt = \beta. \tag{46.b}$$

The functions $\psi_0(.)$ and $\chi_0(.)$ are defined as Eqs.(5.c) and (5.d). Similarly, the constant β is given as

$$\beta = \frac{P - M}{P} E_h(\chi_0). \tag{47}$$

The quantity P is the number of data (i.e. sampling points) to be processed, while M is the number of coefficients to be estimated which depends on different orthogonal functions and different users' requirements. The symbol $E_h(\chi_0)$ stands for the expectation of χ_0 with respect to the Gaussian probability density function and is the same as the one in Eq.(7).

The problem of robust estimation of coefficients is now reduced to a corresponding optimization problem. Utilizing the ordinary iterative Gauss-Newton approach, robust estimates $\hat{\boldsymbol{a}}$ and $\hat{\sigma}$ can be obtained by solving Eq.(46) simultaneously.

Denote the residual as

$$r(t,\boldsymbol{a}) = f(t) - \boldsymbol{a}^T \boldsymbol{g}_{(M)}(t). \tag{48}$$

The main idea of this method is to apply the ordinary iterative Gauss-Newton approach to the suggested criterion (45) with the alternation that the residual $r(t,\boldsymbol{a})$ is replaced by its modified residual $r_*(t,\boldsymbol{a})$. The modified residual is a metrically Winsorized version $r_*(t,\boldsymbol{a})$ and is defined as

$$r_*(t,\boldsymbol{a}) = \psi_0\left(\frac{r(t,\boldsymbol{a})}{\sigma^{(k)}}\right)\sigma^{(k)} \tag{49}$$

where $\sigma^{(k)}$ is an estimate of the residual variance σ at step k.

Furthermore, suppose that

$$\frac{d\psi_0(x)}{dx} = 1 \tag{50}$$

when $|x| < c$, $1 < c < 2$ is an adjusted constant. Therefore, the derivative of $r_*(t,\boldsymbol{a})$ with respect to the coefficient vector $\boldsymbol{a}$ is

$$\frac{\partial r_*(t,\boldsymbol{a})}{\partial \boldsymbol{a}} = -\boldsymbol{g}_{(M)}(t). \tag{51}$$

Suppose a reasonable good estimate $\boldsymbol{a}^{(k)}$ at step k has been obtained appropriately. Then expanding $r_*(t,\boldsymbol{a})$ as the first order Taylor series gives

$$r_*(t,\boldsymbol{a}) \approx r_*(t,\boldsymbol{a}^{(k)}) - \boldsymbol{g}_{(M)}^T(t)\,\Delta\boldsymbol{a} \tag{52}$$

with

$$\Delta\boldsymbol{a} = \boldsymbol{a} - \boldsymbol{a}^{(k)}. \tag{53}$$

Considering the convexity of $\rho_0(x)$, the problem of minimizing (45) with respect to coefficients is reduced to solving Eq.(46.a). That is

$$\int_0^T \boldsymbol{g}_{(M)}(t)\left(r_*(t,\boldsymbol{a}^{(k)}) - \boldsymbol{g}_{(M)}^T(t)\,\Delta\boldsymbol{a}\right)dt = 0. \tag{54}$$

It is interesting to note that the solution of Eq.(54) is simply the optimal solution of the following "mean-square error" criterion, that is

$$M.S.E.(r_*,\Delta\boldsymbol{a}) = \int_0^T \left(r_*(t,\boldsymbol{a}^{(k)}) - \boldsymbol{g}_{(M)}^T(t)\,\Delta\boldsymbol{a}\right)^2 dt. \tag{55}$$

Therefore, bearing in mind the orthogonality of Walsh functions, the solution $\Delta\boldsymbol{a}$ of Eq.(54), or (55) is

$$\Delta\boldsymbol{a} = \frac{1}{T}\int_0^T r_*(t,\boldsymbol{a}^{(k)})\,\boldsymbol{g}_{(M)}(t)\,dt. \tag{56}$$

Furthermore, the robust estimate of the coefficient vector $\boldsymbol{a}$ turns out to be

$$\boldsymbol{a}^{(k+1)} = \boldsymbol{a}^{(k)} + q\,\Delta\boldsymbol{a} \tag{57}$$

where $0<q<2$ is an arbitrary relaxation factor, and $\Delta\boldsymbol{a}$ is provided by Eq.(56).

So far, it has been assumed that the residual variance is known beforehand so that the influence function can be determined properly. Unfortunately, this is not true in most cases. An estimate of the residual variance, which can be obtained by solving Eq.(46.b), is needed. Suppose that the trial values $\boldsymbol{a}^{(k)}$ and $\sigma^{(k)}$ at step k are available. Then, similar to the method developed in the last section, an iterative formula is derived for generating the variance estimate $\sigma^{(k+1)}$. That is

$$\left(\sigma^{(k+1)}\right)^2 = \frac{1}{\beta}\int_0^T \chi_0\left(\frac{r(t,\boldsymbol{a}^{(k)})}{\sigma^{(k)}}\right) dt \left(\sigma^{(k)}\right)^2. \tag{58}$$

It will be proved, in lemma 5, that the estimate of the residual variance given by Eq.(58) minimizes the robust criterion (45). This is, as we pointed out earlier, the one of main advantages of this method over the previous robust methods.

Combining the above equations together, the whole algorithm can be described by the following procedure.

1. Let $\boldsymbol{a}^{(k)}$ and $\sigma^{(k)}$ be reasonably good trial values of $\boldsymbol{a}$ and σ;
2. Calculate $r_{\bullet}(t,\boldsymbol{a}^{(k)})$ from Eq.(49);
3. Solve Eq.(56) for $\boldsymbol{\Delta a}$;
4. Compute the new estimate of the coefficient vector $\boldsymbol{a}^{(k+1)}$, at step $k+1$, from (57);
5. Update the estimate of the residual variance $\sigma^{(k+1)}$ at step $k+1$ by (58);
6. Stop iterating if the coefficients change by less than δ times the estimate of the residual variance, i.e. if

$$|\boldsymbol{\Delta a}| < \delta\,\sigma^{(k+1)} \tag{59}$$

for all m=0, 1, ..., M-1. Otherwise, let $\boldsymbol{a}^{(k)}=\boldsymbol{a}^{(k+1)}$ and $\sigma^{(k)}=\sigma^{(k+1)}$ and go back to the step 2 until the iteration termination criterion (59) is satisfied.

Like the procedure in section III, before the method can be implemented, two points should be specified. The first point is how to select the cost function $\rho_0(x)$ described in Eq.(45). The general definition of the robust cost function will follow the one discussed in Eq.(23) in the last section. In particular, if we substitute the residual of coefficient estimates in Eq.(48) to Eq.(23.b), the following influence function will be obtained

$$\psi_0\left(r(t,\boldsymbol{a})\right) = \begin{cases} r(t,\boldsymbol{a}) & |r(t,\boldsymbol{a})| \le c\,\sigma \\ c\,\sigma\, sign\left(r(t,\boldsymbol{a})\right) & |r(t,\boldsymbol{a})| > c\,\sigma. \end{cases} \tag{60}$$

Clearly, the turning point of influence functions depends upon the residual variance if the constant c is appropriately selected.

As to another point, we should select proper initial values $\boldsymbol{a}^{(0)}$ and $\sigma^{(0)}$ before starting the iteration. Similar to the initial values in the last section, we will choose the initial value of the coefficient vector $\hat{\boldsymbol{a}}$ as $\hat{\boldsymbol{a}}_{MSE}$ given in Eq.(40), i.e.

$$\boldsymbol{a}^{(0)} = \hat{\boldsymbol{a}}_{MSE}. \tag{61.a}$$

The estimate of the residual variance is the median absolute deviation of the residual

$$\sigma^{(0)} = med\left(\left|r_{\bullet}(t,\boldsymbol{a}^{(0)})\right|\right). \tag{61.b}$$

Huber has proved that the initial value of the residual variance in (61.b) is more robust

[1].

IV.C. Convergence Analysis

To investigate the attribute of convergence of the robust iterative least squares method with modified residuals, several lemmas, followed by a convergence theorem are proved. In fact, they are similar to those in the preceding section. Following the notations in the last section, we have the following lemmas and theorem.

Lemma 5

Based on the assumptions in lemma 1, if

$$\left(\sigma^{(k+1)}\right)^2 = \frac{1}{\beta}\int_0^T \chi_0\left(\frac{r(t,\boldsymbol{a}^{(k)})}{\sigma^{(k)}}\right) dt \left(\sigma^{(k)}\right)^2$$

then, the following inequality is true

$$J\left(\boldsymbol{a}^{(k)},\sigma^{(k)}\right) - J\left(\boldsymbol{a}^{(k)},\sigma^{(k+1)}\right) \geq \beta\,\frac{\left(\sigma^{(k+1)} - \sigma^{(k)}\right)^2}{\sigma^{(k)}}.$$

In particular, J is strictly decreased unless Eq.(46.b) is already satisfied. ∎

Proof: See Appendix VII.B.

Lemma 6

Suppose that the assumptions in lemma 1 are true. If

$$\boldsymbol{a}^{(k+1)} = \boldsymbol{a}^{(k)} + q\,\boldsymbol{\Delta\hat{a}}$$

where $\boldsymbol{\Delta\hat{a}}$ is determined by Eq.(56), then

$$\begin{aligned} J\left(\boldsymbol{a}^{(k)},\sigma^{(k)}\right) - J\left(\boldsymbol{a}^{(k+1)},\sigma^{(k)}\right) &\geq \frac{q(2-q)T}{2\sigma^{(k)}}\|\boldsymbol{\Delta\hat{a}}\|^2 \\ &= \frac{(2-q)T}{2\sigma^{(k)}q}\|\boldsymbol{a}^{(k+1)} - \boldsymbol{a}^{(k)}\|^2 \end{aligned}$$

with the norm defined as

$$\|\boldsymbol{\Delta\hat{a}}\|^2 = \left(\boldsymbol{\Delta\hat{a}}\right)^{\mathrm{T}}\left(\boldsymbol{\Delta\hat{a}}\right).$$

In particular, unless Eq.(46.a) is already satisfied, J is strictly decreased. ∎

Proof: See Appendix VII.B.

Lemma 7

Based on the assumptions in lemma 1, the sets of the form

$$\mathfrak{C}_b = \left\{(\boldsymbol{a},\sigma) \mid \sigma \geq 0,\ J(\boldsymbol{a},\sigma) \leq b < \infty\right\}$$

are compact. ∎

Proof: See Appendix VII.B.

Theorem 2

Based on the assumptions of lemmas 1 and 5 to 7, assume that the time function $g(t)$ is absolutely integrable, then, the following statements are true:

(1) The sequence $(\boldsymbol{a}^{(k)},\sigma^{(k)})$ in $\mathfrak{C}_b$ has at least one accumulation point $(\hat{\boldsymbol{a}},\hat{\sigma})$;

(2) Every accumulation point $(\hat{\boldsymbol{a}},\hat{\sigma})$ with $\hat{\sigma}>0$ is a solution of Eq.(46) and minimizes Eq.(45). ∎

Proof: See Appendix VII.B.

IV.D. Results of Simulation

To illustrate the developed robust method, expansions of time functions with Walsh functions are considered.

Example 1

Consider a linear time invariant system of order 2

$$\begin{bmatrix} \dot{x}_1(t) \\ \dot{x}_2(t) \end{bmatrix} = \begin{bmatrix} 0 & 1 \\ -1.4 & -0.7 \end{bmatrix} \begin{bmatrix} x_1(t) \\ x_2(t) \end{bmatrix} + \begin{bmatrix} 1.0 \\ 2.0 \end{bmatrix} u(t). \tag{62}$$

Two sets of output data of Eq.(62) are generated using the input signal $u(t)=e^{-0.5t}$, the sampling interval is 0.01, and the final time T is 2.56. The first set data is noise-free data which is exactly the solution of Eq.(62). Another set is generated by

$$\begin{bmatrix} f_1(t) \\ f_2(t) \end{bmatrix} = \begin{bmatrix} x_1(t) \\ x_2(t) \end{bmatrix} + \begin{bmatrix} e_1(t) \\ e_2(t) \end{bmatrix} \tag{63}$$

where $\{e_i(t)\ i=1,2\}$ are additive noise, $\{x_i(t)\ i=1,2\}$ and $\{f_i(t)\ i=1,2\}$ are noise-free and contaminated output sequences, respectively. The noise sequences belong to the $\mathcal{P}_\epsilon$ in Eq.(10) with $\sigma_1^2=0.05$, $\sigma_2^2=5.0$, and $\epsilon=0.1$.

Table III
Coefficient estimates of Walsh functions for state 1 in Ex.1

Coefficients	Coefficient estimates		
	non-robust	robust	noise-free
a_0	1.133906	1.142383	1.148705
a_1	-0.327333	-0.376386	-0.362364
a_2	-0.180764	-0.165641	-0.164561
a_3	-0.290793	-0.226435	-0.236028
a_4	-0.002437	-0.074453	-0.070256
a_5	-0.082327	-0.102991	-0.107990
a_6	-0.024185	-0.039902	-0.048408
a_7	-0.071663	0.017127	0.023191
a_8	-0.188251	-0.045157	-0.030726
a_9	-0.015600	-0.053502	-0.054132
a_{10}	-0.041589	-0.026631	-0.027828
a_{11}	-0.187109	0.022240	0.012570
a_{12}	-0.028917	-0.006446	-0.010991
a_{13}	-0.025102	0.003447	0.013633
a_{14}	0.060025	0.003126	0.005030
a_{15}	0.118455	0.017118	-0.001808

Table IV
Coefficient estimates of Walsh functions for state 2 in Ex.1

Coefficients	Coefficient estimates		
	non-robust	robust	noise-free
a_0	-0.088449	-0.082070	-0.072809
a_1	0.419667	0.472632	0.475237
a_2	0.282306	0.161377	0.150780
a_3	-0.070726	-0.160315	-0.162125
a_4	0.063385	0.073039	0.068024
a_5	-0.130456	-0.085762	-0.078196
a_6	-0.051751	-0.047558	-0.046219
a_7	-0.132186	-0.085536	-0.085848
a_8	0.131840	0.032456	0.032005
a_9	-0.149989	-0.048040	-0.036124
a_{10}	-0.090027	-0.028842	-0.022174
a_{11}	0.027922	-0.033945	-0.045889
a_{12}	-0.006091	-0.011724	-0.005530
a_{13}	-0.184944	-0.024431	-0.020936
a_{14}	0.008746	-0.004635	-0.014869
a_{15}	0.103084	0.003156	-0.000403

To implement the simulation, the quantity δ is taken as 1.0×10^{-6}, the constant c is chosen as 1.5, and the relaxation factor q is selected as 1. The initial values are chosen by Eq.(61). In addition, to demonstrate the advantage of the proposed method, the coefficient error norm is introduced, that is

$$\textit{Coefficient Error Norm} = \frac{\|\hat{\boldsymbol{a}} - \boldsymbol{a}_0\|}{\|\boldsymbol{a}_0\|} \tag{64}$$

where the Euclidean norm is used. In the above equation, $\boldsymbol{a}_0$ implies the estimate of the coefficient vector of orthogonal function series in the noise-free case and $\hat{\boldsymbol{a}}$ indicates the robust, or non-robust (from Eq.(40)) coefficient estimates. Obviously, the norm is similar to that defined in Eq.(28) except for different vectors.

Robust coefficient estimates of 16 Walsh functions for $x_1(t)$ are obtained after 23 iterations while 20 iterations are used for $x_2(t)$. The estimates of coefficients of Walsh functions are tabulated in Tables III and IV.

The approximated (noise-free, robust, non-robust) solutions and the exact solutions of the outputs are plotted in figures after obtaining the coefficient estimates. It is easily seen, from Figs. 3 and 4, that the non-robust (*M.S.E.*) approximation severely deviates from the exact solution when the series contains outliers. On the contrary, the robust (RILSMMR) approximation approaches the exact solution with better accuracy. It can also be observed, from Figs. 5 and 6, there is almost no difference between the robust and noise-free approximations although the time series contains not only normal noise but also outliers. In addition, from Fig. 7, it is a straightforward conclusion that the coefficient error norm decreases rapidly as the number of iterations increases.

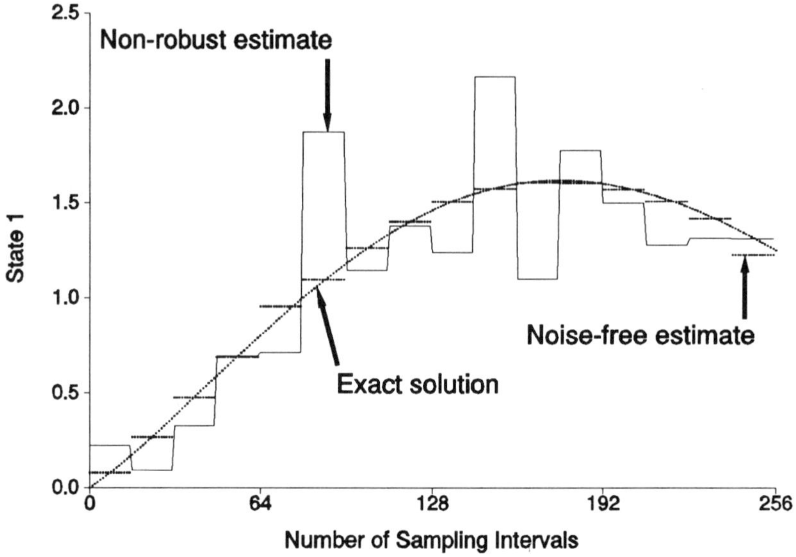

Figure 3. The comparison of exact solution with noise-free and non-robust approximations using Walsh functions

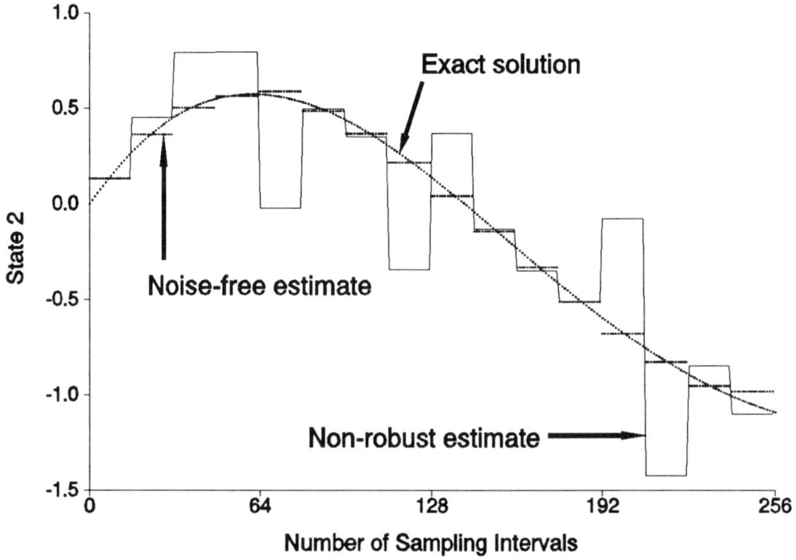

Figure 4. The comparison of exact solution with noise-free and non-robust approximations using Walsh functions

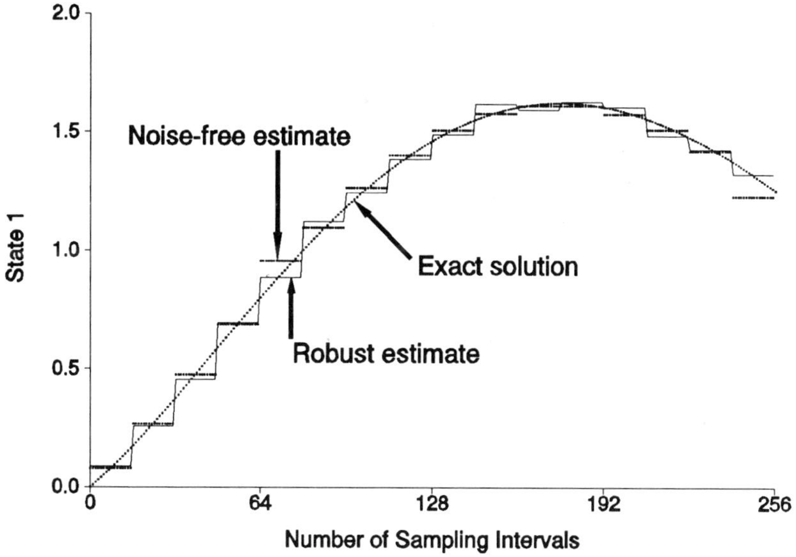

Figure 5. The comparison of exact solution with noise-free and robust approximations using Walsh functions

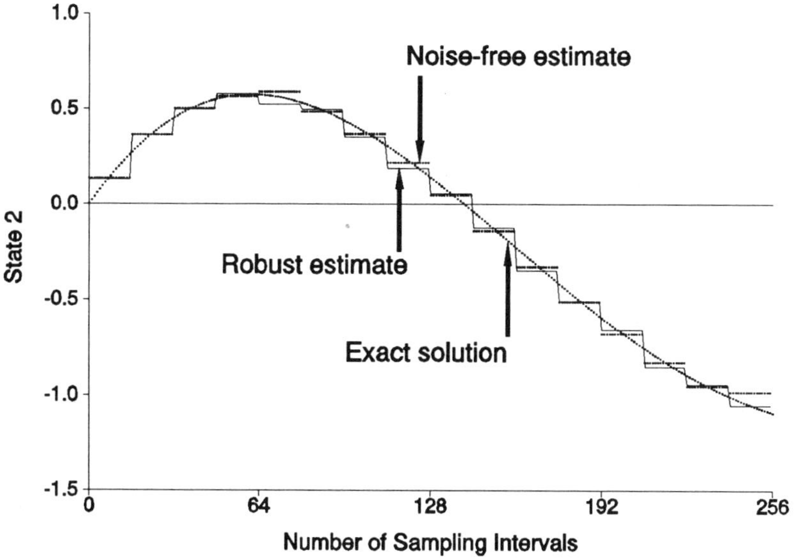

Figure 6. The comparison of exact solution with noise-free and robust approximations using Walsh functions

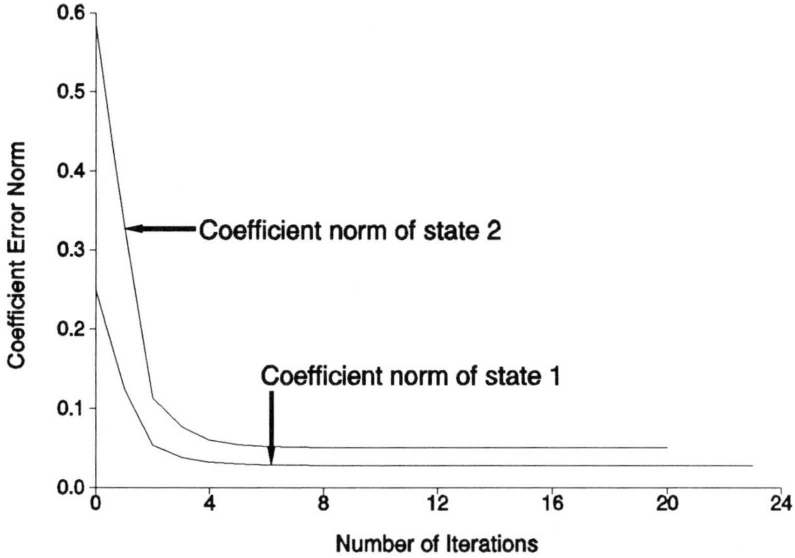

Figure 7. Illustrations of convergence and robustness of robust coefficient estimates of Walsh functions using the RILSMMR

Example 2

Consider a scalar bilinear system as follows

$$\begin{aligned} \dot{x}(t) &= -2x(t) + 2x(t)u(t) + 3u(t) \\ y(t) &= x(t) + e(t). \end{aligned} \tag{65}$$

The variables $x(t)$ and $y(t)$ are noise-free output and noisy output sequences, respectively. Again, the additive noise introduced here belongs to the so-called "ϵ-contaminated family". The variances σ_1^2 and σ_2^2 of normal noise and outliers are taken as 0.05 and 5.0, respectively. The probability of outliers occurring is 0.1.

Like example 1, two sets of data of Eq.(65) are generated using the input signal $u(t)=e^{-0.5t}$, the sampling interval is 0.01, and the final time T is 2.56. The first set is noise-free data which is exactly the solution of Eq.(65). For implementation, the initial values and the constants δ, c, and q are selected as the same as those in example 1.

After 25 iterations, the robust estimates of 16 Walsh functions are obtained by using the suggested expansion approach. They, together with non-robust and noise-free estimates, are tabulated in Table V.

The approximated (noise-free, robust, and non-robust) solutions and the exact solutions are plotted in Figs. 8 and 9. Again, from the figures, it is self-evident to see the advantage of the robust method over the non-robust one. There is almost no difference between the robust and noise-free approximations. Also, the robustness and convergence of the developed method are confirmed by observing Fig. 10.

Clearly, these results in the two examples confirm that convergence and robustness of the coefficient estimate are satisfied when the proposed method is used.

Table V
Coefficient estimates of Walsh functions for the state in Ex.2

Coefficients	Coefficient estimates		
	non-robust	robust	noise-free
a_0	1.353832	1.360691	1.360797
a_1	-0.061909	-0.110985	-0.096947
a_2	-0.140861	-0.122048	-0.124658
a_3	-0.394730	-0.329271	-0.340141
a_4	0.018687	-0.058707	-0.054001
a_5	-0.134189	-0.157497	-0.159855
a_6	-0.064418	-0.077629	-0.088642
a_7	-0.158497	-0.064811	-0.063643
a_8	-0.180048	-0.038276	-0.022523
a_9	-0.040219	-0.081098	-0.078751
a_{10}	-0.062492	-0.046549	-0.048779
a_{11}	0.142649	-0.017719	-0.031899
a_{12}	-0.034798	-0.013933	-0.016866
a_{13}	-0.042439	-0.015365	-0.003705
a_{14}	0.049668	-0.002723	-0.005325
a_{15}	0.112790	0.015533	-0.007474

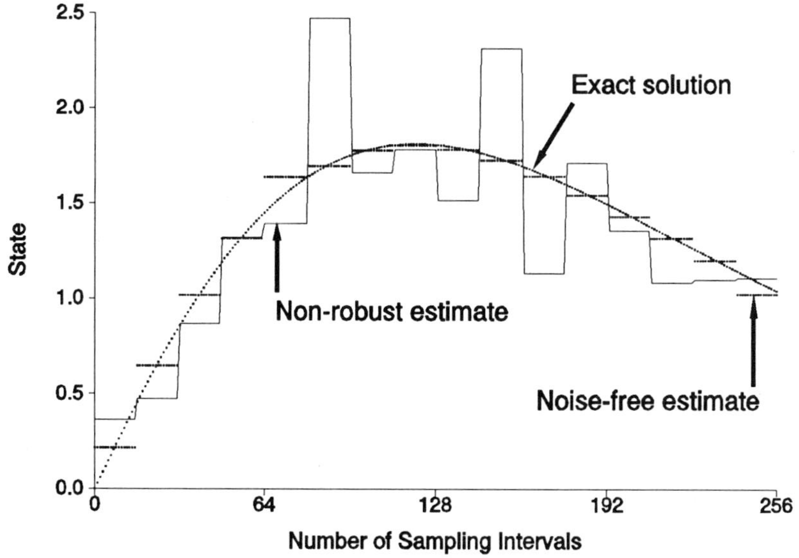

Figure 8. The comparison of exact solution with noise-free and non-robust approximations using Walsh functions

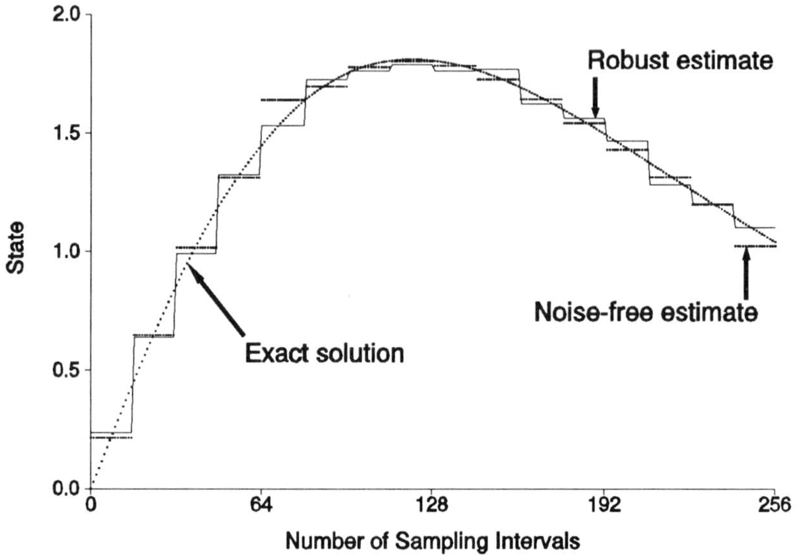

Figure 9. The comparison of exact solution with noise-free and robust approximations using Walsh functions

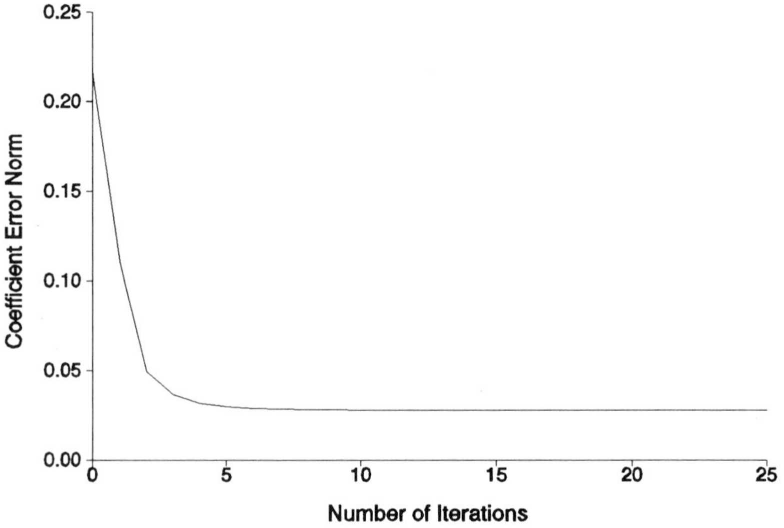

Figure 10. Illustrations of convergence and robustness of robust coefficient estimates of Walsh functions using the RILSMMR

V. ROBUST IDENTIFICATION OF CONTINUOUS-TIME SYSTEMS

Until now, most robust estimation methods are dedicated to the robust identification problem of discrete-time systems. From the point of view of continuous-time system identification, those methods can be classified into robust indirect approaches. Very few attempt has been made to develop a direct approach for robust identification of continuous-time systems. The only attempt, so far, is due to an authors' previous paper [16][2] and will be described in this section.

V.A. Introduction

As a set of orthogonal functions, Walsh functions have been widely applied to many areas, especially in identification and optimal control of systems. Using Walsh functions, a typical paper [17] from Karanam etc., has been provided for bilinear system identification. The most attractive attribute of this approach is that the identification problem can be reduced into an algebraic form by employing the prominent properties of Walsh functions. These properties are, for instance, the group property of Walsh function multiplications, the operational matrix concept of Walsh function integrations and the permutation characteristic of the product matrix of two Walsh function vectors. However, like other methods, the approach is limited when input-output data of systems contain not only small noise or measurement errors but also outliers.

In this section, a robust approach of using Walsh functions for system identification, which has been proposed in [16], will be described. The robust approach discussed in the preceding section is utilized to reduce the effect of noise, especially outliers, on the coefficient estimates of Walsh function expansions of input-output series. Consequently, parameter estimates of systems can be obtained directly by using the corresponding robust estimates of coefficients. It will be shown that the convergence analysis of the developed procedure is based on the convergence proof in the last section. The simulation results given in this chapter will illustrate the robustness and convergence of the developed method.

V.B. Walsh Functions

It is well known that it is convenient to define Walsh functions in terms of Rademacher functions [18]. That is, Walsh functions $\{g_m(t), m=1, 2, \dots\}$ are given as

$$\begin{aligned} g_0(t) &= 1 \\ g_m(t) &= h_{m_1}(t)h_{m_2}(t)\cdots h_{m_r}(t) \end{aligned} \tag{66.a}$$

with $m = 2^{m_1} + 2^{m_2} + \dots + 2^{m_r}$ and Rademacher functions h_i, $i=m_1, m_2, \dots, m_r$, are

$$h_m(t) = \begin{cases} 1 & 0 \le t < \dfrac{T}{2^{(m+1)}} \\ -1 & \dfrac{T}{2^{(m+1)}} \le t < \dfrac{T}{2^m}. \end{cases} \tag{66.b}$$

The non-negative integers m_i are uniquely determined by $m_{i+1} < m_i$.

---

[2] Figures 11 and 12, lemmas 1, 5 to 7, and theorem 2 have been reproduced from reference 16, by kind permission of Control Theory and Advanced Technology.

As a complete set of orthogonal functions, Walsh functions allow a representation of every absolutely integrable function $f(t)$ in the form in Eq.(37). For the sake of completeness, we shall briefly describe several important properties which will be utilized when deriving the identification procedure.

(1). Integration property:

$$\int_0^t g_{(M)}(\tau)\, d\tau \approx TP_{(M)} g_{(M)}(t) \tag{67}$$

with

$$g_{(M)}^T = \left[g_0(t) \;\; g_1(t) \;\; \cdots \;\; g_{M-1}(t) \right] \tag{68}$$

and the matrix $P_{(M)}$ is termed an "operational matrix". A general expression for the operational matrix $P_{(M)}$ of arbitrary order M can be found in [17]. That is

$$P_{(1)} = 0.5;$$
$$P_{(M)} = \begin{bmatrix} P_{(M/2)} & -\dfrac{1}{2M} I_{(M/2)} \\ \dfrac{1}{2M} I_{(M/2)} & 0_{(M/2)} \end{bmatrix} \tag{69}$$

for $M=2^k$; $k=1, 2, \ldots$. The matrices $I_{(M/2)}$ and $0_{(M/2)}$ are the identity and null matrices of order $M/2$.

(2). Group property:

The group property of Walsh functions has been exploited in [17] and has been used for solving the identification problem for classes of nonlinear systems with perfect information, such as variable-structure systems and systems with polynomial-type nonlinearities. Concisely, these main results are introduced below.

Denote the matrix

$$G_{(M)} = g_{(M)}(t) g_{(M)}^T(t) \qquad t \in [0, T). \tag{70}$$

It has been shown that the matrix $G_{(M)}(t)$ is reducible to a simple algebraic form and is symmetric, with $g_i(t)g_j(t)$, $i, j=0, 1, \ldots, M\text{-}1$, as a typical element. Employing the multiplication rule of Walsh functions, the matrix can be represented in the following form

$$G_{(M)}(t) = \left[g_{(M)}(t) \;\; \Lambda_1^{(M)} g_{(M)}(t) \;\; \Lambda_2^{(M)} g_{(M)}(t) \;\; \cdots \;\; \Lambda_{M-1}^{(M)} g_{(M)}(t) \right] \tag{71}$$

with the $\Lambda_i^{(M)}$ matrices are defined as

$$\Lambda_i^{(M)} = \begin{bmatrix} \Lambda_i^{(M/2)} & 0_{(M/2)} \\ 0_{(M/2)} & \Lambda_i^{(M/2)} \end{bmatrix};$$
$$\Lambda_{i+(M/2)}^{(M)} = \begin{bmatrix} 0_{(M/2)} & \Lambda_i^{(M/2)} \\ \Lambda_i^{(M/2)} & 0_{(M/2)} \end{bmatrix} \tag{72}$$

for $i=0, 1, 2, \ldots, (M/2\text{-}1)$ and $\Lambda_0^{(M/2)} = I_{(M/2)}$.

Based on the properties above, a lemma, which is useful for developing the robust approach of system identification, has been derived in [17]. The lemma is:

lemma 8

$$g_{(M)}(t)g_{(M)}^T(t)w = \Lambda_w g_{(M)}(t)$$

where

$$\Lambda_w = \begin{bmatrix} w & \Lambda_1^{(M)}w & \cdots & \Lambda_{M-1}^{(M)}w \end{bmatrix}$$

and w is any vector with an appropriate dimension.

■

Proof: A proof of the lemma can be found in [17].

V.C. System Identification with Walsh Functions

The Walsh function approach to solve the problem of system identification has been proposed for over 15 years. For the sake of convenience, a brief review and description of this approach is stated first. In fact, the robust approach to be described in the next section follows the same procedure except that a different method is used to expand the very noisy data.

Consider a class of bilinear continuous-time systems described as below

$$\begin{cases} \dot{X}(t) = AX(t) + \sum_{k=1}^{q} C_k X(t) u_k(t) + BU(t) \\ Y(t) = rX(t) \end{cases} \tag{73}$$

where $X(t)$, $Y(t) \in \Re^n$, $U(t) \in \Re^q$; A, B, and C_k are parameter matrices of appropriate dimensions, and r is a known nonzero scalar-valued function. Furthermore, suppose that $\bar{a}_i$, $\bar{b}_j$, and $\bar{c}_i(k)$ are the column vectors of matrices A, B, and C_k, respectively. The variable y_i is the i-th component of vector Y and u_j the j-th component of vector U.

Integrating the first equation of Eq.(73) gives $X(t)$, which may be substituted in the second equation of Eq.(73) in order to derive the input-output relationship of the system. That is

$$Y(t) = Y(0) + \int_0^t \sum_{i=1}^{n} \bar{a}_i y_i(\tau)\, d\tau + \int_0^t \sum_{j=1}^{q} \bar{b}_j r u_j(\tau)\, d\tau + \int_0^t \sum_{k=1}^{q} \sum_{i=1}^{n} \bar{c}_i(k) y_i(\tau) u_k(\tau)\, d\tau. \tag{74}$$

The problem of identifying parameter vectors $\bar{a}_i$, $\bar{b}_j$, and $\bar{c}_i(k)$ for $i=1, ..., n$, j and $k=1, ..., q$ in the above equation can now be performed in two relatively independent steps. The first step is to approximate the known input $u(t)$ and output $y(t)$ functions by their expansions of $M=2^k$ (for some integer k) Walsh functions. That is

$$y_i(t) \approx p_i^T g_{(M)}(t) \qquad i = 1 \cdots n \tag{75.a}$$

$$u_j(t) \approx z_j^T g_{(M)}(t) \qquad j = 1 \cdots q \tag{75.b}$$

where the column vectors p_i and z_j correspond to the coefficient estimates of the i-th output series and j-th input series, respectively. Substituting (75) into (74) leads to

$$\begin{aligned} Lg_{(M)}(t) \approx{} & Y(0) + \sum_{i=1}^{n} \bar{a}_i p_i^T \int_0^t g_{(M)}(\tau)\, d\tau + r\sum_{j=1}^{q} \bar{b}_j z_j^T \int_0^t g_{(M)}(\tau)\, d\tau \\ & + \sum_{k=1}^{q} \sum_{i=1}^{n} \bar{c}_i(k) p_i^T \int_0^t g_{(M)}(\tau) g_{(M)}^T(\tau) z_k\, d\tau. \end{aligned} \tag{76}$$

Here L represents the matrix formed by p_i^T $i=1, ..., n$ as row vectors, i.e.

$$L^T = [p_1 \;\cdots\; p_n]. \tag{77}$$

Using lemma 8 and the properties given in section V.B., Eq.(76) can be rewritten as

$$\begin{aligned} L g_{(M)}(t) \approx\ & Y(0) + T\sum_{i=1}^{n} \bar{a}_i p_i^T P_{(M)} g_{(M)}(t) + rT\sum_{j=1}^{q} \bar{b}_j z_j^T P_{(M)} g_{(M)}(t) \\ & + T\sum_{k=1}^{q}\sum_{i=1}^{n} \bar{c}_i(k) p_i^T \Lambda_{z_k} P_{(M)} g_{(M)}(t). \end{aligned} \tag{78}$$

It is now a relatively simple computational problem to identify $\bar{a}_i$, $\bar{b}_j$, and $\bar{c}_i(k)$ from the known r and the robust estimates p_i, $i=1, \ldots, n$, and z_j, $j=1, \ldots, q$.

V.D. Robust Identification with Walsh Functions

As mentioned earlier, the traditional identification method using Walsh functions is very sensitive to outliers, and even to normal noise contained in the input-output data. To develop a direct robust method for system identification using Walsh functions, the conventional non-robust approach should be modified.

The basic idea of deriving the robust approach for system identification using Walsh functions is to introduce a general robust criterion so that robust estimates of expansion coefficients of a Walsh series approximating very noisy data can be obtained. Thus, noise or outliers disturbing the input-output series will have little effect on the approximation accuracy of Walsh function expansions of the series. Consequently, robust estimates of parameters can be easily obtained by using the robust expansions.

In the following part of this subsection, we shall present the robust approach in four steps. In the first two steps, we shall describe the direct robust approach for identification of continuous-time systems. And then, a convergence analysis and simulation results with the proposed approach will be provided in order to show the advantages of the method.

V.D.1. Robust Estimation of Expansion Coefficients

The approach to be used for solving the problem of robust estimation of expansion coefficients has been discussed in detail in section IV. We can simply follow the procedure suggested in that section to estimate all expansion coefficients of Walsh functions for given noisy data.

Without loss of generality, the coefficient vector $\boldsymbol{a}$ denotes a vector of expansion coefficients for either input or output series if input or output series contains noise, or outliers. Otherwise, only the traditional expansion approach is needed and estimates of the coefficients will be given by Eq.(40).

As to the additional considerations, selections of the cost function $\rho_0(.)$ and the initial values $\boldsymbol{a}^{(0)}$ and $\sigma^{(0)}$ may follow the description in Eqs.(23) and (61), respectively.

V.D.2. Robust Identification of Systems

Once robust estimates of the expansion coefficients for input-output series are obtained using the suggested approach, robust identification of systems turns out to be a relatively simple computational problem.

Suppose that the robust estimates of coefficients for j-th input or i-th output time series are z_j, $j=1, ..., q$ and p_i, $i=1, ..., n$, respectively. After the vectors z_j and p_i in Eq.(76) are replaced by the robust estimates, linearly independent equations corresponding to the number of unknown parameters can be generated. Then, these equations can easily be solved for the parameters.

In general, there are two ways to calculate the unknown parameters. One is to simply calculate parameters from M equations which can be obtained by equating the components of M Walsh functions $\{g_m(t), m=0, 1, ..., M\text{-}1\}$ on the both sides of Eq.(76).

Specially, Eq.(76) can be rewritten as

$$
\begin{aligned}
\left\{\begin{bmatrix} p_1^{\mathrm{T}} \\ p_2^{\mathrm{T}} \\ \vdots \\ p_n^{\mathrm{T}} \end{bmatrix} - \begin{bmatrix} p_1^{\mathrm{T}}(0) \\ p_2^{\mathrm{T}}(0) \\ \vdots \\ p_n^{\mathrm{T}}(0) \end{bmatrix}\right\} \boldsymbol{g}_{(\mathrm{M})}(t) = {} & \left\{\begin{bmatrix} a_{11} \\ a_{21} \\ \vdots \\ a_{n1} \end{bmatrix} p_1^{\mathrm{T}} + \cdots + \begin{bmatrix} a_{1n} \\ a_{2n} \\ \vdots \\ a_{nn} \end{bmatrix} p_n^{\mathrm{T}}\right\} \boldsymbol{P}_{(\mathrm{M})}\boldsymbol{g}_{(\mathrm{M})}(t)\,T \\
& + r\left\{\begin{bmatrix} b_{11} \\ b_{21} \\ \vdots \\ b_{n1} \end{bmatrix} z_1^{\mathrm{T}} + \cdots + \begin{bmatrix} b_{1q} \\ b_{2q} \\ \vdots \\ b_{nq} \end{bmatrix} z_q^{\mathrm{T}}\right\} \boldsymbol{P}_{(\mathrm{M})}\boldsymbol{g}_{(\mathrm{M})}(t)\,T \\
& + \left\{\begin{bmatrix} c_{11}(1) \\ c_{21}(1) \\ \vdots \\ c_{n1}(1) \end{bmatrix} p_1^{\mathrm{T}} + \cdots + \begin{bmatrix} c_{1n}(1) \\ c_{2n}(1) \\ \vdots \\ c_{nn}(1) \end{bmatrix} p_n^{\mathrm{T}}\right\} \Lambda_{z_1} \boldsymbol{P}_{(\mathrm{M})}\boldsymbol{g}_{(\mathrm{M})}(t)\,T \\
& + \cdots + \\
& + \left\{\begin{bmatrix} c_{11}(q) \\ c_{21}(q) \\ \vdots \\ c_{n1}(q) \end{bmatrix} p_1^{\mathrm{T}} + \cdots + \begin{bmatrix} c_{1n}(q) \\ c_{2n}(q) \\ \vdots \\ c_{nn}(q) \end{bmatrix} p_n^{\mathrm{T}}\right\} \Lambda_{z_q} \boldsymbol{P}_{(\mathrm{M})}\boldsymbol{g}_{(\mathrm{M})}(t)\,T.
\end{aligned}
\tag{79}
$$

For the k-th equation of Eq.(79), equating the coefficients of $\boldsymbol{g}_{(\mathrm{M})}(t)$ gives

$$p_k - p_k(0) = \Phi\, \theta_k T \tag{80}$$

where the matrix Φ and vector θ are

$$\Phi = \begin{bmatrix} \varphi_{\mathrm{p}} & \varphi_{\mathrm{z}} & \varphi_{z_1\mathrm{p}} & \cdots & \varphi_{z_q\mathrm{p}} \end{bmatrix} \tag{81.a}$$

$$\theta_k^{\mathrm{T}} = \begin{bmatrix} \boldsymbol{a}_k & \boldsymbol{b}_k & \boldsymbol{c}_k(1) & \cdots & \boldsymbol{c}_k(q) \end{bmatrix} \tag{81.b}$$

with

$$\left.\begin{aligned}\varphi_{\mathrm{p}} &= \left[\boldsymbol{P}^{\mathrm{T}}_{(\mathrm{M})}p_1 \;\cdots\; \boldsymbol{P}^{\mathrm{T}}_{(\mathrm{M})}p_{\mathrm{n}}\right]\\ \varphi_{\mathrm{z}} &= \left[r\boldsymbol{P}^{\mathrm{T}}_{(\mathrm{M})}z_1 \;\cdots\; r\boldsymbol{P}^{\mathrm{T}}_{(\mathrm{M})}z_{\mathrm{q}}\right]\\ \varphi_{z_1\mathrm{p}} &= \left[\boldsymbol{P}^{\mathrm{T}}_{(\mathrm{M})}\Lambda^{\mathrm{T}}_{z_1}p_1 \;\cdots\; \boldsymbol{P}^{\mathrm{T}}_{(\mathrm{M})}\Lambda^{\mathrm{T}}_{z_1}p_{\mathrm{n}}\right]\\ &\cdots\\ \varphi_{z_q\mathrm{p}} &= \left[\boldsymbol{P}^{\mathrm{T}}_{(\mathrm{M})}\Lambda^{\mathrm{T}}_{z_q}p_1 \;\cdots\; \boldsymbol{P}^{\mathrm{T}}_{(\mathrm{M})}\Lambda^{\mathrm{T}}_{z_q}p_{\mathrm{n}}\right]\end{aligned}\right\} \tag{82.a}$$

$$\left.\begin{aligned}\boldsymbol{a}_{\mathrm{k}} &= \left[a_{\mathrm{k1}} \;\cdots\; a_{\mathrm{kn}}\right]\\ \boldsymbol{b}_{\mathrm{k}} &= \left[b_{\mathrm{k1}} \;\cdots\; b_{\mathrm{kn}}\right]\\ \boldsymbol{c}_{\mathrm{k}}(1) &= \left[c_{\mathrm{k1}}(1) \;\cdots\; c_{\mathrm{kn}}(1)\right]\\ &\cdots\\ \boldsymbol{c}_{\mathrm{k}}(q) &= \left[c_{\mathrm{k1}}(q) \;\cdots\; c_{\mathrm{kn}}(q)\right]\end{aligned}\right\} \tag{82.b}$$

for $k=1, ..., n$.
Clearly, one straightforward solution of Eq.(80) is the least squares estimate, i.e.

$$\hat{\theta}_{\mathrm{k}} = \frac{1}{T}(\Phi^{\mathrm{T}}\Phi)^{-1}\Phi^{\mathrm{T}}(p_{\mathrm{k}} - p_{\mathrm{k}}(0)) \qquad k = 1 \;\cdots\; n. \tag{83}$$

These should be robust estimates of the parameters if the Walsh function expansions of input-output series are robust.

It is also possible to estimate the parameter vector from the linearly independent equations generated by Eq.(76) using properly selected and independent sampling points. In this way, this robust approach can be extended to time varying systems.

V.D.3. Convergence Analysis

Convergence of the proposed procedure is studied in this part. From the procedure described in the previous section, it is clear that convergence of identified parameters depends only upon the convergence of the estimated coefficients. Therefore, convergence of the proposed method can be proved by showing local convergence of the robust estimates of coefficients.

Local convergence of the robust coefficient estimates has been proved and confirmed in section IV.C. It, therefore, can directly be concluded that the suggested method for robust system identification has satisfactory convergence.

V.D.4. Results of Simulations

Two examples are provided to illustrate the advantage of the developed method over the traditional one.

Example 3.

Consider the following linear time-invariant system of order 2

$$\begin{bmatrix}\dot{x}_1(t)\\ \dot{x}_2(t)\end{bmatrix} = \begin{bmatrix}0 & 1\\ -1.4 & -0.7\end{bmatrix}\begin{bmatrix}x_1(t)\\ x_2(t)\end{bmatrix} + \begin{bmatrix}1.0\\ 2.0\end{bmatrix}u(t) \tag{84.a}$$

$$\begin{bmatrix} f_1(t) \\ f_2(t) \end{bmatrix} = \begin{bmatrix} x_1(t) \\ x_2(t) \end{bmatrix} + \begin{bmatrix} e_1(t) \\ e_2(t) \end{bmatrix}. \tag{84.b}$$

The variables $\{x_i(t), i=1,2\}$ and $\{y_i(t), i=1,2\}$ are the noise-free output and contaminated output sequences respectively. And $\{e_i(t), i=1,2\}$ are the additive noise and belong to $\wp_\epsilon$ in Eq.(10) with $\sigma_1^2=0.05$, $\sigma_2^2=5.0$, and $\epsilon=0.1$.

The matrix Φ and the k-th parameter vector θ_k are defined as

$$\begin{aligned} \Phi &= \begin{bmatrix} P_{(M)}^T p_1 & P_{(M)}^T p_2 & r P_{(M)}^T z \end{bmatrix} \\ \theta_k^T &= \begin{bmatrix} a_{k1} & a_{k2} & b_k \end{bmatrix} \end{aligned} \tag{85}$$

for $k=1$ and 2. Clearly, they are particular cases of Eqs.(81) and (82).

Two sets of data, $\{x_i(t), i=1,2\}$ and $\{y_i(t), i=1,2\}$ $t\in[0, 2.56)$, are generated using the input signal $u(t)=e^{-0.5t}$, the sampling interval is 0.01.

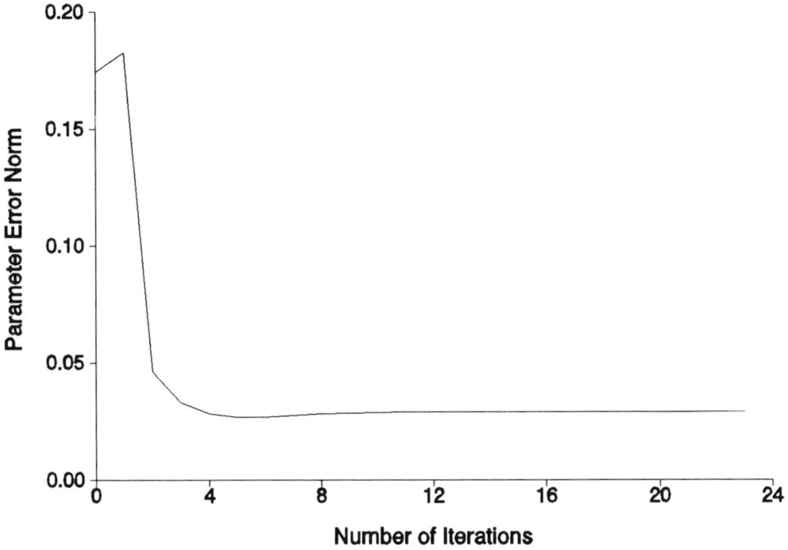

Figure 11. Illustrations of convergence and robustness of parameter estimates of linear systems by the robust Walsh function approach

For noise-free data, only the traditional method (i.e. non-robust method) is used [17]. For the very noisy data $\{y_i(t), i=1,2, t\in[0, 2.56)\}$, both non-robust and robust (i.e. the proposed approach) methods are applied. To implement the robust method, the quantity δ is taken as 1.0×10^{-6}, the constant c is chosen as 1.5, and the relaxation factor q is selected as 1. The initial values are chosen by Eq.(61). To show the robustness of the proposed method, the parameter error norm is taken as that in Eq.(28) except for a different definition of true parameter vector θ_0. Here, θ_0 is the vector of true parameters, and $\hat{\theta}$ is the vector of parameter estimates when the noise-free, or noisy data is used.

Table VI.
Parameter estimates of linear systems obtained by robust and non-robust Walsh function approaches

Parameters	Parameter estimates			True parameters
	non-robust	robust	noise-free	
a_{11}	0.0021	0.0314	0.0011	0.0
a_{12}	1.1941	1.0444	1.0698	1.0
a_{21}	-1.2525	-1.3981	-1.4266	-1.4
a_{22}	-0.4004	-0.7343	-0.7868	-0.7
b_1	0.9391	0.9520	0.9712	1.0
b_2	1.6785	1.9725	2.0618	2.0

After iterations (23 for state 1, 20 for state 2), robust coefficient estimates of 16 Walsh functions are obtained. Using the sequence of coefficient estimates generated during the iterations, the corresponding sequence of parameter estimates can be produced as well as the sequence of parameter error norms.

From the results tabulated in Table VI, it is evident that the robust estimates of parameters are much better than the non-robust ones and approach those in noise-free case very well. In addition, from Fig. 11, the robustness and convergence of the proposed method are confirmed. Clearly, the presented method is much superior to the original one.

Example 4.

In this example, a class of scalar bilinear systems is considered. The system is

$$\begin{aligned} \dot{x}(t) &= -2x(t) + 2x(t)u(t) + 3u(t) \\ y(t) &= x(t) + e(t). \end{aligned} \tag{86}$$

Here, the variables $x(t)$ and $y(t)$ are the noise-free output and noisy output sequences, respectively. The additive noise introduced here belongs to the so-called "ϵ-contaminated family". The variances σ_1^2 and σ_2^2 of normal noise and outliers are taken as 0.05 and 5.0, respectively. The probability of occurrence of outliers is 0.1.

Again, the matrix Φ and vector θ are given as

$$\begin{aligned} \Phi &= \left[\boldsymbol{P}^{\mathrm{T}}_{(\mathrm{M})}p \quad \boldsymbol{P}^{\mathrm{T}}_{(\mathrm{M})}z \quad \boldsymbol{P}^{\mathrm{T}}_{(\mathrm{M})}\Lambda^{\mathrm{T}}_{z}p \right] \\ \theta^{\mathrm{T}}_{\mathrm{k}} &= \left[a \quad b \quad c \right]. \end{aligned} \tag{87}$$

For the sake of comparison, two sets of data, $x(t)$ and $y(t)$, $t \in [0, 2.56)$ are generated using the input signal $u(t) = e^{-0.5t}$ and the sampling interval 0.01.

As with example 3, only the traditional method is applied [17] for expanding the noise-free data, $x(t)$, with 16 Walsh functions. For the contaminated data containing not only normal noise but also outliers, both robust and non-robust methods are used to estimate the coefficients of 16 Walsh functions. To implement the robust method, the quantity δ is taken as 1.0×10^{-6}, the constant c is chosen as 1.5, and the relaxation factor q is selected as 1. The initial values are chosen using Eq.(61). The parameter error norm used here for showing the robustness of the proposed method is selected in Eq.(28) with θ_0 defined as the true parameter vector.

Table VII.
Parameter estimates of bilinear systems obtained by robust and non-robust Walsh function approaches

Parameters	Parameter estimates			True parameters
	non-robust	robust	noise-free	
a	-2.5075	-1.9469	-2.0346	-2.0
c	3.0952	1.9233	2.0280	2.0
b	2.6589	2.9594	3.0091	3.4

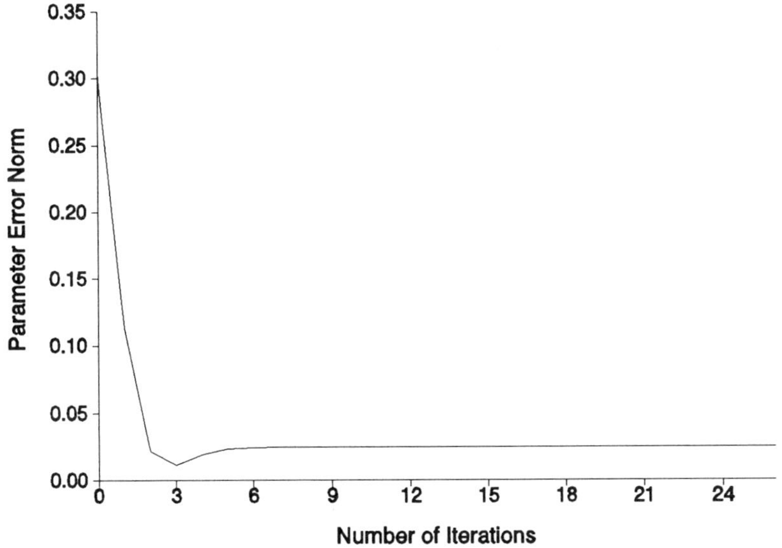

Figure 12. Illustrations of convergence and robustness of parameter estimates of bilinear systems by the robust Walsh function approach

Using the suggested approach in this section, the robust estimates of coefficients are obtained after 25 iterations. Consequently, the sequence of parameter estimates are generated as well as the sequence of parameter error norms by using the corresponding sequence of coefficient estimates obtained during the iteration.

The results of simulation are listed in Table VII and the parameter error norm is plotted in Fig. 12. It is straightforward to confirm the robustness and convergence of the well developed method. Once again, it can be observed that the presented method is much superior to original approach.

Note that the proposed method remains valid for non-scalar bilinear systems because the same procedure is applied.

VI. CONCLUSIONS

The main contribution of this chapter is development of a robust off-line method for parameter estimation. The method is called Robust Iterative Least Squares Method with Modified Residuals. Applications of the method to different estimation problems have been thoroughly discussed.

First of all, a detailed discussion about robust criterions and their effect on robust identification of systems is included. Subsequently, a class of general criterions possessing desirable features is suggested and used for solving the robust identification problem. To minimize the suggested cost functions, the ordinary iterative Gauss-Newton approach is applied except that the residual is replaced by their metrically Winsorized versions.

The key point of the proposed method is replacing the commonly used least squares kind criterion by a general class of robust criterions. It has been shown that the developed method can provide robust estimates of parameters and the estimate of the residual variance simultaneously. Therefore, the proposed method is proved superior to not only the non-robust methods but also the earlier robust ones. Its advantage over the other is that the estimate of residual variance, hence the turning point of influence function, can be obtained directly. Consequently, minimization of the robust criterion is guaranteed and the robustness of parameter estimates is achieved.

The first application of this method is robust identification of both linear and bilinear discrete-time systems. In section III, convergence analysis is included, and theorem 1 on local convergence is proved. The simulation results included in section III show the advantages of the proposed method over non-robust methods. It can be concluded that this method is more useful and practical in dealing with outliers compared to the previous robust and non-robust ones. Therefore, it can be widely applied in real-life situations.

As the second application of the method, the problem of robust estimation of expansion coefficients of any orthogonal series has been successfully solved. The traditional mean-square error criterion, which has been usually minimized in order to approximate a time series with any orthogonal function series, has been replaced by a general robust criterion. To illustrate the derived method, two simulation examples have been given. In both examples, time functions are approximated with Walsh function expansions. The contaminated time functions not only contain noise but also a certain amount of outliers. The results of the simulations show that the robust estimates can well approximate the true values (noise-free coefficients). A local convergence analysis has been theoretically proved.

It is well known that the expansion, using orthogonal functions, of given time series is the very basis of most applications of orthogonal functions in different areas. Therefore, using the presented method, it is possible to utilize orthogonal functions for solving many practical problems which were earlier considered intractable. It should be emphasized once more that the method, in principle, can be utilized for coefficient estimation of other orthogonal functions, for example, block-pulse functions, for any given time functions. The only requirement is that the function be absolutely integrable in an interval $[0,T)$, which should always be satisfied in a variety of applications.

In section V, a direct robust approach for continuous-time system identification using Walsh functions has been proposed, based on robust estimates of expansion coefficients of Walsh functions. The convergence of the proposed procedure is proved through local convergence of coefficient estimates. The simulation results of both linear and bilinear systems presented in section V have shown that robustness of parameter estimates and convergence are always guaranteed due to the use of robust coefficient estimates of Walsh functions. Clearly, this direct robust method for continuous-time system identification is superior to the traditional approach when the data contain noise, especially outliers. Therefore, it has a potential for applications in real-life situations.

It should be noted that the principle used in section V can also be extended to other possible applications of Walsh functions. In particular, extensions to identification of other systems, such as time-varying, distributed systems, etc. are straightforward. In addition, as mentioned earlier, similar applications can be easily performed when other orthogonal functions are utilized. A typical example is development of the direct robust identification approach using block-pulse functions for continuous-time systems.

VII. APPENDICES

VII.A. Convergence Proofs of Lemmas 1 to 4 and Theorem 1

VII.A.1 Proof of Lemma 1:

Put

$$\begin{aligned} \ddot{\rho}_{01}(x) &= \ddot{\rho}_0(x) \qquad \textit{for} \quad |x| \le x_0 \\ \ddot{\rho}_{02}(x) &= \ddot{\rho}_0(x) \qquad \textit{for} \quad |x| > x_0. \end{aligned} \tag{A.1}$$

Then, integrating $\ddot{\rho}_{01}(x)$, once and twice, gives

$$\begin{aligned} \dot{\rho}_{01}(x) &= k_1 x + i_{11} \\ \rho_{01}(x) &= \frac{1}{2} k_1 x^2 + i_{11} x + i_{12} \end{aligned} \tag{A.2}$$

where i_{11} and i_{12} are integral constants to be determined.

Since we have $\rho_0(0)=0$ and $\dot{\rho}_0(0)=0$, then,

$$i_{11} = 0 \qquad \textit{and} \qquad i_{12} = 0$$

Eq.(A.2) can be rewritten

$$\begin{aligned} \dot{\rho}_{01}(x) &= k_1 x \\ \rho_{01}(x) &= \frac{1}{2} k_1 x^2. \end{aligned} \tag{A.3}$$

Similarly, we may have

$$\begin{aligned} \dot{\rho}_{02}(x) &= k_2 x + i_{21} \\ \rho_{02}(x) &= \frac{1}{2} k_2 x^2 + i_{21} x + i_{22} \end{aligned} \tag{A.4}$$

where i_{21} and i_{22} are integral constants to be determined.

Considering assumptions (1) and (2) leads to

$$\dot{\rho}_{01}(x_0) = \dot{\rho}_{02}(x_0)$$
$$\rho_{01}(x_0) = \rho_{02}(x_0).$$

Therefore, we have

$$i_{21} = (k_1 - k_2)x_0$$
$$i_{22} = \frac{1}{2}(k_2 - k_1)x_0^2.$$

Substituting the equation above into (A.4) gives

$$\begin{aligned}\dot{\rho}_{02}(x) &= k_2 x + (k_1 - k_2)x_0 \\ \rho_{02}(x) &= \frac{1}{2}k_2 x^2 + (k_1 - k_2)x_0 x + \frac{1}{2}(k_2 - k_1)x_0^2.\end{aligned} \tag{A.5}$$

After substituting (A.3) and (A.5) into (5.d), $\chi_0(x)$ can be given

$$\chi_0(x) = \begin{cases} \frac{1}{2}k_1 x^2 & |x| \le x_0 \\ \frac{1}{2}\left(k_2 x^2 + (k_1 - k_2)x_0^2\right) & |x| > x_0. \end{cases} \tag{A.6}$$

Now, it is straightforward to confirm the conclusion from the following

$$\ddot{\rho}_0(x) - \frac{2}{x^2}\chi_0(x) = \begin{cases} 0 & |x| \le x_0 \\ -(k_1 - k_2)\frac{x_0^2}{x^2} & |x| > x_0. \end{cases} \tag{A.7}$$

■

VII.A.2 Proof of Lemma 2:

To prove the lemma, let us first construct a simple "comparison function" $Q(\sigma)$ that agrees with $J(\theta^{(m)},\sigma)$ at $\sigma=\sigma^{(m)}$, that lies wholly above $J(\theta^{(m)},.)$, and that reaches its minimum at $\sigma^{(m)}$, namely,

$$Q(\sigma) = J\left(\theta^{(m)},\sigma^{(m)}\right) + \beta\left(\sigma - \sigma^{(m)}\right) + \frac{1}{N}\sum_{t=1}^{N}\chi_0\left(\frac{r(t)}{\sigma^{(m)}}\right)\left[\frac{(\sigma^{(m)})^2}{\sigma} - \sigma^{(m)}\right]. \tag{A.8}$$

Obviously, $Q(\sigma^{(m)})=J(\theta^{(m)},\sigma^{(m)})$. The derivatives of $Q(\sigma)$ and $J(\theta^{(m)},\sigma)$ with respect to σ are

$$\frac{dQ(\sigma)}{d\sigma} = -\frac{1}{N}\sum_{t=1}^{N}\chi_0\left(\frac{r(t)}{\sigma^{(m)}}\right)\left(\frac{\sigma^{(m)}}{\sigma}\right)^2 + \beta \tag{A.9.a}$$

$$\frac{\partial J(\theta^{(m)},\sigma)}{\partial\sigma} = -\frac{1}{N}\sum_{t=1}^{N}\chi_0\left(\frac{r(t)}{\sigma}\right) + \beta. \tag{A.9.b}$$

Hence, they agree at $\sigma=\sigma^{(m)}$. Define

$$f(z) = Q\left(\frac{1}{z}\right) - J\left(\theta^{(m)},\frac{1}{z}\right) \qquad z > 0. \tag{A.10}$$

The first derivative of $f(z)$ with respect to z is

$$\frac{df(z)}{dz} = \frac{1}{N}\sum_{t=1}^{N}\left[\frac{\chi_0\left(r(t)z^{(m)}\right)}{\left(z^{(m)}\right)^2} - \frac{\chi_0\left(r(t)z\right)}{z^2}\right] \tag{A.11}$$

with $z^{(m)} = 1/\sigma^{(m)}$. Differentiating (A.11) again gives

$$\frac{d^2f(z)}{dz^2} = -\frac{1}{N}\sum_{t=1}^{N}\frac{\left(r(t)\right)^2}{z}\left[\ddot{\rho}_0\left(r(t)z\right) - 2\frac{\chi_0\left(r(t)z\right)}{\left(r(t)z\right)^2}\right]. \tag{A.12}$$

Using lemma 1 gives

$$\frac{d^2f(z)}{dz^2} \geq 0$$

for all $z>0$. Since

$$f(z^{(m)}) = 0 \qquad and \qquad \frac{df(z)}{dz}\Big|_{z=z^{(m)}} = 0$$

it follows that $f(z)\geq 0$ for all $z>0$, hence

$$Q(\sigma) \geq J(\theta^{(m)},\sigma) \tag{A.13}$$

for all $\sigma>0$.

From (A.9.a), $Q(\sigma)$ reaches its minimum at $\sigma^{(m+1)}$ which has already been given in (21). Therefore, a simple calculation, using Eq.(21), leads to

$$\begin{aligned} Q(\sigma^{(m+1)}) &= J\left(\theta^{(m)},\sigma^{(m)}\right) + \beta\left(\sigma^{(m+1)} - \sigma^{(m)}\right) + \beta\left(\frac{\sigma^{(m+1)}}{\sigma^{(m)}}\right)^2\left[\frac{\left(\sigma^{(m)}\right)^2}{\sigma^{(m+1)}} - \sigma^{(m)}\right] \\ &= J\left(\theta^{(m)},\sigma^{(m)}\right) - \beta\frac{\left(\sigma^{(m+1)} - \sigma^{(m)}\right)^2}{\sigma^{(m)}}. \end{aligned}$$

Substituting the above equation into (A.13) leads to the conclusions in lemma 2.

■

VII.A.3 Proof of Lemma 3:

As in the proof of lemma 2, a comparison function that agrees with J at $\theta^{(m)}$, that lies wholly above J, and that reaches its minimum at $\theta^{(m+1)}$, is considered. Put

$$O(\Delta\theta) = J\left(\theta^{(m)},\sigma^{(m)}\right) + \frac{1}{2\sigma^{(m)}N}\sum_{t=1}^{N}\left\{\left(r_*(t) - \phi^T(t)\Delta\theta\right)^2 - \left(r_*(t)\right)^2\right\}. \tag{A.14}$$

The first derivatives of $O(\Delta\theta)$ and $J(\theta^{(m)}+\Delta\theta,\sigma^{(m)})$ with respect to $\Delta\theta$ are

$$\frac{\partial O(\Delta\theta)}{\partial\Delta\theta_k} = -\frac{1}{\sigma^{(m)}N}\sum_{t=1}^{N}\left(r_*(t) - \phi^T(t)\Delta\theta\right)\phi_k(t) \tag{A.15.a}$$

$$\frac{\partial J\left(\theta^{(m)}+\Delta\theta,\sigma^{(m)}\right)}{\partial\Delta\theta_k} = -\frac{1}{N}\sum_{t=1}^{N}\psi_0\left(\frac{y(t) - \phi^T(t)\theta^{(m+1)}}{\sigma^{(m)}}\right)\phi_k(t) \tag{A.15.b}$$

where $\phi_k(t)$ is the k-th element in the vector $\phi(t)$ and $k=1, ..., P$.

If $\Delta\theta=0$, $O(\Delta\theta)$ and $J(\theta^{(m)}+\Delta\theta,\sigma^{(m)})$ then have the same value and the same first derivative.

The matrix of second order derivatives of the difference between two functions is given

$$\frac{\partial^2\big(O(\Delta\theta)-J(\theta^{(m)}+\Delta\theta,\sigma^{(m)})\big)}{\partial\Delta\theta_j\,\partial\Delta\theta_k} = \frac{1}{\sigma^{(m)}N}\sum_{t=1}^{N}\phi_j(t)\phi_k(t)\left\{1-\psi_0'\Big(\frac{r(t)-\phi^T(t)\Delta\theta}{\sigma^{(m)}}\Big)\right\}. \quad (A.16)$$

Clearly, it is positive semidefinite, hence

$$O(\Delta\theta) \geq J\big(\theta^{(m)}+\Delta\theta,\sigma^{(m)}\big) \quad (A.17)$$

for all $\Delta\theta$.

From (A.15.a), it is easy to derive the estimate $\Delta\hat{\theta}$ such that $O(\Delta\theta)$ reaches its minimum at $\theta^{(m+1)}=\theta^{(m)}+\Delta\hat{\theta}$. The expression of $\Delta\hat{\theta}$ has already been given in (19). Substituting (19) into the second term of (A.14) gives

$$\begin{aligned}
&\sum_{t=1}^{N}\left\{\big(r_*(t)-\phi^T(t)\Delta\theta\big)^2-\big(r_*(t)\big)^2\right\}\\
&= -2\sum_{t=1}^{N}r_*(t)\phi^T(t)\Delta\theta + \sum_{t=1}^{N}\big(\phi^T(t)\Delta\theta\big)^2\\
&= -2\Delta\hat{\theta}^T\Phi^T(N)\boldsymbol{r}_* + \Delta\hat{\theta}^T\big(\Phi^T(N)\Phi(N)\big)\Delta\hat{\theta}\\
&= -\big(\Phi^T(N)\boldsymbol{r}_*\big)^T\big(\Phi^T(N)\Phi(N)\big)^{-1}\big(\Phi^T(N)\boldsymbol{r}_*\big)\\
&= -\Delta\hat{\theta}^T\big(\Phi^T(N)\Phi(N)\big)\Delta\hat{\theta}
\end{aligned}$$

with $\Phi(N)$ and $\boldsymbol{r}_*$ defined as before.

Define $\|x\|^2=x^T[\Phi^T(N)\Phi(N)]x$, then, the minimum value of $O(\Delta\theta)$, after using the above equation, is given

$$O(\Delta\hat{\theta}) = J\big(\theta^{(m)},\sigma^{(m)}\big) - \frac{1}{2\sigma^{(m)}N}\|\Delta\hat{\theta}\|^2. \quad (A.18)$$

As a function of q,

$$O(q\Delta\hat{\theta}) - J\big(\theta^{(m)},\sigma^{(m)}\big)$$

is quadratic, vanishes at $q=0$, has the minimum at $q=1$, and must vanish again at $q=2$ for reasons of symmetry. Hence, the following equation can be obtained

$$O(\Delta\hat{\theta}) - J\big(\theta^{(m)},\sigma^{(m)}\big) = -\frac{q(2-q)}{2\sigma^{(m)}N}\|\Delta\hat{\theta}\|^2.$$

Finally, the proof of lemma 3 is completed by substituting the above equation into (A.17).

■

VII.A.4 Proof of Lemma 4:

Based on the definition of the sets $\mathfrak{C}_b$, they are obviously closed since J is continuous. Also, we have $\sigma\leq b/\beta$, i.e. σ is bounded. From the assumptions in this lemma, $\|\theta\|$ must also be bounded (otherwise at least one of the $\phi^T(t)\theta$ would be unbounded on $\mathfrak{C}_b$, hence $J(\theta^{(m)},\sigma^{(m)})$ would be unbounded). Therefore, the sets $\mathfrak{C}_b$ are compact.

■

VII.A.5 Proof of Theorem 1:

From lemma 4, the compactness of the sets $\mathfrak{C}_b$ directly implies (1).

To prove (2), assume $\hat{\sigma}$. Given a sequence $(\theta^{(m)},\sigma^{(m)})$, let $(\theta^{(m_l)},\sigma^{(m_l)})$ be a subsequence in $(\theta^{(m)},\sigma^{(m)})$, converging towards $(\hat{\theta},\hat{\sigma})$. Then, from lemmas 2 and 3, we have

$$J\left(\theta^{(m_l)},\sigma^{(m_l)}\right) \geq J\left(\theta^{(m_l)},\sigma^{(m_l+1)}\right) \geq J\left(\theta^{(m_l+1)},\sigma^{(m_l+1)}\right).$$

The two outer members of this inequality tend to $J(\hat{\theta},\hat{\sigma})$ since $(\theta^{(m_l)},\sigma^{(m_l)})$ is a converging subsequence. Hence

$$J\left(\theta^{(m_l)},\sigma^{(m_l)}\right) - J\left(\theta^{(m_l)},\sigma^{(m_l+1)}\right) \geq \beta \frac{\left(\sigma^{(m_l+1)} - \sigma^{(m_l)}\right)^2}{\sigma^{(m_l)}}$$

converges to 0. In particular, it follows that

$$\left[\frac{\sigma^{(m_l+1)}}{\sigma^{(m_l)}}\right]^2 = \frac{1}{N\beta}\sum_{t=1}^{N} \chi_0\left(\frac{r(t)}{\sigma^{(m_l)}}\right)$$

converges to 1, that means

$$\lim_{m_l\to\infty}\frac{1}{N}\sum_{t=1}^{N} \chi_0\left(\frac{r(t)}{\sigma^{(m_l)}}\right) = \beta.$$

Thus, Eq.(5.b) is satisfied.

Similarly, we obtain from lemma 3 that

$$J\left(\theta^{(m_l)},\sigma^{(m_l)}\right) - J\left(\theta^{(m_l+1)},\sigma^{(m_l)}\right) \geq \frac{q(2-q)}{2\sigma^{(m_l)}N}\,\|\Delta\hat{\theta}\|^2$$

$$= \frac{q(2-q)}{2\sigma^{(m_l)}N}\left(\Phi^{T}(N)\boldsymbol{r}_*\right)^{T}\left(\Phi^{T}(N)\Phi(N)\right)^{-1}\left(\Phi^{T}(N)\boldsymbol{r}_*\right)$$

tends to 0, in particular, since $[\Phi^T(N)\Phi(N)]^{-1}>0$, then,

$$\Phi^{T}(N)\boldsymbol{r}_* = \sum_{t=1}^{N}\phi(t)r_*(t) \to 0.$$

That is

$$\sigma^{(m_l)}\sum_{t=1}^{N}\psi_0\left(\frac{r(t)}{\sigma^{(m_l)}}\right)\phi(t) \to 0.$$

Hence in the limit

$$\frac{1}{N}\sum_{t=1}^{N}\phi(t)\psi_0\left(\frac{\hat{r}(t)}{\hat{\sigma}^{(m_l)}}\right) = 0$$

where $\hat{r}(t)=y(t)-\phi^T(t)\hat{\theta}$. Therefore, Eq.(5.a) holds. In view of the convexity of J, every solution of (5.a) and (5.b) minimizes Eq.(4). It leads to the end of the proof.

∎

VII.B. Convergence Proofs of Lemmas 5 to 7 and Theorem 2

VII.B.1 Proof of Lemma 5:

Similar to the proof of lemma 2, first of all, we shall construct a "comparison function" $Q(\sigma)$ that lies wholly above $J(\boldsymbol{a}^{(k)},.)$, reaches its minimum at $\sigma^{(k)}$, and that agrees with $J(\boldsymbol{a}^{(k)},\sigma)$ at $\sigma=\sigma^{(k)}$. That is

$$Q(\sigma) = J\left(\boldsymbol{a}^{(k)},\sigma^{(k)}\right) + \beta\left(\sigma - \sigma^{(k)}\right) + \int_0^T \chi_0\left(\frac{r(t,\boldsymbol{a}^{(k)})}{\sigma^{(k)}}\right)dt \left[\frac{(\sigma^{(k)})^2}{\sigma} - \sigma^{(k)}\right]. \tag{A.19}$$

Clearly, we can see that $Q(\sigma^{(k)})=J(\boldsymbol{a}^{(k)},\sigma^{(k)})$. The derivatives of $Q(\sigma)$ and $J(\boldsymbol{a}^{(k)},\sigma)$ with respect to σ are

$$\frac{dQ(\sigma)}{d\sigma} = -\int_0^T \chi_0\left(\frac{r(t,\boldsymbol{a}^{(k)})}{\sigma^{(k)}}\right)\left(\frac{\sigma^{(k)}}{\sigma}\right)^2 dt + \beta \tag{A.20.a}$$

$$\frac{\partial J(\boldsymbol{a}^{(k)},\sigma)}{\partial\sigma} = -\int_0^T \chi_0\left(\frac{r(t,\boldsymbol{a}^{(k)})}{\sigma}\right)dt + \beta. \tag{A.20.b}$$

They are equal if we put $\sigma=\sigma^{(k)}$. For convenience, define $f(z)$ as

$$f(z) = Q\left(\frac{1}{z}\right) - J\left(\boldsymbol{a}^{(k)},\frac{1}{z}\right) \qquad z > 0. \tag{A.21}$$

Considering Eq.(A.20), the first derivative of $f(z)$ with respect to z is

$$\frac{df(z)}{dz} = \int_0^T \left[\frac{\chi_0\left(r(t,\boldsymbol{a}^{(k)})z^{(m)}\right)}{\left(z^{(k)}\right)^2} - \frac{\chi_0\left(r(t,\boldsymbol{a}^{(k)})z\right)}{z^2}\right]dt \tag{A.22}$$

with $z^{(k)}=1/\sigma^{(k)}$. Differentiating (A.22) again gives

$$\frac{d^2f(z)}{dz^2} = -\int_0^T \frac{\left(r(t,\boldsymbol{a}^{(k)})\right)^2}{z}\left[\ddot{\rho}_0\left(r(t,\boldsymbol{a}^{(k)})z\right) - 2\frac{\chi_0\left(r(t,\boldsymbol{a}^{(k)})z\right)}{\left(r(t,\boldsymbol{a}^{(k)})z\right)^2}\right]dt.$$

Considering the conclusion in lemma 1 gives

$$\frac{d^2f(z)}{dz^2} \geq 0$$

for all $z>0$. Since

$$f(z^{(k)}) = 0 \qquad \textit{and} \qquad \frac{df(z)}{dz}\Big|_{z=z^{(k)}} = 0$$

it follows that $f(z)\geq 0$ for all $z>0$. Therefore, the following inequality is true

$$Q(\sigma) \geq J(\boldsymbol{a}^{(k)},\sigma) \tag{A.23}$$

for all $\sigma>0$.

From (A.20.a), we can clearly observe that $Q(\sigma)$ reaches its minimum at $\sigma^{(k+1)}$ which has already been given in (58). Furthermore, combining the above equations with Eq.(58) leads to

$$\begin{aligned} Q(\sigma^{(k+1)}) &= J\left(\boldsymbol{a}^{(k)},\sigma^{(k)}\right) + \beta\left(\sigma^{(k+1)} - \sigma^{(k)}\right) + \beta\left(\frac{\sigma^{(k+1)}}{\sigma^{(k)}}\right)^2\left[\frac{\left(\sigma^{(k)}\right)^2}{\sigma^{(k+1)}} - \sigma^{(k)}\right] \\ &= J\left(\boldsymbol{a}^{(k)},\sigma^{(k)}\right) - \beta\frac{\left(\sigma^{(k+1)} - \sigma^{(k)}\right)^2}{\sigma^{(k)}}. \end{aligned}$$

Now it is very straightforward to have the conclusions in this lemma after substituting the above equation into (A.23).

■

VII.B.2 Proof of Lemma 6:

To prove the lemma, we shall first introduce a comparison function that agrees with J at $\boldsymbol{a}^{(k)}$, that lies wholly above J, and that reaches its minimum at $\boldsymbol{a}^{(k+1)}$. Put

$$O(\boldsymbol{\Delta a}) = J(\boldsymbol{a}^{(k)},\sigma^{(k)}) + \frac{1}{2\sigma^{(k)}}\int_0^T\left\{\left(r_*(t,\boldsymbol{a}^{(k)}) - \boldsymbol{g}_{(M)}^T(t)\boldsymbol{\Delta a}\right)^2 - \left(r_*(t,\boldsymbol{a}^{(k)})\right)^2\right\}dt. \quad \text{(A.24)}$$

Taking the first derivatives of $O(\boldsymbol{\Delta a})$ and $J(\boldsymbol{a}^{(k)}+\boldsymbol{\Delta a},\sigma^{(k)})$ with respect to $\boldsymbol{\Delta a}_k$ gives

$$\frac{\partial O(\boldsymbol{\Delta a})}{\partial \Delta a_k} = -\frac{1}{\sigma^{(k)}}\int_0^T\left(r_*(t,\boldsymbol{a}^{(k)}) - \boldsymbol{g}_{(M)}^T(t)\,\boldsymbol{\Delta a}\right)g_k(t)\,dt \quad \text{(A.25.a)}$$

$$\frac{\partial J(\boldsymbol{a}^{(k)}+\boldsymbol{\Delta a},\sigma^{(k)})}{\partial \Delta a_k} = -\int_0^T\psi_0\left(\frac{f(t) - \boldsymbol{g}_{(M)}^T(t)\left(\boldsymbol{a}^{(k)} + \boldsymbol{\Delta a}\right)}{\sigma^{(k)}}\right)g_k(t)\,dt. \quad \text{(A.25.b)}$$

In the above equation, $g_k(t)$ is the k-th element in the vector $\boldsymbol{g}_{(M)}(t)$ and $k=1, ..., M\text{-}1$.

Again, $O(\boldsymbol{\Delta\theta})$ and $J(\boldsymbol{a}^{(k)}+\boldsymbol{\Delta\theta},\sigma^{(k)})$ have the same value and the same first derivative if $\boldsymbol{\Delta a}=0$.

The second order derivative matrix of the difference between two functions is given as

$$O(\boldsymbol{\Delta a}) \geq J(\boldsymbol{a}^{(k)}+\boldsymbol{\Delta a},\sigma^{(k)}) \qquad \textit{for all } \boldsymbol{\Delta a}. \quad \text{(A.26)}$$

with $j,k = 1, ..., M\text{-}1$. Clearly, it is positive semidefinite, hence

$$\frac{\partial^2\left(O(\boldsymbol{\Delta a})-J(\boldsymbol{a}^{(k)}+\boldsymbol{\Delta a},\sigma^{(k)})\right)}{\partial \Delta a_j\,\partial \Delta a_k} = \frac{1}{\sigma^{(k)}}\int_0^T g_j(t)g_k(t)\left[1 - \psi_0'\left(\frac{r(t,\boldsymbol{a}^{(k)}) - \boldsymbol{g}_{(M)}^T(t)\boldsymbol{\Delta a}}{\sigma^{(k)}}\right)\right]dt \quad \text{(A.27)}$$

Now, from (A.25.a), it is straightforward to conclude that $O(\boldsymbol{\Delta a})$ reaches its minimum at $\boldsymbol{a}^{(k+1)}=\boldsymbol{a}^{(k)}+\boldsymbol{\Delta\hat{a}}$ if the estimate $\boldsymbol{\Delta\hat{a}}$ is given in (56). Substituting (56) into the second term of (A.24) gives

$$\begin{aligned}
&\int_0^T\left\{\left(r_*(t,\boldsymbol{a}^{(k)}) - \boldsymbol{g}_{(M)}^T(t)\boldsymbol{\Delta\hat{a}}\right)^2 - \left(r_*(t,\boldsymbol{a}^{(k)})\right)^2\right\}dt \\
&= -2\int_0^T r_*(t,\boldsymbol{a}^{(k)})\boldsymbol{g}_{(M)}^T(t)\,\boldsymbol{\Delta\hat{a}}\,dt + \int_0^T\left(\boldsymbol{g}_{(M)}^T(t)\,\boldsymbol{\Delta\hat{a}}\right)^2 dt \\
&= -2\int_0^T r_*(t,\boldsymbol{a}^{(k)})\boldsymbol{g}_{(M)}^T(t)\,dt\,\boldsymbol{\Delta\hat{a}} + \boldsymbol{\Delta\hat{a}}^T\int_0^T \boldsymbol{g}_{(M)}(t)\boldsymbol{g}_{(M)}^T(t)\,dt\,\boldsymbol{\Delta\hat{a}} \\
&= -T\boldsymbol{\Delta\hat{a}}^T\boldsymbol{\Delta\hat{a}}
\end{aligned}$$

with $r_*(t,\boldsymbol{a}^{(k)})$ defined as before.

Define $\|x\|^2=x^Tx$, and then, using the above equation yields the following minimum value of $O(\boldsymbol{\Delta a})$

$$O(\boldsymbol{\Delta\hat{a}}) = J(\boldsymbol{a}^{(k)},\sigma^{(k)}) - \frac{T}{2\sigma^{(k)}}\|\boldsymbol{\Delta\hat{a}}\|^2. \quad \text{(A.28)}$$

As a function of q,

$$O(q\,\boldsymbol{\Delta\hat{a}}) - J(\boldsymbol{a}^{(k)},\sigma^{(k)})$$

is quadratic, vanishes at $q=0$, has the minimum at $q=1$, and must vanish again at $q=2$

for reasons of symmetry. Hence, the following equation can be obtained

$$O(q\,\Delta\hat{\boldsymbol{a}}) - J\left(\boldsymbol{a}^{(k)},\sigma^{(k)}\right) = -\frac{Tq(2-q)}{2\sigma^{(k)}}\|\Delta\hat{\boldsymbol{a}}\|^2.$$

Finally, substituting the above equation into (A.27) can lead to the conclusions in this lemma.

■

VII.B.3 Proof of Lemma 8:

Like the proof in lemma 4, the sets $\mathfrak{C}_b$ are obviously closed since J is continuous if we consider the definition of the sets. Since we have $\sigma \leq b/\beta$, i.e. σ is bounded. Based on the assumptions in lemma 1, $\psi_0(x)$ is bounded for any x and $r_*(t,\boldsymbol{a})$ is bounded. It leads to that $\|\boldsymbol{a}\|$ is bounded. Therefore, the sets $\mathfrak{C}_b$ are compact.

■

VII.B.4 Proof of Theorem 2:

The compactness of the sets $\mathfrak{C}_b$, i.e. conclusion (1), can be directly obtained if we use lemma 8.

Without loss of generality, we assume $\hat{\sigma}$. Furthermore, given a sequence $(\boldsymbol{a}^{(k)},\sigma^{(k)})$, let $(\boldsymbol{a}^{(k_l)},\sigma^{(k_l)})$ be a converging subsequence in $(\boldsymbol{a}^{(k)},\sigma^{(k)})$ and converge towards $(\hat{\boldsymbol{a}},\hat{\sigma})$. Then, using lemmas 6 and 7, we have

$$J\left(\boldsymbol{a}^{(k_l)},\sigma^{(k_l)}\right) \geq J\left(\boldsymbol{a}^{(k_l)},\sigma^{(k_l+1)}\right) \geq J\left(\boldsymbol{a}^{(k_l+1)},\sigma^{(k_l+1)}\right).$$

Since $(\boldsymbol{a}^{(k_l)},\sigma^{(k_l)})$ is a converging subsequence, the two outer members of this inequality tend to $J(\hat{\boldsymbol{a}},\hat{\sigma})$. Therefore,

$$J\left(\boldsymbol{a}^{(k_l)},\sigma^{(k_l)}\right) - J\left(\boldsymbol{a}^{(k_l)},\sigma^{(k_l+1)}\right) \geq \beta\,\frac{\left(\sigma^{(k_l+1)} - \sigma^{(k_l)}\right)^2}{\sigma^{(k_l)}}$$

converges to 0. It simply implies that

$$\left(\frac{\sigma^{(k_l+1)}}{\sigma^{(k_l)}}\right)^2 = \frac{1}{\beta}\int_0^T \chi_0\left(\frac{r(t,\boldsymbol{a}^{(k_l)})}{\sigma^{(k_l)}}\right)dt$$

converges to 1, that means

$$\lim_{k_l\to\infty}\int_0^T \chi_0\left(\frac{r(t,\boldsymbol{a}^{(k_l)})}{\sigma^{(k_l)}}\right) = \beta.$$

Thus, Eq.(46.b) is satisfied.

Similarly, from lemma 8, we can have that

$$J\left(\boldsymbol{a}^{(k_l)},\sigma^{(k_l)}\right) - J\left(\boldsymbol{a}^{(k_l+1)},\sigma^{(k_l)}\right) \geq \frac{Tq(2-q)}{2\sigma^{(k_l)}}\|\Delta\hat{a}\|^2$$

tends to 0. It indicates that $\|\Delta\hat{\boldsymbol{a}}\|^2$ tends to 0, i.e. $\Delta\hat{\boldsymbol{a}}$ tends to 0. In particular,

$$\frac{1}{T}\int_0^T r_*(t,\boldsymbol{a}^{(k)})g_{(M)}(t)\,dt \to 0.$$

Therefore, in the limit Eq.(46.a) is satisfied, that is

$$\int_0^T \psi_0\left(\frac{f(t) - \boldsymbol{a}^T \boldsymbol{g}_{(M)}(t)}{\sigma}\right)\boldsymbol{g}_{(M)}(t) = 0.$$

In view of the convexity of J, every solution of Eq.(46) minimizes Eq.(45). It ends the proof of theorem 2.

■

VIII. REFERENCES

1. P.J. Huber, *Robust Statistics*, John Wiley & Sons, New York (1981).

2. C.I. Masreliez and R.D. Martin, "Robust Bayesian estimation for a linear model and robustifying the Kalman filter", *IEEE Trans. Automat. Contr.*, **AC-22**(3), pp.361-371 (1977).

3. S.S. Stanković and B.D. Kovacević, "Analysis of robust stochastic approximation algorithms for process identification", *Automatica*, **22**(4), pp.483-488 (1986).

4. S.C. Puthenpura and N.K. Sinha, "Modified maximum likelihood method for the robust estimation of system parameters from very noisy data", *Automatica*, **22**(2), pp.231-235 (1986).

5. S.C. Puthenpura and N.K. Sinha, "Robust identification from impulse and step response", *IEEE Trans. on Industrial Electronics*, **IE-34**(3), pp.366-370 (1987).

6. S.C. Puthenpura and N.K. Sinha, "Robust instrumental variables method for system identification", *Control Theory and Advanced Technology*, **1**(3), pp.175-188 (1985).

7. S.C. Puthenpura, N.K. Sinha, and O.P. Vidal, "Application of M-estimation i robust recursive system identification", *IFAC Symp. on Stochastic Control*, pp.23-30 (1985).

8. H. Dai and N.K. Sinha, "Robust recursive least squares method with modified weights for bilinear system identification", *IEE Proceedings-D Control Theory and Applications*, **136**(3), pp.122-126 (1989).

9. H. Dai and N.K. Sinha, "Robust combined estimation of states and parameters of bilinear systems", *Automatica*, **25**(4), pp.613-616 (1989).

10. H. Dai and N.K. Sinha, "Robust recursive instrumental variable method with modified weights for bilinear system identification", *IEEE Trans. on Industrial Electronics*, **38**(1), pp.1-7 (1991).

11. H. Dai and N.K. Sinha, "Robust recursive output error method for bilinear system identification", *Proc. 8th IFAC Symp. Ident. Syst. Par. Est.*, **2**, pp.1141-1146 (1988).

12. H. Dai and N.K. Sinha, "A Robust off-line method for system identification: robust iterative least squares method with modified residuals", to appear at *ASME J. of Dynamic. Syst., Measu., and Control.*

13. L. Ljung, *System Identification: Theory for the User*, Prentice-Hall, Inc., New Jersey (1987).

14. H. Dai and N.K. Sinha, "Robust approximation of time series with Walsh functions", *Electronic Letters*, **25**(8), pp.527-529 (1989).

15. H. Dai and N.K. Sinha, "Robust coefficient estimation of Walsh functions", *IEE Proc.-D Control Theory and Applications*, **137**(6), pp.357-363 (1990).

16. H. Dai and N.K. Sinha, "Robust identification of systems with Walsh functions", *Control Theory and Applications*, **6**(4), pp.633-654 (1990).

17. V.R. Karanam, P.A. Frick, and R.R. Mohler, "Bilinear system identification by Walsh functions", *IEEE Trans. Automat. Contr.*, **AC-23**(4), pp.709-713 (1978).

18. N.J. Fine, "On the Walsh functions", *Trans. Amer. Math. Soc.*, **65**, pp.372-414 (1949).

Loop Transfer Recovery For General Nonminimum Phase Discrete Time Systems

Part 1: Analysis

Ben M. Chen

Department of Electrical Engineering
State University of New York at Stony Brook
Stony Brook, New York 11794-2350

Ali Saberi

School of Electrical Engineering and Computer Science
Washington State University
Pullman, Washington 99164-2752

Peddapullaiah Sannuti

Department of Electrical and Computer Engineering
P.O. Box 909
Rutgers University
Piscataway, New Jersey 08855-0909

Yacov Shamash

College of Engineering and Applied Sciences
State University of New York at Stony Brook
Stony Brook, New York 11794

I. INTRODUCTION

In recent years, a method of multivariable feedback control system design using so called LQG/LTR techniques has gained significance. As is known, many performance and robust stability objectives can be cast in terms of maximum magnitude or maximum singular values of some particular *closed-loop* transfer functions, e.g., sensitivity and complementary sensitivity functions at certain points in a closed-loop. Such magnitude or singular value requirements on some *closed-loop* transfer functions can be directly determined by corresponding singular values of certain related *open-loop* transfer functions. Thus equivalently, the design specifications can be prescribed in terms of some *open-loop* transfer functions. In prescribing the *open-loop* transfer functions, the point at which loop is broken can be the input or output or any arbitrary point of the given plant. Here we deal with the situation when the loop is broken at the input point of the plant. Then the design methodology of LQG/LTR can essentially be partitioned into two steps. The first step involves "loop shaping" utilizing a state feedback control law so that the resulting open-loop transfer function when the loop is broken at the input point of the plant meets the design specifications. The resulting open-loop transfer function is called the target loop transfer function. The second step, called loop transfer recovery (LTR), involves the design of an output feedback control law such that the resulting open-loop transfer function would be either exactly or approximately the same as the target open-loop transfer function. In other words, the idea of LTR is to come up with a measurement feedback compensator, typically observer based, to recover either exactly or asymptotically a specific open-loop transfer function prescribed in terms of a state feedback gain.

Ever since the seminal works of [14] and [9], LTR has been the subject of a number of authors including [2], [3], [4], [5], [6], [7], [10], [12], [13], [15], [17], [18], [19], [21], [23], [24], [25], [29], [31], [32], [34] and [35]. Both continuous and discrete systems have been treated earlier. Recently Chen, Saberi and Sannuti in [6] analyzed in depth the mechanism of LTR for continuous systems. The analysis given there considers four main issues. The first issue is concerned with what can and what cannot be achieved for a given system and for an arbitrarily specified target loop transfer function. On the other hand, the second issue is concerned with the development of necessary or/and sufficient conditions a target loop has to satisfy so that it can either exactly or asymptotically be recovered for a given system while the third issue is concerned with the development of necessary or/and sufficient conditions on a given system such that it has at least one recoverable target loop. The fourth issue deals with a generalization of all the above three issues when recovery is required over a subspace of the control space. It concerns with generalizing the traditional LTR concept to sensitivity recovery over a subspace and deals with method(s) to test whether projections of target and achievable sensitivity and complementary sensitivity functions onto a given subspace match each other or not. Such an analysis pinpoints the limitations of the given system for the recovery of arbitrarily specified target loops via either current or prediction estimator based controllers. These limitations are the consequences of the structural properties (i.e., finite and infinite zero structure, and invertibility) of the given system. Also, the conditions required on a target loop transfer function for its recoverability, turn out to be constraints on its finite and infinite zero structure as related to the corresponding structure of the given system. Furthermore, the analysis given in [6] discovers a multitude of ways in which freedom exists to shape the loops in a desired way as close as possible to the target shapes. Also, possible pole zero cancellations between the eigenvalues of the controller and the input or/and output decoupling zeros of the given system are characterized. Next, regarding the design of controllers for LTR, [7] developed three methods of observer based controller design. The first method is an asymptotic time-scale and eigenstructure assignment (ATEA) method while the other two are optimization based, one minimizing the H_∞ norm and the other H_2 norm of a so called recovery matrix. The analysis as well as design methods as given in [6] and [7] are fairly complete for general nonminimum phase nonstrictly proper plants of continuous type.

In contrast to the continuous systems, the results available for discrete systems are relatively few. In order to facilitate the discussion of the available results, let us first recall that for discrete systems there exist three different types of observer based controllers; namely, 'prediction estimator', and full or reduced order type 'current estimator' based controllers. In the case of continuous systems, as is well known, any arbitrary target loop transfer function is asymptotically recoverable provided that the given system is left invertible and of minimum phase. However, this is not necessarily so for discrete systems as discussed first by Goodman [12]. Using prediction estimator based controllers, Goodman characterized the so called recovery matrix and showed that in general it cannot be rendered zero even for square minimum phase strictly proper systems. He showed further that prediction estimator when its gain is calculated via Kalman filter formalism in which the covariance of the fictitious input noise is arbitrarily increased to infinity, minimizes the H_2 norm of the recovery matrix. Later on, Maciejowski [15] continued the study of LTR for square minimum phase strictly proper discrete systems using current estimators. Although Maciejowski studied the recovery at the output point of the plant, his results when translated to recovery at the input point of the plant, imply that recovery of an arbitrarily specified target loop transfer function is possible for the class of systems he considered, namely, strictly proper square minimum phase systems having only infinite zeros of order one. Also, Maciejowski [15] as well as Ishihara and Takeda [13] observe that it is impossible in general to have either exact or asymptotic LTR when the plant is either nonminimum phase or when prediction estimator is used even if it has all infinite zeros of order one. Realizing that in general LTR for discrete systems is not feasible, Niemann and Sogaard-Andersen [19] consider square strictly proper systems with a prediction estimator, and develop a parameterization of exactly recoverable target loop transfer functions in terms of system zeros and associated zero directions. Recently, Zhang and Freudenberg [35], considering only square strictly proper plants having only infinite zeros of order one, study the LTR mechanism at the output point of the plant. They develop explicit expressions for the recovery error and the resulting sensitivity function when prediction as well as current estimator based controllers are used and when optimization is used to minimize the H_2 norm of a recovery matrix (for precise definition of recovery matrix, see Lemma 1). The analysis of LTR done so far on discrete systems, as summarized above, although presents some glimpses of what is happening in some special cases, it does not reveal a total picture of LTR mechanism for general discrete systems. In fact, it is fair to say that no systematic analysis of all the issues involved in LTR exists to date for general discrete systems, and whatever is available is far away from being complete. For example, as pointed out by Maciejowski [15], most practical discrete systems have a direct feed through from inputs to outputs and thus are non-strictly proper. Yet no work to date deals with non-strictly proper discrete systems. Similarly, as shown in Astrom et al [1], sampling of continuous systems most often introduces unstable invariant zeros in the resulting discrete systems.

Yet, even for strictly proper discrete systems, the results showing the effects of unstable invariant zeros on LTR are to a great extent incomplete; just to mention a few, no characterization of recoverable target loops of a given system exists, similarly analysis for recovery in any given subspace of control space is nonexistent. Similarly, regarding design for LTR, while partial results are available based on minimization of H_2 norm of a certain recovery matrix [12], [35], no methods of H_∞ norm minimization and eigenstructure assignment are yet available. Thus the intent of this two part paper is to present both systematic analysis as well as design tools for LTR of general nonminimum phase nonstrictly proper discrete plants. Part 1 of the paper deals with complete analysis. All the four issues mentioned earlier in connection with continuous systems are reexamined for discrete systems. The analysis and the method of presentation given in Part 1 unifies the discussion regarding all the three controllers, namely, 'prediction estimator' based, and full or reduced order 'current estimator' based controllers. Part 2 of the paper [8] deals with design where both eigenstructure assignment method and optimization based methods in which either H_∞ or H_2 norm of certain recovery matrix is minimized, are developed.

The analysis and design aspects presented in this two part paper reveal both similarities as well as fundamental differences between continuous and discrete systems. One fundamental difference which we want to emphasize here is this. In discrete systems, as is well known, in order to preserve stability, all the closed-loop eigenvalues must be restricted to lie within the unit circle in complex plane. This implies that unlike continuous case which permits both finite as well as asymptotically infinite eigenvalue assignment, in the discrete case one is restricted to only finite eigenvalue assignment. This restriction leads to several important differences in connection with LTR between continuous and discrete systems. To quote one such difference, let us recall that asymptotic recovery in the case of continuous systems allows assignment of both asymptotically finite as well as infinite observer eigenvalues by using high observer gains, where as exact recovery allows only assignment of finite observer eigenvalues. Thus, in continuous systems, there exists target loops which are only asymptotically recoverable but not exactly recoverable. On the other hand, in discrete systems, since both asymptotic as well as exact recovery involves only finite eigenvalue assignment, every asymptotically recoverable target loop is also exactly recoverable. This implies that one needs to talk about just recovery rather than emphasizing exact or asymptotic recovery. However, in optimization based design methods, such as H_∞ norm minimization, one some times ends up in suboptimal designs which correspond to only asymptotic recovery. In that case, a distinction can be made between exact and asymptotic recovery.

Throughout the paper, A' denotes the transpose of A, A^H denotes the complex conjugate transpose of A, I denotes an identity matrix while I_k denotes the identity matrix of dimension $k \times k$. $\lambda(A)$ and $\text{Re}[\lambda(A)]$ respectively denote the set of eigenvalues and real parts of eigenvalues of A. Similarly, $\sigma_{max}[A]$ and $\sigma_{min}[A]$ respectively denote the maximum and minimum singular values of A. Ker $[V]$ and Im $[V]$ denote respectively the kernel and the image of V. $\mathbf{C}^\odot$ denotes the set of complex numbers inside the open unit circle while $\mathbf{C}^\otimes$ is the complementary set of $\mathbf{C}^\odot$. Also, $\mathcal{R}_p$ denotes the sub-ring of all proper rational functions of z while the set of matrices of dimension $l \times q$ whose elements belong to $\mathcal{R}_p$ is denoted by $\mathcal{M}^{l\times q}(\mathcal{R}_p)$. Given a discrete transfer function $G(z)$, we define the discrete frequency response $G^*(j\omega)$ as $G(e^{j\omega T})$ where T is the sampling period of the discrete-time system. An asymptotically stable matrix is the one whose eigenvalues are all in $\mathbf{C}^\odot$.

II. PROBLEM FORMULATION

In this section, we formulate the LTR problem in precise mathematical terms. Let us consider a nonstrictly proper discrete-time system Σ,

$$x(k+1) = Ax(k) + Bu(k)\,, \quad y(k) = Cx(k) + Du(k) \tag{1}$$

where the state vector $x \in \mathbf{R}^n$, output vector $y \in \mathbf{R}^p$ and input vector $u \in \mathbf{R}^m$. Without loss of generality, assume that $[B', D']'$ and $[C, D]$ are of maximal rank. Let us also assume that Σ is stabilizable and detectable. Let F be a full state feedback gain matrix such that (a) the closed-loop system is asymptotically stable, i.e., eigenvalues of $A - BF$ lie inside the unit circle, and (b) the open-loop transfer function when the loop is broken at the input point of the given system meets the given frequency dependent specifications. The state feedback control is

$$u(k) = -Fx(k) \tag{2}$$

and the loop transfer function evaluated when the loop is broken at the input point of the given system, the so called target loop transfer function, is

$$L_t(z) = F\Phi B \tag{3}$$

where $\Phi = (zI_n - A)^{-1}$. The corresponding target sensitivity and complementary sensitivity functions are

$$S_t(z) = [I_m + F\Phi B]^{-1} \quad \text{and} \quad T_t(z) = I_m - S_t(z). \tag{4}$$

Arriving at an appropriate value for F is concerned with the issue of loop shaping which is an engineering art and often includes the use of linear quadratic regulator (LQR) design in which the cost matrices are used as free design parameters to generate the target loop transfer function $L_t(z)$ and thus the desired sensitivity and complementary sensitivity functions. The next step of design is to recover the target loop using only a measurement feedback controller. This is the problem of loop transfer recovery (LTR) and is the focus of this paper. To explain it clearly, consider the configuration of Figure 1 where $\mathrm{C}(z)$ and $P(z)$,

$$P(z) = C\Phi B + D,$$

are respectively the transfer functions of a controller and of the given system. Given $P(z)$ and a target loop transfer function $L_t(z)$, one seeks then to design a $\mathrm{C}(z)$ such that the loop recovery error $E(z)$,

$$E(z) \equiv L_t(z) - \mathrm{C}(z)P(z),$$

is either exactly or approximately equal to zero in the frequency region of interest while guaranteeing the stability of the resulting closed-loop system. The notion of achieving exact LTR (ELTR) corresponds to $E(z) = 0$ for all z. In the case of asymptotic recovery, one normally parameterizes the controller $\mathrm{C}(z)$ in terms of a scalar tuning parameter σ and thus obtains a family of controllers $\mathrm{C}(z, \sigma)$. We say asymptotic LTR (ALTR) is achieved if $\mathrm{C}(z, \sigma)P(z) \to L_t(z)$ pointwise in z as $\sigma \to \infty$. Achievability of ALTR enables the designer to choose a member of the family of controllers that corresponds to a particular value of σ which achieves a desired level of recovery. We now consider the following definitions in order to impart precise meanings to ELTR and ALTR:

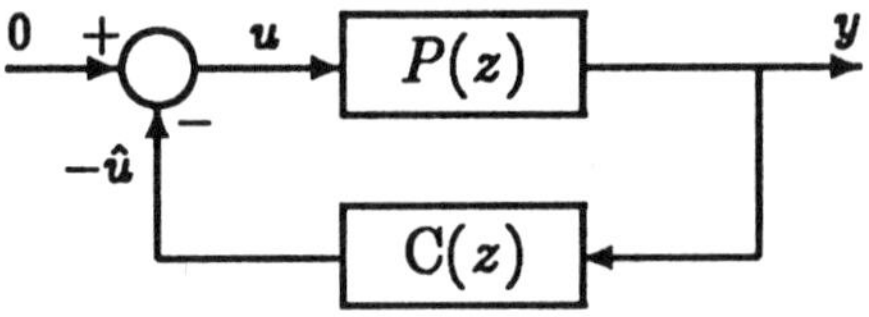

Figure 1: Plant—Controller closed-loop configuration.

Definition 1. *The set of admissible target loops* $\mathbf{T}(\Sigma)$ *of a system* Σ *is defined by*

$$\mathbf{T}(\Sigma) = \{L_t(z) \in \mathcal{M}^{m\times m}(\mathcal{R}_p) \mid L_t(z) = F\Phi B \text{ and } \lambda(A - BF) \in \mathbb{C}^{\odot}\}.$$

Definition 2. *A target loop transfer function* $L_t(z) \in \mathbf{T}(\Sigma)$ *is said to be exactly recoverable (ELTR) if there exists a* $\mathrm{C}(z) \in \mathcal{M}^{m\times p}(\mathcal{R}_p)$ *such that (i) the closed-loop system comprising of* $\mathrm{C}(z)$ *and* $P(z)$ *as in the configuration of Figure 1 is asymptotically stable, and (ii)* $\mathrm{C}(z)P(z) = L_t(z)$.

Definition 3. *A target loop transfer function* $L_t(z) \in \mathbf{T}(\Sigma)$ *is said to be asymptotically recoverable (ALTR) if there exists a parameterized family of controllers* $\mathrm{C}(z,\sigma) \in \mathcal{M}^{m\times p}(\mathcal{R}_p)$, *where* σ *is a scalar parameter taking positive values, such that (i) the closed-loop system comprising of* $\mathrm{C}(z,\sigma)$ *and* $P(z)$ *as in the configuration of Figure 1 is asymptotically stable for all* $\sigma > \sigma^*$, *where* $0 \le \sigma^* < \infty$, *and (ii)* $\mathrm{C}(z,\sigma)P(z) \to L_t(z)$ *pointwise in* z *as* $\sigma \to \infty$. *Moreover, the limits, as* $\sigma \to \infty$, *of all the eigenvalues of the closed-loop system should remain in* $\mathbb{C}^{\odot}$.

As mentioned earlier, it turns out that for discrete systems in contrast with continuous systems, every asymptotically recoverable target loop can also be exactly recoverable and vice versa. One might then wonder why one needs to distinguish between ELTR and ALTR. This is perhaps, as will be seen in Part 2 of the paper, even for the case when ELTR can be achieved, some optimization based design methods, such as H_∞ norm minimization, typically end up in suboptimal designs which correspond to asymptotic recovery. But in this Part 1 of the paper which is mainly concerned with analysis, we will not hereafter distinguish between the notions of exact and asymptotic recovery. Also, we will not parameterize a controller in terms of a tunable parameter σ in an attempt to achieve whatever can be achieved asymptotically rather than exactly. We maintain that such a parameterization can always be done if one chooses so. We have the following additional definitions.

Definition 4. *A target loop transfer function $L_t(z)$ belonging to $\mathbf{T}(\Sigma)$ is said to be recoverable if $L_t(z)$ is either exactly or asymptotically recoverable.*

Definition 5. *The set of recoverable target loops for the given system Σ is denoted by $\mathbf{T}^{R}(\Sigma)$* †.

Next, in view of the definition of sensitivity function $S_t(z)$ as in (4), it is simple to note that the recovery of target loop $L_t(z)$ implies the recovery of $S_t(z)$ and vice versa.

As mentioned in introduction, the purpose of this Part 1 paper is to do in depth analysis of LTR mechanism in general discrete systems. As in the case of continuous systems, the analysis of LTR mechanism carried out here concentrates on four fundamental issues. The first issue is concerned with what can and what cannot be achieved for a given system without taking into account any specific target loop transfer function, i.e., the target loop transfer function is considered as arbitrarily given. On the other hand, the second issue is concerned with the development of necessary or/and sufficient conditions a target loop has to satisfy so that it can either exactly or asymptotically be recovered for a given system. The third issue deals with the development of necessary or/and sufficient conditions on a given system such that it has at least one recoverable target loop. The fourth issue concerns with a generalization of all the above three issues when recovery is required over a subspace of the control space. To be exact, it concerns with generalizing the traditional LTR concept to sensitivity recovery over a subspace and deals with method(s) to test whether projections of target and achievable sensitivity and complementary sensitivity functions onto a given subspace match each other or not. As in the case of continuous systems, the analysis presented here shows some fundamental limitations of

† In continuous time systems, we defined three sets; the set of exactly recoverable target loops $\mathbf{T}^{ER}(\Sigma)$, the set of recoverable target loops $\mathbf{T}^{R}(\Sigma)$, and the set of target loops which are recoverable but not exactly recoverable $\mathbf{T}^{AR}(\Sigma)$. But in discrete case, owing to the fact that every asymptotically recoverable target loop is exactly recoverable and vice versa, we need to define only the set of recoverable target loops $\mathbf{T}^{R}(\Sigma)$.

the given system as a consequence of its structural properties, namely finite and infinite zero structure and invertibility. It also discovers a multitude of ways in which freedom exists to shape the recovery error in a desired way.

The rest of the paper is organized as follows. Section III recalls a special coordinate basis (s.c.b) of [27] and [28] which displays explicitly both the finite as well as infinite zero structure of the given system. In Section IV, the structural details of three observer based controllers, namely prediction, full and reduced order current estimator based controllers, are discussed. Also, in Section IV, some preliminary analysis is given showing that the required LTR analysis for all the three controllers considered here can be unified into a single mathematical frame work. Section V deals with all the issues of LTR analysis, while Section VI dualizes the results of Section V for the case when the target loops are specified at the plant output point. Finally, Section VII draws the conclusions of our work.

III. PRELIMINARIES

As in the case of LTR analysis of continuous systems, finite and infinite zero structures of both the given discrete system and the target loop transfer function play a dominant role in the recovery analysis as well as design. Keeping this in mind, we recall in this section a special coordinate basis (s.c.b) of a linear time invariant system [27], [28]. Such a s.c.b has a distinct feature of explicitly displaying the finite and infinite zero structure of a given system. Consider the system Σ characterized by (A, B, C, D). It is simple to verify that there exist non-singular transformations U and V such that

$$UDV = \begin{bmatrix} I_{m_0} & 0 \\ 0 & 0 \end{bmatrix}, \tag{5}$$

where m_0 is the rank of matrix D. Hence hereafter, without loss of generality, it is assumed that matrix D has the form given on the right hand side of (5).

One can now rewrite the system of (1) as,

$$\begin{cases} x(k+1) = A\, x(k) + [B_0 \quad B_1] \begin{pmatrix} u_0(k) \\ u_1(k) \end{pmatrix}, \\ \begin{pmatrix} y_0(k) \\ y_1(k) \end{pmatrix} = \begin{bmatrix} C_0 \\ C_1 \end{bmatrix} x(k) + \begin{bmatrix} I_{m_0} & 0 \\ 0 & 0 \end{bmatrix} \begin{pmatrix} u_0(k) \\ u_1(k) \end{pmatrix}, \end{cases} \tag{6}$$

where the matrices B_0, B_1, C_0 and C_1 have appropriate dimensions. In what follows, whenever there is no ambiguity, in order to avoid the notational clutter, the running time index k will be omitted. We have the following theorem.

Theorem 1 (SCB). *Consider the system Σ characterised by the matrix quadruple (A, B, C, D). There exist nonsingular transformations Γ_1, Γ_2 and Γ_3, an integer $m_f \le m - m_0$, and integer indexes q_i, $i = 1$ to m_f, such that*

$$x = \Gamma_1 \tilde{x} \ , \ y = \Gamma_2 \tilde{y} \ , \ u = \Gamma_3 \tilde{u}$$

$$\tilde{x} = [x_a', \, x_b', \, x_c', \, x_f']' \ , \ x_a = [(x_a^-)', (x_a^+)']'$$

$$x_f = [x_1', \, x_2', \, \cdots, x_{m_f}']'$$

$$\tilde{y} = [y_0', y_f', y_b']' \ , \ y_f = [y_1, \, y_2, \, \cdots, y_{m_f}]'$$

$$\tilde{u} = [u_0', u_f', u_c']' \ , \ u_f = [u_1, \, u_2, \, \cdots, u_{m_f}]'$$

and

$$x_a^-(k+1) = A_{aa}^- x_a^-(k) + B_{0a}^- y_0(k) + L_{af}^- y_f(k) + L_{ab}^- y_b(k) \tag{7}$$

$$x_a^+(k+1) = A_{aa}^+ x_a^+(k) + B_{0a}^+ y_0(k) + L_{af}^+ y_f(k) + L_{ab}^+ y_b(k) \tag{8}$$

$$x_b(k+1) = A_{bb} x_b(k) + B_{0b} y_0(k) + L_{bf} y_f(k) \ , \ y_b = C_b x_b \tag{9}$$

$$\begin{aligned} x_c(k+1) = & A_{cc} x_c(k) + B_c u_c(k) + B_{0c} y_0(k) \\ & + L_{cb} y_b(k) + L_{cf} y_f(k) + B_c[E_{ca}^- x_a^-(k) + E_{ca}^+ x_a^+(k)] \end{aligned} \tag{10}$$

$$y_0 = C_{0a}^- x_a^- + C_{0a}^+ x_a^+ + C_{0b} x_b + C_{0c} x_c + C_{0f} x_f + u_0 \tag{11}$$

and for each $i = 1$ to m_f,

$$\begin{aligned} x_i(k+1) = & A_{q_i} x_i(k) + L_{i0} y_0(k) + L_{if} y_f(k) + B_{q_i} \Big[u_i(k) \\ & + E_{ia} x_a(k) + E_{ib} x_b(k) + E_{ic} x_c(k) + \sum_{j=1}^{m_f} E_{ij} x_j(k) \Big] \end{aligned} \tag{12}$$

$$y_i = C_{q_i} x_i \ , \ y_f = C_f x_f . \tag{13}$$

Here the states x_a^-, x_a^+, x_b, x_c and x_f are respectively of dimension n_a^-, n_a^+, n_b, n_c and $n_f = \sum_{i=1}^{m_f} q_i$ while x_i is of dimension q_i for each

$i=1$ to m_f. The control vectors u_0, u_f and u_c are respectively of dimension m_0, m_f and $m_c = m - m_0 - m_f$ while the output vectors y_0, y_f and y_b are respectively of dimension $p_0 = m_0$, $p_f = m_f$ and $p_b = p - p_0 - p_f$. The matrices A_{q_i}, B_{q_i} and C_{q_i} have the following form:

$$A_{q_i} = \begin{bmatrix} 0 & I_{q_i-1} \\ 0 & 0 \end{bmatrix}, \quad B_{q_i} = \begin{bmatrix} 0 \\ 1 \end{bmatrix}, \quad C_{q_i} = [1, 0, \cdots, 0]. \tag{14}$$

(Obviously for the case when $q_i = 1$, we have $A_{q_i}= 0$, $B_{q_i}= 1$ and $C_{q_i}= 1$.) Furthermore, we have $\lambda(A_{aa}^-) \in \mathbf{C}^{\circleddash}$, $\lambda(A_{aa}^+) \in \mathbf{C}^{\otimes}$, the pair (A_{cc}, B_c) is controllable and the pair (A_{bb}, C_b) is observable. Also, assuming that the variables x_i, $i = 1$ to m_f, are arranged such that $q_i \le q_{i+1}$, the matrix L_{if} has the particular form,

$$L_{if} = [L_{i1}, L_{i2}, \cdots, L_{i\,i-1}, 0, 0, \cdots, 0].$$

Also, the last row of each L_{if} is identically zero.

Proof : This follows from Theorem 2.1 of [27] and [28]. ■

We can rewrite the s.c.b given by Theorem 1 in a more compact form,

$$\tilde{A} := \Gamma_1^{-1}(A - B_0C_0)\Gamma_1 = \begin{bmatrix} A_{aa}^- & 0 & L_{ab}^-C_b & 0 & L_{af}^-C_f \\ 0 & A_{aa}^+ & L_{ab}^+C_b & 0 & L_{af}^+C_f \\ 0 & 0 & A_{bb} & 0 & L_{bf}C_f \\ B_cE_{ca}^- & B_cE_{ca}^+ & L_{cb}C_b & A_{cc} & L_{cf}C_f \\ B_fE_a^- & B_fE_a^+ & B_fE_b & B_fE_c & A_f \end{bmatrix},$$

$$\tilde{B} := \Gamma_1^{-1}[\,B_0 \quad B_1\,]\,\Gamma_3 = \begin{bmatrix} B_{0a}^- & 0 & 0 \\ B_{0a}^+ & 0 & 0 \\ B_{0b} & 0 & 0 \\ B_{0c} & 0 & B_c \\ B_{0f} & B_f & 0 \end{bmatrix},$$

$$\tilde{C} := \Gamma_2^{-1}\begin{bmatrix} C_0 \\ \\ C_1 \end{bmatrix}\Gamma_1 = \begin{bmatrix} C_{0a}^- & C_{0a}^+ & C_{0b} & C_{0c} & C_{0f} \\ 0 & 0 & 0 & 0 & C_f \\ 0 & 0 & C_b & 0 & 0 \end{bmatrix},$$

$$\tilde{D} := \Gamma_2^{-1}D\Gamma_3 = \begin{bmatrix} I_{m_0} & 0 & 0 \\ 0 & 0 & 0 \\ 0 & 0 & 0 \end{bmatrix}.$$

In what follows, we state some important properties of the s.c.b which are pertinent to our present work.

Property 1. *The given system Σ is right invertible if and only if x_b and hence y_b are nonexistent ($n_b = 0$, $p_b = 0$), left invertible if and only if x_c and hence u_c are nonexistent ($n_c = 0$, $m_c = 0$), invertible if and only if both x_b and x_c are nonexistent. Moreover, Σ is degenerate if and only if it is neither left nor right invertible.*

Property 2. *We note that (A_{bb}, C_b) and (A_{q_i}, C_{q_i}) form observable pairs. Unobservability could arise only in the variables x_a and x_c. In fact, the system Σ is observable (detectable) if and only if (A_{obs}, C_{obs}) is an observable (detectable) pair, where*

$$A_{obs} = \begin{bmatrix} A_{aa} & 0 \\ B_c E_{ca} & A_{cc} \end{bmatrix}, \quad A_{aa} = \begin{bmatrix} A_{aa}^- & 0 \\ 0 & A_{aa}^+ \end{bmatrix}, \quad C_{obs} = \begin{bmatrix} C_{0a} & C_{0c} \\ E_a & E_c \end{bmatrix},$$

$$C_{0a} = [C_{0a}^-, C_{0a}^+], \quad E_a = [E_a^-, E_a^+], \quad E_{ca} = [E_{ca}^-, E_{ca}^+].$$

Similarly, (A_{cc}, B_c) and (A_{q_i}, B_{q_i}) form controllable pairs. Uncontrollability could arise only in the variables x_a and x_b. In fact, Σ is controllable (stabilizable) if and only if (A_{con}, B_{con}) is a controllable (stabilizable) pair, where

$$A_{con} = \begin{bmatrix} A_{aa} & L_{ab} C_b \\ 0 & A_{bb} \end{bmatrix}, \quad B_{con} = \begin{bmatrix} B_{0a} & L_{af} \\ B_{0b} & L_{bf} \end{bmatrix},$$

$$B_{0a} = \begin{bmatrix} B_{0a}^- \\ B_{0a}^+ \end{bmatrix}, \quad L_{ab} = \begin{bmatrix} L_{ab}^- \\ L_{ab}^+ \end{bmatrix}, \quad L_{af} = \begin{bmatrix} L_{af}^- \\ L_{af}^+ \end{bmatrix}.$$

Property 3. *Invariant zeros of Σ are the eigenvalues of A_{aa}. Moreover, the minimum phase (or stable) and the nonminimum phase (or unstable) invariant zeros of Σ are the eigenvalues of A_{aa}^- and A_{aa}^+, respectively.*

If all the invariant zeros of a system Σ are in $\mathbf{C}^{\odot}$, i.e., if all the invariant zeros of Σ are stable, then we say Σ is of minimum phase, otherwise Σ is said to be of nonminimum phase.

There are interconnections between the s.c.b and various invariant and almost invariant geometric subspaces. To show these interconnections, we introduce the following geometric subspaces of Σ.

Definition 6. *For a given system Σ characterized by a matrix quadruple (A, B, C, D), we define*

1. *$\mathcal{V}^g(\Sigma)$ to be the maximal subspace of $\mathbf{R}^n$ which is $(A-BF)$-invariant and contained in* Ker $(C-DF)$ *such that the eigenvalues of $(A-BF)|\mathcal{V}^g$ are contained in $\mathbf{C}_g \subseteq \mathbf{C}$ for some F.*

2. *$\mathcal{S}^g(\Sigma)$ to be the minimal $(A-KC)$-invariant subspace of $\mathbf{R}^n$ containing in* Im $(B-KD)$ *such that the eigenvalues of the map which is induced by $(A-KC)$ on the factor space $\mathbf{R}^n/\mathcal{S}^g$ are contained in $\mathbf{C}_g \subseteq \mathbf{C}$ for some K.*

For the cases that $\mathbf{C}_g = \mathbf{C}$, $\mathbf{C}_g = \mathbf{C}^{\odot}$ and $\mathbf{C}_g = \mathbf{C}^{\otimes}$, we replace the index g in $\mathcal{V}^g$ and $\mathcal{S}^g$ by '$*$', '$-$' and '$+$', respectively.

Various components of the state vector of s.c.b have the following geometrical interpretations.

Property 4.

1. $x_a^- \oplus x_a^+ \oplus x_c$ *spans* $\mathcal{V}^*(\Sigma)$.
2. $x_a^- \oplus x_c$ *spans* $\mathcal{V}^-(\Sigma)$.
3. $x_a^+ \oplus x_c$ *spans* $\mathcal{V}^+(\Sigma)$.
4. $x_c \oplus x_f$ *spans* $\mathcal{S}^*(\Sigma)$.
5. $x_a^- \oplus x_c \oplus x_f$ *spans* $\mathcal{S}^+(\Sigma)$.
6. $x_a^+ \oplus x_c \oplus x_f$ *spans* $\mathcal{S}^-(\Sigma)$.

IV. DIFFERENT CONTROLLER STRUCTURES

In this section, we consider three different controller structures used commonly in discrete systems. All three controllers are observer based, but the type of observer (or state estimator) used in each one is structurally different. The estimators considered here are (1) prediction estimator, (2) current estimator and (3) reduced order estimator. Both prediction estimator and current estimator are full order observers. The reduced order estimator is a current estimator but uses the reduced order observer. The prediction estimator estimates the state $x(k+1)$ based on the measurements $y(k)$ up to and including the (k)-th instant, where as the current estimator estimates $x(k+1)$ based on the measurements $y(k+1)$ up to and including the $(k+1)$-th instant. Since in the prediction estimator based controller, the current estimated value of control does not depend on the most current value of the measurement, it might not be as accurate as it could be in the current estimator based controller. However, the prediction estimator based controller could avail itself the entire sampling period to do the required computations and hence is commonly used when the needed computations are excessive. In contrast, when the needed computations can be done in a short time compared to the sampling period, current estimator based controller can easily be used. We note that prediction estimator forces an inherent time delay which otherwise is absent in the structure of controller. As can be expected, the three different controllers have different capabilities regarding LTR; but as will be seen shortly there exists a common mathematical machinery to analyze them under a single frame work. In the sections to follow, we will systematically do LTR analysis using a generic controller which could be any one of these three controllers. In such an analysis, we shall use the following notation :

$C(z) :=$ The transfer function of the controller,

$L(z) := C(z)P(z) =$ The achieved loop transfer function,

$S(z) := [I_m + L(z)]^{-1} =$ The achieved sensitivity function,

$T(z) := I_m - S(z) =$ The achieved complimentary sensitivity function,

$E(z) := L_t(z) - L(z) =$ Loop recovery error,

$M(z) :=$ The recovery matrix (to be defined later on),

$M^0(z) :=$ A part of the recovery matrix $M(z)$ that can be rendered zero,

$M^e(z) :=$ A part of the recovery matrix $M(z)$ that cannot be rendered zero and hence termed as recovery error matrix,

$\mathbf{T}^{\mathrm{R}}(\Sigma) :=$ The set of either exactly or asymptotically recoverable target loops for Σ.

The above notation applies to a generic controller; however, whenever we refer to a particular controller, we use appropriate subscripts to identify them. Subscripts p, c and r are used respectively to represent prediction, current, and reduced order estimator based controllers. For example, $L_p(z)$, $M_c^e(z)$ and $\mathbf{T}_r^{\mathrm{R}}(\Sigma)$ denote respectively the achieved loop transfer function with a prediction estimator based controller, the recovery error matrix when a current estimator based controller is used, and the set of recoverable target loops for Σ using a reduced order estimator based controller.

We now proceed to give the structural details of the controllers considered here.

Prediction Estimator Based Controller :

The dynamic equations of the controller are

$$\begin{cases} \hat{x}(k+1) = A\hat{x}(k) + Bu(k) + K_p[y(k) - C\hat{x}(k) - Du(k)], \\ u(k) = \hat{u}(k) = -F\hat{x}(k), \end{cases} \tag{15}$$

where K_p is the gain chosen so that $A - K_pC$ is asymptotically stable. The transfer function of the controller is

$$\mathbf{C}_p(z) = F[zI_n - A + BF + K_pC - K_pDF]^{-1}K_p. \tag{16}$$

Current Estimator Based Controller :

Here without loss of generality, we assume that the matrices C and D are in the form,

$$C = \begin{bmatrix} C_0 \\ C_1 \end{bmatrix} \quad \text{and} \quad D = \begin{bmatrix} D_0 \\ 0 \end{bmatrix}, \tag{17}$$

where D_0 is of maximal rank, i.e., $\text{rank}(D) = \text{rank}(D_0) = m_0$. Thus, the output y can be partitioned as,

$$\begin{bmatrix} y_0(k) \\ y_1(k) \end{bmatrix} = \begin{bmatrix} C_0 \\ C_1 \end{bmatrix} x(k) + \begin{bmatrix} D_0 \\ 0 \end{bmatrix} u(k).$$

The dynamic equations of the controller are

$$\begin{cases} \hat{x}(k+1) = A\hat{x}(k) + Bu(k) + K_c\left(\begin{bmatrix} y_0(k) \\ y_1(k+1) \end{bmatrix} - C_c\hat{x}(k) - D_cu(k)\right), \\ \hat{u}(k) = u(k) = -F\hat{x}(k), \end{cases} \tag{18}$$

where

$$C_c = \begin{bmatrix} C_0 \\ C_1A \end{bmatrix} \quad \text{and} \quad D_c = \begin{bmatrix} D_0 \\ C_1B \end{bmatrix}, \tag{19}$$

and where the gain K_c is chosen so that $A - K_cC_c$ is asymptotically stable. The transfer function from $-u$ to y that results in using the current estimator is then given by

$$\mathbf{C}_c(z) = F\left[zI_n - A + K_cC_c + BF - K_cD_cF\right]^{-1} K_c \begin{bmatrix} I_{m_0} & 0 \\ 0 & zI_{p-m_0} \end{bmatrix}. \tag{20}$$

Perhaps, some explanation regarding the structure of the current estimator (18) is in order. It is a generalization for nonstrictly proper systems of the existing current estimator given in Franklin et al [11]. Here we note that $y_0(k)$ and $y_1(k)$ together form the measurement vector $y(k)$. In view of (17), $y_0(k)$ depends on the control $u(k)$ explicitly, where as $y_1(k)$ does not depend on any control at the instant k. In order that the controller be physically realizable, in arriving at $\hat{x}(k+1)$, the current estimator utilizes $y_1(k+1)$ which is a part of the measurement at instant $k+1$, and $y_0(k)$ which is a part of the measurement at instant k. If the given system Σ is strictly proper, $y_0(k)$ is nonexistent and the current estimator (18) coalesces with that given in Franklin *et al* [11], Maciejowski [15] and Zhang and Freudenberg [35].

Reduced Order Estimator Based Controller :

Again, without any loss of generality but for simplicity of presentation, it is assumed that the matrices C and D are transformed into the form,

$$C = \begin{bmatrix} 0 & C_{02} \\ I_{p-m_0} & 0 \end{bmatrix} \quad \text{and} \quad D = \begin{bmatrix} D_0 \\ 0 \end{bmatrix}. \tag{21}$$

Then Σ can be partitioned as follows,

$$\begin{cases} \begin{pmatrix} x_1(k+1) \\ x_2(k+1) \end{pmatrix} = \begin{bmatrix} A_{11} & A_{12} \\ A_{21} & A_{22} \end{bmatrix} \begin{pmatrix} x_1(k) \\ x_2(k) \end{pmatrix} + \begin{bmatrix} B_{11} \\ B_{22} \end{bmatrix} u(k), \\ \begin{pmatrix} y_0(k) \\ y_1(k) \end{pmatrix} = \begin{bmatrix} 0 & C_{02} \\ I_{p-m_0} & 0 \end{bmatrix} \begin{pmatrix} x_1(k) \\ x_2(k) \end{pmatrix} + \begin{bmatrix} D_0 \\ 0 \end{bmatrix} u(k). \end{cases} \tag{22}$$

We observe that $y_1 = x_1$ is already available and need not be estimated. Thus we need to estimate only the state variable x_2. We first rewrite the state equation for x_1 in terms of the output y_1 and state x_2 as follows,

$$y_1(k+1) = A_{11}y_1(k) + A_{12}x_2(k) + B_{11}u(k). \tag{23}$$

Since $y_1(k+1)$ and $y_1(k)$ are known, (23) can be rewritten as

$$\tilde{y}_1(k) = A_{12}x_2(k) + B_{11}u(k) = y_1(k+1) - A_{11}y_1(k). \tag{24}$$

Thus, observation of x_2 is made via (24) as well as by

$$y_0(k) = C_{02}x_2(k) + D_0u(k).$$

Now, a reduced order system suitable for estimating the state x_2 is given by

$$\begin{cases} x_2(k+1) = & A_r\, x_2(k) + B_r\, u(k) + A_{21}\, y_1(k), \\ \begin{pmatrix} y_0(k) \\ \tilde{y}_1(k) \end{pmatrix} = & y_r(k) = C_r\, x_2(k) + D_r\, u(k) \end{cases} \tag{25}$$

where

$$A_r = A_{22}\ ,\quad B_r = B_{22}\ ,\quad C_r = \begin{bmatrix} C_{02} \\ A_{12} \end{bmatrix}\ ,\quad D_r = \begin{bmatrix} D_0 \\ B_{11} \end{bmatrix}. \tag{26}$$

Based on equation (25), we can construct a reduced order estimate of the state x_2 as,

$$\begin{aligned} \hat{x}_2(k+1) = A_r\hat{x}_2(k) + B_r u(k) + A_{21}y_1(k) \\ + K_r\left[y_r(k) - C_r\hat{x}_2(k) - D_r u(k)\right], \end{aligned} \tag{27}$$

where K_r, the reduced order estimator gain matrix, is chosen such that $A_r - K_r C_r$ is asymptotically stable. Since $\hat{x}_2(k+1)$ depends on the measurement $y_1(k+1)$, the reduced order estimator (27) belongs to the class of current estimators. For the purpose of implementing it, (27) can be rewritten by partitioning $K_r = [K_{r0},\ K_{r1}]$ in conformity with y_0 and $\tilde{y}_1$ and by defining the following variable $v(k)$,

$$v(k) = \hat{x}_2(k) - K_{r1}y_1(k). \tag{28}$$

Then the reduced order estimator based controller is given by

$$\begin{cases} v(k+1) = (A_r - K_r C_r)v(k) + (B_r - K_r D_r)u(k) + G_r y(k), \\ u(k) = \hat{u}(k) = -F_1 x_1(k) - F_2\hat{x}_2(k) = -F_2 v(k) - [0,\ F_1 + F_2 K_{r1}]y(k), \end{cases} \tag{29}$$

where

$$F = [F_1,\ \ F_2],\qquad G_r = [K_{r0},\ A_{21} - K_{r1}A_{11} + (A_r - K_r C_r)K_{r1}]. \tag{30}$$

The transfer function from $-u$ to y that results in using the reduced order estimator is then given by

$$\begin{aligned} \mathbf{C}_r(z) = F_2[zI - A_r + K_r C_r + B_r F_2 - K_r D_r F_2]^{-1} \\ \cdot \left(G_r - (B_r - K_r D_r)[0,\ F_1 + F_2 K_{r1}]\right) + [0,\ F_1 + F_2 K_{r1}]. \end{aligned} \tag{31}$$

Proposition 1. *For the case when Σ is right invertible and the matrix D is of maximal rank, all the controllers considered here, namely, the prediction, current and reduced order estimator based controllers, are one and the same.*

Proof : When Σ is right invertible and matrix D is of maximal rank, we note that in current estimator, $D = D_0$, C_1 is empty, $C_c = C_0 = C$ and $D_c = D_0 = D$. On the other hand, in reduced order estimator, we have $A_{22} = A$, $B_{22} = B$, $C_{02} = C$, $A_r = A_{22} = A$, $B_r = B_{22} = B$, $C_r = C_{02} = C$, $D_r = D_0 = D$. Using these facts, it is easy to verify the above proposition. ∎

We now proceed to do some preliminary analysis of the loop recovery error $E(z)$. It turns out that the expression, $E(z) = L_t(z) - L(z)$, is not well suited for loop transfer recovery analysis. Realising this, for the class of systems he considered, Goodman [12] related $E(z)$ to a matrix $M(z)$, hereafter called the recovery matrix. The following lemma generalizes Goodman's result for general nonstrictly proper systems and for all the three controllers considered here.

Lemma 1. *Let a system Σ be stabilizable and detectable. Also, let $L_t(z) = F\Phi B$ be an admissible target loop, i.e., $L_t(z) \in \mathrm{T}(\Sigma)$. Then the loop recovery error $E(z)$ between the target loop transfer function $L_t(z)$ and that realized by any one of the controllers described earlier, can be written in the form,*

$$E(z) = M(z)[I_m + M(z)]^{-1}(I_m + F\Phi B). \tag{32}$$

Furthermore, for all $\omega \in \Omega$,

$$E^*(j\omega) = 0 \text{ if and only if } M^*(j\omega) = 0$$

where Ω is the set of all $0 \leq |\omega| \leq \pi/T$ for which $L_t^*(j\omega)$ and $L^*(j\omega) = \mathbf{C}^*(j\omega)P^*(j\omega)$ are well defined (i.e., all the required inverses exist). The expression for the recovery matrix $M(z)$ depends on the controller used. In particular, for each one of the controllers considered earlier, we have the following expressions,

$$M_p(z) = F(zI_n - A + K_pC)^{-1}(B - K_pD), \tag{33}$$

$$M_c(z) = F\left(zI_n - A + K_cC_c\right)^{-1}\left(B - K_cD_c\right), \tag{34}$$

$$M_r(z) = F_2(zI - A_r + K_rC_r)^{-1}(B_r - K_rD_r). \tag{35}$$

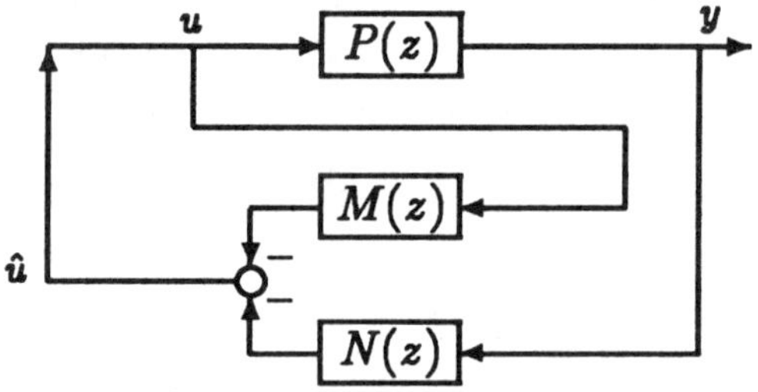

Figure 2: Plant and controller configuration.

Proof : See Appendix A. ∎

A physical interpretation of the the recovery matrix $M(z)$ can be given. To do so, one can view the controller as a device having two inputs, (1) the plant input u and (2) the plant output y as shown in Figure 2. Then, $-M(z)$ is the transfer function from the plant input point to the controller output point while $\tilde{M}(z)$ is the transfer function from the plant input point to the estimated state $\hat{x}$. That is, one can write

$$\hat{U}(z) = -F\hat{X}(z) = -M(z)U(z) - N(z)Y(z), \tag{36}$$

and

$$\hat{X}(z) = \tilde{M}(z)U(z) + \tilde{N}(z)Y(z). \tag{37}$$

Here, depending on the controller used, the expressions for $M(z)$ are as in (33), (34) and (35). Also, $\tilde{M}(z)$ is such that $M(z) = F\tilde{M}(z)$. Moreover, depending on the controller used, the expressions for $N(z)$ and $\tilde{N}(z)$ are as given below.

$$N_p(z) = F(zI - A + K_pC)^{-1}K_p \ , \ N_p(z) = F\tilde{N}_p(z)\,, \tag{38}$$

$$N_c(z) = (zI - A + K_cC_c)^{-1}K_c \begin{bmatrix} I & 0 \\ 0 & zI \end{bmatrix} \ , \ N_c(z) = F\tilde{N}_c(z)\,,$$

$$N_r(z) = F_2(zI - A_r + K_rC_r)^{-1}G_r + [0,\ F_1 + F_2K_{r1}] \ , \ N_r(z) = F\tilde{N}_r(z)\,.$$

In view of the above expressions, Lemma 1 implies that whenever LTR is achieved by the controller, the controller output does not entail any feedback from the plant input point. On the other hand, the state estimate $\hat{x}$ in general depends on the plant input. The significance of Lemma 1 can be seen in two ways. It converts the LTR analysis problem into a study of conditions under which the recovery matrix $M(z)$ can be rendered zero. Also, it unifies the study of $M(z)$ for all the three controllers into a single mathematical framework. To see this explicitly, let us define the auxiliary systems Σ_c and Σ_r which are respectively characterized by the matrix quadruples (A, B, C_c, D_c) and(A_r, B_r, C_r, D_r). Then we have the following observation.

Observation 1.

1. *The LTR mechanism for a given system Σ using a current estimator based controller can be studied using the auxiliary system Σ_c and a a prediction estimator based controller constructed for it.*

2. *The LTR mechanism for a given system Σ using a reduced order estimator based controller can be studied using the auxiliary system Σ_r and a prediction estimator based controller constructed for it where in F_2 takes the place of F.*

In view of Lemma 1 and Observation 1, our study of LTR for all the three controllers is unified and reduces to the study of an appropriate recovery matrix $M(z)$. In order to further cement such a unification, we need to investigate the relationship between the structural properties of Σ_c and Σ, as well as between those of Σ_r and Σ. The following propositions delineates such relationships.

Proposition 2.

1. *Σ_c is of (non-) minimum phase if and only if Σ is of (non-) minimum phase.*

2. *Σ_c is stabilizable and detectable if and only if Σ is stabilizable and detectable.*

3. *Invariant zeros of Σ_c contain invariant zeros of Σ and $z = 0$.*

4. *Orders of infinite zeros of Σ_c are reduced by one from those of Σ.*

5. *Σ_c is left invertible if and only if Σ is left invertible.*

6. *$\mathcal{V}^+(\Sigma_c) = \mathcal{V}^+(\Sigma)$.*

7. *$\mathcal{S}^-(\Sigma_c) = \mathcal{S}^-(\Sigma) \cap \{x \mid Cx \in \mathrm{Im}\,(D)\}$.*

8. $\mathcal{S}^-(\Sigma_c) = \emptyset$ *if and only if* Σ *is left invertible and of minimum phase with no infinite zeros of order higher than one.*

Proof : See Appendix B. ∎

Proposition 3.

1. Σ_r *is of (non-) minimum phase if and only if* Σ *is of (non-) minimum phase.*
2. Σ_r *is detectable if and only if* Σ *is detectable.*
3. *Invariant zeros of* Σ_r *are the same as those of* Σ.
4. *Orders of infinite zeros of* Σ_r *are reduced by one from those of* Σ.
5. Σ_r *is left invertible if and only if* Σ *is left invertible.*
6. $\begin{pmatrix} 0_{(p-m_0)\times(n-p+m_0)} \\ I_{(n-p+m_0)} \end{pmatrix} \mathcal{V}^+(\Sigma_r) = \mathcal{V}^+(\Sigma)$.
7. $\begin{pmatrix} 0_{(p-m_0)\times(n-p+m_0)} \\ I_{(n-p+m_0)} \end{pmatrix} \mathcal{S}^-(\Sigma_r) = \mathcal{S}^-(\Sigma) \cap \{x \mid Cx \in \mathrm{Im}\,(D)\}$.
8. $\mathcal{S}^-(\Sigma_r) = \emptyset$ *if and only if* Σ *is left invertible and of minimum phase with no infinite zeros of order higher than one.*

Proof : See [26]. ∎

Remark 1. *For a left invertible minimum phase system* Σ *with* $D = 0$, *it is simple to see that*

$$\mathcal{S}^-(\Sigma_c) = \mathcal{S}^-(\Sigma_r) = \mathcal{S}^-(\Sigma) \cap \{x \mid Cx \in \mathrm{Im}\,(D)\} = \emptyset$$

if and only if CB *is of maximal rank. Also, for a nonstrictly proper SISO system* Σ,

$$\mathcal{S}^-(\Sigma) = \mathcal{S}^-(\Sigma_c) = \mathcal{S}^-(\Sigma_r) = \mathcal{S}^-(\Sigma) \cap \{x \mid Cx \in \mathrm{Im}\,(D)\} = \emptyset$$

if and only if it is of minimum phase.

V. GENERAL LTR ANALYSIS

This section deals with the general analysis of LTR mechanism using any one of the three controllers discussed in the last section. Notationally, in all our general discussions here, we deal with the given system Σ characterized by the quadruple (A, B, C, D) and the prediction estimator based controller in which K_p is the observer gain. In view of Observation 1, all the general discussions presented here can be particularized to current and reduced order estimator based controllers with appropriate notational changes. In all our main theorems, we will however explicitly point out the capabilities of each controller as they could be different for each case.

As is evident from Lemma 1, the nucleus of LTR analysis is the study of $M_p(z)$ to ascertain how and when it can or cannot be rendered zero. The required study of $M_p(z)$ can be undertaken in two ways, with or without the prior knowledge of F that prescribes the target loop transfer function $L_t(z)$. Note that the study of $M_p(z)$ without the prior knowledge of F imitates the traditional LQG design philosophy in which the two tasks of obtaining F and K_p are separated. Keeping this in mind, our goal in the first subsection to follow is to study $M_p(z)$ without taking into account any specific characteristics of F. The second subsection, devoted to LTR analysis while taking into account appropriate characteristics of F, complements the analysis of the first subsection. Decomposing $M_p(z)$ as $F\tilde{M}_p(z)$, the study of $M_p(z)$ without knowing F is the same one as the study of $\tilde{M}_p(z)$. A detailed study of $\tilde{M}_p(z)$ leads to a fundamental Lemma 2 involving with an eigenstructure assignment to the observer dynamic matrix $A - K_pC$ by an appropriate design of K_p. This Lemma 2 reveals the limitations of the given system as a consequence of its structural properties in recovering an arbitrary target loop transfer function via the given controller structure. Thus it leads to Theorem 2 which, for each controller, shows under what conditions on Σ the set of recoverable target loops $\mathbf{T}^R(\Sigma)$ is equal to the set of admissible target loops $\mathbf{T}(\Sigma)$. Most of the results available so far in the literature can then be seen to be special cases of Theorem 2. Also, Lemma 2 enables one to decompose $\tilde{M}_p(z)$ into two essential parts, $\tilde{M}_p^0(z)$ and $\tilde{M}_p^e(z)$. The first part $\tilde{M}_p^0(z)$ can be rendered zero by an appropriate eigenstructure assignment to $A - K_pC$, while the second part $\tilde{M}_p^e(z)$ in general cannot be rendered zero, by any means, although our analysis of $\tilde{M}_p^e(z)$ reveals a multitude of ways by which it can be shaped. The decomposition of $\tilde{M}_p(z)$ into two parts and the subsequent analysis of each part forms the core of entire analysis given throughout this paper. In particular, it leads to several important results given in this section. For example, Theorem 3 characterizes the loop transfer function as well as the sensitivity and complementary sensitivity functions achievable by the considered controller. On the other hand, Theorem 4 shows the subspace $\mathcal{S}^e \in \mathbf{R}^m$ in which $\tilde{M}_p^e(z)$ can be rendered zero, i.e, the projections of the target and achievable sensitivity and complementary sensitivity functions onto $\mathcal{S}^e$ can match each other. Next, in Subsection B, Theorem 5 develops the necessary and sufficient conditions a target loop transfer function $L_t(z)$ has to satisfy so that it can be recoverable for the given system Σ. On the other hand, Theorem 6 develops the necessary and sufficient conditions on Σ so that it has at least one recoverable target loop transfer function. Subsection C generalizes the results of Subsections A and B when recovery is important over a prescribed subspace of the control space. Furthermore, our analysis in this section reveals the mechanism of pole-zero cancellation between the controller eigenvalues and the input or output decoupling zeros of Σ.

A. Recovery Analysis For An Arbitrary Target Loop

In this subsection, we consider that the target loop transfer function $L_t(z) = F\Phi B$ is arbitrary. That is, we do not take into account any specific characteristics of $L_t(z)$ in analyzing the LTR mechanism. As mentioned before, we will focus our attentions on the prediction estimator based controller with gain K_p. Then, as implied by Lemma 1, $\tilde{M}_p(z)$ as given below forms the basis of our study,

$$\tilde{M}_p(z) = (zI_n - A + K_pC)^{-1}(B - K_pD). \tag{39}$$

It is evident that the gain K_p is the only free design parameter in $\tilde{M}_p(z)$. First of all, in order to guarantee the closed-loop stability, K_p must be such that $A - K_pC$ is an asymptotically stable matrix. The remaining freedom in choosing K_p can then be used for the purpose of achieving LTR. We note that exact loop transfer recovery (ELTR) is possible for an arbitrary F if and only if

$$\tilde{M}_p^*(j\omega) = (e^{j\omega T}I_n - A + K_pC)^{-1}(B - K_pD) \equiv 0.$$

However, due to the nonsingularity of $(e^{j\omega T}I_n - A + K_pC)^{-1}$, the fact that $\tilde{M}_p^*(j\omega) \equiv 0$ implies that $B - K_pD \equiv 0$. Hence, rendering all the parts of $\tilde{M}_p^*(j\omega)$ zero is possible only for a very restrictive class of systems. In general only certain parts of $\tilde{M}_p^*(j\omega)$ can be rendered zero. To proceed with our analysis, for clarity of presentation we will temporarily assume that $A - K_pC$ is nondefective. This allows us to expand $\tilde{M}_p(z)$ and hence $M_p(z)$ in a dyadic form,

$$\tilde{M}_p(z) = \sum_{i=1}^{n} \frac{\tilde{R}_i}{z - \lambda_i} \tag{40}$$

where the residue $\tilde{R}_i$ is given by

$$\tilde{R}_i = W_i V_i^{H}[B - K_p D]. \tag{41}$$

Here W_i and V_i are respectively the right and left eigenvectors associated with an eigenvalue λ_i of $A - K_pC$ and they are scaled so that $WV^H = V^H W = I_n$ where

$$W = [W_1, W_2, \cdots, W_n] \quad \text{and} \quad V = [V_1, V_2, \cdots, V_n]. \tag{42}$$

Remark 2. *The assumption that K_p is selected so that $A - K_pC$ is nondefective is not essential. However, it simplifies our presentation. A removal of this condition necessitates the use of generalized right and left eigenvectors of $A - K_pC$ instead of the right and left eigenvectors W_i and V_i and consequently the expansion of $\tilde{M}_p(z)$ requires a double summation in place of the single summation used in (40).*

We are looking for conditions under which the i-th term of $\tilde{M}_p(z)$ in (40) can be made zero for each $i = 1$ to n. There is only one possibility in discrete-time LTR to do so, namely, assigning λ_i to any location in $\mathbb{C}^{\odot}$ while simultaneously rendering the corresponding residue $\tilde{R}_i$ zero. In other words, such a possibility corresponds to appropriate finite eigenstructure assignment to $A - K_pC$ to render $\tilde{R}_i$ zero. In continuous systems, besides the above possibility, there exists another possibility, namely, assigning λ_i asymptotically large in the negative half s-plane so that a term of the type

$$\frac{\tilde{R}_i}{s - \lambda_i}$$

tends to zero as $\lambda_i \to \infty$. This possibility deals with an infinite eigenstructure assignment to $A - K_pC$. The possibility of assigning an infinite eigenstructure, however, does not exist in discrete systems since λ_i is restricted to $\mathbf{C}^{\odot}$ in order to guarantee the stability of the resulting closed-loop system. Given the fact that $|\lambda_i|$ cannot go to ∞, it is easy to observe that the *notions of exact* LTR (ELTR) *and asymptotic* LTR (ALTR) *in discrete-time systems are equivalent in the sense that any target loop that is asymptotically recoverable is also exactly recoverable and vice versa.* Because of this, throughout this paper, whenever we talk about recovery, we mean both exact and asymptotic recovery. For example, whenever we say that an admissible target loop is recoverable, we mean by it that the specified target loop is exactly as well as asymptotically recoverable as stated in definition 4. Nevertheless, we will in Part 2 of this paper, distinguish between ELTR and ALTR. This is because, as we mentioned in the introduction, some optimization based design methods such as H_∞ norm minimization methods some times lead to suboptimal designs that correspond to asymptotic recovery. To be specific, in optimization methods, one normally generates a sequence of observer gains by solving parameterized algebraic Riccati equations. As the parameter tends to a certain value, the corresponding sequence of H_∞ norms of the resulting recovery matrices tends to a limit which is the infimum of the H_∞ norm of the recovery matrix over the set of all possible observer gains. A suboptimal solution is obtained when one selects an observer gain corresponding to a particular value of the parameter. On the other hand, in eigenstructure assignment methods, the required observer gain is obtained without solving any parameterized equations. Thus some times the observer gain K_p is designed as a function of a parameter and some other times independent of it.

The following lemma answers the question of how many residues $\tilde{R}_i$ can be rendered zero by an appropriate finite eigenstructure assignment of $A - K_pC$.

Lemma 2. *Let λ_i and V_i be an eigenvalue and the corresponding left eigenvector of $A - K_pC$ for any gain K_p such that $A - K_pC$ is asymptotically stable. Then the maximum possible number of $\lambda_i \in \mathbf{C}^{\odot}$ which satisfy the condition $V_i^{\mathrm{H}}[B - K_pD] = 0$ is $n_a^- + n_b$. A total of n_a^- of these λ_i coincide with the system invariant zeros which are in $\mathbf{C}^{\odot}$ (the so called stable or minimum phase invariant zeros) and the remaining n_b eigenvalues can be assigned arbitrarily to any locations in $\mathbf{C}^{\odot}$. All the eigenvectors V_i that correspond to these $n_a^- + n_b$ eigenvalues span the subspace $\mathbf{R}^n / S^-(\Sigma)$. Moreover, the n_a^- eigenvectors V_i which correspond to the eigenvalues which* *coincide with the system invariant zeros in $\mathbf{C}^{\odot}$ coincide with the corresponding left state zero directions and span the subspace $\mathcal{V}^*(\Sigma)/\mathcal{V}^+(\Sigma)$.*

Proof : See Appendix B of [6]. ∎

Remark 3. *Instead of rendering the $n_a^- + n_b$ residues $\tilde{R}_i$ mentioned in Lemma 2 exactly zero, if one prefers, they can be rendered asymptotically zero as certain parameter tends to a particular limit. In that case n_a^- eigenvalues coincide asymptotically with the n_a^- minimum phase invariant zeros while the corresponding eigenvectors in the limit coincide with the corresponding left state zero directions and span the subspace $\mathcal{V}^*(\Sigma)/\mathcal{V}^+(\Sigma)$. As stated earlier, in this part of paper, we will not distinguish between such exact and asymptotic assignments.*

Lemma 2 points out that there are altogether $n_a^- + n_b$ eigenvalues which can be assigned inside $\mathbf{C}^{\odot}$ so that the corresponding terms of $\tilde{M}_p(z)$ in its dyadic expansion (40) are zero. This fact leads to some structural conditions on Σ so that any arbitrary admissible target loop can be recovered. This is explored in the following theorem.

Theorem 2. *Consider a stabilizable and detectable system Σ. Depending upon the controller used, we have the following characterization of Σ so that any arbitrary admissible target loop can be recovered.*

1. Prediction estimator based controller: *Any arbitrary admissible target loop is recoverable, i.e., $\mathbf{T}_p^{\mathrm{R}}(\Sigma) = \mathbf{T}(\Sigma)$, if and only if Σ is left invertible and of minimum phase with no infinite zeros (i.e., D is maximal rank).*

2. Current estimator based controller: *Any arbitrary admissible target loop is recoverable, i.e., $\mathbf{T}_c^{\mathrm{R}}(\Sigma) = \mathbf{T}(\Sigma)$, if and only if Σ is left invertible and of minimum phase with no infinite zeros of order higher than one.*

3. Reduced order estimator based controller: *Any arbitrary admissible target loop is recoverable, i.e., $\mathbf{T}_r^{\mathrm{R}}(\Sigma) = \mathbf{T}(\Sigma)$, if and only if Σ is left invertible and of minimum phase with no infinite zeros of order higher than one.*

Proof : Let us take the case of a prediction estimator based controller. The fact that Σ is left invertible and of minimum phase with no infinite zeros implies that $n_a^+ = n_c = n_f = 0$. Thus $n_a^- + n_b = n$. Hence the result follows from (32) and Lemma 2. Conversely, it is simple to see that the recoverability of all the admissible target loops implies that $V_i^{\mathrm{H}}(B - K_p D) = 0$, $i = 1, \cdots, n$. Then by Lemma 2, we know that this is possible only when $n_a^- + n_b = n$. Hence, $n_a^+ = n_c = n_f = 0$, and thus Σ is left invertible and of minimum phase with no infinite zeros. Now, for the case of current and reduced order observer based controllers, in view of Propositions 2 (i.e., item 8) and 3 (i.e., item 8), we note that $n_a^+ + n_c + n_f$ corresponding to both Σ_c and Σ_r is equal to zero if and only if Σ is left invertible and of minimum phase with no infinite zeros of order higher than one. ∎

We have the following interesting special case of Theorem 2.

Corollary 1. *Let Σ be left invertible and of minimum phase with $D = 0$ and CB of maximal rank. Then all infinite zeros of Σ are of order one, and hence it follows from Theorem 2 that $\mathrm{T}_c^{\mathrm{R}}(\Sigma) = \mathrm{T}_r^{\mathrm{R}}(\Sigma) = \mathrm{T}(\Sigma)$, i.e., all the admissible target loops are recoverable by appropriate current and reduced order estimator based controllers; but not in general by a prediction estimator based controller.*

Proof : It is obvious. ∎

The above result was obtained earlier by Maciejowski [15] and Zhang and Freudenberg [35] for the case of a current estimator based controller.

As is evident by Theorem 2, the required structural conditions for recovery of any arbitrary admissible target loop are very stringent, and call for $n_a^- + n_b$ to be equal to the dimension n of Σ. To see what is and what is not feasible when $n_a^- + n_b \neq n$, and to emphasize explicitly the behavior of each term of $\tilde{M}_p(z)$, let us partition the dyadic expansion (40) of $\tilde{M}_p(z)$ into three parts, each part having a particular type of characteristics,

$$\tilde{M}_p(z) = \tilde{M}_p^-(z) + \tilde{M}_p^b(z) + \tilde{M}_p^e(z), \tag{43}$$

where

$$\tilde{M}_p^-(z) = \sum_{i=1}^{n_a^-} \frac{\tilde{R}_i^-}{z - \lambda_i^-}, \quad \tilde{M}_p^b(z) = \sum_{i=1}^{n_b} \frac{\tilde{R}_i^b}{z - \lambda_i^b}$$

and

$$\tilde{M}_p^e(z) = \sum_{i=1}^{n_a^+ + n_c + n_f} \frac{\tilde{R}_i^e}{z - \lambda_i^e}.$$

In the above partition, appropriate superscripts $-$, b, and e are added to $\tilde{R}_i$ and λ_i in order to associate them respectively with $\tilde{M}_p^-(z)$, $\tilde{M}_p^b(z)$, and $\tilde{M}_p^e(z)$. Next, define the following sets where $n^e = n_a^+ + n_c + n_f$:

$$\Lambda^- = \{\lambda_i^- ; i=1 \text{ to } n_a^-\},\ V^- = \{V_i^- ; i=1 \text{ to } n_a^-\},\ W^- = \{W_i^- ; i=1 \text{ to } n_a^-\}$$

$$\Lambda^b = \{\lambda_i^b ; i=1 \text{ to } n_b\},\quad V^b = \{V_i^b ; i=1 \text{ to } n_b\},\quad W^b = \{W_i^b ; i=1 \text{ to } n_b\}$$

$$\Lambda^e = \{\lambda_i^e ; i=1 \text{ to } n_e\},\quad V^e = \{V_i^e ; i=1 \text{ to } n_e\},\quad W^e = \{W_i^e ; i=1 \text{ to } n_e\}.$$

We now note that various parts of $\tilde{M}_p(z)$ have the following interpretation:

1. $\tilde{M}_p^-(z)$ contains n_a^- terms. The n_a^- eigenvalues of $A - K_pC$ represented in it form a set Λ^-. In accordance with Lemma 2, there exists a gain K_p such that $\tilde{M}_p^-(z)$ can be rendered identically zero by assigning the elements of Λ^- to coincide with the system minimum phase invariant zeros while the corresponding set of left eigenvectors V^- coincides with the the corresponding set of left state zero directions.

2. $\tilde{M}_p^b(z)$ contains n_b terms. The n_b eigenvalues of $A - K_pC$ represented in it form a set Λ^b. In accordance with Lemma 2, there exists a gain K_p such that $\tilde{M}_p^b(z)$ can be rendered zero by assigning the elements of Λ^b to arbitrary locations in $\mathbb{C}^{\odot}$.

3. $\tilde{M}_p^e(z)$ contains $n^e = n_a^+ + n_c + n_f$ terms. The n_e eigenvalues of $A - K_pC$ represented in $\tilde{M}_p^e(z)$ form a set Λ^e. In view of Lemma 2, $\tilde{M}_p^e(z)$ cannot in general be rendered zero by any assignment of Λ^e and the associated sets of right and left eigenvectors W^e and V^e.

Since both $\tilde{M}_p^-(z)$ and $\tilde{M}_p^b(z)$ can be rendered zero, for future use, we can combine them into one term,

$$\tilde{M}_p^0(z) = \tilde{M}_p^-(z) + \tilde{M}_p^b(z).$$

We define likewise, $\Lambda^0 = \Lambda^- \cup \Lambda^b$, $W^0 = W^- \cup W^b$, $V^0 = V^- \cup V^b$. Thus $\tilde{M}_p(z)$ can be rewritten as

$$\tilde{M}_p(z) = \tilde{M}_p^0(z) + \tilde{M}_p^e(z). \tag{44}$$

Since $\tilde{M}^e_p(z)$ cannot in general be rendered zero, it can be termed as recovery error matrix.

As the above discussion indicates, Lemma 2 forms the heart of the underlying mechanism of discrete-time LTR. It shows clearly what is and what is not feasible under what conditions. Although it does not directly provide methods of obtaining the gain K_p, it provides structural guide lines as to how certain eigenvalues and eigenvectors are to be assigned while indicating a multitude of ways in which freedom exists in assigning the other eigenvalues and eigenvectors of $A - K_pC$. These guidelines, in turn, can appropriately be channeled to come up with a design method to obtain an appropriate gain K_p. Leaving aside now the methods of design which will be discussed systematically in Part 2 of the paper, let us at this stage simply define the following sets of gains:

Definition 7. *Consider the system Σ. Let $\mathcal{K}^*_p(\Sigma)$ be a set of gains $K_p \in \mathbf{R}^{n\times p}$ such that (1) $A - K_pC$ is asymptotically stable, and (2) $\tilde{M}^0_p(z)$ is zero. In a similar manner, define $\mathcal{K}^*_c(\Sigma_c)$ and $\mathcal{K}^*_r(\Sigma_r)$ for systems Σ_c and Σ_r.*

As mentioned earlier, we do not parameterize here the gain K_p in terms of a tunable parameter σ. We deal only with a fixed gain K_p. In case if one deals with asymptotic recovery and thus with a sequence of controller gains $K_p(\sigma)$ for different values of σ, the set of recoverable gains is also parameterized and hence can be written as $\mathcal{K}^*_p(\Sigma, \sigma)$. In that case, one defines $\mathcal{K}^*_p(\Sigma, \sigma)$ as a set of gains $K_p(\sigma) \in \mathbf{R}^{n\times p}$ such that (1) $A - K_p(\sigma)C$ is asymptotically stable for all $\sigma > \sigma^*$ where $0 \leq \sigma^* < \infty$, (2) the limits, as $\sigma \to \infty$, of all the eigenvalues of $A - K_p(\sigma)C$ remain in $\mathbf{C}^{\odot}$, and (3) $\tilde{M}^0_p(z)$ is either identically zero or asymptotically zero. Similarly, $\mathcal{K}^*_c(\Sigma_c, \sigma)$ and $\mathcal{K}^*_r(\Sigma_r, \sigma)$ are defined for systems Σ_c and Σ_r. Such a characterization is useful in design methods dealt with in Part 2 of the paper.

It is obvious that the sets of gains defined above are nonempty. We note also that whenever K_p is chosen as an element of $\mathcal{K}^*_p(\Sigma)$, the resulting error in the recovery matrix $M_p(z)$ is $M^e_p(z) = F\tilde{M}^e_p(z)$. As such $M^e_p(z)$ is called hereafter as the *'recovery error matrix'*.

Theorem 3 given below characterizes the achieved loop transfer function as well as sensitivity and complementary sensitivity functions.

Theorem 3. *Let the given system Σ be stabilizable and detectable. Also, let $L_t(z) = F\Phi B$ be an admissible target loop, i.e., $L_t(z) \in \mathbf{T}(\Sigma)$. Then for a prediction estimator based controller with estimator gain $K_p \in \mathcal{K}_p^*(\Sigma)$, we have*

$$E_p(z) = M_p^e(z)[I_m + M_p^e(z)]^{-1}[I_m + L_t(z)], \tag{45}$$

$$S_p(z) = S_t(z)[I_m + M_p^e(z)], \tag{46}$$

$$T_p(z) = T_t(z) - S_t(z)M_p^e(z), \tag{47}$$

and

$$\frac{\mid \sigma_i[S_p^*(j\omega)] - \sigma_i[S_t^*(j\omega)] \mid}{\sigma_{max}[S_t^*(j\omega)]} \leq \sigma_{max}[M_p^{e*}(j\omega)], \tag{48}$$

$$\frac{\mid \sigma_i[T_p^*(j\omega)] - \sigma_i[T_t^*(j\omega)] \mid}{\sigma_{max}[S_t^*(j\omega)]} \leq \sigma_{max}[M_p^{e*}(j\omega)]. \tag{49}$$

The above results are true for current and reduced order estimator based controllers as well provided the subscript p is changed to c and r, and quadruple (A, B, C, D) is changed to (A, B, C_c, D_c) and (A_r, B_r, C_r, D_r) respectively. Also, in the case of reduced order estimator based controller, F in (45) to (49) is to be replaced by F_2.

Proof : It follows from Lemma 2. ∎

Remark 4. *Theorem 2 is a special case of the above theorem. To see this, let us examine first the case when a prediction estimator based controller is used. If the given system Σ is left invertible and of minimum phase with no infinite zeros, then the recovery error matrix $\tilde{M}_p^e(z)$ is nonexistent and hence $E_p(z)$ can be rendered zero for all $z \in \mathbb{C}$. Similarly, if Σ is left invertible and of minimum phase with no infinite zeros of order higher than one, then $\tilde{M}_c^e(z)$ and $\tilde{M}_r^e(z)$ are nonexistent and hence the exact recovery is achievable by using either current or reduced order estimator based controllers. Thus, for the special cases considered in Theorem 2, the results of Theorem 3 are reduced to those of Theorem 2.*

Remark 5. *Theorem 3 also holds if we use the estimator gain $K_p(\sigma) \in \mathcal{K}_p^*(\Sigma, \sigma)$. However, in this case, the equalities in (45) to (47) should be replaced by pointwise convergences in z as $\sigma \to \infty$.*

As implied by Theorem 3, the recovery error matrix $M_p^e(z)$ plays a dominant role in the recovery process and hence it should be shaped to yield as best as possible the desired results. Shaping $M_p^e(z)$ involves selecting the set of eigenvalues Λ^e represented in $M_p^e(z)$ and the associated set of right and left eigenvectors W^e and V^e. Such a selection can be done in a number of ways subject to the constraints imposed in selecting the eigenvectors [16]. Since there is an ample amount of freedom in selecting Λ^e, W^e and V^e, there exists a set of admissible recovery error matrices, and such a set can be denoted as $\mathcal{M}_p^e(z)$. Similarly, sets $\mathcal{M}_c^e(z)$ and $\mathcal{M}_r^e(z)$ can be formulated for the current and reduced order estimator based controllers. Among the design methods to be discussed in Part 2 of the paper, the eigenstructure assignment methods make use of the available flexibility, and are capable of attaining any given $M_p^e(z) \in \mathcal{M}_p^e(z)$ while making $F\tilde{M}_p^0(z) = 0$. On the other hand, in optimization design methods, the optimal solution would render $F\tilde{M}_p^0(z) = 0$ while minimizing the H_∞ or H_2 norm of the recovery error matrix $M_p^e(z)$. Thus, $F\tilde{M}_p^0(z)$ is always rendered zero but the attained $M_p^e(z)$ varies from one design method to the other.

In multivariable systems, one interesting aspect of Theorem 3 is that there could exist a subspace of the control space in which $\tilde{M}_p^e(z)$ can be rendered zero. To pinpoint this, let

$$e_i = [B - K_p D]' V_i \ , \quad V_i \in V^e, \tag{50}$$

and let $\mathcal{E}^e$ be the subspace of $\mathbf{R}^m$,

$$\mathcal{E}^e = \mathrm{Span}\{e_i \mid V_i \in V^e\}. \tag{51}$$

Let the dimension of $\mathcal{E}^e$ be m_e. Now let

$$\mathcal{S}^e = \text{orthogonal complement of } \mathcal{E}^e \text{ in } \mathbf{R}^m. \tag{52}$$

Let P^e be the orthogonal projection matrix onto $\mathcal{S}^e$. Then the following theorem pinpoints the directional behavior of $\tilde{M}_p(z)$ and consequently the behavior of $S(z)$ and $T(z)$.

Theorem 4. *Let Σ be stabilisable and detectable, and $L_t(z)$ be a member of the set of admissible target loops $\mathrm{T}(\Sigma)$. Then for any $K_p \in \mathcal{K}_p^*(\Sigma)$, the corresponding prediction estimator based controller satisfies*

$$\tilde{M}_p(z) P^e = 0,$$

$$S_p(z) P^e = S_t(z) P^e$$

$$T_p(z)P^s = T_t(z)P^s,$$

where P^s is the orthogonal projection onto $\mathcal{S}^s \in \mathbf{R}^m$ as given in (52). The above results are true for current and reduced order estimator based controllers as well provided the subscript $_p$ is changed to $_c$ and $_r$, and quadruple (A, B, C, D) is changed to (A, B, C_c, D_c) and (A_r, B_r, C_r, D_r) respectively.

Proof : In view of the definitions of the matrix P^s and the subspaces $\mathcal{E}^s$ and $\mathcal{S}^s$, Theorem 3 implies the results of Theorem 4. ∎

It is interesting to note that although the projections of $S_p(z)$ and $S_t(z)$ and hence those of $T_p(z)$ and $T_t(z)$ onto $\mathcal{S}^s$ are equivalent, it need not be true that the projections of achieved and target loops, $\mathbf{C}_p(z)P(z)$ and $L_t(z)$, onto $\mathcal{S}^s$ are equivalent. That is, in general, $\mathbf{C}_p(z)P(z)P^s \neq L_t(z)P^s$. This is illustrated by the following example.

Example 1 : Consider a non-strictly proper discrete-time system characterized by

$$A = \begin{bmatrix} 3 & 0 \\ 0 & 2 \end{bmatrix}, \quad B = C = D = \begin{bmatrix} 1 & 0 \\ 0 & 1 \end{bmatrix},$$

which is invertible with two nonminimum phase invariant zeros at $z = 1$ and $z = 2$. Let the target loop $L_t(z)$ and target sensitivity function $S_t(z)$ be specified by

$$F = \begin{bmatrix} 3 & 0 \\ 0 & 2 \end{bmatrix}.$$

Let the prediction estimator gain be given by

$$K_p = \begin{bmatrix} 5 & -4 \\ 1 & 0 \end{bmatrix}.$$

Then, it is easy to calculate that

$$L_t(z) = F(zI - A)^{-1}B = \frac{\begin{bmatrix} 3z-6 & 0 \\ 0 & 2z-6 \end{bmatrix}}{z^2 - 5z + 6},$$

$$S_t(z) = [I + L_t(z)]^{-1} = \frac{\begin{bmatrix} z-3 & 0 \\ 0 & z-2 \end{bmatrix}}{z},$$

$$\mathbf{C}_p(z)P(z) = \frac{\begin{bmatrix} 15z^3 - 72z^2 + 108z - 48 & -12z^3 + 48z^2 - 36z \\ 2z^3 - 8z^2 + 8z & -16z^2 + 64z - 48 \end{bmatrix}}{z^4 - 15z^3 + 64z^2 - 100z + 48}$$

and

$$S_p(z) = \frac{\begin{bmatrix} z^3 - 15z^2 + 48z - 36 & 12z^2 - 48z + 36 \\ -2z^2 + 8z - 8 & z^3 - 8z + 8 \end{bmatrix}}{z^3}.$$

Now consider a subspace $\mathcal{S}^e$ having the orthogonal projection matrix P^s as

$$P^s = \begin{bmatrix} 0.5 & 0.5 \\ 0.5 & 0.5 \end{bmatrix}.$$

It is now straightforward to verify that

$$S_t(z)P^s = S_p(z)P^s = \frac{\begin{bmatrix} z-3 & z-3 \\ z-2 & z-2 \end{bmatrix}}{2z}.$$

This implies that the projections of the target and the achieved sensitivity functions on to the subspace $\mathcal{S}^e$ are equal to one another. On the other hand, we have

$$L_t(z)P^s = \frac{\begin{bmatrix} 3(z-2) & 3(z-2) \\ 2(z-3) & 2(z-3) \end{bmatrix}}{2(z^2 - 5z + 6)}$$

and

$$\mathbf{C}_p(z)P(z)P^s = \frac{\begin{bmatrix} 3z^3 - 24z^2 + 72z - 48 & 3z^3 - 24z^2 + 72z - 48 \\ 2z^3 - 24z^2 + 72z - 48 & 2z^3 - 24z^2 + 72z - 48 \end{bmatrix}}{2(z^4 - 15z^3 + 64z^2 - 100z + 48)}.$$

Obviously, $L_t(z)P^s \neq \mathbf{C}_p(z)P(z)P^s$. That is, the projections of the target and the achieved loop transfer functions on to the subspace $\mathcal{S}^e$ do not match each other. □

In view of the directional behavior of the recovery error matrix $\tilde{M}_p^e(z)$ as given by Theorem 4, one could try to shape it in a particular way so as to obtain the recovery of sensitivity and complementary sensitivity functions in certain desired directions or one could try to shape $\tilde{M}_p^e(z)$ so that the subspace $\mathcal{S}^e$ has as large a dimension as possible, i.e., the subspace $\mathcal{E}^e$ has as small a dimension as possible. In this regard, we note that we have already selected Λ^0 and the corresponding sets of eigenvectors V^0 so that $\tilde{M}_p^0(z)$ is zero. We also note that although all the $n_a^+ + n_c + n_f$ vectors $V_i \in V^e$ can be selected to be linearly independent, the corresponding $e_i \equiv [B - K_pD]'V_i$ need not be linearly independent. In fact for a given $e \neq 0$, the equation

$$e = [B - K_pD]'V,$$

has $n - m + 1$ linearly independent solutions for V. Of course, not all such $n-m+1$ vectors could be admissible eigenvectors of $A - K_pC$ for different eigenvalues of it inside $\mathbf{C}^{\odot}$, and moreover some or all of these $n - m + 1$ vectors could also be linearly dependent on already selected eigenvectors in the set V^0. Thus the problem of shaping $\mathcal{E}^e$ is to find an admissible set of eigenvalues λ_i and vectors e_i, $i = 1$ to $n_a^+ + n_c + n_f$, which are not necessarily linearly independent, but the associated eigenvectors V_i of $A - K_pC$ satisfying $e_i = [B - K_pD]'V_i$, $i = 1$ to $n_a^+ + n_c + n_f$, together with the vectors in the set V^0 form n linearly independent vectors. This problem of selecting an admissible set (λ_i, e_i) is very much related to the traditional problem of distributing the modes of a closed-loop system to various output components by an appropriate selection of the closed-loop eigenstructure. This traditional problem of 'shaping the output response characteristics' of a closed-loop system has been studied for continuous systems first by Moore [16] and Shaked [30] and more recently by Sogaard-Andersen [33] although to this date there exists no systematic design procedure.

The above discussion focuses on how to shape the subspace $\mathcal{S}^e$ in which $\tilde{M}_p(z)$, $S_t(z)$ and $T_t(z)$ are recovered. A practical problem of interest could be to achieve recovery of $\tilde{M}_p(z)$, $S_t(z)$ and $T_t(z)$ in a prescribed subspace $\mathcal{S}^e$. We will discuss this aspect of the problem in Subsection C.

We will next examine the open-loop eigenvalues of the prediction estimator based controller $\mathbf{C}_p(z)$ and the mechanism of pole zero cancellation between the controller eigenvalues and the input or output decoupling zeros [22] of the system. It is important to know the eigenvalues of $\mathbf{C}_p(z)$ as they are included among the invariant zeros of the closed-loop system and hence affect the performance of it, e.g., command following. The controller transfer function is given by (16) while the eigenvalues of it are

$$\lambda[A - K_pC - BF + K_pDF].$$

To study the nature of these eigenvalues, let

$$\det[zI_n - A + K_pC] = \phi^0(z)\phi^e(z)$$

where $\phi^0(z)$ and $\phi^e(z)$ are polynomials in z whose zeros are the eigenvalues of $A - K_pC$ that belong to the sets Λ^0 and Λ^e respectively. Also, let

$$F\tilde{M}_p^e(z) = \frac{R^e(z)}{\phi^e(z)} \tag{53}$$

where $R^e(z)$ is a polynomial matrix in z. Now consider the following when $K_p \in \mathcal{K}_p^*(\Sigma)$:

$$\begin{aligned}
&\det[zI_n - A + K_pC + BF - K_pDF] \\
&\quad = \det[zI_n - A + K_pC]\det[I_n + (zI_n - A + K_pC)^{-1}(B - K_pD)F] \\
&\quad = \phi^0(z)\phi^e(z)\det[I_m + F(zI_n - A + K_pC)^{-1}(B - K_pD)] \\
&\quad = \phi^0(z)\phi^e(z)\det[I_m + F\tilde{M}_p(z)] \\
&\quad = \phi^0(z)\phi^e(z)\det[I_m + F\tilde{M}_p^e(z)] \\
&\quad = \phi^0(z)\phi^e(z)\det[I_m + \frac{R^e(z)}{\phi^e(z)}] \\
&\quad = \phi^0(z)\frac{\det[I_m\phi^e(z) + R^e(z)]}{[\phi^e(z)]^{m-1}}.
\end{aligned} \tag{54}$$

We note that the observer can be designed such that $\phi^0(z)$ and $\phi^e(z)$ are coprime. Thus the open-loop eigenvalues of the controller (16) are the zeros of $\phi^0(z)$ and

$$\frac{\det[I_m\phi^e(z) + R^e(z)]}{[\phi^e(z)]^{m-1}}.$$

Thus Λ^0 is contained among the eigenvalues of the controller. Although Λ^0 is in $\mathbf{C}^{\odot}$, there is no guarantee that the zeros of

$$\frac{\det[I_m\phi^e(z) + R^e(z)]}{[\phi^e(z)]^{m-1}}$$

are in $\mathbf{C}^{\odot}$. Hence the controller may or may not be open-loop stable. However, it is obvious to see that if recovery is achieved, i.e., if $F\tilde{M}_p^e(z) = 0$ and thus $R^e(z) = 0$, then the eigenvalues of the prediction estimator based controller are given by the roots of $\phi^0(z)\phi^e(z) = 0$. As the roots of $\phi^0(z)\phi^e(z) = 0$ are inside the unit circle, whenever recovery is achieved, the controller is open-loop stable. This is also apparent from equation (36). Note that in this case $\mathbf{C}_p(z) = N_p(z)$ where $N_p(z)$ is as in (38).

In general, the loop transfer function $\mathbf{C}_p(z)P(z)$ has $2n$ eigenvalues, n of them coming from the given system Σ and the other n from the controller. However, there are several cancellations among the input or output decoupling zeros [22] of $\mathbf{C}_p(z)P(z)$ and the controller eigenvalues. The following Lemma 3 which is a generalization of a similar one in Goodman [12], explores such a cancellation.

Lemma 3. *Let λ be an eigenvalue of $A - K_pC$ with the corresponding left eigenvector V such that $V^H[B - K_pD] = 0$. Then λ is an eigenvalue of $A - K_pC - BF + K_pDF$ with the corresponding left eigenvector as V. Moreover, λ cancels an input decoupling zero of $\mathbf{C}_p(z)P(z)$.*

Proof : See Appendix C. ∎

Thus, in view of Lemma 2, the above lemma implies that whatever may be the matrix F, if the controller is appropriately designed, there are $n_a^- + n_b$ cancellations among the eigenvalues of the controller and the input decoupling zeros of $\mathbf{C}_p(z)P(z)$. As will be seen in the next subsection, there may be additional cancellations if F satisfies certain properties.

Remark 6. *Equation (54) and Lemma 3 are equally true for current as well as reduced order estimator based controllers. In these cases, notationally the quadruple (A, B, C, D), F and $\mathbf{C}_p(z)$ are to be replaced respectively by (A, B, C_c, D_c), F and $\mathbf{C}_c(z)$ for a current estimator based controller, and by (A_r, B_r, C_r, D_r), F_2 and $\mathbf{C}_r(z)$ for a reduced order estimator based controller.*

B. Analysis For Recoverable Target Loops

In the previous subsection, loop transfer recovery analysis is conducted without taking into account any knowledge of F. It involves essentially the study of the matrix $\tilde{M}_p(z)$ or $M_p(z)$ to ascertain when it can or cannot be rendered zero. This subsection complements the analysis of Subsection A by taking into account the knowledge of F. Obviously then, the analysis of this subsection is a study of $M_p(z) = F\tilde{M}_p(z)$. One of the important questions that needs to be answered here is as follows. What class of target loops can be recovered for the given system? As it forms a coupling between the analysis and design, characterization of $L_t(z)$ to determine whether it can be recovered or not for the given system, plays an extremely important role. That is, although the physical tasks of designing F and K_p are separable, one can benefit enormously by knowing ahead what kind of target loops are recoverable. The necessary and sufficient conditions developed here on $L_t(z)$ for its recoverability, turn out to be constraints on the finite and infinite zero structure of $L_t(z)$ as related to the corresponding structure of Σ. An interpretation of these conditions reveals that recovery of $L_t(z)$ for general nonminimum phase systems is possible under a variety of conditions.

Another important question that arises before one undertakes formulating any target loop transfer function $L_t(z)$ for a given system Σ is as follows. What are the necessary and sufficient conditions on Σ so that it has at least one recoverable target loop? An answer to this question obviously helps a designer to remodel the given plant if necessary by appropriately modifying the number or type of plant inputs or/and outputs. To answer the question posed, we develop here an auxiliary system Σ^{au} of Σ for each one of the three controllers considered here, and show that the set of recoverable target loops is nonempty if and only if Σ^{au} is stabilizable by a static output feedback controller. A close look at this condition reveals a surprising necessary condition, namely, strong stabilizability of Σ is necessary for it to have at least one recoverable target loop.

Finally, another aspect of the analysis given here shows the mechanism of pole-zero cancellation between the controller eigenvalues and the input or output decoupling zeros of Σ for the case when F is known.

We proceed now to give the following result regarding the recoverability of a target loop transfer function $L_t(z) = F\Phi B$ for the given system Σ.

Theorem 5. *Consider a stabilizable and detectable discrete-time system Σ characterized by a matrix quadruple (A, B, C, D), which is not necessarily left invertible and not necessarily of minimum phase. Then, an admissible target loop transfer function $L_t(z)$ of Σ, i.e., $L_t(z) \in \mathrm{T}(\Sigma)$, is recoverable if and only if the following condition is satisfied, depending on the controller used.*

1. *For a prediction estimator based controller, the condition is that $\mathcal{S}^-(\Sigma) \subseteq \mathrm{Ker}\,(F)$.*

2. *For a current estimator based controller, the condition is that $\mathcal{S}^-(\Sigma) \cap \{x \mid Cx \in \mathrm{Im}\,(D)\} \subseteq \mathrm{Ker}\,(F)$.*

3. *For a reduced order estimator based controller, the condition is that $\mathcal{S}^-(\Sigma) \cap \{x \mid Cx \in \mathrm{Im}\,(D)\} \subseteq \mathrm{Ker}\,(F)$.*

Thus the set of recoverable target loops under each controller is characterised as follows:

1. Prediction estimator based controller :

$$\mathbf{T}_p^{\mathrm{R}}(\Sigma) = \{L_t(z) \in \mathbf{T}(\Sigma) \mid \mathcal{S}^-(\Sigma) \subseteq \mathrm{Ker}\,(F)\}.$$

2. Current estimator based controller :

$$\mathbf{T}_c^{\mathrm{R}}(\Sigma) = \{L_t(z) \in \mathbf{T}(\Sigma) \mid \mathcal{S}^-(\Sigma) \cap \{x \mid Cx \in \mathrm{Im}\,(D)\} \subseteq \mathrm{Ker}\,(F)\}.$$

3. Reduced order estimator based controller :

$$\mathbf{T}_r^{\mathrm{R}}(\Sigma) = \{L_t(z) \in \mathbf{T}(\Sigma) \mid \mathcal{S}^-(\Sigma) \cap \{x \mid Cx \in \mathrm{Im}\,(D)\} \subseteq \mathrm{Ker}\,(F)\}.$$

Proof : For the case of a prediction estimator based controller, the result follows easily from Theorem 5.1 of [6] or equivalently from Theorem 3.3 of [24] with obvious notational changes. For the current and reduced order estimator based controllers, the results follow from Propositions 2 and 3 and Theorem 5.1 of [6]. ∎

Remark 7. *We note that* $\mathbf{T}_p^{\mathrm{R}}(\Sigma) \subseteq \mathbf{T}_c^{\mathrm{R}}(\Sigma) = \mathbf{T}_r^{\mathrm{R}}(\Sigma)$.

Remark 8. *Let Σ be a strictly proper minimum phase system having infinite zeros of order one, i.e., CB of maximal rank. Then, it is shown in Goodman [12] that a target loop $L_t(z) = F\Phi B$ is recoverable by a prediction estimator based controller if $FB = 0$. This result can easily be deduced from Theorem 5.*

Several interpretations emerge from the recoverability conditions on the target loops given in Theorem 5. In fact the constraints given in Theorem 5 are nothing more than constraints on the finite and infinite zero structure and invertibility properties of $L_t(z)$. Some interesting interpretations in this regard can easily be exemplified as follows.

1. If Σ is not left invertible, any recoverable $L_t(z)$ is not left invertible. On the other hand, left invertibility of Σ does not necessarily imply that a recoverable $L_t(z)$ is left invertible. That is, whenever Σ is left invertible, a recoverable $L_t(z)$ could be either left invertible or not left invertible.

2. Any left invertible and recoverable $L_t(z)$ must contain the unstable invariant zero structure of Σ. A recoverable but not left invertible $L_t(z)$ does not necessarily contain the unstable invariant zero structure of Σ (see Example 2).

Example 2 : Consider a non-strictly proper discrete-time system with sampling period $T = 1$, and characterized by

$$A = \begin{bmatrix} -1 & -1 & -1 \\ 0 & 0 & 0 \\ 1 & 1 & 1 \end{bmatrix}, \quad B = \begin{bmatrix} 1 & 0 \\ 0 & 0 \\ 0 & 1 \end{bmatrix},$$

$$C = \begin{bmatrix} -1 & -1 & -1 \\ 1 & 1 & -1 \end{bmatrix}, \quad D = \begin{bmatrix} 1 & 0 \\ 0 & 1 \end{bmatrix}.$$

This system is invertible and is of nonminimum phase as its invariant zeros are at $\{0,\ 0,\ 2\}$. Let the target loop $L_t(z)$ and the target sensitivity function $S_t(z)$ be specified by

$$F = \begin{bmatrix} 0.5 & 0.5 & 0 \\ -0.5 & -0.5 & 0 \end{bmatrix}.$$

The triple (A, B, F) forms a minimum phase and right invertible system and hence it does not contain the unstable invariant zero structure of Σ. However, for this example, it can easily be seen that

$$\mathcal{S}^-(\Sigma) = \mathcal{S}^-(\Sigma) \cap \{x \mid Cx \in \operatorname{Im}(D)\} = \begin{bmatrix} 0 \\ 0 \\ 1 \end{bmatrix}.$$

Since $\mathcal{S}^-(\Sigma)$ is contained in $\operatorname{Ker}(F)$, in accordance with Theorem 5, the given target loop is recoverable by all the three controllers considered in this paper. In fact, since Σ is invertible with D being of maximal rank, all these controllers are one and the same (see Proposition 1). Thus we can conclude that a recoverable $L_t(z)$ need not contain the unstable invariant zero structure of Σ. The following is the prediction estimator gain which achieves ELTR,

$$K_p = \begin{bmatrix} 1 & 0 \\ 0 & 0 \\ -1 & 0 \end{bmatrix}.$$

□

Our aim next is to develop the conditions on Σ so that the set of recoverable target loops is nonempty. We have the following theorem.

Theorem 6. *Consider a stabilizable and detectable system Σ characterized by a matrix quadruple (A, B, C, D), which is not necessarily of minimum phase and which is not necessarily left invertible. Let $\overline{C}_p$ and $\overline{C}_c$ be any full rank matrices of dimensions $(n_a^- + n_b) \times n$ and $(n_a^- + n_b + m_f) \times n$, respectively, such that*

1. $\text{Ker}\,(\overline{C}_p) = \mathcal{V}^+(\Sigma)$, *and*
2. $\text{Ker}\,(\overline{C}_c) = \mathcal{S}^-(\Sigma) \cap \{x \mid Cx \in \text{Im}\,(D)\}$.

Also, let $\overline{C}_r = \overline{C}_c$. Define three auxiliary systems:

1. *Σ_p^{au} characterized by the matrix triple $(A, B, \overline{C}_p)$,*
2. *Σ_c^{au} characterized by the matrix triple $(A, B, \overline{C}_c)$, and*
3. *Σ_r^{au} characterized by the matrix triple $(A, B, \overline{C}_r)$.*

Then we have the following results depending upon the controller used :

1. Prediction estimator based controller: $\mathbf{T}_p^R(\Sigma)$ *is nonempty if and only if Σ_p^{au} is stabilizable by a static output feedback controller.*
2. Current estimator based controller: $\mathbf{T}_c^R(\Sigma)$ *is nonempty if and only if Σ_c^{au} is stabilizable by a static output feedback controller.*
3. Reduced order estimator based controller: $\mathbf{T}_r^R(\Sigma)$ *is nonempty if and only if Σ_r^{au} is stabilizable by a static output feedback controller.*

Proof : See Appendix D. ∎

Theorem 6 gives the necessary and sufficient conditions under which the set of recoverable target loops for each controller is nonempty. However, the conditions given there are not conducive to any intuitive feelings. The following corollary gives a necessary condition which is surprising as well as intuitively appealing.

Corollary 2. *The strong stabilizability of the given system Σ is a necessary condition for it to have at least one recoverable target loop under any of the three controllers discussed here.*

Proof : See Appendix E. ∎

Corollary 2 tells us that any given system Σ must be strongly stabilizable in order to have at least one recoverable target loop. On the other hand, as seen from Theorem 6, strong stabilizability of Σ alone is not sufficient for $\mathbf{T}^{\mathrm{R}}(\Sigma)$ to be nonempty. The following example illustrates this.

Example 3 : Consider a non-strictly proper discrete-time system Σ characterized by

$$A=\begin{bmatrix}1&0\\0&1\end{bmatrix},\quad B=\begin{bmatrix}1&0\\0&1\end{bmatrix},\quad C=\begin{bmatrix}0&1\\-1&0\end{bmatrix},\quad D=\begin{bmatrix}1&0\\0&1\end{bmatrix}.$$

This system is invertible with two unstable invariant zeros at $\{1+i,\ 1-i\}$. Also, this system is strongly stabilizable as it can be stabilized by the following stable output feedback compensator,

$$\mathbf{C}(z)=\frac{1}{z^2+z+0.5}\begin{bmatrix}-0.25 & -0.5(z+0.5)\\ 0.5(z+0.5) & -0.25\end{bmatrix}.$$

However, it is simple to verify that for this system,

$$\mathcal{S}^-(\Sigma)=\mathcal{S}^-(\Sigma)\cap\{x\mid Cx\in\operatorname{Im}(D)\}=\mathbf{R}^2.$$

Hence, it follows from Theorem 5 that any recoverable target loop $L_t(z)$ must have the following form of F,

$$F=\begin{bmatrix}0&0\\0&0\end{bmatrix}.$$

But it is trivial to see that such a target loop is not admissible. Thus, this system has no recoverable target loop although it is strongly stabilizable.

□

We now proceed to discuss the possible pole-zero cancellations for the recoverable target loops. First, we need the following lemma which is a generalization of a similar one in Goodman [12].

Lemma 4. *Let λ be an eigenvalue of $A-K_pC$ with the corresponding right eigenvector W such that $FW=0$. Then λ is an eigenvalue of $A-K_pC-BF+K_pDF$ with the corresponding right eigenvector as W. Moreover, λ cancels an output decoupling zero of $\mathbf{C}_p(z)$.*

We have the following theorem.

Theorem 7. *If $E_p^*(j\omega) = 0$ for all $0 \leq |\omega| < \infty$, i.e., ELTR is achieved, then every eigenvalue of $A - K_pC - BF + K_pDF$ cancels either an output decoupling zero of $\mathbf{C}_p(z)$ or an input decoupling zero of $\mathbf{C}_p(z)P(z)$.*

Proof : We note that $E^*(j\omega) = 0$ for all $0 \leq |\omega| < \infty$ if and only if either $FW_i = 0$ or $V_i^H[B - K_pD] = 0$ or both. Hence the result follows from Lemmas 3 and 4. ∎

Remark 9. *Lemma 4 and Theorem 7 are equally true for current and reduced order estimator based controllers. In these cases, notationally the quadruple (A, B, C, D), F and $\mathbf{C}_p(z)$ are to be replaced respectively by (A, B, C_c, D_c), F and $\mathbf{C}_c(z)$ for a current estimator based controller, and by (A_r, B_r, C_r, D_r), F_2 and $\mathbf{C}_r(z)$ for a reduced order estimator based controller.*

In view of Lemmas 3 and 4, and Theorem 7, whenever recovery of a given target loop occurs, there are n cancellations among the eigenvalues of the controller and the output decoupling zeros of $\mathbf{C}_p(z)$ or the input decoupling zeros of $\mathbf{C}_p(z)P(z)$.

C. Recovery Analysis in A Given Subspace

In the last two subsections, we discussed recovery of a target loop transfer function $L_t(z) = F\Phi B$ when the recovery is required over the entire control space $\mathbf{R}^m$ and when the knowledge of state feedback gain F is either unknown or known. The traditional LTR problem as treated in there, concentrates on recovering a open-loop transfer function $L_t(z)$ which has been formed to take into account the given design specifications. Actually, design specifications are normally formulated in terms of certain required closed-loop sensitivity and complementary sensitivity functions, $S_t(z) = [I_m + L_t(z)]^{-1}$ and $T_t(z) = I_m - S_t(z)$. In LQG/LTR design philosophy, these given specifications are reflected in formulating an open-loop transfer function called the target loop transfer function. As discussed earlier, this aspect of determining a target loop transfer function is a first step in LQG/LTR design and falls in the category of loop shaping. Generating a target loop transfer function $L_t(z)$ at the present time is an engineering art and often involves the use of linear quadratic design in which the cost matrices are used as free design parameters to obtain the state feedback gain F and thus to obtain $L_t(z) = F\Phi B$ and $S_t(z) = [I_m + F\Phi B]^{-1}$. In the second step of design, the so called loop transfer recovery (LTR) design, $L_t(z)$ is recovered using a measurement feedback controller. Obviously, in the traditional LTR design where recovery is required over the entire control space $\mathbf{R}^m$, the recovery of $L_t(z)$ implies the recovery of the corresponding sensitivity function $S_t(z)$ and hence the recovery of the complementary sensitivity function $T_t(z)$. Conversely, in a similar manner, the recovery of $S_t(z)$ or equivalently that of $T_t(z)$, implies the recovery of $L_t(z)$. In other words, when recovery is required over the entire control space $\mathbf{R}^m$, recovering a certain target loop transfer function is equivalent to recovering a certain target sensitivity function. Thus, without loss of any freedom, historically, recovery of a target loop transfer function has been sought.

As seen in earlier sections, loop transfer recovery in the entire control space $\mathbf{R}^m$ is not possible in general. This may force a designer to seek recovery only in a chosen subspace $\mathcal{S}$ of the control space $\mathbf{R}^m$. In that case, it is natural to think of recovering the projections of both the target loop $L_t(z)$ and the sensitivity function $S_t(z)$ onto $\mathcal{S}$. However, as seen in Example 3, one may obtain the projections of achieved and target sensitivity functions onto $\mathcal{S}$ matching each other, but the projection of the correspondingly achieved loop transfer function may or may not match that of the target loop. This implies that the designer may have to choose between matching the projections onto $\mathcal{S}$ of (1) the achieved and the target sensitivity functions, and (2) the achieved and the target loop transfer functions. Since, most often design specifications are given in terms of sensitivity functions, it is natural to choose matching the projections onto $\mathcal{S}$ of the achieved and the target sensitivity functions. In view of this, in this section, we focus on the recovery of sensitivity functions over a subspace. For the case when $\mathcal{S}$ equals $\mathbf{R}^m$, obviously the sensitivity recovery formulation of this section coincides with the conventional LTR formulation. Thus this section can indeed be viewed as a generalization of the notion of traditional LTR to cover recovery over either the entire or any specified subspace $\mathcal{S}$ of the control space $\mathbf{R}^m$.

A brief outline of this subsection is as follows. At first, precise definitions dealing with the sensitivity recovery problem are given. Then, Lemma 5 is developed generalizing Lemma 1. It formulates the condition for the recoverability of a sensitivity function in $\mathcal{S}$ in terms of a matrix $M^s(z)$. Next,

Theorem 8 specifies the required conditions on Σ so that sensitivity recovery in $\mathcal{S}$ is possible for any arbitrarily specified target sensitivity function $S_t(z)$. Similarly, Theorem 9 specifies the necessary and sufficient conditions for the recoverability of a sensitivity function when the knowledge of F is known. On the other hand, Theorem 10 establishes the necessary and sufficient conditions so that the sets of recoverable sensitivity functions of the given system Σ for a specified subspace $\mathcal{S}$, is nonempty. An important aspect of recovery analysis in a subspace is to determine the maximum possible dimension of a recoverable subspace $\mathcal{S}$. This is discussed at the end of this section.

We have the following formal definitions.

Definition 8. *The set of admissible target sensitivity functions* $\mathbf{S}(\Sigma)$ *for a given system* Σ *is defined as follows:*

$$\mathbf{S}(\Sigma) := \{ S_t(z) \in \mathcal{M}^{m\times m}(\mathcal{R}_p) \mid S_t(z) = [I_m + L_t(z)]^{-1},\ L_t(z) \in \mathbf{T}(\Sigma) \}.$$

Definition 9. *Given* $S_t(z) \in \mathbf{S}(\Sigma)$ *and a subspace* $\mathcal{S} \in \mathbf{R}^m$, *we say* $S_t(z)$ *is recoverable in the subspace* $\mathcal{S}$ *if there exists a controller having the transfer function* $\mathbf{C}(z)$ *such that (i) the closed-loop system comprising of the controller and the plant is asymptotically stable, and (ii)* $S(z)P^s = S_t(z)P^s$, *where* $S(z)$ *is the achieved sensitivity function and* P^s *is the orthogonal projection matrix onto* $\mathcal{S}$.

Definition 10. *The set of recoverable* $S_t(z) \in \mathbf{S}(\Sigma)$ *in the given subspace* $\mathcal{S}$ *is denoted by* $\mathbf{S}^R(\Sigma, \mathcal{S})$.

As usual, subscripts p, c and r are respectively used to distinguish the above sets ($\mathbf{S}(\Sigma)$ and $\mathbf{S}^R(\Sigma, \mathcal{S})$) for prediction, current and reduced order estimator based controllers. Also, we note that the above definitions 8 to 10 are natural extensions of the corresponding definitions given earlier. In fact, the definitions 8 to 10 generalize the concept of recovery to a subspace and enable us to reanalyze all the results of the previous two subsections to cover the recovery in a given subspace $\mathcal{S}$.

The following lemma is analogous to Lemma 1.

Lemma 5. *Let the given system* Σ *be stabilizable and detectable. Also, let* $L_t(z) = F\Phi B$ *be an admissible target loop, i.e.,* $L_t(z) \in \mathbf{T}(\Sigma)$. *Then* $E^s(z)$, *the projection onto a given subspace* $\mathcal{S} \in \mathbf{R}^m$ *of the error between* *the achieved sensitivity function $S(z)$ and the target sensitivity function $S_t(z)$, is given by*

$$E^s(z) = [I_m + F\Phi B]^{-1} M^s(z) \tag{55}$$

where

$$M^s(z) = M(z)P^s. \tag{56}$$

Furthermore, for all $\omega \in \Omega$,

$$E^{s*}(j\omega) = 0 \quad \text{if and only if} \quad M^{s*}(j\omega) = 0,$$

where Ω is the set of all $0 \leq |\omega| \leq \pi/T$ for which $S_t^*(j\omega)$ and $S^*(j\omega)$ are well defined (i.e., all the required inverses exist). Here the expression for $M(z)$ depends on the controller used and in particular for each one of the three controllers considered in this paper, the needed expressions are as in (33), (34), and (35) with an appropriate suffix added to $M(z)$.

Proof : It is obvious. ∎

To proceed with the recovery analysis, let V^s be a matrix whose columns form an orthogonal basis of $\mathcal{S} \in \mathbf{R}^m$. Assume that the columns of V^s are scaled so that the norm of each column is unity. Let $P^s = V^s(V^s)'$ be the unique orthogonal projection matrix onto $\mathcal{S}$. Then, define three auxiliary systems Σ_p^s, Σ_c^s and Σ_r^s characterized, respectively, by the quadruples (A, BV^s, C, DV^s), (A, BV^s, C_c, D_cV^s) and $(A_r, B_rV^s, C_r, D_rV^s)$. Now treating each auxiliary system as the given system, one can re-discuss here mutatis mutandis all the results of Subsections A and B. In particular, we have the following theorems.

Theorem 8. *Consider a stabilizable and detectable discrete-time system Σ characterized by a matrix quadruple (A, B, C, D), which is not necessarily of minimum phase and which is not necessarily left invertible. Let V^s be a matrix whose columns form an orthogonal basis of a given subspace $\mathcal{S} \in \mathbf{R}^m$. Then any arbitrary admissible sensitivity function $S_t(z)$ of Σ, i.e., $S_t(z) \in \mathbf{S}(\Sigma)$ is recoverable in $\mathcal{S}$ if and only if the following condition is satisfied depending upon the controller used.*

1. Prediction estimator based controller: *Any arbitrary admissible sensitivity function $S_t(z)$ of Σ is recoverable if and only if the auxiliary system Σ_p^s is left invertible and of minimum phase with no infinite zeros (i.e., DV^s is of maximal rank).*

2. Current estimator based controller: *Any arbitrary admissible sensitivity function $S_t(z)$ of Σ is recoverable if and only if the auxiliary system Σ_c^s is left invertible and of minimum phase with no infinite zeros (i.e., $D_c V^s$ is of maximal rank).*

3. Reduced order estimator based controller: *Any arbitrary admissible sensitivity function $S_t(z)$ of Σ is recoverable if and only if the auxiliary system Σ_r^s is left invertible and of minimum phase with no infinite zeros (i.e., $D_r V^s$ is of maximal rank).*

Proof : The results are obvious in view of Theorem 2 and Lemma 5. ∎

Theorem 8 is concerned with the recovery analysis when F is arbitrary or unknown. As in Subsection B, one can formulate the recovery conditions for a known F as follows.

Theorem 9. *Consider a stabilizable and detectable system Σ characterized by a matrix quadruple (A, B, C, D), which is not necessarily of minimum phase and which is not necessarily left invertible. Let V^s be a matrix whose columns form an orthogonal basis of a given subspace $\mathcal{S} \in \mathbf{R}^m$. Then an admissible sensitivity function $S_t(z)$ of Σ, i.e., $S_t(z) \in \mathsf{S}(\Sigma)$, is recoverable in $\mathcal{S}$ if and only if the following condition is satisfied depending on the controller used.*

1. *For a prediction estimator based controller, the condition is that*

$$\mathcal{S}^-(\Sigma_p^s) \subseteq \operatorname{Ker}(F).$$

2. *For a current estimator based controller, the condition is that*

$$\mathcal{S}^-(\Sigma_c^s) \subseteq \operatorname{Ker}(F).$$

3. *For a reduced order estimator based controller, the condition is that*

$$\begin{pmatrix} 0 \\ I \end{pmatrix} \mathcal{S}^-(\Sigma_r^s) \subseteq \operatorname{Ker}(F).$$

Thus the set of recoverable sensitivity functions in the given subspace $\mathcal{S}$ is characterized as follows:

1. **Prediction estimator based controller:**

$$\mathbf{S}_p^{\mathrm{R}}(\Sigma, \mathcal{S}) = \{\, S_t(z) \in \mathbf{S}(\Sigma) \mid \mathcal{S}^-(\Sigma_p^s) \subseteq \mathrm{Ker}\,(F)\,\}.$$

2. **Current estimator based controller:**

$$\mathbf{S}_c^{\mathrm{R}}(\Sigma, \mathcal{S}) = \{\, S_t(z) \in \mathbf{S}(\Sigma) \mid \mathcal{S}^-(\Sigma_c^s) \subseteq \mathrm{Ker}\,(F)\,\}.$$

3. **Reduced order estimator based controller:**

$$\mathbf{S}_r^{\mathrm{R}}(\Sigma, \mathcal{S}) = \left\{\, S_t(z) \in \mathbf{S}(\Sigma) \mid \begin{pmatrix} 0 \\ I \end{pmatrix} \mathcal{S}^-(\Sigma_r^s) \subseteq \mathrm{Ker}\,(F) \right\}.$$

Proof : The proof is a consequence of Theorem 5. ∎

Remark 10. *If the given system Σ is strictly proper, i.e., $D = 0$, then it is simple to verify that*

$$\mathcal{S}^-(\Sigma_c^s) = \begin{pmatrix} 0 \\ I \end{pmatrix} \mathcal{S}^-(\Sigma_r^s) = \mathcal{S}^-(\Sigma_p^s) \cap \{\, x \mid Cx \in \mathrm{Im}\,(DV^s)\,\}.$$

This is not true in general for non-strictly proper systems.

In what follows, we give a necessary and sufficient condition under which $\mathbf{S}^{\mathrm{R}}(\Sigma, \mathcal{S})$ is non-empty for the given subspace $\mathcal{S} \in \mathbf{R}^m$. We have the following theorem.

Theorem 10. *Consider a stabilizable and detectable system Σ characterized by a matrix quadruple (A, B, C, D), which is not necessarily of minimum phase and which is not necessarily left invertible. Let V^s be a matrix whose columns form an orthogonal basis of a given subspace $\mathcal{S} \in \mathbf{R}^m$. Let $\overline{C}_p^s$, $\overline{C}_c^s$ and $\overline{C}_r^s$ be any full rank matrices such that*

1. $\mathrm{Ker}\,(\overline{C}_p^s) = \mathcal{S}^-(\Sigma_p^s)$,
2. $\mathrm{Ker}\,(\overline{C}_c^s) = \mathcal{S}^-(\Sigma_c^s)$, *and*
3. $\mathrm{Ker}\,(\overline{C}_r^s) = \begin{pmatrix} 0 \\ I \end{pmatrix} \mathcal{S}^-(\Sigma_r^s)$.

Define three auxiliary systems:

1. Σ_p^{aus} *characterised by the matrix triple* $(A, B, \overline{C}_p^s)$,

2. *Σ_c^{aus} characterised by the matrix triple $(A, B, \overline{C}_c^s)$, and*

3. *Σ_r^{aus} characterised by the matrix triple $(A, B, \overline{C}_r^s)$.*

Then we have the following results depending upon the controller used :

1. Prediction estimator based controller: $\mathbf{S}_p^R(\Sigma, \mathcal{S})$ *is nonempty if and only if Σ_p^{aus} is stabilisable by a static output feedback controller.*

2. Current estimator based controller: $\mathbf{S}_c^R(\Sigma, \mathcal{S})$ *is nonempty if and only if Σ_c^{aus} is stabilisable by a static output feedback controller.*

3. Reduced order estimator based controller: $\mathbf{S}_r^R(\Sigma, \mathcal{S})$ *is nonempty if and only if Σ_r^{aus} is stabilisable by a static output feedback controller.*

Proof : The results are a consequence of Theorem 6. ∎

An important aspect that arises when one is interested in the recovery analysis in a subspace is to determine the maximum possible dimension of a recoverable subspace $\mathcal{S}$. In this regard, our goal in what follows, analogous to continuous systems, is to prove that whatever may be the given target loop transfer function and whatever may be the number of unstable invariant zeros, there exists at least one $m-1$ dimensional subspace $\mathcal{S}$ of $\mathbf{R}^m$ which is always recoverable provided that the given system Σ satisfies some conditions. In what follows, for simplicity of presentation, we will make a technical assumption that all the unstable invariant zeros of Σ have geometric multiplicity equal to unity. We have the following theorem.

Theorem 11. *Let the given system Σ be left invertible with unstable invariant zeros having geometric multiplicity equal to unity. Then, there exists at least one $m-1$ dimensional subspace $\mathcal{S}$ of $\mathbf{R}^m$ such that any admissible target sensitivity function $S_t(z)$ of Σ, i.e., $S_t(z) \in \mathbf{S}(\Sigma)$, is recoverable in $\mathcal{S}$ provided the following condition is satisfied depending on the controller used.*

1. *For a prediction estimator based controller, the condition is that D be of maximal rank, i.e., Σ has no infinite zeros.*

2. *For a current estimator based controller, the condition is that Σ has no infinite zeros of order higher than one.*

3. ***For a reduced order estimator based controller, the condition is that Σ has no infinite zeros of order higher than one.***

Proof : See Appendix F. ∎

VI. DUALITY OF LTR BETWEEN THE INPUT AND OUTPUT BREAK POINTS

The target open-loop transfer functions can be designed when the loop is broken at either the input or the output point of the plant depending upon the given specifications. We have analyzed so far LTR recovery at the input point (LTRI), using any one of prediction, current or reduced order estimator based controllers. Now we like to consider LTR recovery when the loop is broken at the output point (LTRO). For continuous systems, such a method was introduced earlier by Kwakernaak [14] in connection with sensitivity recovery. LTRO is used when the designer specifications and the modelling of uncertainties are reflected at the output point of the plant. In the literature, it is commonly said that LTR recovery at the input and output points (LTRI and LTRO) are dual to one another. This duality is well understood in the case of prediction estimator based controllers. That is, in the case of LTRO, the first step is to design a prediction estimator based controller, via loop shaping techniques, whose loop transfer function meets the design specifications. The next step is to recover the transfer function of prediction estimator based controller via LTR technique. However, this kind of duality is not well understood when current or reduced order estimator based controllers are used. The confusion in the literature [15,35] arises because the duality is sought between the plant and the controller. The given plant Σ and the controller are not necessarily dual to one another whenever any controller other than prediction estimator based is used. An appropriate subsystem of Σ, such as Σ_c or Σ_r, has to be constructed, and then the duality has to be sought between the controller and the subsystem Σ_c or Σ_r rather than between the controller and the given plant Σ. That is, duality has to be sought in the loop transfer recovery analysis or controller design methodology rather than between the given plant and the controller. In order to avoid any confusion, we give below a formal step by step algorithm to show how duality arises for LTR recovery at the input and output points.

1. **Let a plant Σ be characterized by the quadruple (A, B, C, D). Also, let $P(z)$ be the transfer function of Σ,**

$$P(z) = C(zI_n - A)^{-1}B + D.$$

Let $L_t(z) = C(zI_n - A)^{-1}K$ be an admissible target open-loop transfer function, i.e., $\lambda(A - KC) \in \mathbf{C}^{\odot}$, when the loop is broken at the output point of the given plant. Then, in the configuration of Figure 1, we are seeking a controller $\mathbf{C}(z)$ such that the closed-loop system is asymptotically stable and

$$E^o(z) := L_t(z) - P(z)\mathbf{C}(z) \equiv 0 \quad \text{for all } z.$$

Here the controller $\mathbf{C}(z)$ could be of any type. In particular, it could be either a prediction, or a current, or a reduced order estimator based controller.

2. **Define a dual system Σ_d characterized by (A_d, B_d, C_d, D_d) where**

$$A_d := A', \quad B_d := C', \quad C_d := B', \quad D_d := D'. \tag{57}$$

Note that $P_d(z)$, the transfer function of the dual plant Σ_d is $P'(z)$. Let $L_d(z)$ be defined as

$$L_d(z) := L_t'(s) = F_d(zI_n - A_d)^{-1}B_d \text{ where } F_d := K'. \tag{58}$$

Let $L_d(z)$ be considered as a target loop transfer function for Σ_d when the loop is broken at the input point of Σ_d. Let a measurement feedback controller $\mathbf{C}_d(z)$ be used for Σ_d. Here the controller $\mathbf{C}_d(z)$ could be any one of the three controllers, depending upon how $\mathbf{C}(z)$ is chosen. Then, it is simple to verify that the loop recovery error $E^{id}(z)$ at the input point of Σ_d is

$$E^{id}(z) = L_d(z) - \mathbf{C}_d(z)P_d(z) = [E^o(z)]'$$

provided $\mathbf{C}_d(z) = \mathbf{C}'(z)$.

3. **For the purpose of analysis or design alone, consider the fictitious plant Σ_d and the fictitious target loop transfer function $L_d(z)$ as given in Step 2. Then, it is straightforward to verify that the target loop transfer function $L_t(z)$ is recoverable at the output point of the given**

plant Σ if and only if $L_d(z)$ is recoverable at the input point of the fictitious plant Σ_d. To be specific, construct prediction, current, and reduced order estimator based controllers, namely $\mathbf{C}_{pd}(z)$, $\mathbf{C}_{cd}(z)$, and $\mathbf{C}_{rd}(z)$ respectively as in (16), (20) and (29) while using the parameters A_d, B_d, C_d, D_d and F_d instead of A, B, C, D and F. Let $E_p^{id}(z)$, or $E_c^{id}(z)$ or $E_r^{id}(z)$ be the resulting loop recovery error at the input point of Σ_d when $\mathbf{C}_{pd}(z)$, or $\mathbf{C}_{cd}(z)$, or $\mathbf{C}_{rd}(z)$ is used respectively as a controller for Σ_d. Let

$$\mathbf{C}_p(z) = \mathbf{C}'_{pd}(z)\ ,\ \mathbf{C}_c(z) = \mathbf{C}'_{cd}(z) \text{ and } \mathbf{C}_r(z) = \mathbf{C}'_{rd}(z).$$

Let $E_p^o(z)$ or $E_c^o(z)$ or $E_r^o(z)$ be the loop recovery error at the output point of Σ when $\mathbf{C}_p(z)$, or $\mathbf{C}_c(z)$, or $\mathbf{C}_r(z)$ is used respectively as a controller for Σ. It is then straightforward to show that

$$E_p^o(z) = (\,E_p^{id}(z)\,)'\ ,\ E_c^o(z) = (\,E_c^{id}(z)\,)' \text{ and } E_r^o(z) = (\,E_r^{id}(z)\,)'.$$

Thus all the loop transfer recovery analysis at the input point of Σ_d as done in this chapter can easily be interpreted as the loop transfer recovery analysis at the output point of Σ. In other words, the above step by step discussion clearly shows how duality arises between the loop transfer recovery at the input and the output points whatever may be the type of controller used.

VII. CONCLUSIONS

Here we deal with issues concerning the analysis of loop transfer recovery problem using observer based controllers for general non-strictly proper not necessarily minimum phase discrete time systems. Three different observer based controllers, namely, 'prediction estimator', and full or reduced order type 'current estimator' based controllers , are used. As in our earlier work, all the analysis given here is independent of the methodology by which these controllers are designed. Moreover, the analysis corresponding to all these three controllers is unified into a single mathematical frame work. A fundamental difference between continuous time and discrete time systems is this. In the discrete case, as is well known, in order to preserve stability, all the closed-loop eigenvalues must be restricted to lie within the unit circle in complex plane. This implies that unlike continuous case which permits both finite as well as asymptotically infinite eigenvalue assignment, in the discrete case one is restricted to only finite eigenvalue assignment. Thus, in the continuous case, there exists target loops which are not exactly recoverable, but are asymptotically recoverable by appropriate infinite eigenstructure assignment. On the other hand, in discrete systems, since both asymptotic as well as exact recovery involves only finite eigenstructure assignment, every asymptotically recoverable target loop is also exactly recoverable and vice versa.

There are several fundamental results given here. At first, based on the structural properties of the given system, we decompose the recovery matrix between the target loop transfer function and that that can be achieved by any one of the controllers, into two distinct parts. The first part of recovery matrix can be rendered exactly (or asymptotically, if one chooses so) zero by an appropriate finite eigenstructure assignment of the controller dynamic matrix, while the second part cannot be rendered zero, by any means, although there exists a multitude of ways to shape it. Such a decomposition helps us to discover when and under what conditions, an arbitrarily specified target loop is recoverable by using any one of the three controllers considered. Also, it helps to characterize the recovery error and its singular value bounds, whenever a target loop is not recoverable. Thus it shows the limitations of the given system in recovering the target loop transfer functions as a consequence of its structural properties, namely finite and infinite zero structure and invertibility. The next issue of our analysis concentrates on characterizing the required necessary and sufficient conditions on the target loop transfer functions so that they are recoverable by the considered controller for the given system. As in the case of continuous systems, the conditions developed here on a target loop transfer function for its recoverability, turn out to be constraints on its finite and infinite zero structure as related to the corresponding structure of the given system. Next, necessary and sufficient conditions on the given system are established such that it has at least one recoverable target loop. In this regard, we show that strong stabilizability of the given system is necessary for it to have at least one recoverable target loop by any one of the three controllers considered here. Since recovery in all control loops in general is not feasible, our analysis next, focuses in developing the necessary or/and sufficient conditions under which recovery of target sensitivity and complementary sensitivity functions is possible in any specified subspace of the control space. Inherent in all the issues discussed here is the characterization of the resulting controller eigenvalues and possible pole zero cancellations. Such an investigation is important in view of the fact, controller eigenvalues become the invariant zeros of the closed-loop system and thus affect the performance with respect to command following and other design objectives.

To summarize, the analysis presented here adds a considerable amount of flexibility to the process of design and helps a designer to set meaningful goals at the onset of design. In other words, although the actual physical tasks of first designing a target loop and then designing an observer based controller are separable, one can link these two tasks philosophically by knowing ahead what is feasible and how. In a sequel to this paper, for each one of the controllers considered here, we will present design methodologies which are capable of utilizing the complete freedom a design can have as is discovered here.

Acknowledgement

The work of B. M. Chen and A. Saberi is supported in part by Boeing Commercial Airplane Group and in part by NASA Langley Research Center under grant contract NAG-1-1210. Also, B. M. Chen acknowledges Washington State University OGRD 1991 summer research assistantship.

Appendices

A. Proof of Lemma 1

To simplify and to unify our proof of Lemma 1, we first examine the following Luenberger estimator based controller:

$$\begin{cases} v(k+1) = Lv(k) + Gu(k) + Hy(k), \\ -u(k) = F\hat{x}(k) = Pv(k) + Vy(k), \end{cases} \tag{59}$$

where $v \in \mathbf{R}^r$ with r being the order of the controller. It is well known (see e.g., [20]) that $\hat{x}$ is an asymptotic estimate of the state x provided that (a) L is an asymptotically stable matrix and (b) there exists a matrix $T \in \mathbf{R}^{r \times n}$ satisfying the following conditions:

1. $TA - LT = HC$,

2. $G = TB - HD$,
3. $F = PT + VC$, and
4. $VD = 0$.

Then, following the procedure given in [18], it is simple to show that the loop recovery error realised by such a Luenberger estimator based controller is

$$E(z) = M(z)[I_m + M(z)]^{-1}(I_m + F\Phi B), \tag{60}$$

where

$$M(z) = P(zI - L)^{-1}G. \tag{61}$$

Next, it is straightforward to verify that the prediction estimator based controller of (15) is a special case of Luenberger estimator based controller in (59) with

$$\begin{cases} L = A - K_pC, & G = B - K_pD, & H = K_p, \\ P = F, & V = 0, & T = I. \end{cases}$$

Hence, equation (33) of Lemma 1 follows trivially from (60) and (61).

Similarly, in order to prove that (34) is valid for the current estimator based controller, let us first partition the gain matrix $K_c = [K_0, K_1]$ in conformity with $y_0(k)$ and $y_1(k)$. Also, define a new variable

$$v(k) := \hat{x}(k) - K_1y_1(k).$$

Then it is easy to rewrite the current estimator based controller (18) in terms of the new variable $v(k)$,

$$\begin{cases} v(k+1) = (A - K_cC_c)v(k) + (B - K_cD_c)u(k) \\ \qquad\qquad\qquad + [K_0, (A - K_cC_c)K_1]y(k), \\ -u(k) \ = F\hat{x} = Fv(k) + [0, FK_1]y(k). \end{cases} \tag{62}$$

Again, it is straightforward to verify that the current estimator based controller (62) is a special case of Luenberger estimator based controller in (59) with

$$\begin{cases} L = A - K_cC_c, & G = B - K_cD_c, & H = [K_0, (A - K_cC_c)K_1], \\ P = F, & V = [0, FK_1], & T = I - K_1C_1. \end{cases}$$

Then, (34) follows from (60) and (61).

As expected, the reduced order estimator based controller (29) is also a special case of (59) with

$$\begin{cases} L = A_r - K_r C_r, & G = B_r - K_r D_r, \qquad H = G_r, \\ P = F_2, & V = [0,\ F_1 + F_2 K_{r1}], \quad T = [-K_{r1},\ I]. \end{cases}$$

Then, once again in view of (60) and (61), we get (35). This completes the proof of Lemma 1. ∎

B. Proof of Proposition 2

Without loss of generality but for simplicity of presentation, we assume that the given system Σ is in the form of s.c.b, i.e., it is characterized by the quadruple $(\tilde{A}, \tilde{B}, \tilde{C}, \tilde{D})$ as in Theorem 1. Let us partition $x_f = [(x_{f1})', (x_{f0})', (x_{f2})']'$, where x_{f1} is the part of output associated with infinite zeros of order one, x_{f0} is the rest of output associated with infinite zeros of order higher than one and x_{f2} consists of the state variables corresponding to the rest of infinite zeros. Then it is simple to verify that C_f and B_f have the following forms,

$$C_f = \begin{bmatrix} I_{m_1} & 0 & 0 \\ \\ 0 & I & 0 \end{bmatrix}, \quad B_f = \begin{bmatrix} I_{m_1} & 0 \\ 0 & 0 \\ 0 & B_{f2} \end{bmatrix}.$$

Now, by appropriate permutation transformation of the state variables, we can partition the given system as follows,

$$x = \left[(x_{f1})', (x_{f0})', (x_b)', (x_a^-)', (x_a^+)', (x_c)', (x_{f2})'\right]'$$

and $\tilde{A} = A - B_0 C_0 =$

$$\begin{bmatrix} E_{f11} & E_{f10} & E_{b1} & E_{a1}^- & E_{a1}^+ & E_{c1} & E_{f12} \\ L_{f01} & L_{f00} & 0 & 0 & 0 & 0 & C_{f2} \\ L_{bf1} & L_{bf0} & A_{bb} & 0 & 0 & 0 & 0 \\ L_{af1}^- & L_{af0}^- & L_{ab}^- C_b & A_{aa}^- & 0 & 0 & 0 \\ L_{af1}^+ & L_{af0}^+ & L_{ab}^+ C_b & 0 & A_{aa}^+ & 0 & 0 \\ L_{cf1} & L_{cf0} & B_c E_{cb} & B_c E_{ca}^- & B_c E_{ca}^+ & A_{cc} & 0 \\ A_{f21} & A_{f20} & B_{f2} E_{b2} & B_{f2} E_{a2}^- & B_{f2} E_{a2}^+ & B_{f2} E_{c2} & A_{f22} + B_{f2} E_{f22} \end{bmatrix},$$

$$\tilde{B} = [\,B_0 \quad B_1\,] = \begin{bmatrix} B_{f01} & I_{m_1} & 0 & 0 \\ B_{f00} & 0 & 0 & 0 \\ B_{b0} & 0 & 0 & 0 \\ B_{a0}^- & 0 & 0 & 0 \\ B_{a0}^+ & 0 & 0 & 0 \\ B_{c0} & 0 & 0 & B_c \\ B_{f02} & 0 & B_{f2} & 0 \end{bmatrix},$$

$$\tilde{C} = \begin{bmatrix} C_0 \\ \\ C_1 \end{bmatrix} = \begin{bmatrix} 0 & 0 & 0 & C_{0a}^- & C_{0a}^+ & C_{0c} & C_{0f2} \\ I_{m_1} & 0 & 0 & 0 & 0 & 0 & 0 \\ 0 & I & 0 & 0 & 0 & 0 & 0 \\ 0 & 0 & C_b & 0 & 0 & 0 & 0 \end{bmatrix}$$

and

$$\tilde{D} = \begin{bmatrix} I & 0 & 0 \\ 0 & 0 & 0 \\ 0 & 0 & 0 \end{bmatrix}.$$

Let us define

$$C_c = \begin{bmatrix} C_0 \\ C_1(A - B_0C_0) \end{bmatrix} \quad \text{and} \quad D_c = \begin{bmatrix} I_{m_0} & 0 \\ 0 & C_1B_1 \end{bmatrix}.$$

We note that $C_c = \Gamma C_c$ and $D_c = \Gamma D_c$, where Γ is a nonsingular matrix,

$$\Gamma = \begin{bmatrix} I_{m_0} & 0 \\ -C_0B_1 & I \end{bmatrix}.$$

Thus, establishing the required properties for a system characterized by $(\tilde{A}, \tilde{B}, C_c, D_c)$ is equivalent to doing the same for a system characterized by $(\tilde{A}, \tilde{B}, C_c, D_c)$. We next rewrite C_c and D_c in the form,

$$C_c = \begin{bmatrix} 0 & 0 & 0 & C_{0a}^- & C_{0a}^+ & C_{0c} & C_{0f2} \\ E_{f11} & E_{f10} & E_{b1} & E_{a1}^- & E_{a1}^+ & E_{c1} & E_{f12} \\ L_{f01} & L_{f00} & 0 & 0 & 0 & 0 & C_{f2} \\ C_bL_{bf11} & C_bL_{bf10} & C_bA_{bb} & 0 & 0 & 0 & 0 \end{bmatrix}$$

and

$$D_c = \begin{bmatrix} I_{m_0} & 0 & 0 & 0 \\ 0 & I_{m_1} & 0 & 0 \\ 0 & 0 & 0 & 0 \\ 0 & 0 & 0 & 0 \end{bmatrix}.$$

It is trivial then to verify that system $(\tilde{A}, \tilde{B}, C_c, D_c)$ has the same finite and infinite zero structure, and invertibility property as the system (A_1, B_1, C_1)

does, where $A_1 =$

$$\begin{bmatrix} 0 & 0 & 0 & 0 & 0 & 0 & 0 \\ L_{f01} & L_{f00} & 0 & 0 & 0 & 0 & C_{f2} \\ L_{bf1} & L_{bf0} & A_{bb} & 0 & 0 & 0 & 0 \\ L_{af1}^- & L_{af0}^- & L_{ab}^- C_b & A_{aa}^- & 0 & 0 & 0 \\ L_{af1}^+ & L_{af0}^+ & L_{ab}^+ C_b & 0 & A_{aa}^+ & 0 & 0 \\ L_{cf1} & L_{cf0} & B_c E_{cb} & B_c E_{ca}^- & B_c E_{ca}^+ & A_{cc} & 0 \\ A_{f21} & A_{f20} & B_{f2} E_{b2} & B_{f2} E_{a2}^- & B_{f2} E_{a2}^+ & B_{f2} E_{c2} & A_{f22} + B_{f2} E_{f22} \end{bmatrix}$$

$$B_1 = \begin{bmatrix} 0 & 0 \\ 0 & 0 \\ 0 & 0 \\ 0 & 0 \\ 0 & 0 \\ 0 & B_c \\ B_{f2} & 0 \end{bmatrix},$$

and

$$C_1 = \begin{bmatrix} L_{f01} & L_{f00} & 0 & 0 & 0 & 0 & C_{f2} \\ C_b L_{bf1} & C_b L_{bf0} & C_b A_{bb} & 0 & 0 & 0 & 0 \end{bmatrix}.$$

We note here that $C_{f2} C_{f2}' = I$.

Next, we define a dual system,

$$\bar{x}(k+1) = \bar{A}\bar{x}(k) + \bar{B}\bar{u}(k), \quad \bar{y}(k) = \bar{C}\bar{x}(k),$$

where

$$\bar{A} = A_1' \ , \ \bar{B} = C_1' \ , \ \bar{C} = B_1' \, ,$$

$$\bar{x} = \left[(\bar{x}_0)', (\bar{x}_a^-)', (\bar{x}_a^+)', (\bar{x}_b)', (\bar{x}_f)' \right]' ,$$

$$\bar{u} = [(\bar{u}_f)', (\bar{u}_c)']' \quad \text{and} \quad \bar{y} = [(\bar{y}_f)', (\bar{y}_b)']' .$$

The dynamic equations of the above dual system are given by

$$\bar{x}_0(k+1) = \bar{A}_{00}\bar{x}_0(k) + \bar{A}_{10}\bar{x}_a^-(k) + \bar{A}_{20}\bar{x}_a^+(k) + \bar{A}_{b0}\bar{x}_b(k) + \bar{A}_{f0}\bar{x}_f(k)$$

$$+\bar{K}_f\bar{u}_f(k) + \bar{K}_c\bar{u}_c(k)$$

$$\bar{x}_a^-(k+1) = \bar{A}_{aa}^-\bar{x}_a^-(k) + \bar{L}_{ab}^-\bar{y}_b(k) + \bar{L}_{af}^-\bar{y}_f(k)$$

$$\bar{x}_a^+(k+1) = \bar{A}_{aa}^+\bar{x}_a^+(k) + \bar{L}_{ab}^+\bar{y}_b(k) + \bar{L}_{af}^+\bar{y}_f(k)$$

$$\bar{x}_b(k+1) = \bar{A}_{bb}\bar{x}_b(k) + \bar{L}_{bf}\bar{y}_f(k), \quad \bar{y}_b = \bar{C}_b\bar{x}_b$$

$$\bar{x}_f(k+1) = \bar{A}_f \bar{x}_f(k) + \bar{L}_f \bar{y}_f(k) + \bar{B}_f[\bar{u}_f(k) + \bar{E}_0 \bar{x}_0(k)], \qquad \bar{y}_f = \bar{C}_f \bar{x}_f$$

where

$$\bar{A}_{00} = \begin{bmatrix} 0 & 0 & 0 \\ L_{f01} & L_{f00} & 0 \\ L_{bf1} & L_{bf0} & A_{bb} \end{bmatrix}',$$

$$\bar{A}_{10} = [L^-_{af1},\, L^0_{af0},\, L^-_{ab}C_b]', \quad \bar{A}_{20} = [L^+_{af1},\, L^+_{af0},\, L^+_{ab}C_b]',$$

$$\bar{A}_{b0} = [L_{cf1},\, L_{cf0},\, B_c E_{cb}]', \quad \bar{A}_{f0} = [A_{f21},\, A_{f20},\, B_{f2}E_{b2}]',$$

$$\bar{A}^-_{aa} = (A^-_{aa})', \quad \bar{L}^-_{ab} = (E^-_{ca})', \quad \bar{L}^-_{af} = (E^-_{a2})', \quad \bar{A}^+_{aa} = (A^+_{aa})',$$

$$\bar{L}^+_{ab} = (E^+_{ca})', \quad \bar{L}^+_{af} = (E^+_{a2})', \quad \bar{A}_{bb} = (A_{cc})', \quad \bar{L}_{bf} = (E_{c2})',$$

$$\bar{E}_0 = [0,\, I,\, 0], \quad \bar{A}_f = (A_{ff2})', \quad \bar{C}_f = (B_{f2})', \quad \bar{C}_b = (B_c)',$$

and

$$\bar{L}_f = (E_{f22})', \quad \bar{K}_f = [L_{f01},\, L_{f00},\, 0]', \quad \bar{K}_c = [C_b L_{bf1},\, C_b L_{bf0},\, C_b A_{bb}]'.$$

Next, we will perform some transformations among the state variables in order to bring the new system $(\bar{A}, \bar{B}, \bar{C})$ into the form of s.c.b. Let us first define

$$\bar{\bar{x}}_0 = \bar{x}_0 - \bar{K}_f \bar{B}'_f \bar{x}_f.$$

As $\bar{B}'_f \bar{B}_f = C_{f2}C'_{f2} = I$, it is straightforward to verify that

$$\bar{\bar{x}}_0(k+1) = \tilde{A}_{00}\bar{\bar{x}}_0(k) + \bar{K}_c \bar{u}_c(k) + \bar{A}_{10}\bar{x}^-_a(k) + \bar{A}_{20}\bar{x}^+_a(k) + \bar{A}_{b0}\bar{x}_b(k) + A_{f0}\bar{x}_f(k)$$

where

$$\tilde{A}_{00} = \begin{bmatrix} 0 & 0 & 0 \\ 0 & 0 & 0 \\ L_{bf1} & L_{bf0} & A_{bb} \end{bmatrix}'$$

and A_{f0} is some appropriate matrix. Also,

$$\bar{x}_f(k+1) = \bar{A}_f \bar{x}_f(k) + \bar{L}_f \bar{y}_f(k) + \bar{B}_f[\bar{u}_f(k) + \bar{E}_0 \bar{\bar{x}}_0(k) + \bar{K}_f \bar{B}'_f \bar{x}_f(k)].$$

Then it follows from the results of Sannuti and Saberi (1987) that there exists a nonsingular transformation T such that

$$\left[(\bar{\bar{x}}_0)',\, (\bar{x}^-_a)',\, (\bar{x}^+_a)',\, (\bar{x}_b)',\, (\bar{x}_f)'\right]' = T\left[(\tilde{\bar{x}}_0)',\, (\bar{x}^-_a)',\, (\bar{x}^+_a)',\, (\bar{x}_b)',\, (\bar{x}_f)'\right]'$$

and

$$\tilde{\bar{x}}_0(k+1) = \tilde{A}_{00}\tilde{\bar{x}}_0(k) + \bar{K}_c \bar{u}_c(k) + \bar{A}_{10}\bar{x}^-_a(k) + \bar{A}_{20}\bar{x}^+_a(k) + \tilde{\bar{L}}_{b0}\bar{y}_b(k) + \tilde{\bar{L}}_{f0}\bar{y}_f(k)$$

$$\bar{x}_a^-(k+1) = \bar{A}_{aa}^-\bar{x}_a^-(k) + \bar{L}_{ab}^-\bar{y}_b(k) + \bar{L}_{af}^-\bar{y}_f(k)$$
$$\bar{x}_a^+(k+1) = \bar{A}_{aa}^+\bar{x}_a^+(k) + \bar{L}_{ab}^+\bar{y}_b(k) + \bar{L}_{af}^+\bar{y}_f(k)$$
$$\bar{x}_b(k+1) = \bar{A}_{bb}\bar{x}_b(k) + \bar{L}_{bf}\bar{y}_f(k), \quad \bar{y}_b = \bar{C}_b\bar{x}_b$$
$$\bar{x}_f(k+1) = \bar{A}_f\bar{x}_f(k) + \bar{L}_f\bar{y}_f(k) + \bar{B}_f[\bar{u}_f(k) + \bar{E}_0\bar{\bar{x}}_0(k) + \bar{\bar{E}}_b\bar{x}_b(k) + \bar{\bar{E}}_f\bar{x}_f(k)],$$
$$\bar{y}_f = \bar{C}_f\bar{x}_f.$$

We note that the above system is not in the standard form of s.c.b since we have not separated the new invariant zeros from $\tilde{A}_{00}$ yet. Let us next examine the pair $(\tilde{A}_{00}, \bar{K}_c)$. We have

$$\begin{aligned}
\operatorname{rank}[zI - \tilde{A}_{00} \quad \bar{K}_c] &= \operatorname{rank}\begin{bmatrix} zI - \tilde{A}'_{00} \\ \\ \bar{K}'_c \end{bmatrix} \\
&= \operatorname{rank}\begin{bmatrix} zI & 0 & 0 \\ 0 & zI & 0 \\ -L_{af1} & -L_{bf0} & zI - A_{bb} \\ C_bL_{bf1} & C_bL_{bf0} & C_bA_{bb} \end{bmatrix} \\
&= \operatorname{rank}\begin{bmatrix} zI & 0 & 0 \\ 0 & zI & 0 \\ -L_{af1} & -L_{bf0} & zI - A_{bb} \\ 0 & 0 & zC_b \end{bmatrix}. \qquad (63)
\end{aligned}$$

From (63) we know that the only possibility that causes $[zI - \tilde{A}_{00} \quad \bar{K}_c]$ to drop its rank is $z = 0$. Thus, since the pair (A_{bb}, C_b) is observable, the system $(\bar{A}, \bar{B}, \bar{C})$ has stable invariant zeros at $z = 0$. Hence, it follows from the results of Sannuti and Saberi (1987) that there exists another nonsingular transformation S such that

$$[(\bar{\bar{x}}_0)', (\bar{x}_a^-)', (\bar{x}_a^+)', (\bar{x}_b)', (\bar{x}_f)']'$$
$$= S\left[(\tilde{\bar{x}}_c^-)', (\tilde{\bar{x}}_a^0)', (\bar{x}_a^-)', (\bar{x}_a^+)', (\bar{x}_b)', (\bar{x}_f)'\right]'$$

and

$$\tilde{\bar{x}}_c(k+1) = \tilde{A}_{cc}\tilde{\bar{x}}_c(k) + \tilde{B}_c[\tilde{E}_{ca}^0\tilde{\bar{x}}_a^0(k) + \tilde{E}_{ca}^-\bar{x}_a^-(k) + \tilde{E}_{ca}^+\bar{x}_a^+(k)]$$
$$+\tilde{L}_{cb}\bar{y}_b(k) + \tilde{L}_{cf}\bar{y}_f(k)$$
$$\tilde{\bar{x}}_a^0(k+1) = 0 \cdot \tilde{\bar{x}}_a^0(k) + \tilde{A}_{0-}\bar{x}_a^-(k) + \tilde{A}_{0+}\bar{x}_a^+(k) + \tilde{L}_{ab}^0\bar{y}_b(k) + \tilde{L}_{af}^0\bar{y}_f(k)$$
$$\bar{x}_a^-(k+1) = \bar{A}_{aa}^-\bar{x}_a^-(k) + \bar{L}_{ab}^-\bar{y}_b(k) + \bar{L}_{af}^-\bar{y}_f(k)$$

$$\bar{x}_a^+(k+1) = \bar{A}_{aa}^+ \bar{x}_a^+(k) + \bar{L}_{ab}^+ \bar{y}_b(k) + \bar{L}_{af}^+ \bar{y}_f(k)$$
$$\bar{x}_b(k+1) = \bar{A}_{bb}\bar{x}_b(k) + \bar{L}_{bf}\bar{y}_f(k), \quad \bar{y}_b = \bar{C}_b \bar{x}_b$$

$$\bar{x}_f(k+1) = \bar{A}_f \bar{x}_f(k) + \bar{L}_f \bar{y}_f(k) + \bar{B}_f[\bar{u}_f(k) + \tilde{E}_c \tilde{\bar{x}}_c(k) + \tilde{E}_a^0 \tilde{\bar{x}}_a^0(k)$$
$$+ \tilde{E}_a^- \bar{x}_a^-(k) + \tilde{E}_a^+ \bar{x}_a^+(k) + \tilde{\bar{E}}_b \bar{x}_b(k) + \tilde{\bar{E}}_f \bar{x}_f(k)],$$
$$\bar{y}_f = \bar{C}_f \bar{x}_f,$$

where $(\tilde{A}_{cc}, \tilde{B}_c)$ is a controllable pair, and $\tilde{E}_{ca}^0$, $\tilde{E}_{ca}^-$, $\cdots$, $\tilde{E}_a^+$ are some constant matrices with appropriate dimensions. It is now trivial to see that the above system is in the standard form of s.c.b. Hence all the properties listed in Proposition 2 can be verified easily by the properties of s.c.b and by some simple algebra. ∎

C. Proof of Lemma 3

Noting from Lemma 1 that

$$E_p(z) := L_t(z) - \mathbf{C}_p(z)P(z) = M_p(z)[I + M_p(z)]^{-1}[I + L_t(z)],$$

we obtain,

$$\mathbf{C}_p(z)P(z) = L_t(z) - M_p(z)[I + M_p(z)]^{-1}[I + L_t(z)]$$
$$= [I + M_p(z)]^{-1}[L_t(z) - M_p(z)],$$

from which $\mathbf{C}_p(z)P(z)$ can be interpreted in terms of a block diagram given below.

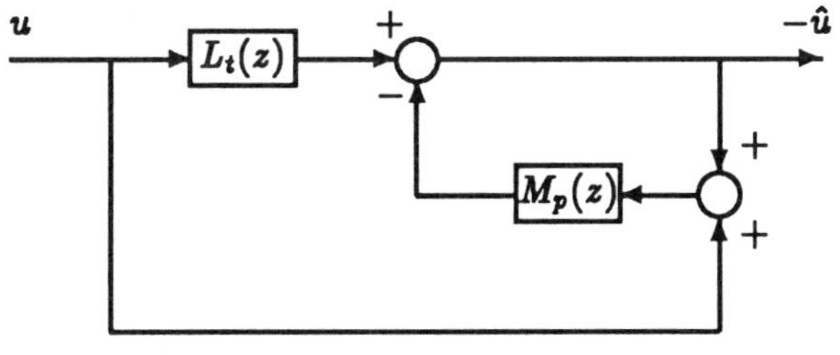

In view of the block diagram, it is straightforward to write a state-space realization of $\mathbf{C}_p(z)P(z)$,

$$\begin{cases} \tilde{x}(k+1) = \begin{bmatrix} A & 0 \\ (B - K_p D)F & A - K_p C - BF + K_p DF \end{bmatrix} \tilde{x}(k) \\ \qquad\qquad + \begin{bmatrix} B \\ B - K_p D \end{bmatrix} u(k), \\ -\hat{u}(k) = [F, \ \ -F]\tilde{x}(k). \end{cases}$$

Let λ be an eigenvalue of $A - K_pC$ and the corresponding left eigenvector V be such that $V^H(B - K_pD) = 0$. It is simple then to verify that

$$[0, \quad V^H] \begin{bmatrix} \lambda I - A & 0 \\ -(B - K_pD)F & \lambda I - A + K_pC + (B - K_pD)F \end{bmatrix} = 0$$

and

$$[0, \quad V^H] \begin{bmatrix} B \\ B - K_pD \end{bmatrix} = 0.$$

This shows that λ is an input decoupling zero of $\mathbf{C}_p(z)P(z)$ and thus the result follows. ∎

D. Proof of Theorem 6

Let us first consider the case of prediction estimator based controller. Without loss of generality we assume that the given system Σ is in the form of s.c.b as in Theorem 1. Now in view of Theorem 5 a recoverable $L_t(z) = F\Phi B$ must satisfy $\mathcal{S}^-(\Sigma) \subseteq \operatorname{Ker}(F)$. This implies that $L_t(z)$ is recoverable if and only if F is of the form,

$$F = \begin{bmatrix} F_{a1}^- & 0 & F_{b1} & 0 & 0 \\ F_{a2}^- & 0 & F_{b2} & 0 & 0 \end{bmatrix}. \tag{64}$$

Thus the fact that the given system has at least one exactly recoverable target loop is equivalent to the existence of some appropriate matrices F_{a1}^-, F_{b1}, F_{a2}^- and F_{b2} such that $A - BF$ is asymptotically stable. Next, in view of the fact that $x_a^+ \oplus x_c \oplus x_f$ spans $\mathcal{S}^-(\Sigma)$, we note that $\overline{C}_p$ as defined in Theorem 6 is of the form,

$$\overline{C}_p = \Gamma \begin{bmatrix} I_{n_a^-} & 0 & 0 & 0 & 0 \\ 0 & 0 & I_{n_b} & 0 & 0 \end{bmatrix}$$

where Γ is any nonsingular matrix of dimension $(n_a^- + n_b) \times (n_a^- + n_b)$. It is now trivial to verify that the existence of a matrix F of the form in (64) such that $A - BF$ is asymptotically stable, is equivalent to the existence of a matrix G of dimension $m \times (n_a^- + n_b)$ such that $A - BG\overline{C}_p$ is asymptotically stable. This is simply due to the fact that $G\overline{C}_p$ has the same structure as F in (64).

The results for current and reduced order estimators follow from similar arguments. This completes the proof of Theorem 6. ∎

E. Proof of Corollary 2

Again, we first consider the case of prediction estimator based controller. If there exists at least one recoverable target loop, i.e., if there exists at least one target loop and a prediction estimator gain K_p such that the corresponding $\tilde{M}_p^e(z) = 0$, then we note from (54) that the eigenvalues of the prediction estimator based controller are given by $\lambda(A - K_pC) \in \mathbf{C}^{\odot}$. Hence the corresponding prediction estimator based controller is asymptotically stable and thus by definition, the given plant is strongly stabilizable. The results for current and reduced order estimators follow from similar arguments. ∎

F. Proof of Theorem 11

Again, we explicitly prove here only the case of prediction estimator based controller. Utilizing Propositions 2 and 3, the results for current and reduced order estimator based controllers can be derived in a similar way.

Let z_i, x_i and w_i, $i = 1$ to n_a^+, be respectively an unstable invariant zero and the associated right state and input zero directions of (A, B, C, D). Since (A, B, C, D) is assumed to be stabilizable and detectable, we have $w_i \neq 0$ for all $i = 1$ to n_a^+. Because if $w_i = 0$, then by definition,

$$(z_iI - A)x_i = Bw_i = 0 \quad \text{and} \quad Cx_i + Dw_i = Cx_i = 0.$$

This implies that z_i is an output decoupling zero of (A, B, C, D). But this contradicts the detectability of (A, B, C, D) as $z_i \in \mathbf{C}^{\otimes}$. Then it follows from Lemma 3.8 of [24] that there exist at least one e such that

$$e'w_i \neq 0 \text{ for all } i = 1 \text{ to } n_a^+.$$

Select e to satisfy the above equation, and then define $\mathcal{S}$ as

$$\mathcal{S} = \text{ The orthogonal complement of the subspace spanned by } e \text{ in } \mathbf{R}^m.$$

It is now trivial to see that $\mathcal{S}$ has a dimension of $m-1$ and that $w_i \notin \mathcal{S}$ for all $i = 1$ to n_a^+. Then it follows from Lemma 6.2 of [6] that the corresponding Σ_p^s is of minimum phase and left invertible. Also, it is simple to verify that Σ_p^s has no infinite zeros, i.e., DV^s is of maximal rank. This in turn implies the results of Theorem 11. ∎

References

[1] K. J. Astrom, P. Hagander and J. Sternby, "Zeros of sampled systems," *Automatica*, 20, pp. 31-38 (1984).

[2] M. Athans, "A tutorial on LQG/LTR methods," *Proceedings of American Control Conference*, Seattle, Washington, pp. 1289-1296 (1986).

[3] B. M. Chen, A. Saberi, P. Bingulac and P. Sannuti, "Loop transfer recovery for non-strictly proper plants," *Control–Theory and Advanced Technology*, Vol. 6, No. 4, pp. 573-594 (1990).

[4] B. M. Chen, A. Saberi and P. Sannuti, "A new stable compensator design for exact and approximate loop transfer recovery," *Automatica*, Vol. 27, No. 2, pp. 257-280 (1991).

[5] B. M. Chen, A. Saberi and P. Sannuti, "Necessary and sufficient conditions for a nonminimum phase plant to have a recoverable target loop — a stable compensator design for LTR," *Automatica*, Vol. 28, No. 3, pp. 493-507 (1992).

[6] B. M. Chen, A. Saberi and P. Sannuti, "Loop transfer recovery for general nonminimum phase non-strictly proper systems — Part 1: analysis," *Control–Theory and Advanced Technology*, Vol. 8, No. 1, pp. 59-100 (1992).

[7] B. M. Chen, A. Saberi and P. Sannuti, "Loop transfer recovery for general nonminimum phase non-strictly proper systems — Part 2: design," *Control–Theory and Advanced Technology*, Vol. 8, No. 1, pp. 101-144 (1992).

[8] B. M. Chen, A. Saberi, P. Sannuti and Y. Shamash, "Loop transfer recovery for general nonminimum phase discrete time systems — Part 2 : design," *Control and Dynamic Systems: Advances in Theory and Applications*, This Volume (1992).

[9] J. C. Doyle and G. Stein, "Robustness with observers," *IEEE Transactions on Automatic Control*, AC-24, No. 4, pp. 607-611 (1979).

[10] J. C. Doyle and G. Stein, "Multivariable feedback design: concepts for a classical / modern synthesis," *IEEE Transactions on Automatic Control*, AC-26, No. 1, pp. 4-16 (1981).

[11] G. F. Franklin, J. D. Powell and M. L. Workman, *Digital control of dynamic systems*, Addison-Wesley, Reading, Massachusetts (1990).

[12] G. C. Goodman, *The LQG/LTR method and discrete-time control systems*, Rep. No. LIDS-TH-1392, MIT, MA (1984).

[13] T. Ishihara and H. Takeda, "Loop transfer recovery techniques for discrete-time optimal regulators using prediction estimators," *IEEE Transactions on Automatic Control*, Vol. AC-31, No. 12, pp. 1149-1151 (1986).

[14] H. Kwakernaak, "Optimal low sensitivity linear feedback systems," *Automatica*, Vol. 5, pp. 279-285 (1969).

[15] J. M. Maciejowski, "Asymptotic recovery for discrete-time systems," *IEEE Transactions on Automatic Control*, Vol. AC-30, No. 6, pp. 602-605 (1985).

[16] B. C. Moore, "On the flexibility offered by state feedback in multivariable systems beyond closed loop eigenvalue assignment," *IEEE Transactions on Automatic Control*, AC-21, pp. 689-692 (1976).

[17] H. H. Niemann and O. Jannerup, "An analysis of pole/zero cancellation in LTR-based feedback design," *Proceedings of 1990 American Control Conference*, San Diego, California, pp. 848-853 (May 1990).

[18] H. H. Niemann, J. Stoustrup and P. Sogaard-Andersen, "General conditions for loop transfer recovery," *Proceedings of American Control Conference*, pp. 333-334 (1991).

[19] H. H. Niemann and P. Sogaard-Andersen, "New results in discrete-time loop transfer recovery," *Proceedings of American Control Conference*, pp. 2483-2489 (1988).

[20] J. O'Reilly, *Observers for linear systems*, Academic Press, London (1987).

[21] D. B. Ridgely and S. S. Banda, *Introduction to robust multivariable control*, Report No. AFWAL-TR-85-3102, Flight Dynamics Laboratories, Air Force Systems Command, Wright – Patterson Air Force Base, Ohio (1986).

[22] H. H. Rosenbrock, *State-space and multivariable theory*, John-Wiley, New York (1970).

[23] G. Stein and M. Athans, "The LQG/LTR procedure for multivariable feedback control design," *IEEE Transactions on Automatic Control*, AC-32, pp. 105-114 (1987).

[24] A. Saberi, B. M. Chen and P. Sannuti, "Theory of LTR for non-minimum phase systems, recoverable target loops, recovery in a subspace — Part 1: analysis," *International Journal of Control*, Vol. 53, No. 5, pp. 1067-1115 (1991).

[25] A. Saberi, B. M. Chen and P. Sannuti, "Theory of LTR for non-minimum phase systems, recoverable target loops, recovery in a Subspace — Part 2: design," *International Journal of Control*, Vol. 53, No. 5, pp. 1117-1160 (1991).

[26] A. Saberi, B. M. Chen and P. Sannuti, *Loop transfer recovery: Analysis and design*, Springer-Verlag, London (1993).

[27] P. Sannuti and A. Saberi, "A special coordinate basis of multivariable linear systems – finite and infinite zero structure, squaring down and decoupling," *International Journal of Control*, Vol. 45, No. 5, pp. 1655-1704 (1987).

[28] A. Saberi and P. Sannuti, "Squaring down of non-strictly proper systems," *International Journal of Control*, Vol. 51, No. 3, pp. 621-629 (1990).

[29] A. Saberi and P. Sannuti, "Observer design for loop transfer recovery and for uncertain dynamical systems," *IEEE Transactions on Automatic Control*, Vol. 35, No. 8, pp. 878-897 (1990).

[30] U. Shaked, "The admissibility of desired transfer-function matrices," *International Journal of Control*, Vol. 25, No. 5, pp. 213-228 (1977).

[31] P. Sogaard-Andersen and H. H. Niemann, "Trade-offs in LTR-based feedback design," *Proceedings of American Control Conference*, Pittsburgh, Pennsylvania, pp. 922-928 (1989).

[32] P. Sogaard-Andersen, "Loop transfer recovery — an eigen-structure interpretation," *Control–Theory and Advanced Technology*, Vol. 5, No. 3, pp 351-365 (1989).

[33] P. Sogaard-Andersen, "An eigenspace and residual approach to transfer function matrix synthesis," *Systems & Control Letters*, 8, pp. 221-223 (1987).

[34] Z. Zhang and J. S. Freudenberg, "Loop transfer recovery for non-minimum phase plants," *IEEE Transactions on Automatic Control*, Vol. 35, No. 5, pp. 547-553 (1990).

[35] Z. Zhang and J. S. Freudenberg, "Loop transfer recovery for non-minimum phase discrete-time systems," *Proceedings of American Control Conference*, Boston, pp. 2214-2219 (1991).

Loop Transfer Recovery For General Nonminimum Phase Discrete Time Systems

Part 2: Design

Ben M. Chen

Department of Electrical Engineering
State University of New York at Stony Brook
Stony Brook, New York 11794-2350

Ali Saberi

School of Electrical Engineering and Computer Science
Washington State University
Pullman, Washington 99164-2752

Peddapullaiah Sannuti

Department of Electrical and Computer Engineering
P.O. Box 909
Rutgers University
Piscataway, New Jersey 08855-0909

Yacov Shamash

College of Engineering and Applied Sciences
State University of New York at Stony Brook
Stony Brook, New York 11794

CONTROL AND DYNAMIC SYSTEMS, VOL. 55

I. INTRODUCTION AND PROBLEM STATEMENT

As discussed earlier in Part 1 [3], the basic loop transfer recovery (LTR) problem is concerned with analyzing and possibly designing a controller which can achieve the same robustness properties as those of a state feedback controller. To be specific, consider a plant Σ,

$$x(k+1) = Ax(k) + Bu(k) \ , \ y(k) = Cx(k) + Du(k) \tag{1}$$

where the state vector $x \in \mathbf{R}^n$, output vector $y \in \mathbf{R}^p$ and input vector $u \in \mathbf{R}^m$. Without loss of generality, assume that $[B' \ D']'$ and $[C \ \ D]$ are of maximal rank. Let us also assume that Σ is stabilizable and detectable. Let the state feedback control law,

$$u = -Fx, \tag{2}$$

be such that (a) the closed-loop system is asymptotically stable, and (b) the open-loop transfer function when the loop is broken at the input point of the plant meets the given frequency dependent specifications. Then $L_t(z)$, $S_t(z)$ and $T_t(z)$, the target loop transfer function, sensitivity and complimentary sensitivity functions are respectively

$$L_t(z) = F\Phi B,$$

$$S_t(z) = [I_m + L_t(z)]^{-1},$$

and

$$T_t(z) = I_m - S_t(z) = [I_m + L_t(z)]^{-1}L_t(z) \tag{3}$$

where $\Phi = (zI_n - A)^{-1}$ and I_m denotes an identity matrix of dimension $m \times m$. We would like to recover $L_t(z)$ using only a measurement feedback controller $\mathrm{C}(z)$. That is, given a target loop transfer function $L_t(z)$ and the plant transfer function P(s),

$$P(z) = C\Phi B + D,$$

we seek to design a controller $\mathrm{C}(z)$ such that the *loop transfer recovery error* $E(z)$,

$$E(z) \equiv L_t(z) - \mathrm{C}(z)P(z), \tag{4}$$

is either exactly or approximately equal to zero in the frequency region of interest while guaranteeing the stability of the resulting closed-loop system.

The notion of achieving exact LTR (ELTR) corresponds to $E(z) = 0$ for all z. In the case of asymptotic recovery, one normally parameterizes the controller $\mathbf{C}(z)$ in terms of a scalar tuning parameter σ and thus obtains a family of controllers $\mathbf{C}(z, \sigma)$. We say asymptotic LTR (ALTR) is achieved if $\mathbf{C}(z, \sigma)P(z) \to L_t(z)$ pointwise in z as $\sigma \to \infty$. Achievability of ALTR enables the designer to choose a member of the family of controllers that corresponds to a particular value of σ which achieves a desired level of recovery.

In Part 1 [3], for general discrete systems, the above LTR problem has been considered and analyzed using any of three different observer or estimator based controllers. The estimators considered there are (1) prediction estimator, (2) current estimator and (3) reduced order estimator. Both prediction estimator and current estimator are full order observers. The reduced order estimator is a current estimator but uses the reduced order observer. The prediction estimator estimates the state $x(k+1)$ based on the measurements $y(k)$ up to and including the (k)-th instant, where as the current estimator estimates $x(k+1)$ based on the measurements $y(k+1)$ up to and including the $(k+1)$-th instant. The analysis of Part 1 corresponding to all these three different estimator based controllers unifies it into a single mathematical frame work. The LTR analysis given there focuses on four fundamental issues, (1) the recoverability of a target loop when it is arbitrarily given, (2) the recoverability of a target loop taking into account its specific characteristics, (3) the establishment of necessary and sufficient conditions on the given system so that it has at least one recoverable target loop transfer function or sensitivity function, and (4) the recoverability of a sensitivity function in a specified subspace of the control space. All this analysis of Part 1 shows some fundamental limitations of the given system as a consequence of its structural properties. Also, Part 1 decomposes the so called recovery matrix into two parts, the first one can always be rendered zero while the other in general cannot be rendered zero and hence can be termed as the ***recovery error matrix***. The analysis of Part 1 also discovers a multitude of ways in which freedom exists to shape the recovery error matrix in a desired way. Thus it helps to set meaningful design goals at the onset of design.

Part 1 also reveals both similarities as well as fundamental differences that arise in LTR analysis of continuous and discrete time systems. A fundamental difference between continuous time and discrete time systems that should be emphasized is this. In the discrete case, as is well known, in order to preserve stability, all the closed-loop eigenvalues must be restricted to lie within the unit circle in complex plane. This implies that unlike continuous case which permits both finite as well as asymptotically infinite eigenvalue assignment to a closed-loop system, in the discrete case one is restricted to only finite eigenvalue assignment. Because of this, in the continuous case, there exists target loops which are not exactly recoverable, but are asymptotically recoverable by appropriate infinite eigenstructure assignment; on the other hand, in discrete systems, since both asymptotic as well as exact recovery involves only finite eigenstructure assignment, every asymptotically recoverable target loop is also exactly recoverable and vice versa. Thus, in discrete systems, one needs to talk about just recovery rather than emphasizing exact or asymptotic recovery.

In this Part 2 of the paper, we consider design of all three, prediction, current and reduced order estimator based controllers for the purpose of loop transfer recovery. For each one of such controllers, after reviewing from Part 1 the necessary design constraints and the available design freedom, three different design techniques are developed. The first one is an eigenstructure assignment scheme, and the other two are optimization based designs. Eigenstructure assignment method yields a controller design which achieves *any chosen* recovery error matrix among a set of admissible recovery error matrices. On the other hand, one of the optimization based design methods leads to a controller that achieves a recovery error matrix having the infimum H_∞ norm, while the other does the same except it achieves a recovery error matrix having the infimum H_2 norm. The eigenstructure assignment method given here is a special case of the asymptotic time-scale and eigenstructure assignment (ATEA) method introduced in [8] and fully developed in [2] in connection with continuous systems. Since in discrete systems, one does not have the option of assigning the asymptotically infinite eigenvalues, no multiple time-scale structure assignment is feasible. The algorithm of ATEA as in [2] when the option of time-scale structure assignment is removed from it, yields a simple design tool for discrete LTR as well. Regarding optimization based design methods, while partial results are available in the literature based on H_2 norm minimization [5], [11], no methods of H_∞ norm minimization are yet available for discrete systems. This paper develops new H_∞ norm minimization methods, and then streamlines and strengthens the available H_2 norm minimization methods. An important difference between ATEA and optimization based designs is this. ATEA is capable of achieving any admissible recovery error matrix where as optimization based methods always lead to a particular recovery error matrix having the infimum H_∞ or H_2 norm depending on the method used.

As mentioned earlier, in discrete systems, when one talks about recoverability, one need not distinguish between the notions of 'exact' and 'asymptotic' recoverabilities as they both imply one and the same. However, as will be seen in the text, optimization based methods of recovery design some times lead only to suboptimal designs. For the case of recoverable target loops, such suboptimal designs yield asymptotic recovery. To be specific, in H_∞-optimization methods, one normally generates a sequence of observer gains by solving parameterized algebraic Riccati equations. As the parameter tends to a certain value, the corresponding sequence of H_∞ norms of the resulting recovery matrices tends to a limit which is the infimum of the H_∞ norm of the recovery matrix over the set of all possible observer gains. Obviously, for the case when the infimum of H_∞ norm of the recovery matrix is zero, the sequence of observer gains thus obtained lead to a suboptimal design that corresponds to asymptotic recovery.

The conventional LTR design task seeks the recovery over the entire control space. As discussed in Part 1, one can also formulate another generalized design task which seeks the recovery only over a specified subspace of the control space. Such a formulation is meaningful especially when recovery over the entire control space is not feasible. All the three design methods developed here can easily be modified to deal with such a generalized design task.

The paper is organized as follows. Section II reviews the necessary design constraints and the available design freedom. Section III develops the general ATEA method of design. Section IV develops optimization based designs. Here two designs are considered; one minimizes the H_∞ norm of a recovery matrix while the other minimizes the H_2 norm of the same. Section V considers the generalized design task of recovering the target sensitivity and complimentary sensitivity functions over a subspace of the control space. All the previous sections consider the case when the target loop transfer function is specified at the input point of the given system. Section VI reformulates the LTR design when a target loop transfer function is specified at the output point of the given system in terms of

LTR design when a target loop transfer function is specified at the input point of a system dual to the given system. Finally, Section VII draws the conclusions of our work.

As in Part 1, throughout this paper, A' denotes the transpose of A, A^H denotes the complex conjugate transpose of A, I denotes an identity matrix while I_k denotes the identity matrix of dimension $k \times k$. $\lambda(A)$ denotes the set of eigenvalues of A. Similarly, $\sigma_{max}[A]$ and $\sigma_{min}[A]$ respectively denote the maximum and minimum singular values of A. Ker $[V]$ and Im $[V]$ denote respectively the kernel and the image of V. $\mathbf{C}^{\odot}$ denotes the set of complex numbers inside the open unit circle while $\mathbf{C}^{\otimes}$ is the complimentary set of $\mathbf{C}^{\odot}$. Given a discrete transfer function $G(z)$, we define the discrete frequency response $G^*(j\omega)$ as $G(e^{j\omega T})$ where T is the sampling period of the discrete-time system. An asymptotically stable matrix is the one whose eigenvalues are all in $\mathbf{C}^{\odot}$.

While discussing the design procedures, we will always use a generic controller which could be based on any one of the three estimators, prediction, current or reduced order. In that case, as in Part 1, we will always use the following notation :

$$
\begin{aligned}
\mathbf{C}(z) &:= \text{The transfer function of the controller,} \\
L(z) &:= \mathbf{C}(z)P(z) = \text{The achieved loop transfer function,} \\
S(z) &:= [I_m + L(z)]^{-1} = \text{The achieved sensitivity function,} \\
T(z) &:= I_m - S(z) = \text{The achieved complimentary sensitivity} \\
&\qquad \text{function,} \\
E(z) &:= L_t(z) - L(z) = \text{Loop recovery error,} \\
M(z) &:= \text{The recovery matrix (to be defined later on),} \\
M^0(z) &:= \text{A part of the recovery matrix } M(z) \text{ that can be} \\
&\qquad \text{rendered zero,} \\
M^e(z) &:= \text{A part of the recovery matrix } M(z) \text{ that cannot be} \\
&\qquad \text{rendered zero and hence termed as recovery error matrix,} \\
\mathbf{T}^R(\Sigma) &:= \text{The set of either exactly or asymptotically recoverable} \\
&\qquad \text{target loops for } \Sigma.
\end{aligned}
$$

Whenever we have a particular controller in mind, we use appropriate subscripts to distinguish them. Subscripts p, c and r are used respectively to represent prediction, current, and reduced order estimator based controllers. For example, $L_p(z)$, $M_c^s(z)$ and $\mathbf{T}_r^{\mathrm{R}}(\Sigma)$ denote respectively the achieved loop transfer function with prediction estimator based controller, the recovery error matrix when a current estimator based controller is used, and the set of recoverable target loops for Σ using reduced order estimator based controllers.

II. CONTROLLER STRUCTURES — DESIGN CONSTRAINTS AND AVAILABLE FREEDOM

In this section, we will recall three different controller structures as well as their design constraints and freedom available for the purpose of achieving LTR. All the three controllers considered are observer based, but the type of observer (or state estimator) used in each one is structurally different. The estimators considered are (1) prediction estimator, (2) current estimator and (3) reduced order estimator. The structural details of the controllers are as follows.

Prediction estimator based controller :

The dynamic equations of the controller are

$$\begin{cases} \hat{x}(k+1) = A\hat{x}(k) + Bu(k) + K_p[y(k) - C\hat{x}(k) - Du(k)], \\ u(k) = \hat{u}(k) = -F\hat{x}(k), \end{cases} \tag{5}$$

where K_p is the gain chosen so that $A - K_pC$ is asymptotically stable. The transfer function of the controller is

$$\mathbf{C}_p(z) = F[zI_n - A + BF + K_pC - K_pDF]^{-1}K_p. \tag{6}$$

Current Estimator Based Controller :

Let us first rewrite the matrices C and D in the form,

$$C = \begin{bmatrix} C_0 \\ C_1 \end{bmatrix} \quad \text{and} \quad D = \begin{bmatrix} D_0 \\ 0 \end{bmatrix}, \tag{7}$$

where D_0 is of maximal rank, i.e., $\mathrm{rank}(D) = \mathrm{rank}(D_0) = m_0$. Thus, the output y can be partitioned as,

$$\begin{bmatrix} y_0(k) \\ y_1(k) \end{bmatrix} = \begin{bmatrix} C_0 \\ C_1 \end{bmatrix} x(k) + \begin{bmatrix} D_0 \\ 0 \end{bmatrix} u(k).$$

The dynamic equations of the controller are

$$\begin{cases} \hat{x}(k+1) = A\hat{x}(k) + Bu(k) + K_c\left(\begin{bmatrix} y_0(k) \\ y_1(k+1) \end{bmatrix} - C_c\hat{x}(k) - D_c u(k)\right), \\ \hat{u}(k) = u(k) = -F\hat{x}(k), \end{cases} \tag{8}$$

where

$$C_c = \begin{bmatrix} C_0 \\ C_1 A \end{bmatrix} \quad \text{and} \quad D_c = \begin{bmatrix} D_0 \\ C_1 B \end{bmatrix}, \tag{9}$$

and where the gain K_c is chosen so that $A - K_cC_c$ is asymptotically stable. The transfer function from $-u$ to y that results in using the current estimator is then given by

$$\mathbf{C}_c(z) = F\left[zI_n - A + K_cC_c + BF - K_cD_cF\right]^{-1} K_c \begin{bmatrix} I_{m_0} & 0 \\ 0 & zI \end{bmatrix}. \tag{10}$$

Reduced Order Estimator Based Controller :

Again, without any loss of generality but for simplicity of presentation, it is assumed that the matrices C and D are transformed into the form,

$$C = \begin{bmatrix} 0 & C_{02} \\ I_{p-m_0} & 0 \end{bmatrix} \quad \text{and} \quad D = \begin{bmatrix} D_0 \\ 0 \end{bmatrix}. \tag{11}$$

Then Σ can be partitioned as follows,

$$\begin{cases} \begin{pmatrix} x_1(k+1) \\ x_2(k+1) \end{pmatrix} = \begin{bmatrix} A_{11} & A_{12} \\ A_{21} & A_{22} \end{bmatrix} \begin{pmatrix} x_1(k) \\ x_2(k) \end{pmatrix} + \begin{bmatrix} B_{11} \\ B_{22} \end{bmatrix} u(k), \\ \begin{pmatrix} y_0(k) \\ y_1(k) \end{pmatrix} = \begin{bmatrix} 0 & C_{02} \\ I_{p-m_0} & 0 \end{bmatrix} \begin{pmatrix} x_1(k) \\ x_2(k) \end{pmatrix} + \begin{bmatrix} D_0 \\ 0 \end{bmatrix} u(k). \end{cases} \tag{12}$$

Since $y_1 = x_1$ is already available, one needs to estimate only the state variable x_2. Then the dynamic equations of the reduced order estimator based controller are as follows:

$$\begin{cases} v(k+1) = (A_r - K_rC_r)v(k) + (B_r - K_rD_r)u(k) + G_ry(k), \\ u(k) = \hat{u}(k) = -F_1x_1(k) - F_2\hat{x}_2(k) = -F_2v(k) - [0, F_1 + F_2K_{r1}]y(k) \end{cases} \tag{13}$$

where the gain K_r is chosen such that $A_r - K_rC_r$ is asymptotically stable, and where

$$A_r = A_{22}, \quad B_r = B_{22}, \quad C_r = \begin{bmatrix} C_{02} \\ A_{12} \end{bmatrix}, \quad D_r = \begin{bmatrix} D_0 \\ B_{11} \end{bmatrix}, \tag{14}$$

$$F = [F_1, \quad F_2], \qquad K_r = [K_{r0}, K_{r1}], \tag{15}$$

$$G_r = [K_{r0}, A_{21} - K_{r1}A_{11} + (A_r - K_rC_r)K_{r1}]. \tag{16}$$

The transfer function from $-u$ to y that results in using the reduced order estimator is then given by

$$\begin{aligned} \mathbf{C}_r(z) &= F_2(zI - A_r + K_rC_r + B_rF_2 - K_rD_rF_2)^{-1} \\ &\cdot \Big(G_r - (B_r - K_rD_r)[0, \ F_1 + F_2K_{r1}]\Big) + [0, \ F_1 + F_2K_{r1}]. \end{aligned} \tag{17}$$

A fundamental result of Part 1, namely Lemma 1, rewrites the loop transfer recovery error $E(z)$ between the target loop transfer function $L_t(z)$ and that realized by any one of the above controllers, in terms of a so called *recovery matrix* $M(z)$. That is,

$$E(z) = M(z)[I_m + M(z)]^{-1}(I_m + F\Phi B). \tag{18}$$

The expression for the recovery matrix $M(z)$ depends on the controller used. In particular, we have the following expressions,

$$M_p(z) = F(zI_n - A + K_pC)^{-1}(B - K_pD), \tag{19}$$

$$M_c(z) = F\left(zI_n - A + K_cC_c\right)^{-1}\left(B - K_cD_c\right), \tag{20}$$

$$M_r(z) = F_2(zI - A_r + K_rC_r)^{-1}(B_r - K_rD_r). \tag{21}$$

It is easy to see that

$$E^*(j\omega) = 0 \text{ if and only if } M^*(j\omega) = 0$$

for all $\omega \in \Omega$, where Ω is the set of all $0 \leq |\,\omega\,| \leq \pi/T$ for which $L_t^*(j\omega)$ and $L^*(j\omega) = \mathbf{C}^*(j\omega)P^*(j\omega)$ are well defined (i.e., all required inverses exist). This implies that the study of LTR can be cast in terms of the study of the recovery matrix $M(z)$. Also, since the expression for $M(z)$ for each controller is structurally similar to those of others, one can unify the LTR analysis and design involving all three different controllers into a single mathematical framework. This is done by defining the auxiliary systems Σ_c and Σ_r which are respectively characterized by the matrix quadruples (A, B, C_c, D_c) and (A_r, B_r, C_r, D_r).

In view of the above discussion, in order to determine the available design freedom for each controller, one needs to study an appropriate recovery matrix $M(z)$. Such a study has been undertaken in Part 1. It is shown there that the recovery matrix $M(z)$ can be decomposed into two parts, $M^0(z)$ that can always be rendered zero and $M^e(z)$ that cannot in general be rendered zero. As such $M^e(z)$ is termed as *recovery error matrix*. Let us present here a brief summary of the analysis given in Part 1. As in Part 1, our general discussion is always in terms of the given system Σ and the prediction estimator based controller. The details for other two controllers are presented only when they need to be emphasized.

Assuming that $A - K_pC$ is nondefective, one can expand the recovery matrix $M_p(z)$ in a dyadic form,

$$M_p(z) = \sum_{i=1}^{n} \frac{R_i}{z - \lambda_i} \tag{22}$$

where the residue R_i is given by

$$R_i = FW_iV_i^{\mathrm{H}}[B - K_pD]. \tag{23}$$

Here W_i and V_i are respectively the right and left eigenvectors associated with an eigenvalue λ_i of $A - K_pC$ and they are scaled so that $WV^{\mathrm{H}} = V^{\mathrm{H}}W = I_n$ where

$$W = [W_1, W_2, \cdots, W_n] \quad \text{and} \quad V = [V_1, V_2, \cdots, V_n]. \tag{24}$$

To review what can and what cannot be rendered zero, let us partition $M_p(z)$ into three parts, each part having a particular type of characteristics,

$$M_p(z) = M_p^-(z) + M_p^b(z) + M_p^e(z), \tag{25}$$

where

$$M_p^-(z) = \sum_{i=1}^{n_a^-} \frac{R_i^-}{z - \lambda_i^-}, \quad M_p^b(z) = \sum_{i=1}^{n_b} \frac{R_i^b}{z - \lambda_i^b},$$

and

$$M_p^e(z) = \sum_{i=1}^{n_a^+ + n_c + n_f} \frac{R_i^e}{z - \lambda_i^e}.$$

In the above partition, appropriate superscripts $-$, b, and e are added to R_i and λ_i in order to associate them respectively with $M_p^-(z)$, $M_p^b(z)$, and $M_p^e(z)$. Next, define the following sets where $n_e = n_a^+ + n_c + n_f$:

$$\Lambda^- = \{\lambda_i^- \,;\, i=1 \text{ to } n_a^-\},\ V^- = \{V_i^- \,;\, i=1 \text{ to } n_a^-\},\ W^- = \{W_i^- \,;\, i=1 \text{ to } n_a^-\}$$

$$\Lambda^b = \{\lambda_i^b \,;\, i=1 \text{ to } n_b\},\quad V^b = \{V_i^b \,;\, i=1 \text{ to } n_b\},\quad W^b = \{W_i^b \,;\, i=1 \text{ to } n_b\}$$

$$\Lambda^e = \{\lambda_i^e \,;\, i=1 \text{ to } n_e\},\quad V^e = \{V_i^e \,;\, i=1 \text{ to } n_e\},\quad W^e = \{W_i^e \,;\, i=1 \text{ to } n_e\}.$$

We now proceed to describe in detail the necessary design constraints and the available design freedom in assigning an appropriate eigenstructure to $A - K_pC$. We do this by considering one part of $M_p(z)$ at a time.

Discussion on $M_p^-(z)$: Consider an arbitrary target loop transfer function $L_t(z)$. The term $M_p^-(z)$ can identically be rendered zero. To accomplish this, the set of n_a^- eigenvalues Λ^- and the corresponding set of left eigenvectors V^- of $A - K_pC$ must be selected to coincide respectively with the set of plant minimum phase invariant zeros and the corresponding left state zero directions of Σ.

Discussion on $M_p^b(z)$: Consider an arbitrary target loop transfer function $L_t(z)$. The term $M_p^b(z)$ can identically be rendered zero. To accomplish this, the set of n_b eigenvalues Λ^b can be assigned arbitrarily in $\mathbb{C}^{\odot}$, while the corresponding set of left eigenvectors V^b of $A - K_pC$ is in the null space of matrix $[B - K_pD]'$.

Discussion on $M_p^e(z)$: In general, for an arbitrary target loop transfer function $L_t(z)$, it cannot be rendered zero either asymptotically or otherwise by any assignment of Λ^e and the associated sets of right and left eigenvectors, W^e and V^e. Note also that the sets of eigenvectors W^e must span the subspace $\mathcal{S}^-(\Sigma)$ [3].

Since both $M_p^-(z)$ and $M_p^b(z)$ can be rendered zero, for future use, we can combine them into one term,

$$M_p^0(z) = M_p^-(z) + M_p^b(z).$$

We define likewise, $\Lambda^0 = \Lambda^- \cup \Lambda^b$, $W^0 = W^- \cup W^b$, $V^0 = V^- \cup V^b$. Similarly, we define the set of residues corresponding to the eigenvalues in Λ^0 as R^0, and the one corresponding to the eigenvalues in Λ^e as R^e. Thus $M_p(z)$ can be rewritten as

$$M_p(z) = M_p^0(z) + M_p^e(z). \tag{26}$$

To summarize the above development, $M_p(z)$ can essentially be decomposed into two parts, $M_p^0(z)$ and $M_p^e(z)$. The first part $M_p^0(z)$ is dependent on Λ^0 a set of eigenvalues, and R^0 the corresponding set of residues. R^0 in turn depends on the sets of right and left eigenvectors, W^0 and V^0. $M_p^0(z)$ can always be rendered zero by choosing appropriately Λ^0, W^0 and

V^0. On the other hand, the second part $M_p^e(z)$ cannot be rendered zero in general for an arbitrary target loop transfer function. Hence, $M_p^e(z)$ can be termed as the ***recovery error matrix***. This recovery error matrix $M_p^e(z)$ is parameterized in terms of Λ^e and R^e, where R^e in turn is parameterized in terms of W^e and V^e. There is complete freedom in choosing the set of eigenvalues Λ^e so that its elements are all with in the unit circle in complex plane, where as the sets of eigenvectors W^e and V^e have to satisfy the well known eigenvector assignment constraints [6]. Also, W^e must span the subspace $\mathcal{S}^-(\Sigma)$. Although W^e and V^e have to satisfy certain constraints, there exists a considerable amount of freedom in selecting them. As such, one can shape the recovery error matrix $M_p^e(z)$ by selecting appropriately Λ^e, W^e and V^e. In other words, for every given system and for each type of controller, there exists a set of admissible recovery error matrices, and such a set can be denoted as $\mathcal{M}^e(\Sigma)$. Thus, notationally, $\mathcal{M}_p^e(\Sigma)$, $\mathcal{M}_c^e(\Sigma)$ and $\mathcal{M}_r^e(\Sigma)$ are respectively the admissible sets of recovery error matrices for prediction, current and reduced order estimator based controllers. Now proceeding with our general discussion for a prediction estimator based controller, one then naturally seeks a design method which leads to a chosen recovery error matrix $M_p^e(z)$ among the set of admissible recovery error matrices $\mathcal{M}_p^e(\Sigma)$. In the following section, we will give an eigenstructure assignment design method capable of achieving any chosen $M_p^e(z) \in \mathcal{M}_p^e(\Sigma)$. In Section IV, we will describe two optimization based design methods, one method leads to a design that yields the infimum H_∞ norm of the recovery error matrix, while the other yields the infimum H_2 norm. We emphasize that the eigenstructure assignment design method can lead to any chosen recovery error matrix, where as the optimization based design methods yield a particular recovery error matrix having either the infimum H_∞ norm or H_2 norm depending on the method used.

The above discussion pertains to the case where the target loop transfer function $L_t(z) = F\Phi B$ is arbitrarily specified. However, as stated in Theorem 5 of Part 1, when specific properties of $L_t(z)$ are taken into account, one can render the recovery error matrix zero provided the given system satisfies the following conditions depending on the controller used:

(1) For a prediction estimator based controller, the condition is that $\mathcal{S}^-(\Sigma) \subseteq \mathrm{Ker}\,(F)$.

(2) For a current estimator based controller, the condition is that $\mathcal{S}^-(\Sigma) \cap \{x \mid Cx \in \mathrm{Im}\,(D)\} \subseteq \mathrm{Ker}\,(F)$.

(3) For a reduced order estimator based controller, the condition is that $\mathcal{S}^-(\Sigma) \cap \{x \mid Cx \in \text{Im}(D)\} \subseteq \text{Ker}(F)$.

Thus under the above conditions, the set of admissible recovery error matrices for a given target loop transfer function contains an element which is identically zero for all z in complex plane. In that case, either eigenstructure assignment method of design or optimization based methods of design can achieve exact loop transfer recovery.

III. DESIGN BY EIGENSTRUCTURE ASSIGNMENT

For continuous time systems, we developed earlier a time-scale and eigenstructure assignment (ATEA) method of LTR design which is capable of exploiting all the available design freedom [8], [2]. In discrete time systems, as explained earlier, one does not deal with asymptotically infinite eigenvalues and as such there is no feasibility of assigning a multiple time scale structure to controller dynamics as in the case of continuous systems. That is, in discrete time systems, one can assign only a finite eigenstructure. As such, the design procedure we propose here is a special case of ATEA in which the option of assigning a chosen time scale structure is removed and hence is some what simpler than that for continuous systems. Although there is no time scale structure assignment, since the method proposed here is a special case of ATEA, we still call it as ATEA design. The present ATEA design method does not call for parameterizing the gain K in terms of a tunable parameter.

The input parameters to ATEA design are the sets of eigenvalues Λ^b and Λ^e, and the residue set R^e which can equivalently be specified in terms of the right and left eigenvectors W^e and V^e. Also, R^0 is to be rendered zero so that $M_p^0(z) = 0$. Note that Λ^b and Λ^e in addition to Λ^- form the eigenvalues of the observer dynamic matrix. Furthermore Λ^e and R^e shape the recovery error matrix $M_p^e(z)$ as desired. Thus the prescription of Λ^e and R^e is equivalent to prescribing a desired $M_p^e(z) \in \mathcal{M}_p^e(\Sigma)$. We now give a step by step ATEA design method of obtaining the observer gain K_p which when used in prediction estimator based controller leads to the prescribed recovery error matrix $M_p^e(z)$. The following steps of the ATEA design algorithm assume that the given system Σ has already been transformed to the form of s.c.b (see, Section III of Part 1 [3]).

Step 1 : This step deals with the assignment of eigenstructure to the subsystem (9) of Part 1. Choose a gain K_p^b such that $\lambda(A_{bb} - K_p^b C_b)$ coincides with the specified set Λ^b. Note that the existence of such a K_p^b is guaranteed by Property 2 of Section III of Part 1 [3]. We also note that the eigenvectors of $A_{bb} - K_p^b C_b$ can be assigned in any chosen way consistent with the freedom available in assigning them [6]. Owing to the properties of s.c.b, ATEA design always results in an eigenvector set V^b corresponding to the eigenvalues Λ^b of the observer, in the null space of $(B - K_p D)'$ so that $M_p^b(z) = 0$.

Step 2 : This step deals with the assignment of eigenstructure to the subsystems (8), (10) and (12) of Part 1. Let A^e and C^e be defined as

$$A^e = \begin{bmatrix} A_{aa}^+ & 0 & L_{af}^+ C_f \\ B_c E_{ca}^+ & A_{cc} & L_{cf} C_f \\ B_f E_a^+ & B_f E_c & A_f \end{bmatrix}, \qquad C^e = \begin{bmatrix} C_{0a}^+ & C_{0c} & C_{0f} \\ 0 & 0 & C_f \end{bmatrix}. \tag{27}$$

The design specifications utilized here are Λ^e and W^e. In view of the s.c.b, W^e is of the form,

$$W^e = [0 \quad (W_{a+}^e)^{\mathrm{H}} \quad 0 \quad (W_c^e)^{\mathrm{H}} \quad (W_f^e)^{\mathrm{H}}]^{\mathrm{H}}.$$

Let $W_e^e = [(W_{a+}^e)^{\mathrm{H}} \quad (W_c^e)^{\mathrm{H}} \quad (W_f^e)^{\mathrm{H}}]^{\mathrm{H}}$. Now select a gain K_p^e such that $\lambda(A^e - K_p^e C^e)$ and the set of right eigenvectors of $A^e - K_p^e C^e$ coincide with the specified Λ^e and W_e^e. Again note that the existence of such a K_p^e is guaranteed by Property 2 of Section III of Part 1. Let us next partition K_p^e as

$$K_p^e = \begin{bmatrix} K_p^{a0+} & K_p^{a1+} \\ K_p^{c0} & K_p^{c1} \\ K_p^{f0} & K_p^{f1} \end{bmatrix}.$$

Step 3 : In this step, K_p^b and K_p^e calculated in Steps 1 and 2 are put together into a composite matrix. Let

$$K_p = \Gamma_1 \begin{bmatrix} B_{0a}^- & L_{af}^- & L_{ab}^- \\ B_{0a}^+ + K_p^{a0+} & K_p^{a1+} & L_{ab}^+ \\ B_{0b} & L_{bf} & K_p^b \\ B_{0c} + K_p^{c0} & K_p^{c1} & L_{cb} \\ B_{0f} + K_p^{f0} & K_p^{f1} & 0 \end{bmatrix} \Gamma_2^{-1}. \tag{28}$$

We have the following theorem.

Theorem 1. *Consider a gain as given by (28). Then we have the following properties:*

1. *The eigenvalues of $A - K_pC$ are given by Λ^-, Λ^b and Λ^e.*
2. *The achieved recovery error matrix coincides with the specified $M_p^e(z)$ for all z in $\mathbf{C}$.*

Proof : It follows from the properties of s.c.b and some simple algebra. ∎

Example 1 : Consider a non-strictly proper discrete-time system Σ with sampling period $T = 1$, and characterized by

$$A = \begin{bmatrix} 0 & 1 & 0 & 0 \\ 0 & 0 & 1 & 1 \\ 1 & 0 & 0.5 & 0 \\ 1 & 0 & 0 & 1.5 \end{bmatrix}, \quad B = \begin{bmatrix} 0 & 0 \\ 0 & 1 \\ 1 & 0 \\ 0 & 0 \end{bmatrix},$$

$$C = \begin{bmatrix} 0 & 0 & 1 & 0 \\ 1 & 0 & 0 & 0 \end{bmatrix}, \qquad D = \begin{bmatrix} 1 & 0 \\ 0 & 0 \end{bmatrix}.$$

Let the target loop $L_t(z)$ and thus the target sensitivity function $S_t(z)$ be specified by

$$F = \begin{bmatrix} 0.00 & 0.0000 & 1 & 0.000 \\ 1.25 & 0.8333 & 1 & 2.875 \end{bmatrix}.$$

It is simple to verify that the given system Σ is invertible, i.e., $n_b = n_c = 0$, with two infinite zeros of order 2 and two invariant zeros at $\{-0.5, 1.5\}$. It can also be verified that the target loop specified by the given F is not recoverable by any of the three controllers being considered in this paper. Let Λ^e and W^e along with the corresponding recovery error matrix $M^e(z)$ be as given below depending on the controller used:

Prediction Estimator Based Controller :

$$\Lambda^e = \{-0.1, 0, 0.1\}\ , \quad W^e = \begin{bmatrix} -0.3684 & -0.3478 & 0.3276 \\ -0.5157 & -0.5217 & 0.5241 \\ 0.0000 & 0.0000 & 0.0000 \\ -0.7735 & -0.7790 & 0.7861 \end{bmatrix}$$

and

$$M_p^e(z) = \frac{1}{z^3 - 0.01z} \begin{bmatrix} 0 & 0 \\ 0 & 0.8333z^2 + 1.25z - 13.41 \end{bmatrix}.$$

Current Estimator Based Controller :

$$\Lambda^e = \{-0.1, 0.1, \} \ , \ W^e = \begin{bmatrix} 0.0000 & 0.0000 \\ -0.5812 & -0.5300 \\ 0.0000 & 0.0000 \\ -0.8137 & -0.8480 \end{bmatrix}$$

and

$$M_c^e(z) = \frac{1}{z^2 - 0.01} \begin{bmatrix} 0 & 0 \\ 0 & 0.8333z - 7.69 \end{bmatrix}.$$

Reduced Order Estimator Based Controller :

$$\Lambda^e = \{-0.1, 0.1, \} \ , \ W^e = \begin{bmatrix} -0.5812 & -0.5300 \\ 0.0000 & 0.0000 \\ -0.8137 & -0.8480 \end{bmatrix}$$

and

$$M_r^e(z) = \frac{1}{z^2 - 0.01} \begin{bmatrix} 0 & 0 \\ 0 & 0.8333z - 7.69 \end{bmatrix}.$$

Then using ATEA algorithm, we obtain the following controllers.

Prediction Estimator Based Controller :

$$\hat{x}(k+1) = \begin{bmatrix} -1.50 & 1.0000 & 0.0 & 0.000 \\ -3.49 & -0.8333 & 0.0 & -1.875 \\ 0.00 & 0.0000 & -0.5 & 0.000 \\ -3.36 & 0.0000 & 0.0 & 1.500 \end{bmatrix} \hat{x}(k) + \begin{bmatrix} 0 & 1.50 \\ 0 & 2.24 \\ 1 & 1.00 \\ 0 & 4.36 \end{bmatrix} y(k),$$

$$-u(k) = \begin{bmatrix} 0.00 & 0.0000 & 1 & 0.000 \\ 1.25 & 0.8333 & 1 & 2.875 \end{bmatrix} \hat{x}(k).$$

The eigenvalues of the above prediction estimator are at $\{-0.5, -0.1, 0, 0.1\}$ while the achieved recovery error matrix M_p^e coincides with the one specified.

Current Estimator Based Controller :

$$v(k+1) = \begin{bmatrix} 0.00 & 0.0000 & 0.0 & 0.000 \\ -1.25 & -2.3333 & 0.0 & -1.875 \\ 1.00 & 0.0000 & -0.5 & 0.000 \\ 1.00 & -2.2400 & 0.0 & 1.500 \end{bmatrix} v(k) + \begin{bmatrix} 0 & 0.00 \\ 0 & -8.95 \\ 1 & 1.00 \\ 0 & 1.00 \end{bmatrix} y(k),$$

$$-u(k) = \begin{bmatrix} 0.00 & 0.0000 & 1 & 0.000 \\ 1.25 & 0.8333 & 1 & 2.875 \end{bmatrix} v(k) + \begin{bmatrix} 0 & 0.00 \\ 0 & 8.94 \end{bmatrix} y(k).$$

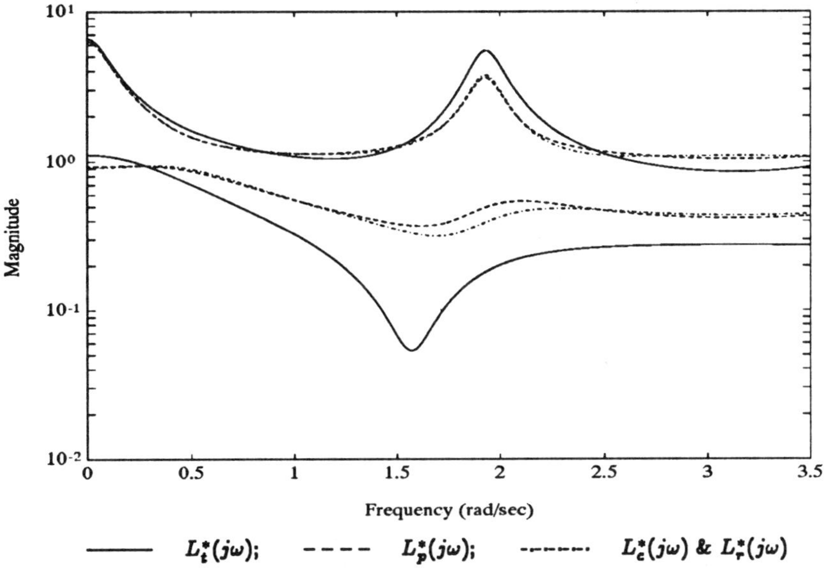

Figure 1: The max. and min. singular values of target and achieved loops.

The eigenvalues of the above current estimator are at $\{-0.5, -0.1, 0, 0.1\}$ while the achieved recovery error matrix M_c^e coincides with the one specified.

Reduced Order Estimator Based Controller :

$$v(k+1) = \begin{bmatrix} -2.3333 & 0.0 & -1.875 \\ 0.0000 & -0.5 & 0.000 \\ -2.2400 & 0.0 & 1.500 \end{bmatrix} v(k) + \begin{bmatrix} 0 & -8.95 \\ 1 & 1.00 \\ 0 & 1.00 \end{bmatrix} y(k),$$

$$-u(k) = \begin{bmatrix} 0.0000 & 1 & 0.000 \\ 0.8333 & 1 & 2.875 \end{bmatrix} v(k) + \begin{bmatrix} 0 & 0.00 \\ 0 & 8.94 \end{bmatrix} y(k).$$

The eigenvalues of the above reduced order estimator are at $\{-0.5, -0.1, 0.1\}$ while the achieved recovery error matrix M_r^e coincides with the one specified.

The plots of maximum and minimum singular values of the target and the achieved loop transfer function via all the three controllers are shown in Figure 1. Also, the plots of maximum singular values of $M_p^*(j\omega)$, $M_c^*(j\omega)$ and $M_r^*(j\omega)$ are shown in Figure 2, while the plots of maximum singular

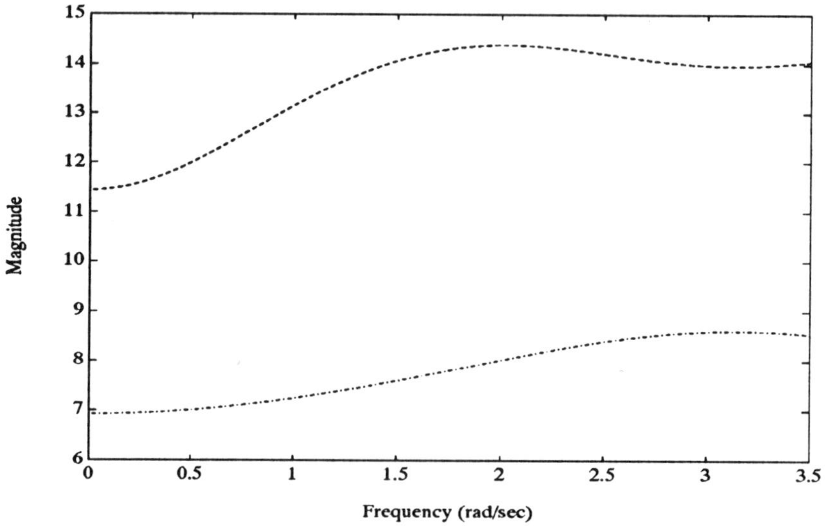

- - - - $\sigma_{max}[M_p^*(j\omega)]$; -·-·-·-· $\sigma_{max}[M_c^*(j\omega)]$ & $\sigma_{max}[M_r^*(j\omega)]$

Figure 2: The max. singular values of recovery matrices.

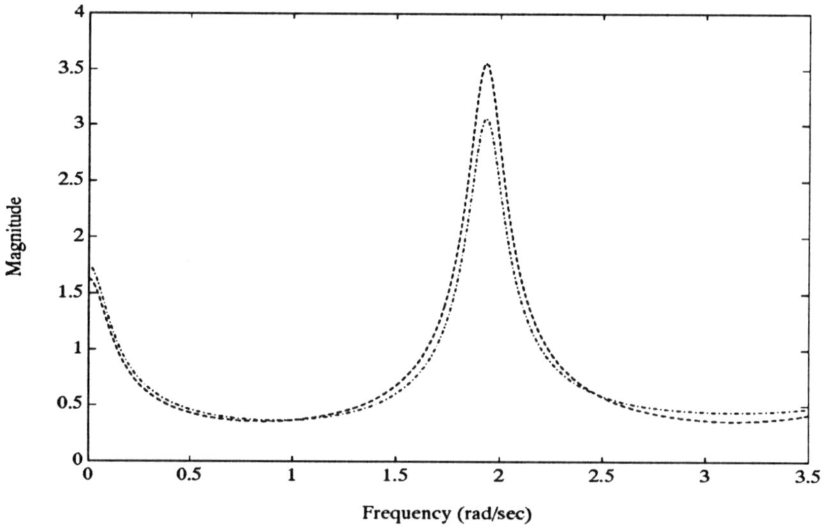

- - - - $\sigma_{max}[E_p^*(j\omega)]$; -·-·-·-· $\sigma_{max}[E_c^*(j\omega)]$ & $\sigma_{max}[E_r^*(j\omega)]$

Figure 3: The max. singular values of loop transfer recovery errors.

values of $E_p^*(j\omega)$, $E_c^*(j\omega)$ and $E_r^*(j\omega)$ are shown in Figure 3. Clearly, as expected, the above current and reduced order estimators yield the same performance in the sense that the achieved recovery is same. □

IV. OPTIMIZATION BASED DESIGN METHODS

As is clear from Section II, the whole notion of LTR is to render the recovery matrix $M_p(z) = F(zI_n - A + K_pC)^{-1}(B - K_pD)$ small in some sense or other. The ATEA design method of Section III views this task from the perspective of eigenstructure assignment to the controller dynamic matrix. It enables us to design a controller which achieves any specified recovery error matrix among a set of such admissible matrices. An alternative method, as in the case of continuous systems, is to view the controller design for LTR as finding an observer gain K_p which minimizes some (say, either H_∞ or H_2) norm of $M_p(z)$. That is, one can cast the LTR design as a straightforward mathematical optimization problem. A suboptimal or optimal solution to such an optimization problem provides the needed observer gain. Note that when LTR problem is formulated as an optimization problem of minimizing some norm of $M_p(z)$ by appropriate selection of K_p, the optimization method apparently renders $M_p^0(z)$ zero while minimizing the specified norm of $M_p^e(z)$. In contrast to this, eigenstructure assignment method is flexible, and is capable of yielding any recovery error matrix $M_p^e(z) \in \mathcal{M}_p^e(\Sigma)$ while rendering $M_p^0(z)$ zero.

Goodman [5] is the first person who formulated earlier the LTR problem for discrete systems as a H_2 minimization problem of the recovery matrix $M_p(z)$. He considered only strictly proper square invertible minimum phase systems having infinite zeros of order one. Recently, Zhang and Freudenberg [11] considered square strictly proper nonminimum phase systems. They develop explicit expressions for the resulting recovery error matrix and the sensitivity function when prediction as well as current estimator based controllers are used and when optimization is used to minimize the H_2 norm of the recovery matrix. The optimization procedure used by [11] follows along the same lines as that for continuous systems as in [4]. It turns out, as in the continuous case [2], that the controller design based on optimization procedures for LTR of general discrete systems, can be cast as an optimal state feedback design for an auxiliary system related to the given one. Then, following a mass of existing literature on such optimal state feedback designs, especially for continuous systems, one can develop in a straightforward way optimization based design procedures for LTR of general discrete systems. This is what we pursue in this section.

To proceed with, we consider the following auxiliary system,

$$\Sigma^{au1}: \begin{cases} x(k+1) = A'x(k) + C'u(k) + F'w(k), \\ y(k) \quad = x(k), \\ z(k) \quad = B'x(k) + D'u(k). \end{cases} \tag{29}$$

Here w is treated as an exogenous disturbance input to Σ^{au1} and u is the controlling input while the variable z is considered as the controlled output. Then it is trivial to verify that the closed loop transfer function from w to y of Σ^{au1} under a state feedback law $u(k) = -K'_p x(k)$ is given by

$$M'(z) = (B' - D'K'_p)(zI_n - A' + C'K'_p)^{-1}F'. \tag{30}$$

Hence, the minimization of $M_p(z)$ over all the possible stabilizing gains K_p is equivalent to the minimization of $M'(z)$ over all the stabilizing state feedback control laws for Σ^{au1}. As such the design of observer based controllers for LTR is translated to an optimal state feedback controller design.

A. H_∞-optimization Based Algorithm

Throughout this section, we assume that the given system Σ characterized by (A, B, C, D) has no invariant zeros on the unit circle. Denoting γ^* as the infimum of $\|M_p(z)\|_\infty$ over all possible stabilizing gains K_p, we present here a basic algorithm of computing the gain matrix $K_p(\gamma)$ such that the resulting H_∞-norm of the recovery matrix $M_p(z,\gamma)$, is less than a priori given desired scalar $\gamma > \gamma^*$. The algorithm is as follows:

Step 1 : At first we compute nonsingular transformations U and V such that

$$UD'V = \begin{bmatrix} I_{m_0} & 0 \\ 0 & 0 \end{bmatrix}.$$

Then partition

$$C'V = [\tilde{B}_0 \quad \tilde{B}_1] \quad \text{and} \quad UB' = \begin{bmatrix} \tilde{C}_0 \\ \tilde{C}_1 \end{bmatrix}.$$

Following the procedure of constructing a special coordinate basis (s.c.b), see Section III of Part 1 [3], one can readily calculate three nonsingular transformations Γ_s, Γ_i and Γ_o such that

$$\Gamma_s^{-1}(A' - \tilde{B}_0\tilde{C}_0)\Gamma_s = \begin{bmatrix} A_{aa}^- & 0 & L_{ab}^- C_b & 0 & L_{af}^- C_f \\ 0 & A_{aa}^+ & L_{ab}^+ C_b & 0 & L_{af}^+ C_f \\ 0 & 0 & A_{bb} & 0 & L_{bf} C_f \\ B_c E_{ca}^- & B_c E_{ca}^+ & L_{cb} C_b & A_{cc} & L_{cf} C_f \\ B_f E_{fa}^- & B_f E_{fa}^+ & B_f E_{fb} & B_f E_{fc} & A_f \end{bmatrix}, \tag{31}$$

$$\Gamma_s^{-1}[\tilde{B}_0 \quad \tilde{B}_1]\Gamma_i = \begin{bmatrix} B_{0a}^- & 0 & 0 \\ B_{0a}^+ & 0 & 0 \\ B_{0b} & 0 & 0 \\ B_{0c} & 0 & B_c \\ B_{0f} & B_f & 0 \end{bmatrix}, \quad \Gamma_s^{-1}F' = \begin{bmatrix} E_a^- \\ E_a^+ \\ E_b \\ E_c \\ E_f \end{bmatrix}, \tag{32}$$

$$\Gamma_o^{-1} \begin{bmatrix} \tilde{C}_0 \\ \\ \tilde{C}_1 \end{bmatrix} \Gamma_s = \begin{bmatrix} C_{0a}^- & C_{0a}^+ & C_{0b} & C_{0c} & C_{0f} \\ 0 & 0 & 0 & 0 & C_f \\ 0 & 0 & C_b & 0 & 0 \end{bmatrix}, \tag{33}$$

and

$$\Gamma_o^{-1} \begin{bmatrix} I_{m_0} & 0 \\ \\ 0 & 0 \end{bmatrix} \Gamma_i = \begin{bmatrix} I_{m_0} & 0 & 0 \\ 0 & 0 & 0 \\ 0 & 0 & 0 \end{bmatrix}. \tag{34}$$

Next, we define a subsystem Σ^{au2} of the above dual system as,

$$\Sigma^{au2} : \begin{cases} \boldsymbol{x}(k+1) = \boldsymbol{A}\boldsymbol{x}(k) + \boldsymbol{B}\boldsymbol{u}(k) + \boldsymbol{E}\boldsymbol{w}(k), \\ \quad \boldsymbol{y}(k) \quad = \boldsymbol{x}(k), \\ \quad \boldsymbol{z}(k) \quad = \boldsymbol{C}\boldsymbol{x}(k) + \boldsymbol{D}\boldsymbol{u}(k), \end{cases} \tag{35}$$

where

$$\boldsymbol{A} = \begin{bmatrix} A_{aa}^- & 0 & L_{ab}^- C_b & L_{af}^- C_f \\ 0 & A_{aa}^+ & L_{ab}^+ C_b & L_{af}^+ C_f \\ 0 & 0 & A_{bb} & L_{bf} C_f \\ B_f E_{fa}^- & B_f E_{fa}^+ & B_f E_{fb} & A_f \end{bmatrix}, \quad \boldsymbol{B} = \begin{bmatrix} B_{0a}^- & 0 \\ B_{0a}^+ & 0 \\ B_{0b} & 0 \\ B_{0f} & B_f \end{bmatrix}, \tag{36}$$

and

$$\boldsymbol{E} = \begin{bmatrix} E_a^- \\ E_a^+ \\ E_b \\ E_f \end{bmatrix}, \quad \boldsymbol{C} = \begin{bmatrix} 0 & 0 & 0 & 0 \\ 0 & 0 & 0 & C_f \\ 0 & 0 & C_b & 0 \end{bmatrix}, \quad \boldsymbol{D} = \begin{bmatrix} I_{m_0} & 0 \\ 0 & 0 \\ 0 & 0 \end{bmatrix}. \tag{37}$$

Here we note that the system characterized by (A, B, C, D) is left invertible and has no invariant zeros on the unit circle if and only if Σ characterized by (A, B, C, D) has no invariant zeros on the unit circle.

Step 2 : Solve the Riccati equation,

$$P = A'PA + C'C - \begin{bmatrix} B'PA + D'C \\ E'PA \end{bmatrix}'$$
$$\cdot \begin{bmatrix} D'D + B'PB & B'PE \\ E'PB & -\gamma^2 I + E'PE \end{bmatrix}^{-1} \begin{bmatrix} B'PA + D'C \\ E'PA \end{bmatrix},$$

for a positive semi-definite P which satisfies the conditions:

$$D'D + B'PB > 0 \quad \text{and} \quad I - E'PE/\gamma^2 > 0$$

and

$$\lambda \left\{ A - [B \quad E] \begin{bmatrix} D'D+B'PB & B'PE \\ E'PB & -\gamma^2 I + E'PE \end{bmatrix}^{-1} \begin{bmatrix} B'PA + D'C \\ E'PA \end{bmatrix} \right\} \subset \mathbf{C}^{\odot}.$$

We note that such a P always exists and is unique since (A, B, C, D) is left invertible and has no invariant zeros on the unit circle [10].

Step 3 : Compute

$$F_1 = [B'PB + D'D + B'PE(\gamma^2 I - E'PE)^{-1} E'PB]^{-1}$$
$$\cdot [B'PA + D'C + B'PE(\gamma^2 I - E'PE)^{-1} E'PA]$$

and partition it as

$$F_1 = \begin{bmatrix} F_{a0}^- & F_{a0}^+ & F_{b0} & F_{f0} \\ F_{af}^- & F_{af}^+ & F_{bf} & F_{ff} \end{bmatrix}.$$

Then let

$$F(\gamma) = V\Gamma_i \begin{bmatrix} C_{0a}^- + F_{a0}^- & C_{0a}^+ + F_{a0}^+ & C_{0b} + F_{b0} & C_{0c} & C_{0f} + F_{f0} \\ F_{af}^- & F_{af}^+ & F_{bf} & E_{fc} & F_{ff} \\ 0 & 0 & 0 & F_{cc} & 0 \end{bmatrix} \Gamma_s^{-1}$$

where F_{cc} is such that $\lambda(A_{cc} - B_c F_{cc}) \in \mathbf{C}^{\odot}$. Next, choose $K_p(\gamma)$ as

$$K_p(\gamma) = F'(\gamma). \tag{38}$$

We have the following theorem.

Theorem 2. *Let $K_p(\gamma)$ be computed as in (38) and let $M_p(z,\gamma)$ be the resulting recovery matrix. Then, $\|M_p(z,\gamma)\|_\infty$ is strictly less than γ, and tends to γ^* as $\gamma \to \gamma^*$.*

Proof : See Appendix A. ∎

Remark 1. *Note that once the estimator gains $K_p(\gamma^*)$, $K_c(\gamma^*)$ and $K_r(\gamma^*)$ are calculated from the above design procedure, the corresponding recovery error matrices can easily be calculated from the expressions (19), (20) and (21), i.e.,*

$$M_p^e(z) = F[zI_n - A + K_p(\gamma^*)C]^{-1}[B - K_p(\gamma^*)D], \tag{39}$$

$$M_c^e(z) = F\,[zI_n - A + K_c(\gamma^*)C_c]^{-1}\,[B - K_c(\gamma^*)D_c]\,, \tag{40}$$

$$M_r^e(z) = F_2[zI - A_r + K_r(\gamma^*)C_r]^{-1}[B_r - K_r(\gamma^*)D_r]. \tag{41}$$

Moreover, these recovery error matrices have the least H_∞ norm among the sets of the corresponding admissible recovery error matrices.

Example 2 : Consider a discrete-time system Σ given in Astrom et al [1] with sampling period $T = 1$, and characterized by

$$A = \begin{bmatrix} 1.1036 & 1 & 0 \\ -0.4060 & 0 & 1 \\ 0.0498 & 0 & 0 \end{bmatrix}, \quad B = \begin{bmatrix} 0.0803 \\ 0.1544 \\ 0.0179 \end{bmatrix},$$

$$C = [1 \quad 0 \quad 0], \qquad D = 0.$$

Let the target loop $L_t(z)$ and the target sensitivity function $S_t(z)$ be specified by

$$F = [7.1222 \quad 7.5293 \quad 2.7373]\,.$$

It is simple to verify that the given system Σ is invertible with one infinite zero of order one and two invariant zeros at $\{-1.7989, -0.1239\}$. It can also be verified that the target loop specified by the given F is not recoverable either by a prediction or by a current or by a reduced order estimator based controller. The following are the prediction, current and reduced order estimator based controllers obtained by the H_∞-optimization based algorithm. All these controllers achieve the infimum of the H_∞-norm of the corresponding recovery matrices.

Prediction Estimator Based Controller :

$$\hat{x}(k+1) = \begin{bmatrix} -1.5371 & 0.3954 & -0.2198 \\ -1.2039 & -1.1625 & 0.5774 \\ -0.1275 & -0.1348 & -0.0490 \end{bmatrix} \hat{x}(k) + \begin{bmatrix} 2.0688 \\ -0.3017 \\ 0.0498 \end{bmatrix} y(k),$$

$$-u(k) = [7.1222 \quad 7.5293 \quad 2.7373]\,\hat{x}(k).$$

The eigenvalues of the above prediction estimator are placed at $\{0, -0.1239, -0.8413\}$ while the resulting recovery error matrix $M_p^e(z)$ is given by,

$$M_p^e(z) = \frac{1.7834z + 2.1198}{z^2 + 0.8413z}.$$

Current Estimator Based Controller :

$$v(k+1) = \begin{bmatrix} 0.0000 & 0.0000 & 0.0000 \\ -1.5716 & -1.2115 & 0.6046 \\ -0.0777 & -0.1348 & -0.0490 \end{bmatrix} + \begin{bmatrix} 0.0000 \\ -1.7217 \\ -0.0944 \end{bmatrix} y(k),$$

$$-u(k) = [7.1222 \quad 7.5293 \quad 2.7373]\,v(k) + 8.0552y(k).$$

The eigenvalues of the above current estimator are placed at $\{0, 0, -0.1239\}$ while the resulting recovery error matrix $M_c^e(z)$ is given by,

$$M_c^e(z) = \frac{1.1366}{z}.$$

Reduced Order Estimator Based Controller :

$$v(k+1) = \begin{bmatrix} -1.2115 & 0.6046 \\ -0.1348 & -0.0490 \end{bmatrix} v(k) + \begin{bmatrix} -1.7217 \\ -0.0944 \end{bmatrix} y(k),$$

$$-u(k) = [7.5293 \quad 2.7373]\,v(k) + 8.0553y(k).$$

The eigenvalues of the above reduced order estimator are placed at $\{0, -0.1239\}$ while the resulting recovery error matrix $M_r^e(z)$ is given by,

$$M_r^e(z) = \frac{1.1366}{z}.$$

The plots of singular values of the target and the achieved loop transfer function via all the three controllers are shown in Figure 4. Also, the plots of singular values of $M_p^*(j\omega)$, $M_c^*(j\omega)$ and $M_r^*(j\omega)$ are shown in Figure 5, while the plots of singular values of $E_p^*(j\omega)$, $E_c^*(j\omega)$ and $E_r^*(j\omega)$ are shown in Figure 6. Clearly, for this example, the above current and reduced order estimators yield the same performance in the sense that the achieved recovery is same. □

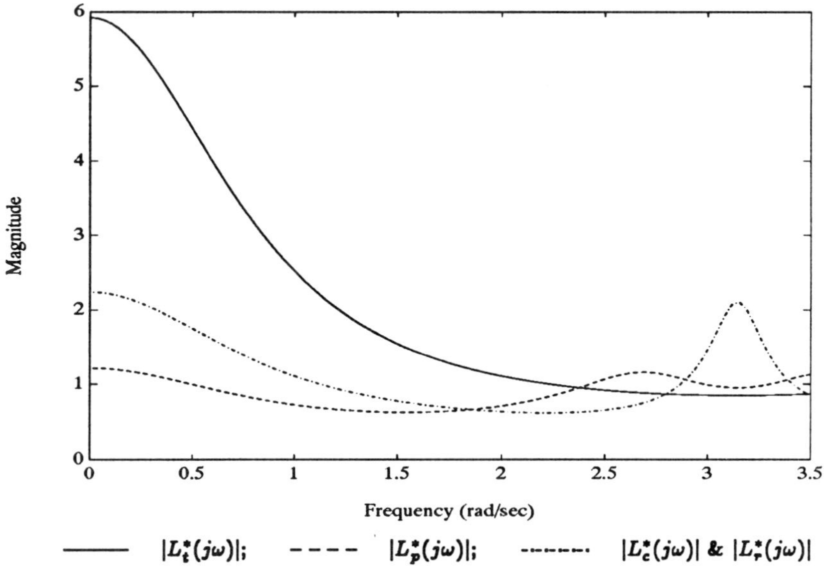

Figure 4: The singular values of target and achieved loops.

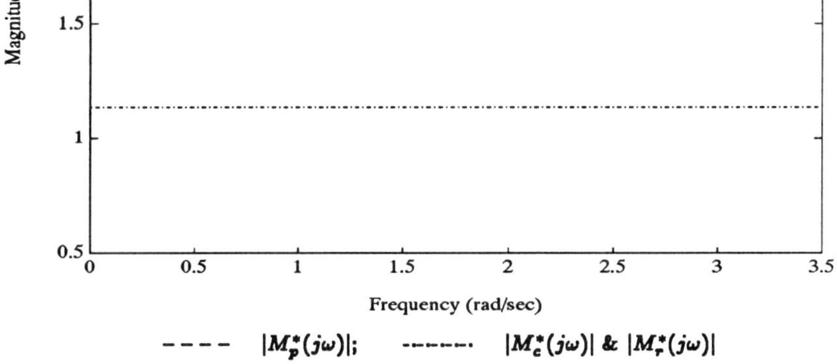

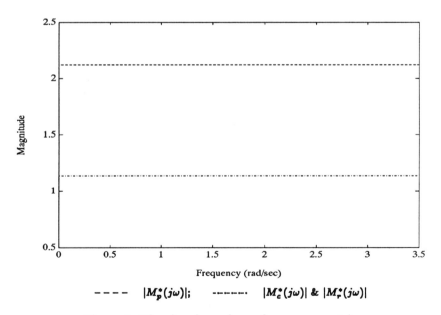

Figure 5: The singular values of recovery matrices.

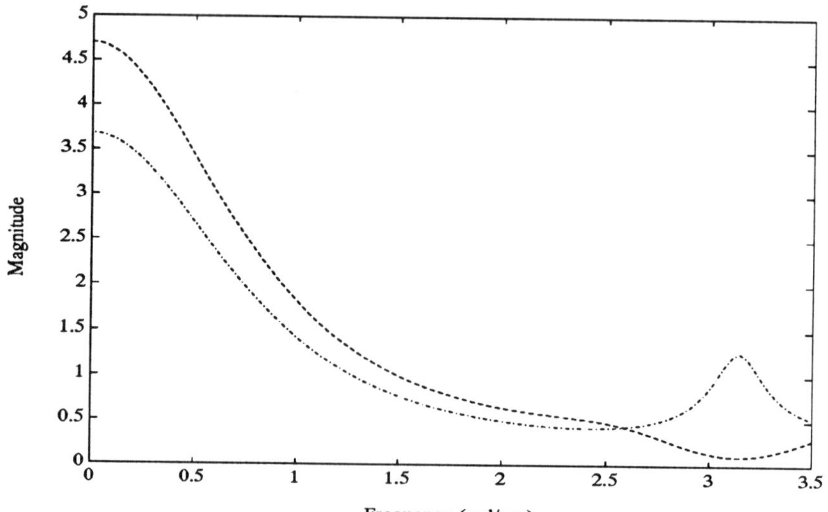

Figure 6: The singular values of loop transfer recovery errors.

B. H_2-optimization Based Algorithm

It is well-known in the literature that the solution of H_2-optimization problem is equivalent to the solution of H_∞-optimization where in γ is set to ∞. Utilizing this fact, in this subsection, we proceed to give an algorithm that minimizes the H_2-norm of $M_p(z)$ over all possible stabilizing gain matrices K_p. To do so, as in the previous subsection, we assume that the given system Σ characterized by (A, B, C, D) has no invariant zeros on the unit circle.

Step 1 : Transform the dual system of (A, B, C, D) in the form of s.c.b, and construct an auxiliary system Σ^{au2} as in (35).

Step 2 : Solve the Riccati equation,

$$P = A'PA + C'C - (B'PA + D'C)'(D'D + B'PB)^{-1}(B'PA + D'C),$$

for a positive semi-definite P which satisfies the conditions:

$$D'D + B'PB > 0$$

and

$$\lambda\{A - B(D'D + B'PB)^{-1}(B'PA + D'C)\} \subset \mathbf{C}^{\odot}.$$

Again, we note that such a P always exists and is unique.

Step 3 : Partition,

$$[B'PB + D'D]^{-1}[B'PA + D'C] = \begin{bmatrix} F_{a0}^{-} & F_{a0}^{+} & F_{b0} & F_{f0} \\ F_{af}^{-} & F_{af}^{+} & F_{bf} & F_{ff} \end{bmatrix}.$$

Step 4 : Let

$$F = V\Gamma_i \begin{bmatrix} C_{0a}^{-} + F_{a0}^{-} & C_{0a}^{+} + F_{a0}^{+} & C_{0b} + F_{b0} & C_{0c} & C_{0f} + F_{f0} \\ F_{af}^{-} & F_{af}^{+} & F_{bf} & E_{fc} & F_{ff} \\ 0 & 0 & 0 & F_{cc} & 0 \end{bmatrix} \Gamma_s^{-1} \tag{42}$$

where F_{cc} is such that $\lambda(A_{cc} - B_c F_{cc}) \in \mathbf{C}^{\odot}$. Next, choose K_p as

$$K_p = F'. \tag{43}$$

We have the following theorem.

Theorem 3. *Let K_p be computed as in (43) and let $M_p(z)$ be the resulting recovery matrix. Then, $\|M_p(z)\|_{H_2}$ is the infimum among all the possible ones.*

Proof : It follows from the well-known relationship between H_∞- and H_2-optimizations. ∎

Remark 2. *Consider the case when the system characterized by the matrix quadruple (A, B, C, D) is invertible with $D = 0$ and $\det(CB) \neq 0$. Then, as determined first by Goodman [5], the gain matrix of (43) reduces to $K_p = [(CB)^{-1}CA]'$ if the system characterized by (A, B, C, D) is of minimum phase. On the other hand, the gain matrix of (43) reduces to $K_p = [(C_m B)^{-1} C_m A]'$ if the system characterized by (A, B, C, D) is of nonminimum phase, where C_m is the minimum phase counterpart of C in the all-pass factorization of (A, B, C, D) [11].*

Remark 3. *Consider a special case of a strictly proper square invertible system with all its infinite zeros of order one. For this special case, and when prediction and current estimator based controllers are used, and moreover* *when the target loop transfer function is specified by breaking the loop at the output point of the given system, Zhang and Freudenberg [11] earlier developed closed-form expressions for the resulting recovery errors when observer gain matrices are calculated using their H_2-optimization procedure. It can easily be shown that for the special cases considered by Zhang and Freudenberg [11], the resulting recovery error matrices when our design procedures are used, are indeed equivalent to the ones given by Zhang and Freudenberg.*

Example 3 : Consider the system and the target loop given in Example 2. The following are the prediction, current and reduced order estimator based controllers obtained by the H_2-optimization based algorithm. All these controllers achieve the infimum of the H_2-norm of the corresponding recovery matrices.

Prediction Estimator Based Controller :

$$\hat{x}(k+1) = \begin{bmatrix} -1.2517 & 0.3954 & -0.2198 \\ -1.1686 & -1.1625 & 0.5774 \\ -0.1275 & -0.1348 & -0.0490 \end{bmatrix} \hat{x}(k) + \begin{bmatrix} 1.7834 \\ -0.3371 \\ 0.0498 \end{bmatrix} y(k),$$

$$-u(k) = [7.1222 \quad 7.5293 \quad 2.7373]\,\hat{x}(k).$$

The eigenvalues of the above prediction estimator are placed at $\{0, -0.1239, -0.5559\}$ while the resulting recovery error matrix $M_p^e(z)$ is given by,

$$M_p^e(z) = \frac{1.7834z + 1.7954}{z^2 + 0.5559z}.$$

Current Estimator Based Controller :

$$v(k+1) = \begin{bmatrix} 0.0000 & 0.0000 & 0.0000 \\ -1.8671 & -1.4313 & 0.7268 \\ -0.1143 & -0.1620 & -0.0339 \end{bmatrix} v(k) + \begin{bmatrix} 0.0000 \\ -2.7901 \\ -0.2268 \end{bmatrix} y(k),$$

$$-u(k) = [7.1222 \quad 7.5293 \quad 2.7373]\,v(k) + 12.4294y(k).$$

The eigenvalues of the above current estimator are placed at $\{0, -0.1239, -0.5559\}$ while the resulting recovery error matrix $M_c^e(z)$ is given by,

$$M_c^e(z) = \frac{0.7854}{z + 0.5559}.$$

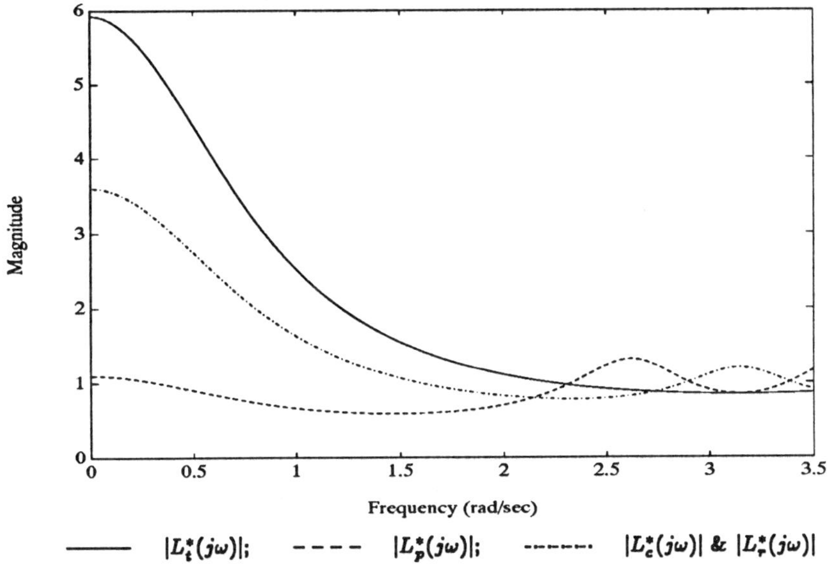

Figure 7: The singular values of target and achieved loops.

Reduced Order Estimator Based Controller :

$$v(k+1) = \begin{bmatrix} -1.4313 & 0.7268 \\ -0.1620 & -0.0339 \end{bmatrix} v(k) + \begin{bmatrix} -2.7901 \\ -0.2268 \end{bmatrix} y(k),$$

$$-u(k) = [\,7.5293 \quad 2.7373\,]\, v(k) + 12.4294 y(k).$$

The eigenvalues of the above reduced order estimator are placed at $\{-0.1239, -0.5559\}$ while the resulting recovery error matrix $M_r^e(z)$ is given by,

$$M_r^e(z) = \frac{0.7854}{z + 0.5559}.$$

The plots of singular values of the target and the achieved loop transfer function via all the three controllers are shown in Figure 7. Also, the plots of singular values of $M_p^*(j\omega)$, $M_c^*(j\omega)$ and $M_r^*(j\omega)$ are shown in Figure 8, while the plots of singular values of $E_p^*(j\omega)$, $E_c^*(j\omega)$ and $E_r^*(j\omega)$ are shown in Figure 9. Clearly, for this example, the above current and reduced order estimators yield the same performance in the sense that the achieved recovery is same. □

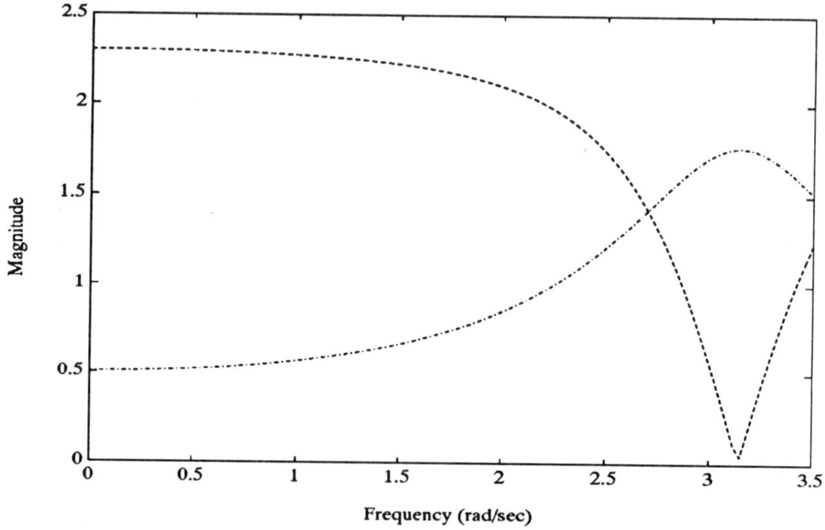

---- $|M_p^*(j\omega)|$; -·-·-· $|M_c^*(j\omega)|$ & $|M_r^*(j\omega)|$

Figure 8: The singular values of recovery matrices.

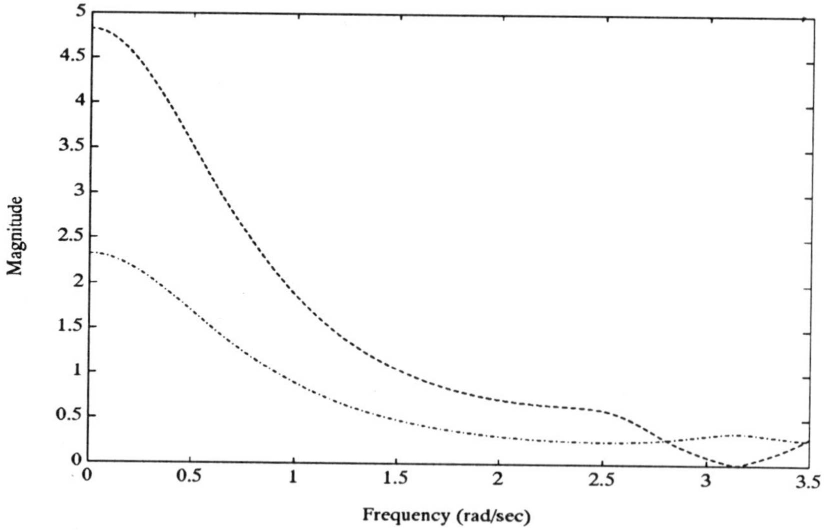

---- $|E_p^*(j\omega)|$; -·-·-· $|E_c^*(j\omega)|$ & $|E_r^*(j\omega)|$

Figure 9: The singular values of loop transfer recovery errors.

Example 4 : Consider a non-strictly proper discrete-time system Σ with sampling period $T = 1$, and characterized by

$$A = \begin{bmatrix} -2.0 & 1.0 & 1 & -1 \\ 1.0 & -0.5 & 0 & 0 \\ 1.0 & 0.0 & 0 & 0 \\ 2.5 & 0.0 & 0 & 2 \end{bmatrix}, \quad B = \begin{bmatrix} 0 & 1 \\ 1 & 0 \\ 0 & 0 \\ 0 & 0 \end{bmatrix},$$

$$C = \begin{bmatrix} 0 & -0.25 & 0 & 0 \\ 1 & 0 & 0 & 0 \end{bmatrix}, \quad D = \begin{bmatrix} 1 & 0 \\ 0 & 0 \end{bmatrix}.$$

Let the target loop $L_t(z)$ and target sensitivity function $S_t(z)$ be specified by

$$F = \begin{bmatrix} 1 & -1 & 1 & 0 \\ -1 & 1 & 1 & 0 \end{bmatrix}.$$

It is simple to verify that the given system Σ is invertible with one infinite zero of order one and three invariant zeros at $\{-0.25,\ 0,\ 2\}$. It can also be verified that

$$\mathcal{S}^-(\Sigma) = \text{span} \begin{bmatrix} 1 & 0 \\ 0 & 0 \\ 0 & 0 \\ 0 & 1 \end{bmatrix} \quad \text{and} \quad \mathcal{S}^-(\Sigma) \cap \{x \,|\, Cx \in \text{Im}\,(D)\} = \text{span} \begin{bmatrix} 0 \\ 0 \\ 0 \\ 1 \end{bmatrix}.$$

Hence, the target loop specified by the given F is not recoverable by prediction estimator based controller, but it is recoverable either by current or by reduced order estimator based controllers. The following are the prediction, current and reduced order estimator based controllers obtained by the H_2-optimization based algorithm. Again, all these controllers achieve the infimum of the H_2-norm of the corresponding recovery matrices.

Prediction Estimator Based Controller :

$$\hat{x}(k+1) = \begin{bmatrix} -0.5 & 0.00 & 0 & -1 \\ 0.0 & -0.25 & 0 & 0 \\ 0.0 & 0.00 & 0 & 0 \\ 3.0 & 0.00 & 0 & 2 \end{bmatrix} \hat{x}(k) + \begin{bmatrix} 0 & -0.5 \\ 1 & 1.0 \\ 0 & 1.0 \\ 0 & -0.5 \end{bmatrix} y(k).$$

$$-u(k) = \begin{bmatrix} 1 & -1 & 1 & 0 \\ -1 & 1 & 1 & 0 \end{bmatrix} y(k).$$

The eigenvalues of the above prediction estimator are placed at $\{0,\ 0,\ -0.25,\ 0.5\}$ while the resulting recovery error matrix $M_p^e(z)$ is given by,

$$M_p^e(z) = \frac{1}{z^2 - 0.5z} \begin{bmatrix} 0 & z-2 \\ 0 & -(z-2) \end{bmatrix}.$$

Current Estimator Based Controller :

$$v(k+1) = \begin{bmatrix} 0 & 0.00 & 0 & 0.0 \\ 1 & -0.25 & 0 & 0.0 \\ 1 & 0.00 & 0 & 0.0 \\ 1 & 0.00 & 0 & 0.5 \end{bmatrix} v(k) + \begin{bmatrix} 0 & 0.00 \\ 1 & 1.00 \\ 0 & 1.00 \\ 0 & 0.25 \end{bmatrix} y(k),$$

$$-u(k) = \begin{bmatrix} 1 & -1 & 1 & 0 \\ -1 & 1 & 1 & 0 \end{bmatrix} v(k) + \begin{bmatrix} 0 & 1 \\ 0 & -1 \end{bmatrix} y(k).$$

The eigenvalues of the above current estimator are placed at $\{0,\ 0,\ -0.25,\ 0.5\}$ while the resulting recovery error matrix $M_c^e(z) \equiv 0$.

Reduced Order Estimator Based Controller :

$$v(k+1) = \begin{bmatrix} -0.25 & 0 & 0.0 \\ 0.00 & 0 & 0.0 \\ 0.00 & 0 & 0.5 \end{bmatrix} v(k) + \begin{bmatrix} 1 & 1.00 \\ 0 & 1.00 \\ 0 & 0.25 \end{bmatrix} y(k).$$

$$-u(k) = \begin{bmatrix} -1 & 1 & 0 \\ 1 & 1 & 0 \end{bmatrix} v(k) + \begin{bmatrix} 0 & 1 \\ 0 & -1 \end{bmatrix} y(k).$$

The eigenvalues of the above reduced order estimator are placed at $\{-0.25,\ 0,\ 0.5\}$ while the resulting recovery error matrix $M_r^e(z) \equiv 0$.

The plots of maximum and minimum singular values of the target and the achieved loop transfer function via all the three controllers are shown in Figure 10. Also, the plot of maximum singular value of $M_p^*(j\omega)$is shown in Figure 11, while the plots of maximum singular value of $E_p^*(j\omega)$ is shown in Figure 12. Clearly, for this example, the above current and reduced order estimators yield exact recovery. □

V. DESIGN FOR RECOVERY OVER A SPECIFIED SUBSPACE

Sections III and IV consider the conventional LTR design problem which seeks the recovery over the entire control space. In this section, given a subspace S of $\mathbf{R}^m$, the interest is in designing a controller so that the achieved and target sensitivity and complimentary sensitivity functions projected onto the subspace S match each other. The conditions under which such a design is possible are given in Part 1. To recapitulate these conditions, let

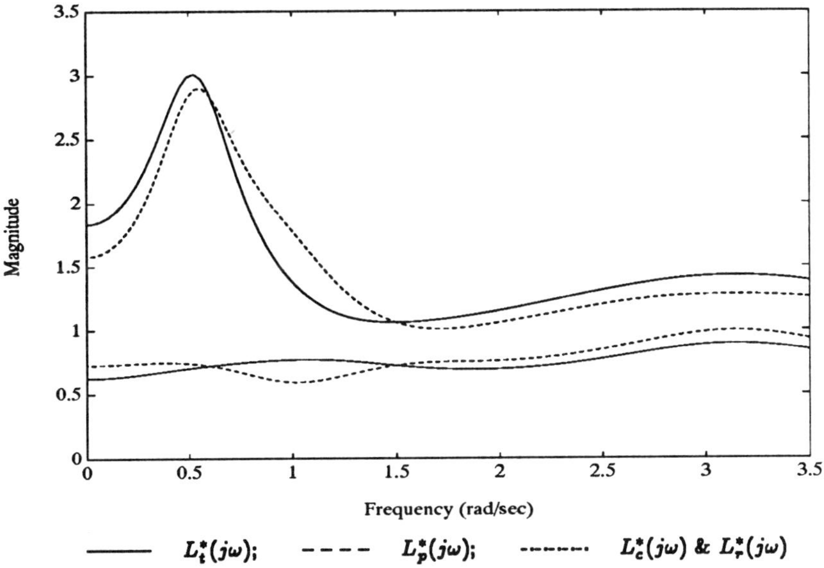

Figure 10: The max. and min. singular values of target and achieved loops.

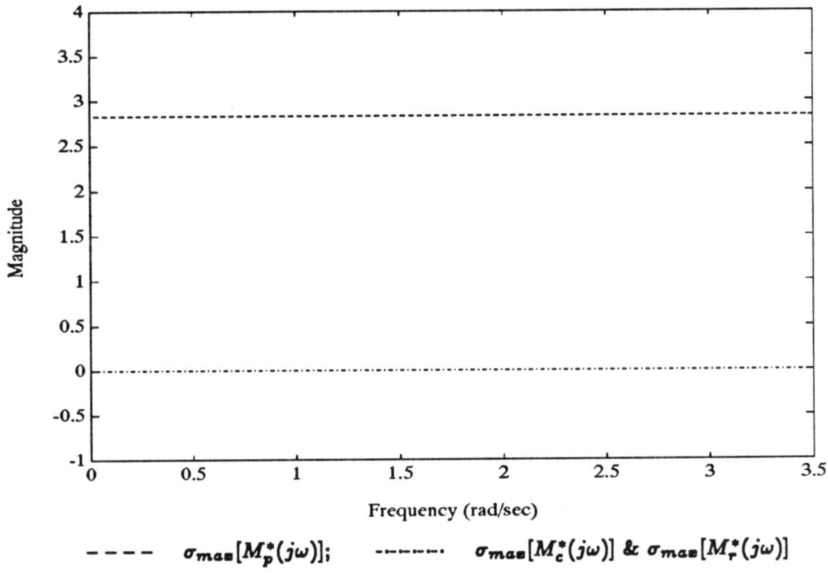

Figure 11: The max. singular values of recovery matrices.

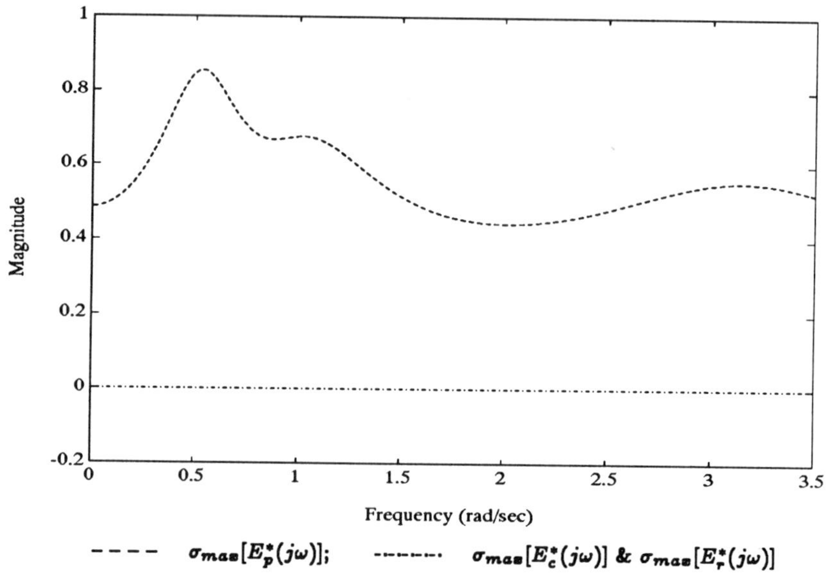

Figure 12: The max. singular values of loop transfer recovery errors.

V^s be a matrix whose columns form an orthogonal basis of $\mathcal{S} \in \mathbf{R}^m$. Assume that the columns of V^s are scaled so that the norm of each column is unity. Let $P^s = V^s(V^s)'$ be the unique orthogonal projection matrix onto $\mathcal{S}$. Then, define three auxiliary systems Σ_p^s, Σ_c^s and Σ_r^s characterized, respectively, by the quadruples (A, BV^s, C, DV^s), (A, BV^s, C_c, D_cV^s) and $(A_r, B_rV^s, C_r, D_rV^s)$. Then the analysis given in Part 1 implies that any admissible and arbitrarily specified sensitivity function (i.e., when F is specified arbitrarily) is recoverable in $\mathcal{S}$ if and only if the following condition is satisfied depending upon the controller used.

(1) Prediction estimator based controller :

Any arbitrary admissible sensitivity function is recoverable if and only if the auxiliary system Σ_p^s is left invertible and of minimum phase with no infinite zeros (i.e., DV^s is of maximal rank).

(2) Current estimator based controller :

Any arbitrary admissible sensitivity function is recoverable if and only if the auxiliary system Σ_c^s is left invertible and of minimum phase with no infinite zeros (i.e., D_cV^s is of maximal rank).

(3) Reduced order estimator based controller :

Any arbitrary admissible sensitivity function is recoverable if and only if the auxiliary system Σ_r^s is left invertible and of minimum phase with no infinite zeros (i.e., $D_r V^s$ is of maximal rank).

The above results are concerned with the recovery of sensitivity function when F is arbitrary or unknown. As is done in Part 1, one can also formulate the recovery conditions for a known F as follows. A known admissible sensitivity function (i.e., when F is known) is recoverable in $\mathcal{S}$ if and only if the following condition is satisfied depending on the controller used.

1. For a prediction estimator based controller, the condition is that $\mathcal{S}^-(\Sigma_p^s) \subseteq \operatorname{Ker}(F)$.
2. For a current estimator based controller, the condition is that $\mathcal{S}^-(\Sigma_c^s) \subseteq \operatorname{Ker}(F)$.
3. For a reduced order estimator based controller, the condition is that $\begin{pmatrix} 0 \\ I \end{pmatrix} \mathcal{S}^-(\Sigma_r^s) \subseteq \operatorname{Ker}(F)$.

Remark 4. *If the given system Σ is strictly proper, i.e., $D = 0$, then it is simple to verify that*

$$\mathcal{S}^-(\Sigma_c^s) = \begin{pmatrix} 0 \\ I \end{pmatrix} \mathcal{S}^-(\Sigma_s^r) = \mathcal{S}^-(\Sigma_p^s) \cap \{ x \mid Cx \in \operatorname{Im}(DV^s) \}.$$

This is not true in general for non-strictly proper systems.

Thus the task of designing a controller for recovery in a subspace collapses to the same task discussed in Sections III and IV except that one needs to use the auxiliary systems Σ_p^s, Σ_c^s and Σ_r^s respectively in place of Σ_p, Σ_c and Σ_r. The following example illustrates this.

Example 5 : Consider a non-strictly proper discrete-time system Σ with sampling period $T = 1$, and characterized by

$$A = \begin{bmatrix} 1 & 4 & 1 & 1 & 1 \\ 4 & 4 & 4 & 4 & 4 \\ 1 & 4 & 5 & 0 & 0 \\ 1 & 4 & 0 & 3 & 0 \\ 1 & 4 & 0 & 0 & 2 \end{bmatrix}, \quad B = \begin{bmatrix} 0 & 1 & 0 \\ 0 & 0 & 1 \\ 1 & 0 & 0 \\ 0 & 0 & 0 \\ 0 & 0 & 0 \end{bmatrix},$$

$$C = \begin{bmatrix} 0 & 0 & 1 & 0 & 0 \\ 1 & 0 & 0 & 0 & 0 \\ 0 & 1 & 0 & 0 & 0 \end{bmatrix}, \quad D = \begin{bmatrix} 1 & 0 & 0 \\ 0 & 0 & 0 \\ 0 & 0 & 0 \end{bmatrix}.$$

Let the target loop $L_t(z)$ and target sensitivity function $S_t(z)$ be specified by

$$F = \begin{bmatrix} 1.0000 & 4.0000 & 5.0000 & 0.0000 & 0.0000 \\ 0.5795 & 3.6504 & 1.0000 & -0.4554 & 1.7594 \\ 5.1676 & 8.3372 & 4.0000 & 9.5197 & 2.4978 \end{bmatrix}.$$

It is simple to verify that the given system Σ has two infinite zeros of order one and three invariant zeros at $\{2, 3, 4\}$. It can also be verified that the target loop specified by the given F is not recoverable either by a prediction or by a current or by a reduced order estimator based controller. Let us consider a subspace $\mathcal{S}$ spanned by

$$V^s = \begin{bmatrix} 1 & 0 \\ 0 & 1 \\ 0 & 0 \end{bmatrix}.$$

It is simple to verify that the given target sensitivity function is recoverable in $\mathcal{S}$ using either a current or a reduced order estimator based controller. The following current and reduced order estimator based controllers obtained by ATEA achieve such a recovery.

Current Estimator Based Controller :

$$v(k+1) = \begin{bmatrix} 0.0000 & 0.0000 & 0 & 0.0000 & 0.0000 \\ 0.0000 & 0.0000 & 0 & 0.0000 & 0.0000 \\ 9.0027 & 33.7281 & 4 & 37.8334 & -10.2964 \\ -5.3994 & -19.7721 & 0 & -27.2535 & 8.2335 \\ 1.8485 & 7.1520 & 0 & 4.0114 & 0.9083 \end{bmatrix} v(k)$$

$$+ \begin{bmatrix} 0 & 0.0000 & 0.0000 \\ 0 & 0.0000 & 0.0000 \\ 1 & 9.0027 & -153.7028 \\ 0 & -5.3994 & 135.5881 \\ 0 & 1.8485 & -14.1746 \end{bmatrix} y(k),$$

$$-u(k) = \begin{bmatrix} 1.0000 & 4.0000 & 5.0000 & 0.0000 & 0.0000 \\ 0.5795 & 3.6504 & 1.0000 & -0.4554 & 1.7594 \\ 5.1676 & 8.3372 & 4.0000 & 9.5197 & 2.4978 \end{bmatrix} v(k)$$

$$+ \begin{bmatrix} 0 & 1.0000 & 38.2712 \\ 0 & 0.5795 & 14.2796 \\ 0 & 5.1676 & -14.6080 \end{bmatrix} y(k).$$

The eigenvalues of the above current estimator are placed at {0, 0.1, 0.2, 0.3}.

Reduced Order Estimator Based Controller :

$$v(k+1)=\begin{bmatrix}4 & 40.9010 & -11.1312\\ 0 & -30.6150 & 9.1483\\ 0 & 4.7193 & 0.7156\end{bmatrix}v(k)+\begin{bmatrix}1 & 9.6516 & -192.8259\\ 0 & -6.1104 & 171.8538\\ 0 & 1.9983 & -20.4207\end{bmatrix}y(k)$$

$$-u(k)=\begin{bmatrix}5 & 0.0000 & 0.0000\\ 1 & -0.4554 & 1.7594\\ 4 & 9.5197 & 2.4978\end{bmatrix}v(k)+\begin{bmatrix}0 & 1.0000 & 41.0500\\ 0 & 0.5795 & 15.3384\\ 0 & 5.1676 & -17.8622\end{bmatrix}y(k).$$

The eigenvalues of the above reduced order estimator are placed at {0, 0.1, 0.2}. □

VI. LTR DESIGN FOR OUTPUT BREAK POINT

All the previous sections consider the case when the target loop transfer function is specified at the input point of the given system. Let us now consider the LTR design when a target loop transfer function $L_t(z) = C(zI - A)^{-1}K$ is specified at the output point of the given system (A, B, C, D). Let $\lambda(A - KC) \in \mathbf{C}^{\odot}$. Then as is discussed in Section VI of Part 1 [3], the above design can be reformulated as the loop transfer recovery problem when a fictitious target loop transfer function $L_d(z) = F_d(zI - A_d)^{-1}B_d$ is specified at the input point of a fictitious dual system Σ_d characterized by (A_d, B_d, C_d, D_d) where $A_d = A'$, $B_d = C'$, $C_d = B'$, $D_d = D'$, and $F_d = K'$. Now, to come up with a controller $\mathbf{C}_d(z)$ for Σ_d, one can utilize any one of the three controllers, namely, the prediction, current and reduced order estimator based controllers. Moreover, the design can be accomplished using any one of the methods developed earlier, namely, the eigenstructure assignment method, the H_∞- or the H_2-optimization based designs. Finally, one needs simply to implement the controller $\mathbf{C}(z) = \mathbf{C}_d'(z)$ to achieve the needed design for the given system.

VII. CONCLUSIONS

This Part 2 of the paper considers three design methods for LTR of general discrete time systems. The first one is an eigenstructure assignment scheme, and the other two are optimization based designs. Eigenstructure assignment method yields a controller design which achieves *any chosen* recovery error matrix among a set of admissible recovery error matrices. On the other hand, one of the optimization based design methods leads to a controller that achieves a recovery error matrix having the infimum H_∞ norm, while the other does the same except it achieves a recovery error matrix having the infimum H_2 norm. Any controller, whether it is prediction or current or reduced order estimator based, can be designed by using any one of the above three design methods. Once the estimator gain is known, the corresponding recovery error matrices can explicitly be calculated in closed form. Besides the conventional LTR design problem which is concerned with the recovery over the entire control space, another generalized recovery design problem where the concern is with the recovery over a specified subspace of the control space is also considered. All the design methods developed here are implemented in a 'Matlab' software package. A number of design examples illustrate several aspects of eigenstructure assignment design as well as optimization based designs.

Acknowledgement

The work of B. M. Chen and A. Saberi is supported in part by Boeing Commercial Airplane Group and in part by NASA Langley Research Center under grant contract NAG-1-1210. Also, B. M. Chen acknowledges Washington State University OGRD 1991 summer research assistantship.

A. Proof of Theorem 2

We need to introduce the following lemma first in order to prove Theorem 2.

Lemma 1. *The following two statements are equivalent:*

1. *There exists an internally stabilizing static state feedback law* $u(k) = -K_p' x(k)$ *such that the closed-loop transfer function from* w *to* z *of* Σ^{au1} *has an* H_∞*-norm less than* 1.

2. *There exists an internally stabilizing static state feedback law* $u(k) = -Fx(k)$ *such that the closed-loop transfer function from* w *to* z *of* Σ^{au2} *has an* H_∞*-norm less than* 1.

Proof of Lemma 1: Without loss of generality, we assume that the system characterized by the quadruple (A', C', B', D') is in the form of s.c.b. Now, let us assume the first statement is true, i.e., there exists a state feedback law $u(k) = -K'x(k)$ such that the resulting closed-loop system of Σ^{au1} is asymptotically stable and the transfer function from w to z has an H_∞-norm less than 1. Partitioning

$$K' = \begin{bmatrix} F_{a0}^- + C_{0a}^- & F_{a0}^+ + C_{0a}^+ & F_{b0} + C_{0b} & F_{c0} + C_{0c} & F_{f0} + C_{0f} \\ F_{af}^- & F_{af}^+ & F_{bf} & F_{cf} & F_{ff} \\ F_{ac}^- & F_{ac}^+ & F_{bc} & F_{cc} & F_{fc} \end{bmatrix},$$

it is trivial to verify that the closed-loop system of Σ^{au1} under the state feedback law $u = -K'x$ is equivalent to the closed-loop system of Σ^{au2} with the following dynamical state feedback,

$$\begin{cases} x_c(k+1) = (A_{cc} - B_c F_{cc}) x_c(k) \\ \quad +[B_c(E_{ca}^- - F_{ac}^-),\ B_c(E_{ca}^+ - F_{ac}^+),\ L_{cb}C_b - B_c F_{bc},\ L_{cf}C_f - B_c F_{fc}] x(k), \\ u(k) = -\begin{bmatrix} F_{c0} \\ F_{cf} \end{bmatrix} x_c(k) - \begin{bmatrix} F_{a0}^- & F_{a0}^+ & F_{b0} & F_{f0} \\ F_{af}^- & F_{af}^+ & F_{bf} & F_{ff} \end{bmatrix} x(k). \end{cases} \tag{44}$$

Obviously, the resulting closed-loop system of Σ^{au2} with the dynamical state feedback law (44) is asymptotically stable and the transfer function from w to z has an H_∞-norm less than 1. Then it follows from Theorem 9.2 of Stoorvogel [10] that there exists a symmetric matrix $P \geq 0$ such that the following conditions hold:

1. $V > 0$ and $R > 0$, where

$$V := DD + B'PB,$$
$$R := I - E'PE + E'PBV^{-1}B'PE.$$

This implies that the matrix $G(P)$ is invertible, where

$$G(P) := \begin{bmatrix} D'D + B'PB & \tilde{B}'PE \\ E'PB & E'PE - I \end{bmatrix}. \tag{45}$$

2. P satisfies the following discrete time ARE:

$$P = A'PA + C'C - \begin{pmatrix} B'PA + D'C \\ E'PA \end{pmatrix}' G(P)^{-1} \begin{pmatrix} B'PA + D'C \\ E'PA \end{pmatrix}. \tag{46}$$

3. The matrix A_{cl} has all its eigenvalues inside the unit circle, where

$$A_{cl} := A - [B \quad E]\, G(P)^{-1} \begin{pmatrix} B'PA + D'C \\ E'PA \end{pmatrix}. \tag{47}$$

Now let us define a cost function,

$$\mathcal{J}(0) := \sup_{w} \inf_{u^+} \{ \|z_{u,w}\|^2 - \|w\|^2 \}, \tag{48}$$

where $u^+ := u|_{[1,\infty)}$. Moreover, we impose an additional constraint $u(0) = 0$. Then the dynamic state feedback induces in closed loop a mapping from w to u, say f. In other words $u = f(w)$. Because we have $x(0) = 0$ we found that $u(0) = [f(w)](0) = 0$. It is clear that

$$\begin{aligned} \mathcal{J}(0) &\le \sup_{w} \{ \|z_{f(w),w}\|^2 - \|w\|^2 \} \\ &\le \sup_{w} \{ (\gamma^2 - 1)\|w\|^2 \} \\ &= 0, \end{aligned}$$

where $\gamma < 1$ is the closed-loop H_∞ norm we obtained via the *dynamic* state feedback control law (44). Also note that the supremum in (48) is finite and is only attained by 0. We know that

$$\sup_{w^+} \inf_{u^+} \{ \|z_{u^+,w^+}\|^2 - \|w^+\|^2 = x'(1)Px(1),$$

where $w^+ := w|_{[1,\infty)}$, as an optimization problem on $[1,\infty)$. Therefore we have

$$\mathcal{J}(0) = \sup_{w(0)} \{ \|z(0)\|^2 - \|w(0)\|^2 + x'(1)Px(1). \tag{49}$$

Moreover, the maximum is uniquely attained by $w(0) = 0$ (uniqueness stems from the uniqueness in maximization in (48)), (49) can be rewritten in the following form

$$\sup_{w(0)} \{ w'(0)[E'PE - I]w(0) \}.$$

Then boundedness and uniqueness of maximum imply

$$E'PE - I < 0.$$

By Theorem 9.4 of Stoorvogel [10] that there exists an internally stabilizing static state feedback law $u(k) = -Fx(k)$ such that the closed-loop transfer function from w to z of Σ^{au2} has an H_∞-norm less than 1.

Conversely, let us assume that the second statement is true, i.e., there exists a state feedback law $u(k) = -Fx(k)$ such that the resulting closed-loop system of Σ^{au2} is asymptotically stable and the transfer function from w to z has an H_∞-norm less than 1. Again, let us partition F as

$$F = \begin{bmatrix} F_{a0}^- & F_{a0}^+ & F_{b0} & F_{f0} \\ F_{af}^- & F_{af}^+ & F_{bf} & F_{ff} \end{bmatrix}.$$

Choosing

$$K' = \begin{bmatrix} F_{a0}^- + C_{0a}^- & F_{a0}^+ + C_{0a}^+ & F_{b0} + C_{0b} & C_{0c} & F_{f0} + C_{0f} \\ F_{af}^- & F_{af}^+ & F_{bf} & E_{fc} & F_{ff} \\ 0 & 0 & 0 & F_{cc} & 0 \end{bmatrix},$$

where F_{cc} is such that $\lambda(A_{cc} - B_c F_{cc}) \in \mathbf{C}^{\odot}$, it is simple to verify that the closed-loop system of Σ^{au1} under the state feedback law $u(k) = -K'x(k)$ is asymptotically stable and the transfer function from w to z has an H_∞-norm less than 1. □

Remark 5. *The above lemma is a generalization of a state feedback H_∞-optimization problem for discrete-time systems given in [10].*

Now, the proof of Theorem 2 follows directly from the above lemma and the results of Stoorvogel [10]. ■

References

[1] K. J. Astrom, P. Hagander and J. Sternby, "Zeros of sampled systems," *Automatica*, 20, pp. 31-38 (1984).

[2] B. M. Chen, A. Saberi and P. Sannuti, "Loop transfer recovery for general nonminimum phase non-strictly proper systems — Part 2: design," *Control–Theory and Advanced Technology*, 8, pp. 59-100 (1992).

[3] B. M. Chen, A. Saberi, P. Sannuti and Y. Shamash, "Loop transfer recovery for general nonminimum phase discrete time systems — Part 1 : analysis," *Control and Dynamic Systems: Advances in Theory and Applications*, This Volume (1992).

[4] J. C. Doyle and G. Stein, "Robustness with observers," *IEEE Transactions on Automatic Control*, AC-24, No. 4, pp. 607-611 (1979).

[5] G. C. Goodman, *The LQG/LTR method and discrete-time control systems*, Rep. No. LIDS-TH-1392, MIT, MA (1984).

[6] B. C. Moore, "On the flexibility offered by state feedback in multivariable systems beyond closed loop eigenvalue assignment," *IEEE Transactions on Automatic Control*, AC-21, pp. 689-692 (1976).

[7] A. Saberi and P. Sannuti, "Squaring down of non-strictly proper systems," *International Journal of Control*, Vol. 51, No. 3, pp. 621-629 (1990).

[8] A. Saberi and P. Sannuti, "Time-scale structure assignment in linear multivariable systems using high-Gain feedback," *International Journal of Control*, Vol. 49, No. 6, pp. 2191-2213 (1989).

[9] P. Sannuti and A. Saberi, "A special coordinate basis of multivariable linear systems – finite and infinite zero structure, squaring down and decoupling," *International Journal of Control*, Vol. 45, No. 5, pp. 1655-1704 (1987).

[10] A. A. Stoorvogel, *The H_∞ control problem: A state space approach*, Ph.D. dissertation, Technical University of Eindhoven, The Netherlands (1990).

[11] Z. Zhang and J. S. Freudenberg, "Loop transfer recovery for nonminimum phase discrete-time systems," *Proceedings of American Control Conference*, Boston, pp. 2214-2219 (1991).

SET–INDUCED NORM BASED ROBUST CONTROL TECHNIQUES

Mario Sznaier

Department of Electrical Engineering
University of Central Florida, Orlando, FL 32816

I. Introduction

A substantial number of control problems can be summarized as the problem of designing a controller capable of achieving acceptable performance under system uncertainty and design constraints. This statement looks deceptively simple, but even in the case where the system under consideration is linear, the problem is far from solved. Generally, the design constraints include both frequency and time domain constraints. The former usually result from performance criteria while the later reflect physical considerations such as the presence of actuator limits or the need to maintain the states of the plant confined to a region where the model used in the design process is valid. Since in general the satisfaction of time–domain constraints cannot be ensured by shaping the Fourier transform of the relevant transfer function, it is clear that frequency domain specifications alone are inadequate to handle these situations. However, even though there currently exist several computationally efficient design techniques able to handle most types of frequency–domain constraints, [1], satisfactory design techniques capable of handling time–domain constraints have started to appear only recently (see for instance [2–3] and references therein). Moreover, all of these schemas assume exact knowledge of the dynamics involved (i.e. exact knowledge of the model). Such an assumption can be too restrictive, ruling out cases where good qualitative models of the plant are

CONTROL AND DYNAMIC SYSTEMS, VOL. 55

available but the numerical values of various parameters are unknown or even change during operation.

On the other hand, during the last decade a considerable amount of time has been spent analyzing the question of whether some relevant properties of a system (most notably asymptotic stability) are preserved under the presence of unknown perturbations. This research effort has led to procedures for designing controllers, termed "robust controllers", capable of achieving desirable properties under various classes of plant perturbations while, at the same time, satisfying frequency–domain constraints. Central to these developments is the Youla parametrization of all stabilizing linear controllers [4] in terms of a free parameter Q (a stable transfer matrix). By using this parametrization, the problem of finding the "best" linear controller can be formulated as the constrained optimization problem of minimizing the weighted norm of a function affine in Q. This optimization problem, which with the proper choice of constraints is convex, has been solved using a number of techniques, including linear programming [5] and non-smooth optimization [6]. In particular, a powerful H_∞ framework has been developed, addressing the issues of robust stability and robust performance in the presence of norm–bounded model uncertainty, by minimizing a weighted supremum–norm (H_∞ norm). Since its introduction, the original formulation of Zames [7] has been substantially simplified, resulting in reliable and efficient computational algorithms for finding solutions ([8–11] and references therein).

The methods based upon the use of the Q parametrization are appealing because they yield the "best" stabilizing linear controller that meets the specifications. Moreover, these methods are efficient, i.e. they are guaranteed to find a solution, provided that a *linear* controller capable of satisfying all the constraints exists. However, being essentially frequency domain based methods, in their present form they can address time–domain constraints in a very conservative way. (For instance, through bounds on the l_1 norm of the relevant transfer function.) Although the use of suitable weighting functions may partially circumvent this problem, we can expect this approach to be inherently conservative, which may preclude finding a satisfactory solution if the constraints are tight.

Recently, some progress has been made towards addressing explicitly time–domain constraints within an H_∞ type framework. Helton and Sideris [12] and Sideris and Rotstein [13] incorporate time–domain constraints constraints over a finite horizon into an H_∞ optimal control problem which is then transformed into a finite dimensional optimization problem. However, at this stage constraints over an infinite horizon can be handled only indirectly.

In this paper, following the spirit of [14–16], we will address time–domain constraints using an operator norm theoretic approach. Specifically, we use a set induced operator norm to define a simple robustness measure that indicates how well the family of systems under consideration satisfies a given set of time–domain constraints. The available degrees of freedom are then used to optimize this measure subject to additional performance specifications. We believe that the results presented here will provide a useful new approach for addressing more realistic control design problems including a combination of time–domain and frequency–domain specifications.

The paper is organized as follows: In section II we introduce the concept of *Constrained Stability* and *Robust Constrained Stability* and we use these concepts to give a formal definition to the *Robust Constrained Control* problem. The main result of this section is a necessary and sufficient condition guaranteeing the constrained stability of a family of systems. In section III we consider the simpler case of linear time–invariant systems subject to additive parametric model uncertainty. We show that in this case, a very simple framework accounts for most of the time–domain based robustness bounds available for discrete–time systems. In section IV we apply the results of section III to the synthesis problem using fixed order controllers. We show that, in the absence of additional performance specifications, our approach yields convex optimization problems. The problem of incorporating additional performance specifications is addressed in section V, where we also extend our results to dynamic uncertainty. The main result of this section shows that, by using a parametrization of all internally stabilizing controllers, H_2 and H_∞ performance criteria can be incorporated into the design, yielding convex optimization problems. Finally, in section VI, we summarize our results and we indicate directions for future research.

II. Problem Formulation and Preliminary Results

In this section we present a formal definition to our problem. We begin by introducing the notation to be used and several required concepts and preliminary results.

2.1 Notation

By L_∞ we denote the Lebesgue space of complex valued transfer matrices which are essentially bounded on the unit circle. H_∞ (H_∞^-) denotes the set of stable (antistable) complex matrices $g(z) \in L_\infty$, i.e analytic in $|z| \geq 1$ ($|z| \leq 1$). $\mathcal{RH}_\infty$ ($\mathcal{RH}_\infty^-$) denotes the subset of H_∞ (H_∞^-) formed by real rational transfer matrices. Throughout the paper we will use packed notation to represent state–space realizations, i.e.

$$G(z) = C(zI - A)^{-1}B + D \stackrel{\Delta}{=} \left(\begin{array}{c|c} A & B \\ \hline C & D \end{array} \right)$$

Finally, for a transfer matrix $G(z)$, $G^{\sim} \stackrel{\Delta}{=} G'(\frac{1}{z})$.

2.2 Preliminary Definitions

- **Def. 1:** Consider the linear, discrete time, autonomous system modeled by the difference equation:

$$\underline{x}_{k+1} = \phi_k \underline{x}_o, \; k = 0, 1 \ldots \; \phi_0 = I_n \qquad (S^a)$$

subject to the constraint $\underline{x} \in \mathcal{G} \subset R^n$, where $\mathcal{G}$ is a compact, convex set containing the origin in its interior; $\phi_k \in R^{n*n}$ and where $\underline{x}$ indicates x is a vector quantity. The system (S^a) is *Constrained Stable* if for *any* point $\tilde{\underline{x}} \in \mathcal{G}$, the trajectory $\underline{x}_k(\tilde{\underline{x}})$ originating at $\tilde{\underline{x}}$ remains in $\mathcal{G}$ for all k.

In the next definition we take into account model uncertainty by extending the concept of *constrained stability* to a family of systems.

- **Def. 2:** Consider the family of linear discrete time systems modeled by the difference equation:

$$\underline{x}_{k+1} = \phi_{k\Delta} \underline{x}_o, \; k = 0, 1 \qquad (S^a_\Delta)$$

where $\phi_{k\Delta}$ belongs to some family $\mathcal{P} \subset R^{n*n}$ described by the parameter Δ which takes values in a set $\mathcal{D}$. The system (S^a) is *Robustly Constrained Stable* with respect to the family $\mathcal{P}$ if every element of $\mathcal{P}$ is constrained–stable.

2.2.1 Constraint Qualification Hypothesis

We proceed now to restrict the class of constraints allowed in our problem. In this paper, we will limit ourselves to constraints of the form:

$$\underline{x} \in \mathcal{G} = \{\underline{x}: G(\underline{x}) \leq \underline{\gamma}\} \tag{1}$$

where $\underline{\gamma} \in R^p, \gamma_i \geq 0$, the inequalities should be interpreted on a component by component sense and where $G: R^n \rightarrow R^p$ is a positive definite sublinear function, i.e. it has the following properties:

$$\begin{aligned} G(\underline{x})_i &\geq 0, \; i = 1 \ldots p \, \forall \, \underline{x} \\ G(\underline{x}) &= 0 \iff \underline{x} = 0 \\ G(\underline{x} + \underline{y})_i &\leq G(\underline{x})_i + G(\underline{y})_i, \; i = 1 \ldots p \, \forall \, \underline{x}, \underline{y} \\ G(\lambda \underline{x}) &= |\lambda| G(\underline{x}) \end{aligned} \tag{2}$$

As examples of constraints that satisfy these conditions we can mention polyhedral regions of the form:

$$G(\underline{x}) = |G\underline{x}|$$

with $G \in R^{p*n}$ with full column rank and where the $|.|$ should be interpreted on a component by component basis; and hyperellipsoidal regions of the form:

$$G(\underline{x}) = (\underline{x}'Q\underline{x}), Q \text{ positive definite}$$

Restricting the constraints to have the form (2), which embodies the features present in a large percentage of realistic problems, introduces more structure into the problem without affecting significantly the number of real-world problems that can be handled by our formalism [15]. In particular, in the next lemma we show that $G(.)$ induces a norm in R^n, which in the sequel we denote as $\|.\|_{\mathcal{G}}$. This norm plays a key role in deriving necessary and sufficient conditions for constrained stability.

- **Lemma 1:** Let:

$$v(x) = \max_{1 \le i \le p} \left\{ \frac{G(\underline{x})_i}{\gamma_i} \right\} = \|W^{-1}G(\underline{x})\|_\infty \triangleq \|\underline{x}\|_\mathcal{G} \tag{3}$$

where $W = diag(\gamma_1, \ldots, \gamma_m)$. Then $v(.)$ defines a norm in R^n

Proof: The proof of the lemma follows by noting that the constraint qualification hypothesis (2) implies that $\|.\|_\mathcal{G}$ satisfies the conditions for a norm in R^n $\diamond$.

Remark 1: Note that the admissible region $\mathcal{G}$ can be characterized as the $\mathcal{G}$-norm unity ball, i.e.: $\mathcal{G} = \{\underline{x}: \|\underline{x}\|_\mathcal{G} \le 1\}$

2.3 Statement of the Problem

Consider the LTI system represented by the following state–space realization:

$$\begin{aligned} \underline{x}_{k+1} &= A\underline{x}_k + B_1\underline{\omega}_k + B_2\underline{u}_k \\ \underline{\zeta}_k &= C_1\underline{x}_k + D_{11}\underline{\omega}_k + D_{12}\underline{u}_k \end{aligned} \tag{S}$$

subject to the constraint:

$$\underline{x}_k \in \mathcal{G} \subset R^n$$

where $D_{12}'D_{12} > 0$, $\underline{x} \in R^n$ represents the state; $\underline{\zeta} \in R^q$ represents the variables subject to performance specifications; $\underline{u} \in R^m$ represents the control input; and where $\underline{\omega} \in R^r$ contains all external inputs of interest including disturbances and commands (note that these signals may include "fictitious" ones, used to evaluate stability in the presence of structured perturbations [17]). Then, the basic problems that we address in this paper are the following:

- **Robust Constrained Control Analysis Problem:**

  Given the nominal system (S) and a feedback control law $\underline{u}_k = F_k(\underline{x}_k)$, determine if the resulting closed–loop system is constrained–stable in the absence of disturbances (i.e. when $\underline{\omega} \equiv 0$). If the nominal closed–loop system is constrained–stable, determine the maximum allowable level of model uncertainty (in the sense of some previously defined norm) such that the constraints are satisfied for any initial condition $\underline{x} \in \mathcal{G}$.

- **Linear Robust Constrained Control Synthesis Problem:**

  Given the system (S), with additional performance specifications of the form:

$$\|W(z)T_{\zeta w}(z)\|_* \leq 1 \qquad (P)$$

  where $\|.\|_*$ indicates a suitable operator norm (such as $\|.\|_2$ or $\|.\|_\infty$), $W(z)$ is a suitable weighting function and where $T_{\zeta w}$ indicates the closed–loop transfer function from the disturbances to the performance variables, find a *linear* controller such that the resulting closed–loop system is robustly constrained stable and satisfies the performance specifications (P).

2.4 Constrained Stability Analysis

In this section we present a *necessary and sufficient* condition for constrained stability of a system. This result is used to define a qualitative robustness measure in terms of the size of the smallest destabilizing perturbation.

- **Lemma 2:** Consider the system (S^a). Let $\Phi \stackrel{\Delta}{=} \{\phi_k\}$ and denote by $\|.\|_{\mathcal{G}}$ the operator norm induced in R^{n*n} by G, i.e. $\|A\|_{\mathcal{G}} = \sup\limits_{\|\underline{x}\|_{\mathcal{G}}=1} \frac{\|A\underline{x}\|_{\mathcal{G}}}{\|\underline{x}\|_{\mathcal{G}}}$. Finally, let $\|\Phi\|_{\mathcal{G}} \stackrel{\Delta}{=} \sup\limits_k \|\phi_k\|_{\mathcal{G}}$. Then the system (S^a) is constrained stable *iff*:

$$\|\Phi\|_{\mathcal{G}} \leq 1 \tag{4}$$

Proof: The proof follows immediately from Def. 1 and Remark 1 by noting that:

$$\begin{aligned} \|\Phi\|_{\mathcal{G}} \leq 1 &\iff \|\phi_k\|_{\mathcal{G}} \leq 1 \ \forall\ k \iff \|\phi_k \tilde{\underline{x}}\|_{\mathcal{G}} \leq 1\ \forall\ k,\ \|\tilde{\underline{x}}\|_{\mathcal{G}} \leq 1 \\ &\iff \|\underline{x}_k(\tilde{\underline{x}})\|_{\mathcal{G}} \leq 1 \forall \tilde{\underline{x}} \in \mathcal{G} \iff \underline{x}_k(\tilde{\underline{x}}) \in \mathcal{G} \forall\ k \end{aligned} \tag{5}$$

where $\underline{x}_k(\tilde{\underline{x}})$ denotes the trajectory that originates in $\tilde{\underline{x}}$ $\diamond$.

Remark 2: The operator norm defined in Lemma 2 can be extended to $\mathcal{RH}_\infty^{n*n}$ as follows: Let $\Phi(z) \in \mathcal{RH}_\infty^{n*n}$ and let $\{\phi_k\} = Z^{-1}\{\Phi(z)\}$. Then we can define:

$$\|\Phi(z)\|_{\mathcal{G}} \stackrel{\Delta}{=} \sup_k \|\phi_k\|_{\mathcal{G}} \tag{6}$$

Note that since $\Phi(z) \in \mathcal{RH}_\infty^{n*n}$, $\{\phi_i\} \in l_1^{n*n}$. Since $\|\phi_k\|_1$ is uniformly bounded with respect to k, it follows from the equivalence of all finite–dimensional matrix norms (Theorem 5.4.4, [18]) that $\|\phi_k\|_{\mathcal{G}}$ is also uniformly bounded. Hence $\|\Phi\|_{\mathcal{G}}$ is finite.

- **Corollary 1:** The system (S_Δ^a) is robustly constrained stable with respect to $\mathcal{D}$ *iff* $\|\Phi_\Delta\|_{\mathcal{G}} \leq 1$ for all $\Delta \in \mathcal{D}$.

This result can be used to define a quantitative measure of robustness in terms of the "size" of the smallest destabilizing perturbation as follows:

- **Def. 3:** Consider the system (S^a). The *constrained stability measure* $\varrho_{\mathcal{G}}^{\mathcal{N}}$ is defined as:

$$\varrho_{\mathcal{G}}^{\mathcal{N}} \stackrel{\Delta}{=} \min_{\Delta \in \mathcal{D}} \{\|\Delta\|_{\mathcal{N}} \colon \|\Phi_{\Delta}\|_{\mathcal{G}} = 1\}$$

where $\|.\|_{\mathcal{N}}$ denotes a suitable operator norm defined in $\mathcal{D}$. In the particular case that the induced operator norm $\|.\|_{\mathcal{G}}$ is used in the set $\mathcal{D}$, we will denote the constrained stability measure as $\varrho_{\mathcal{G}}$.

III. The Additive Real Parametric Uncertainty Case

In this section we consider the simpler case of Linear Time Invariant Systems subject to additive parametric model uncertainty. In this case the family of autonomous systems (S_{Δ}^{a}) can be represented by:

$$\underline{x}_{k+1} = (A + \Delta)\underline{x}_k \qquad (S_p)$$

and the nominal and perturbed transition matrices are given by $\phi_k = A^k$ and $\phi_{k\Delta} = (A+\Delta)^k$ respectively.

- **Lemma 3:** The system (S_p) is constrained stable iff $\|A\|_{\mathcal{G}} \leq 1$.

Proof: Since the induced norm $\|.\|_{\mathcal{G}}$ is submultiplicative, it follows that:

$$\max_k \{\|\phi_k\|_{\mathcal{G}}\} = \max_k \left\{\|A^k\|_{\mathcal{G}}\right\} \leq 1 \iff \|A\|_{\mathcal{G}} \leq 1$$

The lemma follows then immediately from Lemma 2 $\diamond$.

- **Corollary 2:** For the parametric uncertainty case the constrained robustness measure reduces to:

$$\varrho_{\mathcal{G}}^{\mathcal{N}} = \min_{\Delta \in \mathcal{D}} \{ \|\Delta\|_{\mathcal{N}} : \|A + \Delta\|_{\mathcal{G}} = 1 \} \tag{7}$$

Remark 3: Note that the corollary is quite general since in principle no conditions are imposed over the set $\mathcal{D}$. However, in the general case nothing can be stated about the properties of $\varrho_{\mathcal{G}}^{\mathcal{N}}$ which could conceivably be a *non-continuous* function of A. In the sequel we show that under some assumptions that are commonly verified in practice, $\varrho_{\mathcal{G}}^{\mathcal{N}}$ is a *continuous, concave* function of the dynamics matrix A.

- **Theorem 1:** Assume that the perturbation set $\mathcal{D}$ is is a *closed cone* with vertex at the origin [19], (i.e. $\Delta^o \in \mathcal{D} \iff \alpha\Delta^o \in \mathcal{D} \; \forall \, 0 \leq \alpha$). Then $\varrho_{\mathcal{G}}^{\mathcal{N}}$ is a *continuous, concave* function of A.

Proof: The proof of the theorem is given in Appendix A.

Remark 4: Note that the class of sets considered in this theorem includes as a particular case sets of the form:

$$\mathcal{D} = \left\{ \Delta : \Delta = \sum_{1}^{m} \mu_i E_i ; \; \mu_i \geq 0, \; E_i \text{ given} \right\} \tag{8}$$

which has been the object of much interest lately ([20–22] and references therein).

Next, we introduce a *lower bound* of the constrained stability measure. This lower bound is then used to show that our formalism accounts for most of the robustness bounds previously derived for systems of the form (S_p).

- **Lemma 4:**

$$\varrho_{\mathcal{G}} \geq 1 - ||A||_{\mathcal{G}} \tag{9}$$

Proof: Let Δ_1 be such that $||A + \Delta_1||_{\mathcal{G}} = 1$. Then:

$$1 = ||A + \Delta_1||_{\mathcal{G}} \leq ||A||_{\mathcal{G}} + ||\Delta_1||_{\mathcal{G}} \tag{10}$$

or

$$||\Delta_1||_{\mathcal{G}} \geq 1 - ||A||_{\mathcal{G}} \tag{11}$$

Hence:

$$\varrho_{\mathcal{G}} = \min_{\mathcal{D}} ||\Delta_1||_{\mathcal{G}} \geq 1 - ||A||_{\mathcal{G}} \diamond \tag{12}$$

- **Theorem 2:** For the unstructured perturbation case, i.e. the case where $\mathcal{D} \equiv R^{n*n}$, condition (9) is saturated.

Proof: The proof follows from Lemma 4 by noting that for:

$$\Delta^o \triangleq \frac{(1 - ||A||_{\mathcal{G}})A}{||A||_{\mathcal{G}}} \tag{13}$$

we have $||\Delta^o||_{\mathcal{G}} = 1 - ||A||_{\mathcal{G}}$ and $||A + \Delta^o||_{\mathcal{G}} = 1 \diamond$

Remark 5: Note that the results of Lemma 4 and Theorem 2 can be used to find a lower bound for the constrained robustness measure in the general case when an operator norm different from $||.||_{\mathcal{G}}$ is used in the set $\mathcal{D}$. Since all finite dimensional matrix norms are equivalent, it follows that, given any norm $\mathcal{N}$ in the set $\mathcal{D}$, there exist a constant c such that $||.||_{\mathcal{G}} \leq c||.||_{\mathcal{N}}$. Hence $\varrho_{\mathcal{G}}^{\mathcal{N}} \leq \frac{\varrho_{\mathcal{G}}}{c}$.

3.1 Quadratic Constraints Case

In this subsection we particularize our theoretical results for the special case where the constraint region is an hyperellipsoid, i.e.:

$$G(\underline{x}) = (\underline{x}'P\underline{x})^{\frac{1}{2}}\ , \ P \in R^{n*n} \text{ positive definite} \tag{14}$$

We will show that in this case our approach yields a generalization of the well known technique of estimating the robustness measure by using quadratic based Lyapunov functions [23] by obtaining robustness bounds previously derived in this context. Moreover, using our approach we will show that in some cases these bounds give the *actual* value of the constrained stability measure.

- *Example 1*: (unstructured perturbation)

In this case, Theorem 2 yields $\varrho_{\mathcal{G}} = 1 - \|A\|_{\mathcal{G}}$ where:

$$\|A\|_{\mathcal{G}}^2 = \|A\|_P^2 = \max_{\underline{x}} \left(\frac{\underline{x}'A'PA\underline{x}}{\underline{x}'P\underline{x}} \right) \tag{15}$$

Assume that $\varrho_{\mathcal{G}} > 0$. Then, there exists Q positive definite such that:

$$A'PA - P = -Q \tag{16}$$

and:

$$\|A\|_{\mathcal{G}}^2 = \max_{\underline{x}} \left(1 - \frac{\underline{x}'Q\underline{x}}{\underline{x}'P\underline{x}} \right) \leq 1 - \frac{\sigma_{min}(Q)}{\sigma_{Max}(P)} \tag{17}$$

Hence:

$$\varrho_{\mathcal{G}} = 1 - \|A\|_{\mathcal{G}} \geq 1 - \left(1 - \frac{\sigma_{min}(Q)}{\sigma_{Max}(P)} \right)^{\frac{1}{2}} \tag{18}$$

A common technique in state space robust analysis is to obtain robustness bounds from equation (16) ([24–25]). This case can be accommodated by our formalism by recognizing the fact that once P is selected, the system becomes effectively constrained to remain within an hyperellipsoidal region. It has been suggested ([24–25]) that good robustness bounds can be obtained from (18) when P is selected such that $Q = I$. In this case our approach yields:

$$\varrho_{\mathcal{G}} = 1 - \|A\|_{\mathcal{G}} = 1 - \left(1 - \frac{1}{\sigma_{Max}(P)}\right)^{\frac{1}{2}} \tag{19}$$

which coincides with the robustness bound found by Sezer and Siljak [25]. Note however that our derivation shows this bound to be exact for the unstructured perturbation case.

- *Example 2:* (Unstructured perturbation, A semisimple)

Consider the case where A is semisimple, i.e.

$$\begin{aligned} A &= L^{-1}\Lambda L \\ \Lambda &= diag\left\{\begin{pmatrix} \sigma_1 & \omega_1 \\ -\omega_1 & \sigma_1 \end{pmatrix}, \ldots, \begin{pmatrix} \sigma_p & \omega_p \\ -\omega_p & \sigma_p \end{pmatrix}, \sigma_{p+1}, \ldots, \sigma_n\right\} \end{aligned} \tag{20}$$

Then, the maximum of the stability measure, $\varrho_{\mathcal{G}}$, over all possible positive definite matrices P, is achieved for $P = L'L$.

Proof: From (15) and (20) we have:

$$\begin{aligned} \|A\|_P^2 &= \max_{\underline{x}}\left\{\frac{\underline{x}'A'PA\underline{x}}{\underline{x}'P\underline{x}}\right\} \\ &= \max_{\underline{x}}\left\{\frac{\underline{x}'L'L^{-'}A'L'LAL^{-1}L\underline{x}}{\underline{x}'L'L\underline{x}}\right\} \\ &= \max_{\|\underline{y}\|_2=1} \|LAL^{-1}\underline{y}\|_2^2 \\ &= \|LAL^{-1}\|_2^2 = \sigma_{Max}^2(\Lambda) \end{aligned} \tag{21}$$

From (20) it follows that:

$$\sigma_{Max}(\Lambda) = \max_i |\lambda_i^A| = \rho(A) \tag{22}$$

where λ_i^A denotes the eigenvalues of A and $\rho(.)$ denotes the spectral radius. Since the spectral radius is always *smaller* than any other matrix norm [18] we have that:

$$\|A\|_M \geq \rho(A) = \|A\|_{L'L} \tag{23}$$

and therefore:

$$\begin{aligned} \varrho_{L'L} &= 1 - \|A\|_{L'L} \\ &\geq \varrho_M = 1 - \|A\|_M \; \forall \; M \in R^{n*n}, \text{ positive definite} \diamond \end{aligned} \tag{24}$$

3.2 Polyhedral Constraints

In this subsection we consider the case where the region $\mathcal{G}$ is polyhedral, i.e.:

$$G(x) = |G\underline{x}| \tag{25}$$

where $G \in R^{p*n}$, $\text{rank}(G) = n$, and the $|.|$ should be interpreted on a component by component sense. In this case, $\varrho_{\mathcal{G}}^{\mathcal{N}}$ can be efficiently computed by solving p Linear Programming problems as follows:

- **Lemma 5:** Let $\varrho_i^{\mathcal{N}}$ be the solution of the following optimization problem:

$$\varrho_i^{\mathcal{N}} = \min_{\Delta \in \mathcal{D}} \{\|\Delta\|_{\mathcal{N}} : \|W^{-1}(H + \Delta H)W\|_1^{(i)} \geq 1\} \tag{26}$$

where $H \stackrel{\Delta}{=} GA(G'G)^{-1}G'$, $\Delta H \stackrel{\Delta}{=} G\Delta(G'G)^{-1}G'$, and where $\|M\|_1^{(i)}$ indicates the l_1 norm of the ith row of the matrix M. Then:

$$\varrho_{\mathcal{G}}^{\mathcal{N}} = \min_{1 \leq i \leq p} \{\varrho_i^{\mathcal{N}}\} \tag{27}$$

Proof: From the definition of H we have that $GA = HG$. Hence:

$$\|A\underline{x}\|_{\mathcal{G}} = \|W^{-1}GA\underline{x}\|_\infty = \|W^{-1}HG\underline{x}\|_\infty$$

and

$$\|A\|_{\mathcal{G}} = \sup_{\|\underline{x}\|_{\mathcal{G}}=1} \|A\underline{x}\|_{\mathcal{G}} = \sup_{\|\underline{y}\|_\infty=1} \|W^{-1}HW\underline{y}\|_\infty = \|W^{-1}HW\|_\infty$$

Assume now that the Lemma is false and that there exist $\tilde{\varrho}$ and $\tilde{\Delta}$ such that:

$$\|A + \tilde{\Delta}\|_{\mathcal{G}} = 1; \; \|\tilde{\Delta}\|_{\mathcal{N}} = \tilde{\varrho} < \varrho_{\mathcal{G}}^{\mathcal{N}} \tag{28}$$

Since $\|A + \tilde{\Delta}\|_{\mathcal{G}} = 1$ there exists i^o such that $\|W^{-1}(H + \tilde{\Delta}H)W\|_1^{(i^o)} = 1$, $\|W^{-1}(H + \tilde{\Delta}H)W\|_1^{(j)} \leq 1$, $j \neq i^o$, but this implies (eq. (26)) that $\varrho_{i^o}^{\mathcal{N}} \leq \tilde{\varrho}$ which contradicts (28) $\diamond$.

- *Example 3:* (unstructured perturbation)

Consider the following case:

$$A = \begin{pmatrix} 0.8 & 0.5 \\ -0.0208 & 0.5083 \end{pmatrix} \; G = \begin{pmatrix} 1.0 & 2.0 \\ -1.5 & 2.0 \end{pmatrix} \; \underline{\omega} = \begin{pmatrix} 5.0 \\ 10.0 \end{pmatrix}$$

Then, from the definition of H, we have that:

$$H = \begin{pmatrix} 0.7583 & 0.0 \\ -0.4167 & 0.55 \end{pmatrix}, \; \|A\|_{\mathcal{G}} = 0.7583$$

and, from Lemma 5,

$$\varrho_i = \min_{\|\Delta\|_{\mathcal{G}}} \left\{ \|\Delta\|_{\mathcal{G}} \colon \sum_{j=1}^{2} \frac{|H + \Delta|_{ij}\omega_j}{\omega_i} = 1 \right\} \; i = 1,2 \tag{29}$$

Casting the problems (29) into a linear programming form and solving we have that:

$$\varrho_1 = 0.2417, \; \varrho_2 = 0.2417 \text{ and } \varrho_{\mathcal{G}} = \min_{1 \leq i \leq 2} \varrho_i = 0.2417$$

Note that in this case $\varrho_{\mathcal{G}} = 1 - \|A\|_{\mathcal{G}} = 0.2417$ as shown in Theorem 2.

IV. Fixed Order Robust Controllers Design

From Lemma 3 and its corollary, it follows that a full state feedback matrix F such that the constrained stability measure, $\varrho_{\mathcal{G}}^{\mathcal{N}}$, of the closed loop system is maximized can be selected by solving the following optimization problem:

$$\max_{F}\{\varrho_{\mathcal{G}}^{\mathcal{N}}(F)\} \tag{30}$$

subject to:

$$\varrho_{\mathcal{G}}^{\mathcal{N}}(F) \stackrel{\Delta}{=} \min_{\Delta \in \mathcal{D}}\{\|\Delta\|_{\mathcal{N}} : \|A + BF + \Delta\|_{\mathcal{G}} = 1\} \tag{31}$$

Since from Theorem 1, $\varrho_{\mathcal{G}}^{\mathcal{N}}(F)$ is a concave function, (30) has a global optimum. Hence, the problem of finding the *maximally* robust controller leads to convex, albeit non-differentiable, optimization problems, which can be solved using a number of techniques ([26]). However, in most cases maximizing the robustness measure does not necessarily guarantee a design that meets desirable specifications. Moreover, good performance and good robustness properties are usually conflicting design objectives which must be traded–off as shown in Figure 1.

In particular, note the existence of a region where we can achieve the same value of the constrained robustness measure obtained with a maximum–robustness type design, but with a substantially lower cost. Hence, a substantial improvement over a maximum robustness type design can be achieved by adopting the following design philosophy:

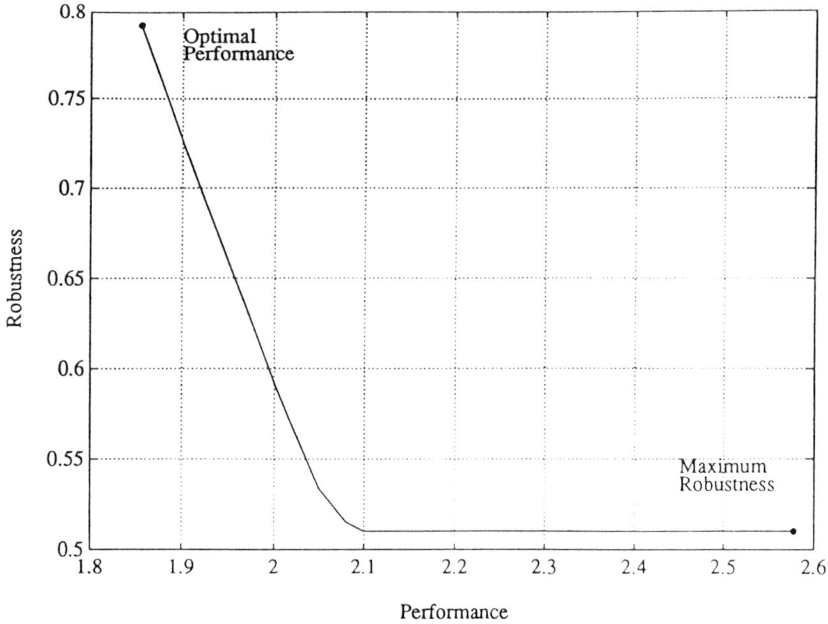

Figure 1. A Typical Robustness Versus Performance Curve

Select a set of specifications. Then use the extra degrees of freedom that may be available in the problem to maximize the robustness measure over the set of all controllers that satisfy the given specifications for the nominal plant.

Thus, the design problem will have the general form of a non-differentiable constrained minimization problem. In the remainder of this section, we address the design problem when the controller is limited to have fixed order. The analysis of the general case of dynamic controllers is postponed until section V. In the case of additive parametric uncertainty and fixed order controllers, the robust constrained control synthesis problem reduces to:

Given: a family of linear time invariant, stabilizable discrete time systems, represented by:

$$\begin{aligned} \underline{x}_{k+1} &= (A+\Delta)\underline{x}_k + B_1\underline{\omega}_k + B_2\underline{u}_k \\ \zeta_k &= C_1\underline{x}_k + D_{11}\underline{\omega}_k + D_{12}\underline{u}_k \end{aligned} \tag{S}$$

where $D_{12}'D_{12} > 0$ and a nominal performance specification of the form $\|T_{\zeta w}\|_* \leq \gamma$

Find: a *constant* feedback matrix F (possibly of the form $F = HK$, H given) such that the nominal closed–loop system is constrained stable and $\varrho_{\mathcal{G}}^{\mathcal{N}}(F)$ is maximized subject to the performance specification constraint.

In the remainder of this section we will concentrate in the H_2 case, i.e. the case where $\underline{\omega}_k$ is assumed to be white gaussian with covariance $\mathbf{E}\{\underline{\omega}_k\underline{\omega}_k'\} = V$ and the performance specification is expressed in term of the 2 norm. Note that in this case we can assume $D_{11} = 0$ without loss of generality since it contributes only with a constant term to $\|T_{zw}\|_2$. Hence, the optimization problem has the following form:

$$\max_F \varrho_{\mathcal{G}}^{\mathcal{N}}(F)$$

subject to:

$$\begin{gathered} Tr\{PV\} \leq \gamma \\ (A+BF)'P(A+BF) - P = -Q - F'RF - F'M' - MF \\ P > 0 \end{gathered} \tag{32}$$

where: $Q = C_1'C_1$, $R = D_{12}'D_{12}$, and $M = C_1'D_{12}$.

Note that although the maximally robust control problem (30) has a unique solution, once performance specifications are imposed, the problem becomes non–convex (since equation (32) is not convex in F) and therefore it may have local extrema. This problem can be partially circumvented by using an homotopy like algorithm as follows:

Controller Design Algorith

Begin:

0) Data: A sequence $\epsilon_i \rightarrow 1$, $\epsilon_o = 0$, $\epsilon_i > \epsilon_j$, $i > j$. Set $i = 0$

1) Solve the maximally robust constrained problem:

$$F_o = \underset{F \in \Re^{m \times n}}{\operatorname{argmax}} \, \varrho_{\mathcal{G}}^{\mathcal{N}} \tag{33}$$

Note that the assumption that $\mathcal{D}$ is a closed–cone guarantees that this problem reduces to the well–posed problem of finding the maximum of a continuous concave function. Let $\gamma_u = Tr\{P_u V\}$ where P_u is the unique positive definite solution to the Lyapunov equation (32).

2) Solve the unconstrained H_2 problem. Let $\gamma_{LQ} = Tr\{P_{LQ} V\}$.

3) Sweep step: Set $i = i + 1$, $\gamma_i = (1 - \epsilon_i)\gamma_u + \epsilon_i \gamma_{LQ}$. Solve problem (32) subject to the modified performance specification:

$$Tr\{PV\} \leq \gamma_i$$

using as initial guess for the solution F_{i-1}.

4) Repeat 3 until a matrix F^o such that the closed–loop system satisfies the given performance specification is found.

End.

- *Example 4:*

Consider the following system:

$$A = \begin{pmatrix} 0 & 1 \\ 0.505 & -0.51 \end{pmatrix} \quad B_1 = I_2 \quad B_2 = \begin{pmatrix} 1 \\ 0 \end{pmatrix}$$

$$C = \begin{pmatrix} \sqrt{0.2} & 0 \\ 0 & 1 \\ 0 & 0 \end{pmatrix} \quad D = \begin{pmatrix} 0 \\ 0 \\ \sqrt{20} \end{pmatrix}$$

$$\mathcal{G} = \{\underline{x}\colon \|\underline{x}\|_2 \leq 1\} \quad \mathbf{E}\{\underline{\omega}_k \underline{\omega}_k'\} = I$$

with the performance specification $\|T_{\zeta w}\|_2^2 \leq 12$. The open–loop system has poles at $s_1 = 0.5$ and $s_2 = -1.01$. Assume that the perturbation set is such that changes the position of the poles while maintaining constant their sum, i. e:

$$\mathcal{D} = \left\{\Delta\colon \Delta = \mu E,\ E \stackrel{\Delta}{=} \begin{pmatrix} 0 & 0 \\ 0 & 1 \end{pmatrix},\ \mu \in \Re\right\} \tag{34}$$

Note that $\|E\|_2 = 1$ hence $\|\Delta\|_2 = |\mu|$.

Step 1: In this case, it is easily seen that the solution to the maximally robust control problem (33) can be computed by solving a matrix dilation problem [27]. Rewrite the dynamics matrix as:

$$A = \begin{pmatrix} x_1 & x_2 \\ a_1 & a_2 \end{pmatrix}$$

where x_i denote elements that can be modified using state–feedback. Since matrix dilations are norm–increasing we have that:

$$\begin{aligned} \|A + \mu E\|_2 &\geq \max\{\|\begin{pmatrix} a_1 & a_2 + \mu \end{pmatrix}\|_2\} \\ &= \sqrt{a_1^2 + (a_2 + \mu)^2} \end{aligned} \tag{35}$$

Define now:

$$\begin{aligned} \mu^0 &= \operatorname{argmin}\{|\mu|, \mu \in \Re: a_1^2 + (a_2 + \mu)^2 = 1\} \\ &= \sqrt{(1 - a_1^2)} - |a_2| \end{aligned} \tag{36}$$

From (35) and (36) it follows that $\|A + \mu^o E\|_2 \geq 1$ which implies that $\varrho_2(F) \leq \mu^o$ for all F. Furthermore, from the definition of μ^o (36) it follows that if F is selected such that $x_1 = x_2 = 0$, then $\varrho_2(F) = \mu^o$. Hence, this choice of F yields the solution to (36). In this particular example we have:

$$F^o = (\,0 \quad 1\,),\ \varrho_2 = 0.3531$$

Step 3: Consider a feedback matrix F and let A_{cl} be the corresponding *closed-loop* matrix, i.e:

$$A_{cl} = A + BF = \begin{pmatrix} a_{11} & a_{12} \\ a_{21} & a_{22} + \mu \end{pmatrix} \tag{37}$$

In order to carry–out the sweep step, we need to compute the corresponding value of the robustness measure. This computation can be performed using standard results on matrix dilations [27] as follows: The set Υ of numbers μ such that $\|A_{cl}\|_2 \leq 1$ can be parametrized as:

$$\Upsilon = \left\{\mu : \mu = -a_{22} - y a_{11} z + (1 - y^2)^{\frac{1}{2}} w (1 - z^2)^{\frac{1}{2}}\right\} \tag{38}$$

where:

$$\begin{aligned} y &= \frac{a_{21}}{(1 - a_{11}^2)^{\frac{1}{2}}} \\ z &= \frac{a_{12}}{(1 - a_{11}^2)^{\frac{1}{2}}} \\ w &\in \Re,\ |w| \leq 1 \end{aligned} \tag{39}$$

From (38) it follow that the constrained stability margin of A_{cl} is given by:

$$\varrho_2(F) = |a_{22} + y a_{11} z - (1 - y^2)^{\frac{1}{2}} w (1 - z^2)^{\frac{1}{2}} \operatorname{sign}(a_{22} + y a_{11} z)|$$

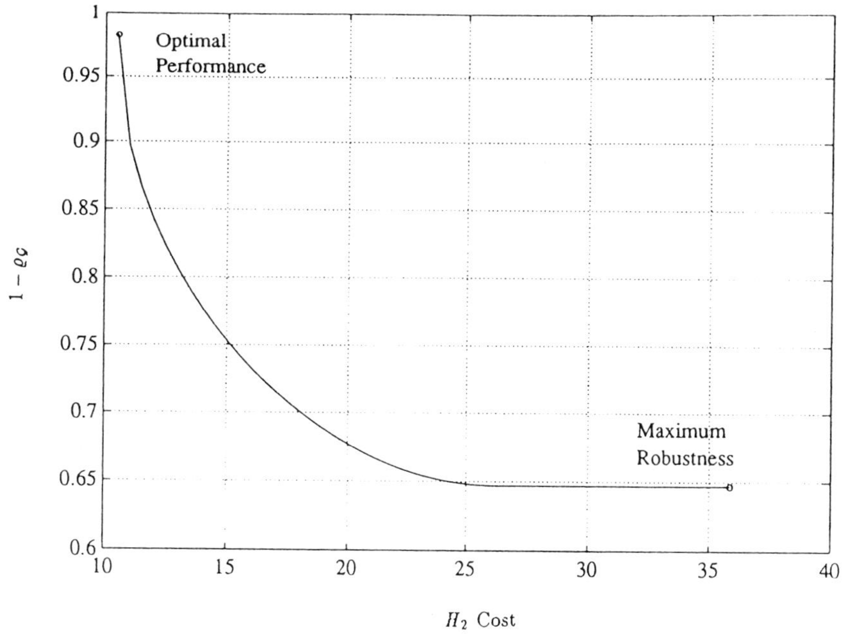

Fig 2. Robustness vs. H_2 cost for Example 4

Figure 2 shows $\varrho_2(F)$ and $||T_{\zeta w}||_2^2$ as the relaxation parameter ϵ is continuously changed from 0 to 1. For $\epsilon = 0$ we recover the maximally robust controller, which yields a robustness measure $\varrho_{\mathbf{max.\ robust}} = 0.3531$ and an H_2 cost $||T_{\zeta w}||_2^2 = 35.8$. For $\epsilon = 1$ we recover the optimal H_2 controller which yields $||T_{\zeta w}||_2^2 = 10.5$, but with a robustness measure $\varrho_{\mathrm{min.\ cost}} = 0.017$ (i.e. the closed–loop system can become constrained unstable for perturbations such that $|\mu| \geq 0.017$. Finally, for $\epsilon = 0.94$ we get $F = (\,-0.0781 \quad 0.3339\,)$, $||T_{\zeta w}||_2^2 = 12.02$ and $\varrho_2(F) = 0.158$ which satisfies the performance specifications.

V. Robust Controller Design using Dynamic Controllers

In the last section, we proposed a simple fixed order controller design method based upon solving a constrained optimization problem. Although this design method has the advantage of yielding simple controllers, the resulting optimization problem is non–convex. Hence the proposed design algorithm could converge to a local rather than global extrema. Moreover, if the performance specifications are tight, there is no guarantee that a low order controller will yield acceptable performance. In this section, we address these problems by using dynamic controllers. Specifically, we will make use of the Youla parametrization of all stabilizing controllers to maximize the robustness measure over the set of all achievable closed–loop plants. Hence, if the problem is feasible, the proposed design algorithm is guaranteed to find a solution. In order to make use of these results, we begin by extending the results of section 2 on constrained stability to a more general uncertainty description.

5.1 Systems with Dynamic Uncertainty

In this section we address the case of multiplicative unstructured dynamic uncertainty. Consider the nominal system (S^a) and let $\Phi^o(z)$ denote its corresponding z–transform. Then, we will assume that the family of systems under consideration is described by:

$$\begin{aligned} \mathcal{P}_\delta &= \left\{ \Phi(z) \colon \Phi(z) = \Phi^o(z)(I_n + \Delta),\ \Delta \in \mathcal{D}_\delta \right\} \\ \mathcal{D}_\delta &= \left\{ \Delta \in \mathcal{RH}_\infty^{n*n} \colon \|\Delta\|_{\mathcal{G},1} \leq \delta \right\} \end{aligned} \qquad (S_d)$$

where

$$||\Delta||_{\mathcal{G},1} \stackrel{\Delta}{=} \sum_{i=0}^{\infty} ||\Delta_i||_{\mathcal{G}} \tag{40}$$

In the next theorem we derive a bound on the $||.||_{\mathcal{G}}$ of the dynamics Φ for all the elements of the family $\mathcal{P}_\delta$ and we show that this bound is tight.

- **Theorem 3:** Consider the family (S_d). Then:

$$||\Phi||_{\mathcal{G}} \leq ||\Phi^o||_{\mathcal{G}}(1+\delta) \tag{41}$$

and there exist at least one $\hat{\Phi} \in \mathcal{P}$ such that (41) is an equality.

Proof: Let $\Psi(z) = \Phi(z)\Delta(z)$. Then:

$$\begin{aligned} ||\Psi_k||_{\mathcal{G}} &= ||\sum_{i=0}^{k} \phi_i \Delta_{k-i}||_{\mathcal{G}} \leq \sum_{i=0}^{k} ||\phi_i||_{\mathcal{G}} ||\Delta_{k-i}||_{\mathcal{G}} \\ &\leq (\sup_k ||\phi_k||_{\mathcal{G}}) \sum_{i=0}^{\infty} ||\Delta_i||_{\mathcal{G}} = ||\Phi||_{\mathcal{G}} ||\Delta||_{\mathcal{G},1} \end{aligned} \tag{42}$$

From (42) it follows that:

$$||\Phi||_{\mathcal{G}} = ||\Phi^o(I+\Delta)||_{\mathcal{G}} \leq ||\Phi^o||_{\mathcal{G}}(1+||\Delta||_{\mathcal{G},1})$$

Note that since $\Delta \in \mathcal{RH}_\infty^{n*n}$, $\{\Delta_i\} \in l_1^{n*n}$ and hence it follows that $||\Delta||_{\mathcal{G},1}$ is finite. Finally, let $\hat{\Delta} \stackrel{\Delta}{=} \delta I_n$. Then $\hat{\Phi} = \Phi^o(I_n + \hat{\Delta}) \in \mathcal{P}$ and $||\hat{\Phi}||_{\mathcal{G}} = ||\Phi^o||_{\mathcal{G}}(1+\delta)\diamond$

- **Corollary 3:** For the family (S_d) the constrained stability measure ϱ reduces to:

$$\varrho \stackrel{\Delta}{=} 1 - ||\Phi^o||_{\mathcal{G}}$$

Remark 6: The uncertainty description of (S_d) includes the parametric uncertainty of section 3 in the sense that if there exist $\tilde{\Delta}$, $\|\tilde{\Delta}\|_{\mathcal{G}} \leq \delta$ such that $\|A + \tilde{\Delta}\|_{\mathcal{G}} = 1$ then $\Delta_o = \frac{\tilde{\Delta}}{z} \in \mathcal{D}_\delta$ and it can be easily shown that $\|\Phi(z)\|_{\mathcal{G}} = \|\Phi^o(z)(I + \Delta_o(z))\|_{\mathcal{G}} = 1$.

5.2 Linear Robust Constrained Control Synthesis

In this section we address the general case of the Linear Robust Constrained Control Synthesis Problem defined in section 2.3. We will show that with an appropriate parametrization of all the achievable closed–loop maps, this problem can be solved by solving a *convex* constrained optimization problem. We begin by introducing a parametrization of all input–output closed–loop mappings achievable through internally stabilizing controllers and by showing that with this parametrization constrained–stability can be checked by checking a *finite number*, m_q, of convex inequalities.

5.2.1 Parametrization of all Closed–Loop Systems Achievable Through Full-State Feedback

- **Lemma 6:** Consider the linear, time invariant discrete time system:

$$\underline{x}_{k+1} = A\underline{x}_k + B_1\underline{\omega}_k + B_2\underline{u}_k,\ k = 0, 1 \ldots \tag{43}$$

with initial condition $\underline{x}(0) = \underline{x}_o$. If the pair (A, B_2) is stabilizable then the set of *all* closed–loop systems achievable by an internally stabilizing full state–feedback proper controller can be represented in terms of a free parameter $\tilde{Q} \in \mathcal{RH}_\infty^{mn}$ as:

$$x(z) = \left(I_n + P_{22}\tilde{Q}\right) P_{21} \begin{pmatrix} \underline{x}_o \\ \underline{\omega}(z) \end{pmatrix} \tag{44}$$

where:

$$
\begin{aligned}
P_{21} &= \left(I_n + (zI_n - A_F)^{-1} A_F \quad (zI_n - A_F)^{-1} B_1 \right) \\
P_{22} &= (zI_n - A_F)^{-1} B_2 \\
A_F &= A + B_2 F
\end{aligned}
\tag{45}
$$

and where F is any matrix that stabilizes (A, B_2).

Proof: The system (S) is equivalent to:

$$
\begin{aligned}
\underline{\tilde{x}}_{k+1} &= A\underline{\tilde{x}}_k + B_1\underline{\omega}_k + B_2\underline{u}_k + A\delta_{k,o}\underline{x}_o \\
\underline{x}_k &= \underline{\tilde{x}}_k + \delta_{ko}\underline{x}_o
\end{aligned}
\tag{S_o}
$$

with initial condition $\underline{\tilde{x}}_o = 0$. Since (A, B_2) is stabilizable, it follows that the system (S_o) is strongly stabilizable [28] (i.e. stabilizable with a stable controller). Hence a parametrization of all stabilizing controllers can be obtained by first pre–stabilizing the plant with a constant state feedback matrix F and then applying the Youla parametrization to the resulting plant. Define:

$$
\underline{v}_k \stackrel{\Delta}{=} \underline{u}_k - F\underline{x}_k \tag{46}
$$

Substituting (46) in (S_o) and taking z–transforms we have:

$$
\begin{aligned}
\underline{x}(z) &= (zI_n - A_F)^{-1} \left(B_1\underline{\omega}(z) + A_F\underline{x}_o + B_2\underline{v}(z) \right) \\
&= P_{21} \begin{pmatrix} \underline{x}_o \\ \underline{\omega}(z) \end{pmatrix} + P_{22}\underline{v}(z)
\end{aligned}
\tag{47}
$$

By using now the Youla parametrization for stable plants [28]we have that *all* controllers that stabilize (S_o) can be parametrized as:

$$
\begin{aligned}
\underline{v}(z) &= K(z)\underline{x}(z) \\
K &= \tilde{Q}\left(I + P_{22}\tilde{Q} \right)^{-1}
\end{aligned}
\tag{48}
$$

where $\tilde{Q} \in \mathcal{RH}_\infty^{mn}$. Hence:

$$
\underline{v} = \tilde{Q}P_{21} \begin{pmatrix} \underline{x}_o \\ \underline{\omega} \end{pmatrix} \tag{49}
$$

and, substituting (49) in (47),

$$\underline{x}(z) = \left(I_n + P_{22}\tilde{Q}\right) P_{21} \begin{pmatrix} \underline{x}_o \\ \underline{\omega}(z) \end{pmatrix} \tag{50}$$

Finally note that the control input is given by:

$$\begin{aligned} \underline{u} &= F\underline{x} + \underline{v} = F\underline{x} + \tilde{Q}P_{21} \begin{pmatrix} \underline{x}_o \\ \underline{\omega} \end{pmatrix} \\ &= \left[\left(\tilde{Q} + F\right) + FP_{22}\tilde{Q}\right] P_{21} \begin{pmatrix} \underline{x}_o \\ \underline{\omega} \end{pmatrix} \diamond \end{aligned} \tag{51}$$

5.2.2 Closed–Loop Constrained Stability Analysis

Setting $\underline{\omega}_k \equiv 0$ in (44) yields:

$$\begin{aligned} x(z) &= \left(I_n + P_{22}\tilde{Q}\right) P_{21} \begin{pmatrix} \underline{x}_o \\ 0 \end{pmatrix} \\ &= \underline{x}_o + (zI_n - A_F)^{-1} \left(A_F + B_2\tilde{Q}\,(zI_n - A_F)^{-1} zI_n\right) \underline{x}_o \end{aligned} \tag{52}$$

Let

$$Q \triangleq \tilde{Q}\left(I_n - \frac{A_F}{z}\right)^{-1} \tag{53}$$

since A_F is stable (53) defines a *bijection* in $\mathcal{RH}_\infty$. Making the change of variable indicated by (53) in (52) yields:

$$\underline{x}(z) = \underline{x}_o + (zI_n - A_F)^{-1} (A_F + B_2Q)\underline{x}_o \tag{54}$$

Let Q_i denote the coefficients of the impulse response of Q, i.e $\{Q_i\} = Z^{-1}\{Q(z)\}$. Then, from (54) we have that:

$$\underline{x}_k = \phi_k \underline{x}_o \tag{55}$$

where:

$$\begin{aligned} \phi_0 &= I_n \\ \phi_k &= A_F^k + \sum_{i=0}^{k-1} A_F^{k-1-i} B_2 Q_i \end{aligned} \tag{56}$$

In the next lemma we show that if Q is approximated by a N_Q^{th} order finite impulse response (FIR) filter, then constrained stability can be checked by checking a finite number of convex constraints.

- **Lemma 7:** Let F be any matrix that stabilizes the pair (A, B_2) and let the corresponding Q be approximated by an N_Q^{th} order FIR, i.e. $Q_i = 0, \ i \geq N_Q$. Then checking constrained–stability can be reduced to checking at most $N_Q + m_q$ inequalities where m_q is a number that depends *only* on F. Moreover these inequalities are *convex* in the variables Q_i.

Proof: Let $\rho(A_F)$ denote the spectral radius of A_F. Since A_F is stable $\rho(A_F) < 1$. Hence there exist a finite dimensional matrix norm such that $||A_F|| < 1$. Since *all* finite dimensional matrix norms are equivalent, there exist a constant c such that $||.||_{\mathcal{G}} \leq c||.||$. Since $||A_F|| < 1$ it follows that there exist m_q such that:

$$
\begin{aligned}
||A_F^k||_{\mathcal{G}} &\leq c||A_F^k|| \leq c||A_F||^k \\
&\leq 1 \, \forall \, k \geq m_q
\end{aligned}
\tag{57}
$$

From (56) it follows that if $Q_i = 0, \ i \geq N_Q$ then:

$$
\begin{aligned}
||\phi_{k+N_Q}||_{\mathcal{G}} &= ||A_F^{k+N_Q} + \sum_{i=0}^{N_Q+k-1} A_F^{N_Q+k-1-i} B_2 Q_i||_{\mathcal{G}} \\
&= ||A_F^{k+N_Q} + \sum_{i=0}^{N_Q-1} A_F^{N_Q+k-1-i} B_2 Q_i||_{\mathcal{G}} \\
&= ||A_F^k \left(A_F^{N_Q} + \sum_{i=0}^{N_Q-1} A_F^{N_Q-1-i} B_2 Q_i \right) ||_{\mathcal{G}} \\
&= ||A_F^k \phi_{N_Q}||_G \leq ||A_F^k||_{\mathcal{G}} ||\phi_{N_Q}||_{\mathcal{G}}
\end{aligned}
\tag{58}
$$

From (58) and Lemma 2 it follows that constrained stability is satisfied *iff* $||\phi_k||_{\mathcal{G}} \leq 1 \ k = 1 \ldots N_Q + m_q - 1$. Finally, since ϕ_k is linear in Q_i it follows that $||\phi_k||_{\mathcal{G}}$ is a convex function of Q_i. Hence the constraints are convex functions of Q_i $\diamond$.

5.2.3 H_2 Performance Criterion

In this section we apply our design procedure to the case where the frequency domain performance specifications have the form of a bound upon $\|T_{\zeta w}\|_2$. Hence the synthesis problem has the form of the following constrained optimization problem:

$$\min_{Q_i \in R^{mn}} \epsilon$$

subject to:

$$\|\phi_k\|_{\mathcal{G}} \leq \epsilon, \; k = 1 \ldots \mathcal{N}_Q + m_q - 1$$
$$\|T_{\zeta w}(z)\|_2 \leq \gamma$$

- **Lemma 8:** Assume that (S) has the following state–space realization:

$$\left(\begin{array}{c|cc} A & B_1 & B_2 \\ \hline C_1 & 0 & D_{12} \\ I_n & 0 & 0 \end{array}\right) \qquad (S)$$

where D_{12} has full column rank (i.e. $D_{12}'D_{12}$ non–singular) and where the pairs (A, B_2) and (C_1, A) are stabilizable and detectable respectively. Note that $D_{11} = 0$ can be assumed without loss of generality since it contributes only with a constant term to $\|T_{\zeta w}\|_2$, and that the rank condition upon D_{12} is required to guarantee that the problem is non–singular. Finally, let F be the unique solution to the unconstrained H_2 problem. Then F stabilizes the pair (A, B_2) and:

$$\begin{aligned} T_{\zeta w} &= G_C B_1 + R_B^{\frac{1}{2}} Q B_1 \\ \|T_{\zeta w}\|_2^2 &= \|G_C B_1\|_2^2 + \|R_B^{\frac{1}{2}} Q B_1\|_2^2 \end{aligned} \qquad (59)$$

where:

$$
\begin{aligned}
G_C &= \left(\begin{array}{c|c} A_F & I_n \\ \hline C_F & 0 \end{array}\right) \\
A_F &= A + B_2 F \\
C_F &= C_1 + D_{12} F \\
R_B &= D'_{12} D_{12} + B'_2 X B_2
\end{aligned}
\tag{60}
$$

and where $X > 0$ is the solution to the Riccati equation associated with the unconstrained H_2 problem.

Proof: The proof of the Lemma, which follows from standard space–state manipulations, is given in Appendix B.

- **Corollary 4:** Let Q_{B_i} denote the coefficients of the impulse response of $R_B^{\frac{1}{2}} Q$. Then, given $\gamma \geq \|G_C B_1\|_2$, all stabilizing controllers yielding $\|T_{\zeta w}\|_2 \leq \gamma$ can be parametrized in terms of Q where $Q_i = R_B{}^{-\frac{1}{2}} Q_{B_i}$ and where Q_{B_i} satisfy the following constraint:

$$
\sum_{i=0}^{i=\infty} \|Q_{B_i} B_1\|_F^2 \leq \gamma^2 - \|G_C B_1\|_2^2 \tag{61}
$$

where $\|.\|_F$ denotes the Frobenius norm. This results follows immediately from (59) by noting that $\|Q(z)\|_2^2 = \sum_{i=0}^{\infty} \|Q_i\|_F^2$

- *Example 5:*

Consider the following system:

$$
\begin{aligned}
&A = \begin{pmatrix} 0.4975 & 0.075 \\ 0.5025 & -1.0075 \end{pmatrix}, \; B_1 = I_2, \; B_2 = \begin{pmatrix} 1 \\ 1 \end{pmatrix} \\
&C_1 = \begin{pmatrix} \sqrt{0.05} & \sqrt{0.05} \\ 0.5 & -0.5 \\ 0 & 0 \end{pmatrix}, \; D_{12} = \begin{pmatrix} 0 \\ 0 \\ \sqrt{20} \end{pmatrix} \\
&G = I_2, \; \underline{\gamma} = (1 \quad 1)', \; \mathcal{G} = |G\underline{x}| \leq \underline{\gamma}, \; \mathbf{E}\{\underline{\omega\omega'}\} = I_2
\end{aligned}
\tag{62}
$$

In this case we have (see eq. B2 in appendix B) that the unconstrained optimal controller and closed-loop dynamics are given by:

$$F = (\,-0.0494 \quad 0.1327\,)\,,\; A_F = \begin{pmatrix} 0.4481 & 0.1402 \\ 0.4531 & -0.8748 \end{pmatrix} \qquad (63)$$

Direct application of Lemma 7 shows that constrained stability requires the satisfaction of $N_Q + 3$ inequalities.

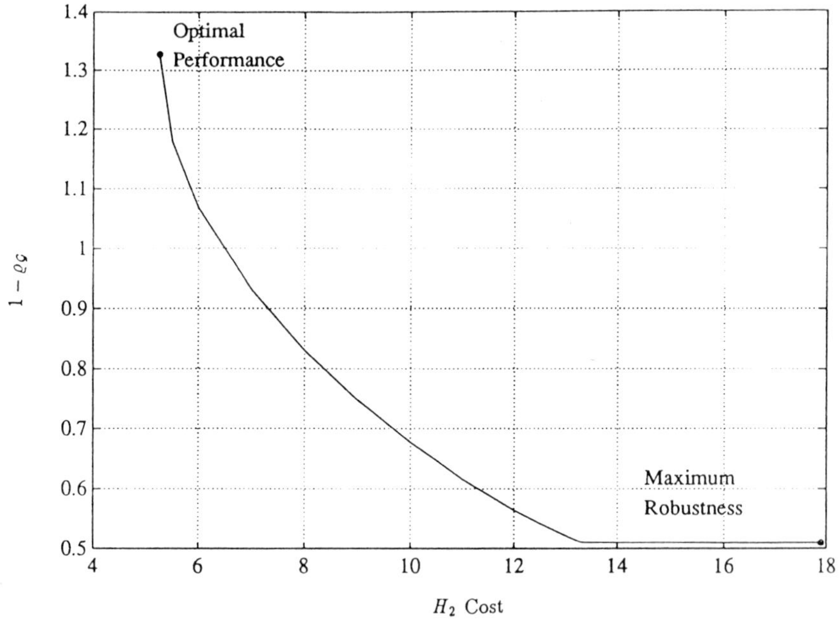

Fig 3. Robustness vs. H_2 cost for Example 5, with $N_Q = 3$.

	$\|\|T_{\zeta w}\|\|_2^2$	$\|\|\Phi\|\|_{\mathcal{G}} = 1 - \varrho$
Open Loop	Unstable	
H_2	5.25	1.328
$\varrho_{opt.}$	17.892	0.51
Norm Based		
N_Q=1	13.5	0.658
N_Q=2		0.51
Norm Based		
N_Q=1	13.0	0.6696
N_Q=2		0.5224
N_Q=3		0.5224
N_Q=20		0.5225
Norm Based		
N_Q=1	10	0.7462
N_Q=2		0.6774
N_Q=3		0.6774
N_Q=20		0.6774
Norm Based		
N_Q=1	6	1.0685
N_Q=2		1.0685
N_Q=3		1.0685
N_Q=20		1.0685

Table 1. Robustness vs. H_2 Cost for Example 5.

Table 1 shows the constrained robustness margin and the H_2 cost for the Norm Based controller for different values of N_Q. H_2 denotes the unconstrained optimal H_2 controller and ϱ_{opt} denotes the controller obtained maximizing ϱ using a linear–programming based procedure. Figure 3 shows a typical robustness vs. performance plot. Note that the proposed approach can provide a substantial reduction of the H_2 cost while maintaining the same robustness level achieved by a maximum constrained–robustness design.

5.2.4 H_∞ Performance Criterion

Similarly to the H_2 case, the case where the performance specifications are given in terms of an H_∞ cost can be stated as follows: Given $\gamma > 0$, find Q such that $\|\Phi\|_{\mathcal{G}}$ is minimized subject to the constraint $\|T_{\zeta w}\|_\infty \leq \gamma$. Thus, the constrained optimization problem has the following form:

$$\min_{Q_i \in R^{mn}} \epsilon$$

subject to:

$$\begin{aligned} \|\phi_k\|_{\mathcal{G}} &\leq \epsilon,\ k = 1 \ldots \mathcal{N}_Q + m_q - 1 \\ \overline{\sigma}(T_{\zeta w}(z)) &\leq \gamma\ \forall z \in \tau \end{aligned} \qquad (P_\infty)$$

where τ denotes the unit circle in the z–plane. Note that this has the form of a semi–infinite optimization problem, since for each $z \in \tau$ there are $N_Q + m_q + 1$ constraints. In [5–6] a similar problem was approximately solved *finite* set of frequency points, hence replacing the semi–infinite constraints by a (large) finite number of single frequency constraints. In this paper, we will use a different approach. We will solve this problem using an algorithm based upon outer approximation methods [29] which yield a sequence of approximations to the solution such that the accumulation point of this sequence is a solution to the original problem. Let $\tau_k \subset \tau$ be a *finite* set and consider the following approximation (P_k) to problem (P_∞)

$$\min_{Q_i \in R^{mn}} \epsilon$$

subject to:

$$\begin{aligned} \|\phi_k\|_{\mathcal{G}} &\leq \epsilon,\ k = 1 \ldots \mathcal{N}_Q + m_q - 1 \\ \overline{\sigma}(T_{\zeta w}(z)) &\leq \gamma\ \forall z \in \tau_k \end{aligned} \qquad (P_k)$$

Then, a solution to P_∞ can be found by solving a sequence of problems of the form (P_k) as follows:

Algorithm $\mathbf{H}_\infty$

0) Set $i = 0$, select an initial set $\tau_o \subset \tau$

1) Solve the approximate problem P_i. Let Q^i be the solution.

2) Compute approximately:

$$z_i = \underset{z \in \tau}{\operatorname{argmax}} \{\overline{\sigma}(T_{\zeta w}(z))\}$$

where $T^i_{\zeta w}(z)$ is given by (B5) evaluated at Q^i.

Let $Z = \{z \in \tau : \overline{\sigma}(T_{\zeta w}(z)) = \overline{\sigma}(T_{\zeta w}(z_i))\}$

3) If $\overline{\sigma}(T_{\zeta w}(z_i)) \leq \gamma$

then: Done

else: Set $\tau_{i+1} = \tau_i \cup Z; i = i + 1$. Go to Step 1

Endif

- *Example 6:*

Consider the system introduced in section 5.2.3 and assume that

$$C_1 = (1 \quad 1); D_{12} = (1)$$

In this case, since $\|A_F\|_{\mathcal{G}} < 1$, constrained stability can be checked by checking only N_Q inequalities. Also note that the norm of the optimal unconstrained H_∞ controller can be found solving a 1–block Nehari approximation problem, as shown in appendix C, yielding an optimal value $\|T_{\zeta w}\|^o_\infty = 1.8543$.

Table 2 shows the constrained robustness margin versus the H_∞ cost for different values of N_Q while figure 4 illustrates a typical robustness versus performance curve. As in the H_2 case, a substantial reduction in the cost can be achieved without sacrificing robustness.

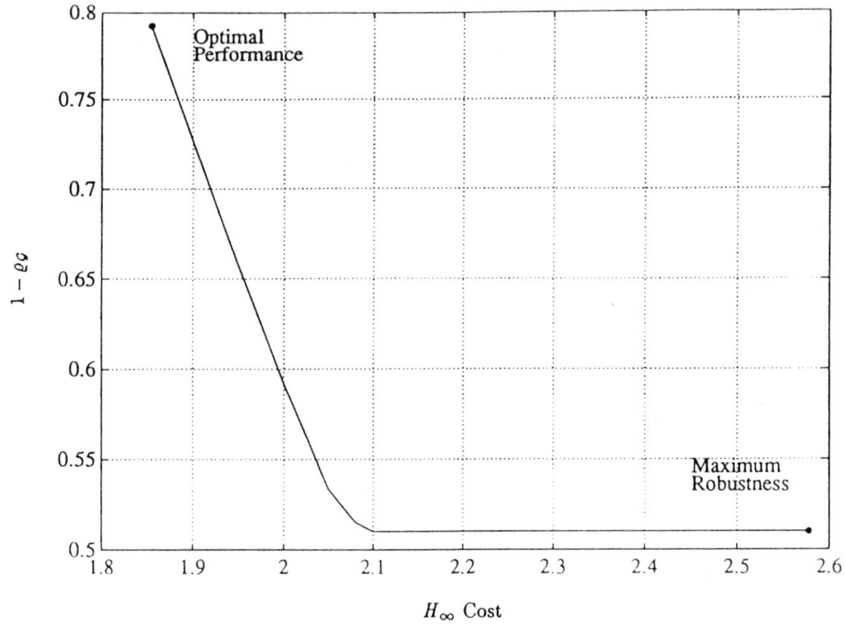

Fig 4. Robustness vs. H_∞ cost for Example 6, with $N_Q = 5$.

VI. Conclusions

Most realistic control problems involve both some type of time–domain constraints and certain degree of model uncertainty. However, the majority of control design methods currently available focus only on one aspect of the problem. Following the spirit of [14–16], in this paper we propose to approach time–domain constraints using an operator norm induced by the constraints to assess the stability properties of a family of systems. Specifically, in section II we introduced a robustness measure that indicates how well the family

	$\|T_{\zeta w}\|_\infty$	$\|\Phi\|_{\mathcal{G}} = 1 - \varrho$
Open Loop	Unstable	
$\varrho_{opt.}$	2.5776	0.51
Norm Based		
N_Q=2	2.1	0.5538
N_Q=3		0.51
Norm Based		
N_Q=2	2.0	0.6489
N_Q=3		0.6163
N_Q=5		0.5922
N_Q=10		0.5812
Norm Based		
N_Q=2	1.95	0.6978
N_Q=3		0.6752
N_Q=5		0.6586
N_Q=10		0.6500
Norm Based		
N_Q=2	1.855	0.7917
N_Q=3		0.7915
N_Q=5		0.7914
N_Q=10		0.7913
Norm Based		
N_Q=2	1.8543	0.7923
N_Q=3		0.7923
N_Q=5		0.7923
N_Q=10		0.7923

Table 2. Robustness vs. H_∞ cost for Example 6.

of systems under consideration satisfies a given set of constraints. In section III we explored the properties of this robustness measure for the case of additive parametric model uncertainty and we showed that our formalism provides a generalization of the well known technique of estimating robustness bounds from the solution of a Lyapunov equation. We then proposed, in section IV, a synthesis procedure for fixed order controllers, based upon maximization of the robustness measure subject to additional performance constraints. This procedure results in static feedback controllers that are easily implementable, but unfortunately, it leads to non–convex optimization problems that might have local extrema. This difficulty was solved in section V, where we extended our results to dynamic uncertainty and proposed a general synthesis procedure for dynamic controllers, based upon maximizing the robustness measure over the set of *all* internally–stabilizing controllers subject to the performance specifications. We showed that this procedure leads to convex optimization problems and we proposed specific algorithms for the cases where the additional performance specifications are formulated in terms of $\|.\|_2$ or $\|.\|_\infty$ of a closed–loop transfer function. The examples in sections IV and V show that our technique yields a significant improvement over a "maximum robustness" type design since the resulting controllers can achieve essentially the same level of "robustness", but with a substantially reduced cost. Furthermore the proposed design paradigm offers a number of significant advantages over the methods currently available:

- Displays, by means of the robustness measure, the trade–offs between model uncertainty and design constraints. For instance, it indicates when an additional system identification effort is required in the predesign phase in order to satisfy the design constraints

- Provides a systematic approach that deals explicitly with frequency and time–domain constraints. Hence we can expect a considerable simplification of the design phase.

- Presents the advantages of modern control techniques, i.e. it is an algorithmic approach, capable of dealing with multivariable systems and guaranteeing optimality in some previously defined sense.

Perhaps the most severe limitation to the theory in its present form arises from the fact that it currently addresses only parametric and unstructured dynamic uncertainty. It is our goal to extend the theory to include *dynamic structured uncertainty*, i.e. the case where the uncertainty has a block diagonal structure in addition to the norm bound introduced in section 5.1 (equation (40)).

Another interesting open research area concerns the numerical properties of the design algorithms and the resulting controllers, in particular their order. Finally, we are currently looking into an alternative to the outer approximation method proposed in section 5.2.4. We have preliminary results showing that the frequency discretization can be replaced by a condition requiring a matrix which is a convex function of the parameters of the controller to be positive definite. Furthermore, the resulting optimization problem can be *decoupled* into a constrained finite dimensional convex optimization, followed by the solution of an *unconstrained* Nehari approximation problem. This research direction is currently being pursued and a future paper reporting these results is planned.

VII. Acknowledgements

The author is grateful to Mr. Hector Rotstein for a discussion on the use of outer approximation methods for solving semi–infinite optimization problems and to Professors M. Damborg at the University of Washington and A. Sideris at Caltech for many suggestions.

VIII. References

[1]. S. Boyd and C. Barrat, "Linear Controller Design: Limits of Performance," Prentice Hall Information and Systems Sciences Series, Englewood Cliffs, New Jersey, 1991.

[2]. M. Vassilaki, J. C. Hennet and G. Bitsoris, "Feedback Control of Discrete–Time Systems under State and Control Constraints", *Int. J. Control,* 47, pp. 1727–1735, 1988.

[3]. M. Sznaier and M. J. Damborg, "Heuristically Enhanced Feedback Control of Constrained Discrete Time Linear Systems," *Automatica,* Vol 26, 3, pp 521–532, 1990.

[4]. D. C. Youla, J. J. Bongiorno and H. A. Jabr, "Modern Wiener–Hopf Design of Optimal Controllers–Part I: The single–input–output case," *IEEE Trans. Autom. Contr.,* Vol 21, 1, pp 3–13. "–Part II: The Multivariable Case," *IEEE Trans. Autom. Contr.,* Vol 21, 3, pp. 319–338, 1976.

[5]. S. Boyd et. al., "A New CAD Method and Associated Architectures for Linear Controllers,"*IEEE Trans. Autom. Contr.,* Vol 33, 3, pp 268–283, 1988.

[6]. E. Polak and S. Salcudean, "On the Design of Linear Multivariable Feedback Systems Via Constrained Nondifferentiable Optimization in H_∞ Spaces," *IEEE Trans. Autom. Contr.,* Vol 34, 3, pp 268–276, 1989.

[7]. G. Zames, "Feedback and Optimal Sensitivity: Model Reference Transformations, Multiplicative Seminorms and Approximate Inverses", *IEEE Trans. Autom. Contr.,* Vol 26, 2, pp. 301–320, 1981.

[8]. B. Francis, "A Course in H_∞ Control Theory," Vol 88 in *Lecture Notes in Control and Information Science,* M. Thoma and A. Wyner, Editors, Springer–Verlag, New–York, 1987.

[9]. B. Francis and J. Doyle, "Linear Control Theory with an H_∞ Optimality Criterion," *SIAM Journal in Control and Optimization,"* Vol 25, 4, pp. 815–844, 1987.

[10]. J. Doyle, K. Glover, P. Khargonekar and B. Francis , "State–Space Solutions to Standard H_2 and H_∞ Control Problems," *IEEE Trans. Autom. Contr.,* Vol 34, 8, pp. 831–846, 1989.

[11]. D. C. McFarlane and K. Glover, "Robust Controller Design Using Normalized Coprime Factor Plant Descriptions," *Vol 138 in Lecture Notes in Control and Information Sciences,* M. Thoma and A. Wyner, Editors, Springer–Verlag, New–York, 1990.

[12]. J. W. Helton and A. Sideris, "Frequency Response Algorithms for H_∞ Optimization with Time Domain Constraints," *IEEE Trans. Autom. Contr.,* Vol 34, 4, pp. 427–434, 1989.

[13]. A. Sideris and H. Rotstein, "H_∞ Optimization with Time Domain Constraints over a Finite Horizon," *Proc. of the 29th IEEE CDC,* Hawaii, Dec 5–7 1990, pp. 1802–1807.

[14]. M. Sznaier, "Norm Based Robust Control of Constrained–Discrete Time Linear Systems," *Proc. of the 29th IEEE CDC,* Hawaii, Dec 5–7, 1990, pp. 1925–1930.

[15]. M. Sznaier, "Norm Based Robust Control of State Constrained Discrete Time Linear Systems," *IEEE Trans. Autom. Contr.,* to appear, 1992.

[16]. M. Sznaier and A. Sideris, "Norm Based Optimally Robust Control of Constrained Discrete Time Linear Systems," *Proc. 1991 ACC,* Boston, Massachusetts, June 23–25, 1991, pp. 2710–2715.

[17]. J. Doyle, "Analysis of Feedback Systems with Structured Uncertainties," *IEE Proc.,* Vol 129, Pt. D, 6, pp. 252–250, 1982.

[18]. R. A. Horn and C. R. Johnson, "Matrix Analysis," Cambridge University Press, 1985.

[19]. D. G. Luenberger, "Optimization by Vector Space Methods," Wiley, New York, 1989.

[20]. R. K. Yedavalli, "Improved Measures of Stability Robustness for Linear State Space Models," *IEEE Trans. Autom Contr.,* Vol AC-30, pp. 557–579, 1985.

[21]. K. Zhou and P. P. Khargonekar, "Stability Robustness Bounds for Linear State–Space Models with Structured Uncertainty," *IEEE Trans. Autom Contr.*, Vol AC-32, N°7, pp 621–623, 1987.

[22]. L. H. Keel et. al., " Robust Control With Structured Perturbations," *IEEE Trans. Autom Contr.*, Vol 33, N°1, pp. 68-77, 1988.

[23]. D. D. Siljak, " Parameter Space Methods for Robust Control Design: A Guided Tour, " *IEEE Trans. Autom Contr.*, Vol 34, N°7, pp. 674–687, 1989.

[24]. R. V. Patel and M. Toda, "Quantitative Measures of Robustness for Multivariable Systems," *Proc. Joint Automat. Contr. Conf.*, San Francisco, CA, 1980, paper TP-8A.

[25]. M. E. Sezer and D. D. Siljak, "Robust Stability of Discrete Systems," *Int. J. Control.*, Vol 48, N°5, pp. 2055–2063, 1988.

[26]. F. H. Clarke, *Optimization and Nonsmooth Analysis*, Canadian Mathematical Society Series of Monographs and Advanced Texts, Wiley, N. Y., 1983.

[27]. K. Zhou and J. Doyle, "Notes on MIMO Control Theory," Lecture Notes, California Institute of Technology, 1990.

[28]. M. Vidyasagar, "Control Systems Synthesis: A Factorization Approach," MIT Press, Cambridge, MA, 1985.

[29]. C. Gonzaga and E. Polak, "On Constraint Dropping Schemes and Optimality Functions for a Class of Outer Approximation Algorithms,"*SIAM J. on Contr. and Opt.*, Vol. 17, 4, pp. 477–493, 1979.

[30]. A. Naylor and G. R. Sell, "Linear Operator Theory in Engineering and Science," Vol 40 in Applied Mathematical Sciences, Springer–Verlag, New York, 1982.

[31]. C. C. Chu, J. Doyle and E. B. Lee, "The General Distance Problem in H_∞ Optimal Control Theory," *Int. J. Control,* Vol 44, 2, pp. 565–596, 1986.

Appendix A: Proof of Theorem 1

We begin by introducing two preliminary results:

- **Lemma 9:** Consider the system (S^a). Assume that the perturbation set D is *a closed cone with vertex at the origin* [19], i.e. $\Delta^o \in \mathcal{D} \iff \alpha\Delta^o \in \mathcal{D}\, \forall\, 0 \leq \alpha$ and that (S^a) is constraint stable (i.e. $\|A\|_{\mathcal{G}} < 1$). Let:

$$\Delta^o = \underset{\Delta \in \mathcal{D}}{\operatorname{argmin}} \{\|\Delta\|_{\mathcal{N}} : \|A + \Delta\|_{\mathcal{G}} = 1\} \tag{A1}$$

and consider a sequence $A^i \to A$ such that $\|A^i\|_{\mathcal{G}} < 1$. Finally, define the sequence λ^i as:

$$\lambda^i = \min_{\lambda \in \Re+} \left\{\lambda : \|A^i + \lambda\Delta^o\|_{\mathcal{G}} = 1\right\} \tag{A2}$$

Then the sequence λ^i has an accumulation point at 1.

Proof: Since $\|A^i\|_{\mathcal{G}} < 1$ and since $\mathcal{D}$ is a closed cone it follows that λ^i is well defined. Furthermore, from (A2) it follows that:

$$\lambda^i \leq \frac{1 + \|A^i\|_{\mathcal{G}}}{\|\Delta^o\|_{\mathcal{G}}} \leq \frac{2}{\|\Delta^o\|_{\mathcal{G}}} \tag{A3}$$

Hence from Bolzanno–Weierstrass' theorem [30] it follows that λ^i has an accumulation point $\tilde{\lambda}$ and that there exist a subsequence $\tilde{\lambda}^i \to \tilde{\lambda}$. Hence:

$$\|A^i + \tilde{\lambda}^i\Delta^o\|_{\mathcal{G}} = 1$$

and since $A^i \to A$ then:

$$\|A + \tilde{\lambda}\Delta^o\|_{\mathcal{G}} = 1i \tag{A4}$$

Assume that $\tilde{\lambda} < 1$ and let $\hat{\Delta} \triangleq \tilde{\lambda}\Delta^o$ Then $\|\hat{\Delta}\|_{\mathcal{N}} < \|\Delta^o\|_{\mathcal{N}}$, $\|A + \hat{\Delta}\|_{\mathcal{G}} = 1$ and $\hat{\Delta} \in \mathcal{D}$ (since $\mathcal{D}$ is a cone) which contradicts (A1). Assume now that $\tilde{\lambda} > 1$. Then, for i large enough, $\tilde{\lambda}^i > 1$, which together with (A2) implies that:

$$\|A^i + \Delta^o\|_{\mathcal{G}} < 1 \tag{A5}$$

and hence:

$$\|A + \Delta^o\|_{\mathcal{G}} < 1 \tag{A6}$$

which contradicts (A1). Therefore $\tilde{\lambda} = 1 \diamond$.

- **Lemma 10:** Let $\rho_1 > 0, \rho_2 > 0$ and $0 \leq \lambda \leq 1$ be given numbers and assume that $\mathcal{D}$ is a cone with vertex at the origin. Consider the following sets:

$$\begin{aligned} \rho_1 B\Delta &= \{\Delta \in \mathcal{D}: \|\Delta\|_{\mathcal{N}} \leq \rho_1\} \\ \rho_2 B\Delta &= \{\Delta \in \mathcal{D}: \|\Delta\|_{\mathcal{N}} \leq \rho_2\} \\ \rho B\Delta &= \{\Delta \in \mathcal{D}: \|\Delta\|_{\mathcal{N}} \leq \rho \stackrel{\Delta}{=} \lambda\rho_1 + (1-\lambda)\rho_2\} \end{aligned} \tag{A7}$$

Then:

$$\rho B\Delta \subseteq \lambda\rho_1 B\Delta + (1-\lambda)\rho_2 B\Delta$$

Proof: Consider any $\Delta^o \in \rho B\Delta$. Then:

$$\begin{aligned} \Delta^o &= \frac{\|\Delta^o\|_{\mathcal{N}}}{\rho}\left[\frac{\rho\Delta^o}{\|\Delta^o\|_{\mathcal{N}}}\right] \\ &= \frac{\|\Delta^o\|_{\mathcal{N}}}{\rho}\left[\lambda\rho_1\frac{\Delta^o}{\|\Delta^o\|_{\mathcal{N}}} + (1-\lambda)\rho_2\frac{\Delta^o}{\|\Delta^o\|_{\mathcal{N}}}\right] \\ &= [\lambda\Delta_1 + (1-\lambda)\Delta_2] \end{aligned} \tag{A8}$$

where:

$$\begin{aligned} \Delta_1 &= \alpha\rho_1\frac{\Delta^o}{\|\Delta^o\|_{\mathcal{N}}} \\ \Delta_2 &= \alpha\rho_2\frac{\Delta^o}{\|\Delta^o\|_{\mathcal{N}}} \\ \alpha &= \frac{\|\Delta^o\|_{\mathcal{N}}}{\rho} \leq 1 \end{aligned} \tag{A9}$$

The proof is completed by noting that from (A9) and the hypothesis it follows that $\Delta_1 \in \rho_1 B\Delta$ and $\Delta_2 \in \rho_2 B\Delta$ $\diamond$.

Proof of Theorem 1

Assume that $\varrho_{\mathcal{G}}^{\mathcal{N}}$ is *not* continuous. Then, given $\epsilon > 0$, for every $\delta > 0$ there exist A_δ such that $||A_\delta - A||_{\mathcal{G}} \leq \delta$ and $|\varrho_{\mathcal{G}}^{\mathcal{N}}(A_\delta) - \varrho_{\mathcal{G}}^{\mathcal{N}}| > \epsilon$. Hence there exist a sequence $A^i \rightarrow A$ such that $\varrho_{\mathcal{G}}^{\mathcal{N}^i} \not\rightarrow \varrho_{\mathcal{G}}^{\mathcal{N}}$. Furthermore, it is easily seen that the sequence $\varrho_{\mathcal{G}}^{\mathcal{N}^i}$ is bounded and therefore is contains a convergent subsequence. It follows that there exist a sequence $A^i \rightarrow A$ such that $\varrho_{\mathcal{G}}^{\mathcal{N}^i} \rightarrow \tilde{\varrho} \neq \varrho_{\mathcal{G}}^{\mathcal{N}}$. Let:

$$\Delta^i = \underset{\Delta \in \mathcal{D}}{\operatorname{argmin}} \left\{ ||\Delta||_{\mathcal{N}} : ||A^i + \Delta||_{\mathcal{G}} = 1 \right\} \tag{A10}$$

From (A10) it follows that $||\Delta^i||_{\mathcal{G}} \leq 1 + ||A^i||_{\mathcal{G}}$. It follows then that the sequence Δ^i is bounded and therefore, since $R^{n \times n}$ with a finite dimensional matrix norm is complete and since $\mathcal{D}$ is a closed set, it has an accumulation point $\tilde{\Delta}$ (Bolzano Weierstrass) and a convergent subsequence $\tilde{\Delta}^i \rightarrow \tilde{\Delta}$ such that $||A + \tilde{\Delta}||_{\mathcal{G}} = 1$. Furthermore, from the definition of Δ^o it follows that

$$\tilde{\varrho} = ||\tilde{\Delta}||_{\mathcal{N}} > ||\Delta^o||_{\mathcal{N}} = \varrho_{\mathcal{G}}^{\mathcal{N}} \tag{A11}$$

Hence, for i large enough,

$$||\tilde{\Delta}^i||_{\mathcal{N}} > ||\Delta^o||_{\mathcal{N}} \tag{A12}$$

Applying Lemma 9, we have that there exist a sequence $\lambda^i \rightarrow 1$ such that:

$$\lambda^i = \min_{\lambda \in \Re^+} \left\{ \lambda : ||A^i + \lambda \Delta^o||_{\mathcal{G}} = 1 \right\} \tag{A13}$$

From (A12) and since $\lambda^i \rightarrow 1$ it follows that for i large enough

$$\begin{aligned} ||\lambda^i \Delta^o||_{\mathcal{N}} &< ||\tilde{\Delta}^i||_{\mathcal{N}} \\ ||A^i + \lambda^i \Delta^o||_{\mathcal{G}} &= 1 \end{aligned} \tag{A14}$$

and, since $\mathcal{D}$ is a cone, $\lambda^i \Delta^o \in \mathcal{D}$, which contradicts (A10). The proof is completed by noting that since *all* finite dimensional matrix norms are equivalent [18] then continuity in the $||.||_{\mathcal{G}}$ norm implies continuity in any other norm defined over $R^{n \times n}$ $\diamond$.

To prove concavity, start by considering a convex linear combination $A = \lambda A_1 + (1-\lambda)A_2$, $\lambda \leq 1$ of given matrices A_1 and A_2. Then, from Lemma 10 it follows that:

$$\max_{\Delta \in \rho B\Delta} \|A + \Delta\|_G \leq \max_{\substack{\Delta_1 \in \rho_1 B\Delta \\ \Delta_2 \in \rho_2 B\Delta}} \|\lambda(A_1 + \Delta_1) + (1-\lambda)(A_2 + \Delta_2\|_{\mathcal{G}}$$

$$\leq \lambda \max_{\Delta_1 \in \rho_1 B\Delta} \|A_1 + \Delta_1\|_{\mathcal{G}} + (1-\lambda) \max_{\Delta_2 \in \rho_2 B\Delta} \|A_2 + \Delta_2\|_{\mathcal{G}} \tag{A15}$$

Consider now the case where $\rho_1 = \varrho_{\mathcal{G}}^{\mathcal{N}}(A_1)$ and $\rho_2 = \varrho_{\mathcal{G}}^{\mathcal{N}}(A_2)$. Then it follows from the definition of $\varrho_{\mathcal{G}}^{\mathcal{N}}$ that both maximizations in the right hand side of (A15) yield 1 and therefore:

$$\max_{\Delta \in \rho B\Delta} \|A + \Delta\|_{\mathcal{G}} \leq 1 \tag{A16}$$

Hence, from the definition of $\varrho_{\mathcal{G}}^{\mathcal{N}}$:

$$\varrho_{\mathcal{G}}^{\mathcal{N}}[\lambda A_1 + (1-\lambda)A_2] \geq \varrho = \lambda \varrho_{\mathcal{G}}^{\mathcal{N}}(A_1) + (1-\lambda)\varrho_{\mathcal{G}}^{\mathcal{N}}(A_2) \diamond$$

Appendix B: Proof of Lemma 8

Let $\underline{z} = C_1\underline{x} + D_{12}\underline{u}$ and consider the Riccati equation associated with the problem of minimizing $\|\underline{z}\|_2^2$:

$$A'XA - X + C_1'C_1 - (B_2'XA + D_{12}'C_1)'(D_{12}'D_{12} + B_2'XB_2)^{-1}(B_2'XA + D_{12}'C_1) = 0 \tag{B1}$$

Define:

$$F \stackrel{\Delta}{=} -(R + B_2'XB_2)^{-1}(D_{12}'C_1 + B_2'XA) \tag{B2}$$

where $R \stackrel{\Delta}{=} D_{12}'D_{12} > 0$. From the hypothesis it follows that (B1) has a unique solution $X = X' > 0$ and that $A_F = A + B_2F$ is asymptotically stable. Note that in terms of the closed loop matrices A_F and C_F, (B1) and (B2) are equivalent to:

$$\begin{aligned} A_F'XA_F - X + C_F'C_F &= 0 \\ RF + B_2'XA_F + D_{12}'C_1 &= \\ D_{12}'C_F + B_2'XA_F &= 0 \end{aligned} \tag{B3}$$

Selecting F as the prestabilizing feedback in (45) we have:

$$\begin{aligned} \underline{y} &= P_{21}\underline{\omega} + P_{22}\underline{v} \\ \underline{z} &= C_F\underline{y} + D_{12}\underline{v} \end{aligned} \tag{B4}$$

From (45), (49) and (53), it follows that:

$$\begin{aligned} T_{\zeta w} &= G_C B_1 + N T_{vw} \\ &= G_C B_1 + N\tilde{Q}P_{21} \\ &= G_C B_1 + N\frac{QB_1}{z} \end{aligned} \tag{B5}$$

where G_C is defined in (60) and where:

$$N = \left(\begin{array}{c|c} A_F & B_2 \\ \hline C_F & D_{12} \end{array}\right) \tag{B6}$$

Next we show that $N^{\sim}G_C \in \mathcal{RH}_2^{\perp}$ and that $N^{\sim}N = R_B \stackrel{\Delta}{=} D_{12}'D_{12} + B_2'XB_2$. Using standard manipulation techniques [27] we have that:

$$N^{\sim}G_C = \left(\begin{array}{cc|c} A_F & 0 & I_n \\ {A_F'}^{-1}C_F'C_F & {A_F'}^{-1} & 0 \\ \hline \left(D_{12} - C_F{A_F'}^{-1}B_2\right)'C_F & -B_2'{A_F'}^{-1} & 0 \end{array}\right) \tag{B7}$$

and the similarity transformation:

$$T = \begin{pmatrix} I_n & 0 \\ X & I_n \end{pmatrix}$$

yields:

$$\begin{aligned} TN^{\sim}G_C T^{-1} &= \left(\begin{array}{cc|c} A_F & 0 & I \\ 0 & {A_F'}^{-1} & X \\ \hline 0 & -B_2'{A_F'}^{-1} & 0 \end{array}\right) \\ &= \left(\begin{array}{c|c} {A_F'}^{-1} & X \\ \hline -B_2'{A_F'}^{-1} & 0 \end{array}\right) \end{aligned} \tag{B8}$$

From (B8) it follows that:

$$||T_{\zeta w}||_2^2 = ||G_C B_1||_2^2 + ||N\frac{QB_1}{z}||_2^2 \tag{B9}$$

Finally, from the stability of A_F and (B3) it follows ([27], Lemma 13.8) that:

$$N^{\sim}N = R + B_2' X B_2 \triangleq R_B$$

Hence, $U \triangleq N R_B^{-\frac{1}{2}}$ is an *inner* function. Since $\frac{1}{z}$ is also inner we have that:

$$||T_{\zeta w}||_2^2 = ||G_C B_1||_2^2 + ||R_B^{\frac{1}{2}} Q B_1||_2^2 \diamond$$

Appendix C. Computation of the Norm of the Optimal Approximation for Example 6

Using the factorization introduced in Lemma 8 we have that:

$$T_{\zeta w} = G_C B_1 + N R_B^{\frac{-1}{2}} \left(\frac{R_B^{\frac{1}{2}} \tilde{Q} B_1}{z} \right) \tag{C1}$$

where $U \triangleq N R_B^{\frac{-1}{2}}$ is inner. Hence, we can find $U_\perp$ such that $(U \; U_\perp)$ is unitary. Since premultiplication by a unitary matrix preserves the ∞ norm we have that:

$$||T_{\zeta w}||_\infty = || \begin{pmatrix} U \\ U_\perp \end{pmatrix}^{\sim} T_{\zeta w}||_\infty = || \begin{pmatrix} U^{\sim} G_C B_1 + R_B^{\frac{1}{2}} Q B_1 \\ U_\perp^{\sim} G_C B_1 \end{pmatrix} ||_\infty \tag{C2}$$

Note that, in general, we have a 4–block general distance problem, which must be solved iteratively using the so called γ–iteration technique [31]. However, Example 6 corresponds to the simpler case of a square, right invertible system. In this case U is unitary and the term B_1 can be absorbed into Q. Hence, the problem can be explicitly solved by solving the following Nehari approximation problem:

$$\begin{aligned} \min_{Q \in \mathcal{RH}_\infty} ||T_{\zeta w}||_\infty &= \min_{Q \in \mathcal{RH}_\infty} ||U^{\sim} G_C B_1 + \frac{R_B^{\frac{1}{2}} Q B_1}{z}||_\infty \\ &= \min_{Q_B \in \mathcal{RH}_\infty} ||zR + Q_B||_\infty \end{aligned} \tag{C3}$$

where:

$$R \triangleq U^{\sim} G_C B_1; \; Q_B \triangleq R_B^{\frac{1}{2}} Q B_1$$

and where we used the fact that $\frac{1}{z}$ is an inner function. From eq. (B8) we have that R has the following state space realization:

$$R = \left(\begin{array}{c|c} {A'_F}^{-1} & X B_1 \\ \hline -R_B^{\frac{-1}{2}} B'_2 {A'_F}^{-1} & 0 \end{array} \right) \qquad (C4)$$

Finally, since the best stable approximation to a given function coincides with the best antistable approximation to its conjugate we have:

$$\min_{Q_B \in RH_\infty} \|T_{\zeta w}\|_\infty = \min_{Q_B \in RH_\infty^-} \|z^{-1} R^{\sim} + Q_B\|_\infty \qquad (C5)$$

Hence, to compute the norm of the best approximation, we need to calculate the maximum Hankel singular value of the stable part of $G \triangleq z^{-1} R^{\sim}$ [8]. Standard state–space manipulations yield:

$$R^{\sim} = \left(\begin{array}{c|c} A_F & B_2 R_B^{\frac{-1}{2}} \\ \hline B'_1 X A_F & B'_1 X B'_2 R_B^{\frac{-1}{2}} \end{array} \right) \; R^{\sim}\frac{1}{z} = \left(\begin{array}{c|c} A_F & B_2 R_B^{\frac{-1}{2}} \\ \hline B'_1 X & 0 \end{array} \right) \qquad (C6)$$

Hence $P_+[G] = G$. Substituting the values of A, F, B_1, B_2, C_1 and D_{12} and computing the observability and controllability grammians yields for the maximum Hankel singular value $\Gamma_H = 1.8543$ $\diamond$.

TECHNIQUES FOR ROBUST NONLINEAR LARGE SCALE SYSTEMS

Bor–Sen Chen

Department of Electrical Engineering
National Tsing–Hua University
Hsin–Chu, Taiwan, Republic of China

Wen–June Wang

Department of Electrical Engineering
National Central University
Chung–Li, Taiwan, Republic of China

I. INTRODUCTION

The demands of today's technology in the planning, design, and realization of sophisticated systems have become increasingly large in scope and complex in structure. It is therefore not surprising that over the past decades or more, many researchers have paid their attention to various problems that arise in connection with systems of this type, which are called "large–scale systems". Numerous examples of large–scale systems provide much challenges to system engineers, which include electric power systems, nuclear reactors, aerospace systems, large electric networks, economic systems, process control systems, chemical and petroleum industries, etc.

CONTROL AND DYNAMIC SYSTEMS, VOL. 55

Generally speaking, the published literature in the investigation of the stability and stabilization problems for large–scale systems can be divided into two categories: the time domain and the frequency domain. In the time domain, the problems have been widely studied by those papers (see [3, 4, 7 ~ 12, 15, 17, 18, 28, 45]). However, in the frequency domain, [7, 15, 16, 30, 37] have studied the stability of large–scale systems by the input–output (I/O) stability concepts. In practical systems, the plants are always nonlinear and/or time–varying. If only linear time–invariant models are considered, nonlinear unmodelled dynamics will occur and can be regarded as nonlinear perturbations. Hence, the robustness problem of the perturbed large–scale systems has also received attention in recent years (see [3, 9, 32, 35, 38]).

In this chapter, we consider the robustness problem in perturbed large–scale systems and give some new and simple results which are different from those papers above. In Section II, we first review some mathematical preliminaries which include some definitions and useful lemmas. In Section III, using the properties of M–matrix and the comparison theorem, we derive a simple time–domain criterion for the robust stability of large–scale systems with linear perturbations. In section IV, we consider a nonlinearly perturbed large–scale system whose interconnection matrix of each subsystem satisfies a certain matching condition. Two approaches are proposed to deal with the robust stability problem. One uses a local constant feedback to stabilize the whole system and finds the tolerable perturbation bound. The other uses the optimal control law to stabilize the whole system and finds the tolerable perturbation bound. In this section Lyapunov stability theorem and properties of matrix norm (or vector norm) are used to derive our results. In section V we consider the nonlinear perturbed large–scale systems without satisfying the matching condition. With the aid of Lyapunov stability theorem, robust stability criterion of the perturbed systems is derived.

From this theorem, the tolerable perturbation bound can be observed easily. It should be noted that from Section III to Section V, we discuss the robustness of perturbed large–scale systems in the time domain. Hence, Lyapunov stability theorem is the main tool for solving the problem. Then, Section VI and Section VII consider the problem from another point of view, in other words, the problem will be studied in the frequency domain.

All practical systems, when driven by sufficiently large signals, exhibit the the phenomenon of saturation due to limitations of the practical capabilities of their components. Many components such as amplifiers have an output proportional to their input for a limited range of input signals. When the input exceeds this range the output tends to become nearly constant as shown in Fig. 1 of Section VI. Some books and papers [26, 41, 42] have studied the system with saturating actuator. In general, these authors analyzed the stability of saturated systems by the circle criterion or the Popov's criterion [14, 22, 27]. In Section VI, we shall study the stability problems of large–scale systems with saturating actuators without using circle criterion or Popov's criterion. Here we use the concept of input–output stability and properties of M–matrix to derive the stability criterion. But this stability criterion must be checked by computer simulation. It is well known that circle and Popov criteria are useful tools for investigating the stability of a nonlinear single feedback system. However, the single–loop feedback systems sometimes may hardly meet certain design specifications. Hence, observer–controller compensators have been widely employed in control systems as they offer more freedom in system design [31, 36, 40]. Unfortunately, it is difficult to apply circle or Popov criteria to treat the nonlinear system with an observer–controller compensators. Consequently, in Section VII, based on input–output stability, M–matrix properties and factorization approach, we derive a new stability criterion for nonlinear perturbed large–scale systems with observer–controller compensators. Furthermore, a simple but useful inequality is derived from the stability criterion and is used in the robust compensator design.

II. MATHEMATICAL PRELIMINARIES

Nomenclature:

$\mathbb{R}^n$	Real vector space of dimension n.
$A^T(x^T)$	Transpose of the matrix A (or the vector x).
$\lambda_i(A)$	The i–th eigenvalue of the matrix A.
$\mathbb{R}^{n\times m}$	Real matrix space of dimension n × m.
$\lambda_m(A)$	minimal eigenvalue of the matrix A.
$\lambda_M(A)$	maximal eigenvalue of the matrix A.

Definition 2.1 [14]: For a vector $X=[x_1, ..., x_n]^T \in \mathbb{R}^n$, we have

$$\|X\|_\infty \triangleq \max_i |x_i|;\ \|X\|_1 \triangleq \sum_i |x_i|;\ \|X\|_2 \triangleq [\sum_i |x_i|^2]^{1/2} \tag{2.1}$$

The corresponding induced matrix norms $\|.\|_i$, i=1, 2, ∞ and matrix measures $\mu_i(.)$, i=1, 2, ∞ are defined below.

Definition 2.2 [14]: For a matrix $A \in \mathbb{R}^{n\times n}$

$$\|A\|_\infty \triangleq \max_i \sum_j |a_{ij}|;\ \|A\|_1 \triangleq \max_j \sum_i |a_{ij}|;\ \|A\|_2 \triangleq [\lambda_M(A^TA)]^{1/2} \tag{2.2}$$

and

$$\left.\begin{aligned} &\mu_\infty(A) \triangleq \max_i \{a_{ii} + \sum_{j\neq i} |a_{ij}|\};\ \mu_1(A) \triangleq \max_j \{a_{jj} + \sum_{j\neq i} |a_{ij}|\} \\ &\mu_2(A) \triangleq (\lambda_M\ [A^T+A])/2 \end{aligned}\right\} \tag{2.3}$$

The following lemma will be useful in the subsequent sections.

Lemma 2.3 [44]: For a matrix $A \in \mathbb{R}^{nxn}$ the following relation holds:

$$\|\exp(At)\|_i \leq \exp\{\mu_i(A)t\} \quad i=1, 2, \infty, t \geq 0 \tag{2.4}$$

Definition 2.4 [15, 18]: A matrix $H \in \mathbb{R}^{nxn}$ with non–positive off–diagonal elements h_{ij} (where $i \neq j$) is called an M–matrix if the leading principal minors of H are all positive.

Lemma 2.5 [15, 17]: Let H be an M–matrix, then the following statements are equivalent.

(a) The principal minors of H are all positive.

(b) The leading principal minors of H are all positive.

(c) The matrix H is non–singular and the elements of H^{-1} are all non–negative.

(d) There is a vector X (or Y) whose elements are all positive such that the elements of HX (or H^TY)are all positive.

(e) The real parts of the eigenvalues of H are all positive.

Lemma 2.6 [2]: A matrix $H=[h_{ij}] \in \mathbb{R}^{nxn}$ is called an M–matrix if $h_{ii}>0 \ \forall\, i$, $h_{ij} \leq 0$ whenever $i \neq j$ and

$$h_{ii} > \sum_{j \neq i}^{n} |h_{ij}| \quad i=1, 2, \ldots, n \tag{2.5}$$

Lemma 2.7 [43]:

For any matrix X and Y with appropriate dimensions, we have

$$X^TY+Y^TX \leq rX^TX+(1/r)Y^TY$$

for any positive constant r.

For discussing the robustness of large–scale systems in frequency domain, the following definitions and lemmas will be used in Section VI and Section VII.

Definition 2.8 [27]: The set of real measurable vector functions of the real variable t defined on $[0, \infty]$ is denoted by $\mathbb{R}^n(0, \infty)$, and

$$L_2^n(0, \infty)=\left\{ f(t)\,|\,f(t)\in\mathbb{R}^n(0, \infty),\ \int_0^\infty f^T(t)f(t)dt<\infty \right\} \tag{2.6}$$

Definition 2.9: Let $d\in(0, \infty)$ and define f_d by

$$f_d(t)=\begin{cases} f(t) & \text{for } t\in[0, d] \\ 0 & \text{for } t>d \end{cases} \tag{2.7}$$

for any $f(t) \in \mathbb{R}^n(0, \infty)$ and let $L_{2e}^n=\{f(t) \mid f(t) \in \mathbb{R}^n(0, \infty),\ f_d \in L_2^n(0, \infty)$ for $d \in (0, \infty)\}$ be an extended norm space. An operator (not necessary linear) $G: L_{2e}^n \to L_{2e}^n$ is causal if for an arbitrary $d>0$.

$$(Gf)_d=(Gf_d)_d, \text{ for all } f\in L_{2e}^n$$

Definition 2.10: The L_2 norm of $f(t)\in L_2^n$ is defined by

$$\|f(.)\|_2=\left[\int_0^\infty f^T(t)f(t)dt \right]^{1/2} \tag{2.8}$$

If G is a linear operator, Gf is defined by

$$Gf=\int_0^t g(t-z)f(z)dz, \quad \text{for } f\in L_{2e}^n \tag{2.9}$$

where g(t) denotes the impulse response of operator G and G(s) is a Laplace form of operator G.

Lemma 2.11 [27]: For a linear bounded operator G, there is a property

$$\|G\| = \|G(jw)\|_\infty \tag{2.10}$$

where $\|G(jw)\|_\infty = \sup_{w} \max_{i} [\lambda_i(G^*(jw)G(jw))]^{1/2}$, for $w \geq 0$, and

$$\|G\| \triangleq \sup_{f_d \neq 0} \frac{\|Gf_d\|}{\|f_d\|}, \ \forall f \in L_{2e}^{n}.$$

Note: The maximum singular value of the matrix G(jw) is denoted by $\|G(jw)\|_s$ or $\bar{\sigma}(G(jw))$, where

$$\|G(jw)\|_s = \max_{i} [\lambda_i(G^*(jw)G(jw))]^{1/2}$$

Hence, it is seen that $\|G(jw)\|_\infty = \sup_{w} \|G(jw)\|_s$ for $w \geq 0$.

Definition 2.12 [14, 16]: An operator G: $L_{2e}^{m} \to L_{2e}^{n}$ is said to be L_2–bounded if

(i) $Gf \in L_2^n$ whenever $f \in L_2^m$ and

(ii) there are nonegative a and b such that

$$\|(Gf)\| \leq a \|f\| + b \tag{2.11}$$

for any $f \in L_2^m$.

Lemma 2.13 [23]: For a stable matrix G(s), $\|G(s)\|_\infty < 1$ is equivalent to

$$I - G^*(s)G(s) > 0, \text{ for } s = jw, \ w\ [0, \infty). \tag{2.12}$$

III. STABILITY CRITERION OF LARGE SCALE SYSTEMS WITH LINEAR PERTURBATIONS

§3–1. Introduction

In this section robust stability criteria for the linearly perturbed large–scale system in the time domain will be derived. We use the properties of M–matrix and the concept of induced matrix norms and matrix measures to obtain the bound of linear perturbation in the large–scale system.

§3–2. Robust Stability Criterion

First, consider a linear differential large–scale system

$$\dot{x}_i(t) = -a_i x_i(t) + \sum_{j=1}^{J} b_{ij} x_j(t) \tag{3.1}$$

$i=1, 2, ..., J$, where a_i and b_{ij} are non–negative constants.

Araki and Mori [33] have given a sufficient condition for the stability of the above systems (3.1).

Lemma 3.1: The large–scale systems (3.1) is asymptotically stable if Q is an M–matrix, where

$$Q = [q_{ij}], \qquad q_{ij} = \begin{cases} a_i - b_{ii}, & i=j \\ -b_{ij}, & i \neq j \end{cases} \tag{3.2}$$

#

This lemma can be extended to the large–scale perturbed systems

$$\dot{x}_i(t) = (A_i + \Delta A_i)x_i + (B_i + \Delta B_i) \sum_{j \neq i}^{J} C_{ij} x_j(t), \quad i=1, ..., J \tag{3.3}$$

It is well known that the solution of (3.3) can be written as

$$x_i(t)=\exp(A_it)x_{io}+\int_0^t \exp(A_i(t-\tau)\times\left\{\Delta A_ix_i(\tau)+(B_i+\Delta B_i)\sum_{j\neq i}^{J} C_{ij}x_j(\tau)\right\}d\tau \tag{3.4}$$

where $x_i(0)=x_{io}$ is the initial condition.

Taking the norms (any norm) of both sides of (3.4)

$$\|x_i(t)\|\leq\|\exp(A_it)\|\ \|x_{io}\|+\int_0^t \|\exp(A_i(t-\tau)\|\times\left\{\|\Delta A_i\|\ \|x_i(\tau)\|+(\|B_i\|+\|\Delta B_i\|)\sum_{j\neq i}^{J}\|C_{ij}\|\times\|x_j(\tau)\|\right\}d\tau$$

Let $w_i(t)\triangleq\|x_i(t)\|$ and use Lemma 2.3, we have

$$w_i(t)\leq\exp[\mu(A_i)t]w_i(0)+\int_0^t \exp[\mu(A_i)(t-\tau)]\times\left\{\|\Delta A_i\|\ w_i(\tau)+\sum_{j\neq i}^{J}(\|B_i\|+\|\Delta B_i\|)\|C_{ij}\|\ w_j(\tau)\right\}d\tau$$

Similarly, we have $z_i(t)$ such that

$$z_i(t)=\exp[\mu(A_i)t]z_i(0)+\int_0^t \exp[\mu(A_i)(t-\tau)]\times\left\{\|\Delta A_i\|\ z_i(\tau)+\sum_{j\neq i}^{J}(\|B_i\|+\|\Delta B_i\|)\|C_{ij}\|\ z_j(\tau)\right\}d\tau \tag{3.5}$$

which is the solution of

$$\begin{aligned}\dot{z}_i(t)&=\mu(A_i)z_i(t)+\left\{\|\Delta A_i\|\ z_i(t)+\sum_{j\neq i}^{J}(\|B_i\|+\|\Delta B_i\|)\|C_{ij}\|\ z_j(t)\right\}\\&=(\mu(A_i)+\|\Delta A_i\|)z_i(t)+\left\{\sum_{j\neq i}^{J}(\|B_i\|+\|\Delta B_i\|)\|C_{ij}\|\right\}z_j(t)\end{aligned} \tag{3.6}$$

Since $z_i(t) \geq \|x_i(t)\|$ (from the comparison theorem of Mori et al [43]), the asymptotical stability of $z_i(t)$ implies that $x_i(t)$ is asymptotically stable. Hence, from (3.1) and Lemma 3.1, we have the following theorem.

Theorem 3.2: The large–scale perturbed system (3.3) is asymptotically stable if P is an M–matrix where

$$P=[p_{ij}]_{i,j=1,\cdots,n};\quad p_{ij}=\begin{cases} -(\mu(A_i)+\|\Delta A_i\|); & i=j \quad (3.7a) \\ -(\|B_i\|+\|\Delta B_i\|)\|C_{ij}\|; & i\neq j \quad (3.7b) \end{cases}$$

#

Remark 1: If P is an M–matrix, (3.7a) (i.e. diagonal elements of P) must be positive. It is easy to see that the large–scale perturbed system (3.3) is robustly stable where each isolated perturbed subsystem $\dot{x}_i(t)=(A_i+\Delta A_i)x_i$ must also be robustly stable.

Remark 2: By Lemma 2.5, if $\|\Delta A_i\| << -\mu(A_i)$ and/or $\|C_{ij}\| << 1$, (3.7) is more easily to be an M–matrix.

§3–3. An Example [40]:

Consider a large–scale system which consists of 3 perturbed subsystems S_1, S_2 and S_3, where

$$S_1:\ A_1=\begin{bmatrix} -8 & 5 \\ 0 & -1 \end{bmatrix},\ B_1=\begin{bmatrix} 1 \\ 0 \end{bmatrix},\ C_{12}^T=\begin{bmatrix} 0.15 \\ 0.05 \end{bmatrix},\ C_{13}^T=\begin{bmatrix} 0 \\ 0 \\ 0 \end{bmatrix}$$

$$S_2:\ A_2=\begin{bmatrix} -3 & -2 \\ 1 & -5 \end{bmatrix},\ B_2=\begin{bmatrix} 2 \\ 1 \end{bmatrix},\ C_{21}^T=\begin{bmatrix} 0.05 \\ 0.04 \end{bmatrix},\ C_{23}^T=\begin{bmatrix} 0.1 \\ 0 \\ 0.04 \end{bmatrix}$$

$$S_3:\ A_3=\begin{bmatrix} -2 & 0 & 0 \\ 0 & -4 & 1 \\ 0 & 0 & -3 \end{bmatrix},\ B_3=\begin{bmatrix} 2 \\ 0 \\ -1 \end{bmatrix},\ C_{31}^T=\begin{bmatrix} 0.07 \\ 0.06 \end{bmatrix},\ C_{32}^T=\begin{bmatrix} 0.2 \\ 0.4 \end{bmatrix}$$

Here we have $\mu_\infty(A_1)=-1$, $\mu_\infty(A_2)=-1$, and $\mu_\infty(A_3)=-2$; $\|B_1\|_\infty=1$, $\|B_2\|_\infty=2$, $\|B_3\|_\infty=2$; and $\|C_{12}\|_\infty=0.2$, $\|C_{13}\|_\infty=0$, $\|C_{21}\|_\infty=0.09$, $\|C_{23}\|_\infty=0.14$, $\|C_{31}\|_\infty=0.13$, and $\|C_{32}\|_\infty=0.6$. According to (3.7), the matrix P is established in the following:

$$P=\begin{bmatrix} -(-1+\|\Delta A\|_\infty) & -0.02(1+\|\Delta B_1\|_\infty) & 0 \\ -0.09(2+\|\Delta B_2\|_\infty) & -(-1+\|\Delta A_2\|_\infty) & -0.14(2+\|\Delta B_2\|_\infty) \\ -0.13(2+\|\Delta B_3\|_\infty) & -0.6(2+\|\Delta B_3\|_\infty) & -(-2+\|\Delta A_3\|_\infty) \end{bmatrix}$$

From Lemma 2.6, it is clear that if

$$0.8>\|\Delta A_1\|_\infty+0.2\|\Delta B_1\|_\infty;\ 0.54>\|\Delta A_2\|_\infty+0.23\|\Delta B_2\|_\infty;$$
$$0.54>\|\Delta A_3\|_\infty+0.73\|\Delta B_3\|_\infty$$

then P is an M–matrix and the large–scale perturbed system is robustly stable.

§3–4. Conclusion

With the aid of some properties of matrix norns and matrix measures as well as the M–matrix, simple criterion of the robust stability for linearly perturbed large–scale systems has been derived. The advantage of our approach is that we can easily obtain the perturbation bounds from the nominal system without solving Lyapunov equation. Furthermore, the perturbation matrix is not necessarily time–invariant.

IV. ROBUSTNESS OF NONLINEARLY PERTURBED LARGE SCALE SYSTEMS WITH LOCAL CONSTANT STATE FEEDBACK

§4–1. Introduction

In this section we desire to design the local state feedback to stabilize the nonlinear perturbed large–scale system and to develop the perturbation bound which can be tolerated robustly. Furthermore, these results are also applicable to the linearly perturbed large–scale system. All our developments are based on the concepts of *scalar Lyapunov stability theorem.*

§4–2. System Description and Problem Formulation

Let S be a large–scale system composed of J interconnected subsystems S_i, i = 1,2, ..., J. Each S_i is described by the equations

$$S_i: \quad \dot{x}_i(t)=A_i x_i(t)+B_i \sum_{j\neq i}^{J} C_{ij}x_j(t)+f_i(x_i)+B_i u_i(t); \tag{4.1}$$

where $A_i \in \mathbb{R}^{n_i \times n_i}$ is not necessary a *Hurwitz matrix*; the pair $\{A_i, B_i\}$ is controllable. B_i, C_{ij} are matrices with appropriate dimension and $f_i(x_i)$ is a nonlinear perturbation vector which is not usually available but only some bound on its function may be evaluated [2, 14].

$$\|f_i(x_i)\| \leq b_i\|x_i\| \tag{4.2}$$

Let

$$u_i(t)=k_i x_i(t), \ i=1, 2, ..., J \tag{4.3}$$

where k_i is either (a): the constant feedback gain matrix, or (b): the optimal feedback gain matrix, for the i–th isolated subsystem. Inserting (4.3) into (4.1) then the closed–loop subsystem (denoted by $\hat{S}_i$) becomes

$$\hat{S}_i: \quad \dot{x}_i(t)=(A_i+B_ik_i)x_i(t)+\sum_{j\neq i}^{J} B_iC_{ij}x_j(t)+f_i(x_i),\ \forall\, i \tag{4.4}$$

Our objective is to determine the allowable perturbation bound of $\hat{S}_i$ corresponding to the above two cases (a) and (b) respectively.

§4–3. Robustness of a Local Constant Feedback Design

The large–scale system consisting of J closed–loop subsystems $\hat{S}_i$ is denoted by $\hat{S}$. According to the results given by [4], we have the local constant state feedback gain

$$k_i = -\frac{h_i}{2} B_i^T \overline{P}_i; \quad i = 1, 2, ..., J \tag{4.5}$$

where h_i is an arbitrary scalar satisfying

$$h_i \geq 1 + \sum_{j\neq i}^{J} ||C_{ij}||_2^2, \quad i = 1, 2, ..., J \tag{4.6}$$

where $\overline{P}_i$ is a symmetric positive definite matrix satisfying the following *Riccati–equation*

$$A_i^T\overline{P}_i+\overline{P}_iA_i-\overline{P}_iB_iB_i^T\overline{P}_i+J\ I=0,\ \forall\, i \tag{4.7}$$

where I is an identity matrix with appropriate dimension. Hence the nominal large–scale system (without $f_i(x_i)$ for (4.4)) is stablilized by local constant feedback (4.5) and (4.6). Define a *scalar Lyapunov function*

$$v_i = x_i^T P_i x_i, \quad i = 1, 2, \ldots, J \tag{4.8}$$

where P_i is a solution of the following equation with a constant $L > J$.

$$A_i^T P_i + P_i A_i - P_i B_i B_i^T P_i + L\, I = 0, \ \forall\, i \tag{4.9}$$

Let the *Lyapunov function* V for the large scale system S be

$$V = \sum_{i=1}^{J} v_i\,, \qquad i.e.\ V = X^T P X \tag{4.10}$$

where $X=[x_1, \ldots, x_J]$ and P=block–diag$[P_1, \ldots, P_J]$. Hence, according to ***Lyapunov's*** method, we have the following result.

Theorem 4.1 : If the perturbation $f_i(x_i)$ of each S_i in (4.1) satisfies (4.2), and

$$b_i \le \frac{L - J}{2\ \lambda_M(P_i)}, \qquad i=1, 2, \ldots, J \tag{4.11}$$

Then the large–scale system $\hat{S}$ described as (4.4) is robustly stable with the local constant state feedback (4.5) where h_i satisfies (4.6).

Proof: Taking the time derivative $\dot{v}_i$ of the function v_i (4.8) with respect to (4.4), we obtain

$$\dot{v}_i = x_i^T[A_i^T P_i + P_i A_i + (B_i k_i)^T P_i + P_i B_i k_i] x_i + 2 f_i^T(x_i) P_i x_i$$

$$+ 2 x_i^T P_i \sum_{i \neq j}^{J} B_i C_{ij} x_j \tag{4.12}$$

According to [1]

$$f_i^T P_i x_i \leq \|f_i\|_2 \|P_i x_i\|_2 \leq b_i \|x_i\|_2 \|P_i x_i\|_2 \leq b_i \|x_i\|_2 \|P_i\|_2 \|x_i\|_2$$

$$= b_i \|P_i\|_2 \|x_i\|_2^2 = b_i x_i^T \lambda_M(P_i) x_i \tag{4.13}$$

Substituting (4.5) into (4.12) yields

$$\dot{V} = \sum_{i=1}^{J} \{x_i^T [A_i^T P_i + P_i A_i] x_i + x_i^T [(-0.5 h_i B_i B_i^T P_i)^T P_i$$

$$-0.5 h_i P_i B_i B_i^T P_i] x_i + \sum_j (B_i C_{ij} x_j)^T P_i x_i$$

$$+ x_i^T P_i (\sum_j B_i C_{ij} x_j) + 2 f_i^T (x_i) P_i x_i\} \tag{4.14}$$

Here, we let $C_{ij}=0$ as $i=j$. From (4.2) and (4.9), we have

$$\dot{V} \leq - \sum_i \{x_i^T [(h_i - 1) P_i B_i B_i^T P_i + L I - 2 b_i \lambda_M(P_i) I] x_i$$

$$-x_i^T P_i (\sum_j B_i C_{ij} x_j) - \sum_j (B_i C_{ij} x_j)^T P_i x_i\} \tag{4.15}$$

Under the assumptions of (4.6) and (4.11), then (4.15) becomes

$$\dot{V} \leq - \sum_i \{x_i^T [\sum_j \|C_{ij}\|_2^2 P_i B_i B_i^T P_i + LI - 2 b_i \lambda_M(P_i) I] x_i$$

$$-\,2\,\|B_i^TP_ix_i\|_2\sum_j\|C_{ij}\|_2\,\|x_j\|_2\}$$

$$\le -\sum_i\sum_j[\|C_{ij}\|_2^2\,\|B_i^TP_ix_i\|_2^2-2\|B_i^TP_ix_i\|_2\,\|C_{ij}\|_2\,\|x_j\|_2]-\sum_j J\,\|x_j\|_2^2$$

$$=-\sum_i\sum_j[\|C_{ij}\|_2^2\,\|B_i^TP_ix_i\|_2^2-2\,\|B_i^TP_ix_i\|_2\,\|C_{ij}\|_2\,\|x_j\|_2+\|x_j\|_2^2]$$

$$=-\sum_i\sum_j[\|C_{ij}\|_2\,\|B_i^TP_ix_i\|_2-\|x_j\|_2]^2\le 0$$

From this inequality, we conclude that the choice of local feedback gain k_i given by (4.5) guarantees that the perturbed large scale system $\hat{S}$ is stable if (4.11) holds. □

Remark 1: A review of the foregoing analysis, it is seen that time invariant of $f_i(x_i)$ is not essential. As long as a time varying function $f_i(x_i,t)$ satisfies (4.2), *Theorem* 4.1 is also applicable.

If we consider the linear perturbations ΔA_i of the system, then (4.1) becomes

$$\dot{x}_i(t)=(A_i+\Delta A_i)x_i+\sum_{j\neq i}^{J}B_iC_{ij}x_j(t)$$

where $\Delta A_i=[\Delta\alpha_i^{jk}]$ and $|\Delta\alpha_i^{jk}|\le\phi_{A_i}$, j, k=1, 2, ..., n_i. Using the same control law (4.5) and (4.6), we have the following results.

Corollary **4.2**: If

$$\phi_{A_i} \leq \frac{L - J}{2\, n_i \lambda_M(P_i)}, \quad i=1, 2, \ldots, J \tag{4.16}$$

then the large–scale system S with linear perturbations ΔA_i can be stabilized robustly by the local constant state feedback (4.5, 4.6). #

Proof: It is easily derived from the proof of Theorem 4.1.

Corollary 4.3: If the perturbation $f_i(x_i)$ of each S_i satisfies (4.2), and

$$b_i \leq \frac{L - \sum_{j=1}^{J} \delta_{ji}^2}{2 \lambda_M(P_i)}, \quad i= 1, 2, \ldots, J \tag{4.17}$$

and P_i is the solution of Ricatti equation (4.9). Then the large–scale system S (4.1) can be stabilized robustly by the local state feedback (4.5) and (4.6) in which

$$h_i \geq 1 + \sum_{j=1}^{J} \epsilon_{ij}^2 \, \|C_{ij}\|_2^2 \tag{4.18}$$

where $\delta_{ij}=1/\epsilon_{ij}$ as well as both δ_{ij} and ϵ_{ij} are arbitrary constants satisfying

$$\sum_j \delta_{ij}^2 < L \tag{4.19}$$

for any choice of positive constant L. #

Proof: From (4.15) and (4.18), it follows that

$$\dot{V} \leq -\sum_i \{x_i^T[\sum_j \epsilon_{ij}^2 ||C_{ij}||_2^2 P_i B_i B_i^T P_i + L\, I - 2\, b_i \lambda_M(P_i) I\,]\, x_i$$

$$-\, 2\, ||B_i^T P_i x_i||_2 \sum_j ||C_{ij}||_2\, ||x_j||_2\}$$

When (4.17) holds, then

$$\dot{V} \leq -\sum_i \{x_i^T[\sum_j \epsilon_{ij}^2 ||C_{ij}||_2^2 P_i B_i B_i^T P_i] x_i$$

$$-\, 2\, ||B_i^T P_i x_i||_2 \sum_j ||C_{ij}||_2\, ||x_j||_2\} - \sum_i x_i^T \sum_j \delta_{ji}^2 x_i$$

$$\leq -\sum_i \sum_j [\epsilon_{ij}^2 ||C_{ij}||_2^2 ||B_i^T P_i x_i||_2^2 - 2||B_i^T P_i x_i||_2 ||C_{ij}||_2 ||x_j||_2 + \delta_{ij}^2 ||x_j||_2^2]$$

$$= -\sum_i \sum_j [\epsilon_{ij} ||C_{ij}||_2 ||B_i^T P_i x_i||_2 - \delta_{ij} ||x_j||_2]^2 \leq 0 \tag{4.20}$$

where $\epsilon_{ij} \cdot \delta_{ij} = 1$ then the system is stable robustly. □

Corollary 4.4: For the same system as (4.1), the perturbation of each S_i satisfies (4.2), where

$$b_i \leq \frac{L - \sum_{j=1}^{J} \delta_{ji}^2 ||C_{ji}||_2^2}{2\lambda_M(P_i)}, \quad \forall i \tag{4.21}$$

Then the large scale system S can be stabilized robustly by the local state feedback (4.5) and (4.6) in which

$$h_i \geq 1 + \sum_{j=1}^{J} \epsilon_{ij}^2 \tag{4.22a}$$

where $\epsilon_{ij}=1/\delta_{ij}$, both ϵ_{ij} and δ_{ij} are arbitrary constants satisfying

$$L>\sum_{j}\delta_{ij}^2\|C_{ij}\|_2^2 \tag{4.22b}$$

for any choice of positive constant L. #

Proof: This proof is similar to that of Corollary 4.3, we omit it here. □

Remark 3: In Corollary 4.3 and Corollary 4.4, it is seen that L can be chosen to be smaller than J as long as (4.19) or (4.22b) holds.

Remark 4: From (4.17 and 4.18) or (4.21 and 4.22a) the allowable perturbation bound b_i depends on the choice of h_i. We will give more discussion by an illustrated example in the next subsection.

Remark 5: Theorem 4.1 is the special case of Corollary 4.3 with $\epsilon_{ij}=\delta_{ij}=1$.

§4–4. An Example [6]:

A perturbed large–scale system S is made of three subsystem

$$S_i\text{: } \dot{x}_i(t)=\begin{bmatrix} 0 & 1 \\ -4 & 4 \end{bmatrix} x_i(t)+\begin{bmatrix} 0 \\ 1 \end{bmatrix} \sum_{j\neq i} C_{ij}x_j(t)+B_iu_i(t)+f_i(x_i) \tag{4.23}$$

where $C_{12}=C_{13}=[0.1 \quad 0.1\ ]$, $C_{21}=C_{23}=[0.05 \quad 0.02\ ]$, $C_{31}=[0.1 \quad 0\ ]$, and $C_{32}=[0.2 \quad 0.1\ ]$

and $f_i(x_i)$ is the nonlinear perturbation satisfying (4.2) for all subsystems. Note that in this example each A_i is not a Hurwitz matrix. Firstly, choosing L= 5, (L > J = 3), the solution of (4.9) is given by

$$P_i = \begin{bmatrix} 37.575 & 0.583 \\ 0.583 & 8.708 \end{bmatrix} \tag{4.24}$$

where $\lambda_M(P_i)=37.587$, i=1, 2, 3. Let the stabilizing local feedback gain be

$$k_i = -\frac{h_i}{2}(0.583, 8.708), \quad i=1, 2, 3 \tag{4.25}$$

where $h_1 > 1.04$, $h_2 > 1.0058$ and $h_3 > 1.06$. Then by (4.6), the nominal system is stabilized. From Theorem 4.1, the allowable perturbation bound is given by

$$b_i \le \frac{L-J}{2\lambda_M(P_i)} = 0.0266 \tag{4.26}$$

i=1, 2, 3. Hence, it is seen from Theorem 4.1, the local feedback control (4.3) and (4.25) can stabilize the perturbed system S robustly as long as (4.26) holds. In order to illustrate the results of Corollaries, we make a comparison between Theorem 4.1 and Corollaries for the same problem in the Example. For simplification and clarity, we choose each element of ϵ_{ij} to be equal and then formulate the following table with L=5.

Table 1

	Theorem 4.1	Corollary 4.3	Corollary 4.4

$\forall\epsilon_{ij}=1/1.2$ $\forall\delta_{ij}=1.2$	$b_i\leq 0.0266$ $h_1\geq 1.04$ $h_2\geq 1.0058$ $h_3\geq 1.06$	$b_i\leq 0.0090$ $h_1\geq 1.0278$ $h_2\geq 1.0040$ $h_3\geq 1.0417$	$b_1\leq 0.0663$ $b_2\leq 0.0652$ $b_3\leq 0.0661$ $h_i\geq 3.08$
$\forall\epsilon_{ij}=1.25$ $\forall\delta_{ij}=0.8$	$b_i\leq 0.0266$ $h_1\geq 1.04$ $h_2\geq 1.0058$ $h_3\geq 1.06$	$b_i\leq 0.0410$ $h_1\geq 1.0625$ $h_2\geq 1.0091$ $h_3\geq 1.09375$	$b_1\leq 0.0664$ $b_2\leq 0.06592$ $b_3\leq 0.06632$ $h_i\geq 5.6875$
$\forall\epsilon_{ij}=2$ $\forall\delta_{ij}=0.5$	$b_i\leq 0.0266$ $h_1\geq 1.04$ $h_2\geq 1.0058$ $h_3\geq 1.06$	$b_i\leq 0.0565$ $h_1\geq 1.16$ $h_2\geq 1.0232$ $h_3\geq 1.24$	$b_1\leq 0.0665$ $b_2\leq 0.0663$ $b_3\leq 0.0653$ $h_i\geq 13$
$\forall\epsilon_{ij}=3$ $\forall\delta_{ij}=1/3$	$b_i\leq 0.0266$ $h_1\geq 1.04$ $h_2\geq 1.0058$ $h_3\geq 1.06$	$b_i\leq 0.0621$ $h_1\geq 1.36$ $h_2\geq 1.0522$ $h_3\geq 1.54$	$b_1\leq 0.0665$ $b_2\leq 0.0664$ $b_3\leq 0.0665$ $h_i\geq 28$

Remark 6: The table illustrates the case described in Remark 4.

Remark 7: It is seen from Table 1, for the same L, a certain relation depending on the choice of ϵ_{ij} (or δ_{ij}) exists between the local feedback gain k_i (or h_i) and permissible perturbation bound b_i.

Roughly speaking, tolerable perturbation is increasing with increasing local feedback gain k_i.

§4–5. Robustness of a Local Optimal Control Design

In this section, the same perturbed large–scale system (4.4) is considered. Assume the performance index for each isolated and nominal subsystem (without C_{ij} and $f_i(x_i)$) to be minimized is

$$J_i=\int_0^{\infty}[x_i^T(t)Q_ix_i(t)+u_i^T(t)R_iu_i(t)]dt,\ \forall\, i \tag{4.27}$$

where the weighting matrices Q_i and R_i are assumed to be positive definite matrices. Thus, the optimal control law corresponding to the performance index (4.27) for the isolated nominal subsystem model is

$$u_i(t)=-R_i^{-1}B_i^TP_ix_i(t); \quad i=1, 2, \ldots, J \tag{4.28}$$

where P_i is symmetric positive definite solution of *Riccati equation*

$$A_i^TP_i+P_iA_i-P_i(B_iR_i^{-1}B_i^T)P_i+Q_i=0 \tag{4.29}$$

Now, the question is that if we use the local optimal control law (4.28) to stabilize the whole perturbed large–scale systems, how large is the tolerable perturbation bound and what is the relationship between allowable perturbation and weighting matrices in the performance index (4.27) for each subsystem ? For simplicity, let the weighting matrices Q_i and R_i have the following forms

$$\left.\begin{aligned} Q_i &= \text{diag}\,[q_i] = q_iI \\ R_i &= \text{diag}\,[r_i] = r_iI \end{aligned}\right\} i=1, 2, \ldots, J \tag{4.30}$$

where q_i is selected as $q_i > M$ and

$$M = \sum_{i=1}^{J} r_i \sum_{j \neq i}^{J} \|C_{ij}\|_2^2 \tag{4.31}$$

for the chosen r_i, $i=1, 2, ..., J$. Using the *Lyapunov function* (4.10), we have the following theorem.

Theorem 4.5: If the perturbation $f_i(x_i)$ of each S_i in (4.1) satisfies (4.2), and

$$b_i \leq \frac{q_i - M}{2\lambda_M(P_i)}, \ i, j=1, 2, ..., J \tag{4.32}$$

where P_i is solved from (4.29). Then the large scale S can be stabilized robustly by the local optimal control law (4.28).   #

Proof: With the aid of (4.28), (4.29) and (4.13), (4.12) becomes

$$\dot{V} \leq \sum_i \{ x_i^T (P_i B_i R_i^{-1} B_i^T P_i - Q_i - 2 P_i B_i R_i^{-1} B_i^T P_i) x_i$$

$$+ \sum_j (B_i C_{ij} x_j)^T P_i x_i + x_i^T P_i \sum_j B_i C_{ij} x_j + 2 f_i^T(x_i) P_i x_i \}$$

$$\leq - \sum_i \{ x_i^T [P_i B_i R_i^{-1} B_i^T P_i + Q_i - 2 b_i \lambda_M(P_i) I] x_i$$

$$- x_i^T P_i \sum_j B_i C_{ij} x_j - \sum_j (B_j C_{ij} x_j)^T P_i x_i \} \tag{4.33}$$

Choose ϵ_i such that

$$\epsilon_i = [r_i \sum_j \|C_{ij}\|_2^2]^{-1}, \text{ i.e., } R_i^{-1} = \epsilon_i \sum_j \|C_{ij}\|_2^2 I \tag{4.34}$$

Substituting (4.34) and (4.32) into (4.33) yields

$$\dot{V} \leq -\sum_i \{x_i^T(\sum_j P_i B_i \epsilon_i \|C_{ij}\|_2^2 B_i^T P_i + M)x_i - 2x_i^T P_i \sum_j B_i C_{ij} x_j\}$$

With (4.31) and (4.34), it follows that $M = \sum_i \epsilon_i^{-1}$, then

$$\dot{V} \leq -\sum_i \{x_i^T(\sum_j P_i B_i \epsilon_i \|C_{ij}\|_2^2 B_i^T P_i + \sum_{j \neq i}^{J} \epsilon_j^{-1})x_i - 2x_i^T P_i \sum_j B_i C_{ij} x_j\}$$

$$\leq -\sum_i \sum_j [\epsilon_i \|C_{ij}\|_2^2 \|B_i^T P_i x_i\|_2^2 - 2\|B_i^T P_i x_i\|_2 \|C_{ij}\|_2 \|x_j\|_2 + \epsilon_i^{-1} \|x_j\|_2^2]$$

$$= -\sum_i \sum_j [\epsilon_i^{1/2} \|C_{ij}\|_2 \|B_i^T P_i x_i\|_2 - \epsilon_i^{-1/2} \|x_j\|_2]^2 \leq 0$$

From this inequality, we conclude that optimal feedback gain k_i given by (4.28) guarantees perturbed large scale system $\hat{S}$ to be stable if (4.32) holds. □

Remark 8 : For the given r_i(or R_i) and q_i(or Q_i), (4.32) also presents the relation between perturbation bound and the weighting matrices.

Example 4.6 [5]:

A perturbed large–scale system S is made of three subsystems

$$S_i:\ \dot{x}_1(t)=\begin{bmatrix} 0 & 1 \\ -2 & 3 \end{bmatrix} x_1(t)+\begin{bmatrix} 1 \\ 0 \end{bmatrix}\Big(\sum_{j\neq 1} C_{1j}x_j(t)+u_1(t)\Big)+f_1(x_1) \qquad (4.35a)$$

$$\dot{x}_2(t)=\begin{bmatrix} 0 & 1 \\ -3 & 3 \end{bmatrix} x_2(t)+\begin{bmatrix} 0 \\ 1 \end{bmatrix}\Big(\sum_{j\neq 2} C_{2j}x_j(t)+u_2(t)\Big)+f_2(x_2) \qquad (4.35b)$$

$$\dot{x}_3(t)=\begin{bmatrix} 0 & 1 \\ -6 & 5.5 \end{bmatrix} x_3(t)+\begin{bmatrix} 1 \\ 0 \end{bmatrix}\Big(\sum_{j\neq 3} C_{3j}x_j(t)+u_3(t)\Big)+f_3(x_3) \qquad (4.35c)$$

where $C_{12} = [\ 0.1 \quad 0.1\ ],\ C_{13} = [\ 0.1 \quad 0.1\ ]$

$C_{21} = [\ 0.2 \quad 0.1\ ],\ C_{23} = [\ 0.03 \quad 0.05\ ]$

$C_{31} = [\ 0.1 \quad 0\ ],\ C_{32} = [\ 0.15 \quad 0.1\ ]$

and $f_i(x_i)$ is the nonlinear perturbation satisfying (4.2) for all subsystems. Now, for performance index (4.27) if we choose $Q_i=3I$ and $R_i = 1$ which satisfy (4.30), since

$$M = \sum_{i=1}^{J} r_i \sum_{j\neq i}^{J} \|C_{ij}\|_2^2=0.1284 \qquad (4.36)$$

Solving (4.29) yields

$$P_1 = \begin{bmatrix} 7.5951 & -13.671 \\ -13.671 & 35.208 \end{bmatrix}, \qquad P_2 = \begin{bmatrix} 21.4555 & 0.4641 \\ 0.4641 & 6.5965 \end{bmatrix}$$

$$P_3 = \begin{bmatrix} 12.7037 & -13.199 \\ -13.199 & 17.964 \end{bmatrix} \qquad (4.37)$$

where $\lambda_M(P_1) = 40.832$, $\lambda_M(P_2) = 21.470$, and $\lambda_M(P_3) = 28.792$. Thus, if we use the local optimal control law (4.28) to stabilize the perturbed system S, where

$$u_1(t)=-[7.5951 \quad -13.6714]x_1(t), \quad u_2(t)=-[0.4641 \quad 6.5956]x_2(t)$$
$$u_3(t)=-[12.704 \quad -13.4862]x_3(t) \tag{4.38}$$

and the allowable perturbation bounds given by (4.32) are

$$b_1 \leq 0.0352, \quad b_2 \leq =0.0669, \quad b_3 \leq =0.0499$$

§4–6. Conclusion

By using the local constant state feedback control law and the local optimal control method, we have derived a sufficient condition for the stabilization of large–scale system with nonlinear perturbations. Since these results have the forms of algebraic criteria, we can estimate the perturbation bounds in advance when we design a system.

V. ROBUSTNESS OF LINEAR QUADRATIC LOCAL STATE FEEDBACK DESIGN IN THE PERTURBED LARGE–SCALE SYSTEMS

§5–1. Introduction

It is seen that the interconnection terms of the considered large–scale systems in Section III and Section IV satisfy the so–called "matching condition", i.e. $\sum_{j\neq i} B_i C_{ij} x_j$. Here we try to study the robustness of the nonlinear perturbed large–scale system S^P which is composed of N subsystems S_i^P as follows

$$S_i^P: \quad \dot{x}_i(t) = A_i x_i(t) + B_i u_i(t) + \sum_{j\neq i}^{N} C_{ij} x_j(t) + f_i(x_i(t)) \tag{5.1}$$

where $f_i(x_i(t))$ is a nonlinear perturbation satisfying (4.2) and (A_i, B_i), $\forall$ i, are all controllable. It is noted that the interconnection term need not satisfy the "matching condition" $C_{ij}=B_i\hat{C}_{ij}$. Assume the performance index for each isolated and nominal subsystem (5.1) (without $\sum_{j\neq i} C_{ij}x_j$ and $f_i(x_i)$) to be minimized is as (4.27). Thus, the optimal control law corresponding to the performance index (4.27) for the isolated nominal subsystem model (5.1) is as (4.28). Now, our interest is to find the condition under which the local optimal control law (4.28) can stabilize the whole perturbed large–scale systems S^P.

§5–2. **Main Result**

The closed–loop perturbed large–scale system is denoted by $\hat{S}^P$ and subsystem $\hat{S}_i^P$ is described as follows

$$\hat{S}_i^P: \quad \dot{x}_i(t)=(A_i-B_iR_i^{-1}B_i^TP_i)x_i(t)+\sum_{j\neq i}^{N} C_{ij}x_j+f_i(x_i),\ i=1,\ ...,\ N, \tag{5.2}$$

We have the following theorem.

Theorem 5.1: Consider the system (5.1) if there exist a_i, i=1, ... N, such that

$$\lambda_m(D_i)\geq\frac{N-1}{a_i}+\sum_{j\neq i} a_j\|P_jC_{ji}\|_2^2+2b_i\lambda_M(P_i); \tag{5.3}$$

where

$$D_i=P_iB_iR_i^{-1}B_i^TP_i+Q_i \tag{5.4}$$

Then the perturbed system S^P can be stabilized robustly by (4.28). #

Proof: Using the same *Lyapunov function* V as (4.10) and with the aid of (4.2), (4.13) and (5.2), we have

$$\dot{V}\leq\sum_i\{x_i^T[-P_iB_iR_i^{-1}B_i^TP_i-Q_i+2b_i\lambda_M(P_i)I]x_i+2\sum_{j\neq i}(C_{ij}x_j)^TP_ix_i\} \quad (5.5)$$

Let $\bar{x}_{ij}=P_iC_{ij}x_j$, and by (5.4) and Lemma 2.7 then

$$\begin{aligned}\dot{V}&\leq\sum_i[x_i^T(-D_i+2b_i\lambda_M(P_i))x_i+2\sum_{j\neq i}(C_{ij}x_j)^TP_ix_i]\\&=\sum_i[-x_i^TD_ix_i+2b_ix_i^T\lambda_M(P_i))x_i+\sum_{j\neq i}(\frac{1}{a_i}x_i^Tx_i+a_i\bar{x}_{ij}^T\bar{x}_{ij})]\\&\leq\sum_i[x_i^T(-\lambda_m(D_i)+2b_i\lambda_M(P_i)+\frac{N-1}{a_i}+\sum_{j\neq i}a_j\|P_jC_{ji}\|_2^2)x_i]\end{aligned}$$

From this inequality, we conclude that (5.3) guarantees $\dot{V}\leq 0$, i.e., the closed–loop large–scale system $\hat{S}^P$ is robustly stable via linear quadratic local state feedback control law (4.28). □

Remark 1: From Theorem 5.1, for a given pair $\{Q_i, R_i\}$ if there exist a_i, i=1, ..., N, such that (5.3) holds and the perturbation satisfies

$$b_i\leq\{\lambda_m(D_i)-\frac{N-1}{a_i}-\sum_{j\neq i}a_j\|P_jC_{ji}\|_2^2\}/2\lambda_M(P_i),\ \forall\ i \quad (5.6)$$

then we know the control law (4.28) can stabilize the S^P. For convenience and without loss of generality, a_i, in (5.3) and (5.6) can be assigned to be 1, ∀ i.

Remark 2: Since $b_i \geq 0$, by (5.6) it need have such a_i, for all i, such that (5.3) holds. If such a_i do not exist to satisfy (5.3) then we do not guarantee the stability of nominal large–scale system, it goes without saying the perturbed large–scale system. Hence, with the same control law (4.28), the tolerable perturbation (5.6) may be different if we choose a different set $\{a_i\}$, i=1, ..., N.

§5–4. Example

Consider the same system of Example 4.6 except that the matching condition is not needed in the interconnection matrices as below

$$C_{12} = \begin{bmatrix} 0.030 & 0.000 \\ 0.000 & 0.020 \end{bmatrix}, \; C_{13} = \begin{bmatrix} 0.080 & 0.000 \\ 0.030 & 0.000 \end{bmatrix}$$

$$C_{21} = \begin{bmatrix} 0.040 & 0.010 \\ 0.020 & 0.000 \end{bmatrix}, \; C_{23} = \begin{bmatrix} 0.010 & 0.000 \\ 0.000 & 0.020 \end{bmatrix}$$

$$C_{31} = \begin{bmatrix} 0.020 & 0.000 \\ 0.010 & 0.000 \end{bmatrix}, \; C_{32} = \begin{bmatrix} 0.000 & 0.020 \\ 0.010 & 0.010 \end{bmatrix}$$

and the nonlinear perturbation $f_i(x_i)$ satisfying (4.2) for all subsystems. If we give $Q_i = \begin{bmatrix} 6 & 0 \\ 0 & 6 \end{bmatrix}$ and $R_i = 1$ to solve (4.29), which yields

$$P_1 = \begin{bmatrix} 8.396 & -16.121 \\ -16.121 & 47.687 \end{bmatrix}, \; P_2 = \begin{bmatrix} 24.849 & 0.873 \\ 0.873 & 7.092 \end{bmatrix}$$

$$P_3 = \begin{bmatrix} 13.617 & -14.953 \\ -14.953 & 22.499 \end{bmatrix} \tag{5.8}$$

where $\lambda_M(P_1) = 53.455$, $\lambda_M(P_2) = 24.892$, and $\lambda_M(P_3) = 33.653$.

Then

$$\lambda_m(D_1)=\lambda_m(D_2)=\lambda_m(D_3)=6 \tag{5.9}$$

The control laws are derived from (4.28)

$$u_1(t)=-[\,8.396 \quad -16.121]\,x_1(t),\quad u_2(t)=-[\,0.873 \quad 7.092]\,x_2(t)$$

and

$$u_3(t)=-[13.617 \quad -14.953]\,x_3(t) \tag{5.10}$$

Hence by Theorem 5.1 with $a_i=1$ the tolerable perturbation bounds are $b_1\leq 0.0268$, $b_2\leq 0.0523$, and $b_3\leq 0.0577$ respectively. Suppose $a_1 = 1.4$, $a_2 = 1.45$, and $a_3 = 1.3$ are chosen, then we have the larger tolerable perturbation $b_1 \leq 0.0274$, $b_2 \leq 0.0537$, and $b_3 \leq 0.0638$. That is to say, it is possible, the same control law may tolerate different perturbation by our different choice of a_i, $i=1, \ldots, N$.

§5–5. Conclusion

Section V has considered the robust stabilization of the unmatched perturbed large–scale system. The sufficient conditions for the stability of local optimal state feedback has been derived and the corresponding tolerable perturbation bound has been obtained. According to (5.3) and (5.6), it can be seen that the strength of interconnections can not be too strong otherwise it is difficult to find a set a_i, $i=1, 2, \ldots, J$, such that Theorem 5.1 holds.

VI. STABILITY OF LARGE–SCALE SYSTEMS WITH SATURATING ACTUATORS

§6–1. Introduction

In the previous sections, we have considered the robust stabilization of large–scale system in the time domain. This chapter is concerned with the stability problem of the large–scale system in the frequency domain. Each subsystem contains a saturating actuator, a linear time invariant plant and a linear time invariant controller. A sufficient condition of stability is given for the system with saturating actuators and an algorithm for synthesizing the stabilizing controllers is also proposed.

§6–2. Preliminaries and System Description

Assume that we have reviewed from Definition 2.8 to Lemma 2.13 of Section II then we continue to study the following.

Definition 6.1 [22]: A continuous nonlinear memoryless causal map N: $L_2^m \to L_2^m$ is said to be inside the sector (s_1, s_2) if N satisfies the following properties:

(a) $Nx(t) = N(x(t)),\ x(t) \in \mathbb{R}^m$

(b) $N(0) = 0$ (6.1)

(c) $\left\| (N(x))_d - \frac{s_1+s_2}{2} x_d \right\| \leq \frac{s_1-s_2}{2} \|x_d\|$ for all $x \in \mathbb{R}^m$, $d \in [0, \infty)$.

From the above definitions, it is clear that

$$\|L_s u - 0.5u\| = \|sat(u) - 0.5u\| \leq 0.5\|u\| \tag{6.2}$$

So, $L_s u$ is inside the sector (0, 1), where sat(u) is a saturating actuator shown in Fig.1 and will be defined later.

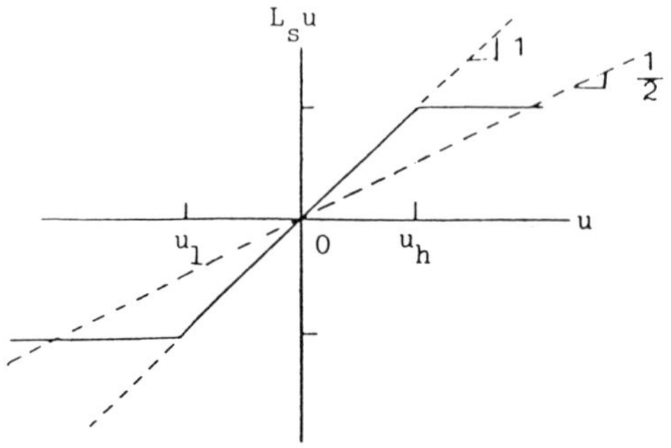

Figure 1. A saturating actuator.

Definition 6.2 [25]: The order of a rational matrix $G_i(s)$ is denoted by $O(G_i(s))=g_i$ if: (a) no entry in $G_i(s)$ grows faster than s^{g_i} as $s \to \infty$ and (b) at least one entry grows exactly like s^{g_i}, where $G_i(s)$ is the Laplace form of the operator G_i.

Now, Consider a large scale system LSS made up of J subsystems; each subsystem contains a saturating actuator and is described as follows (shown in Fig.2)

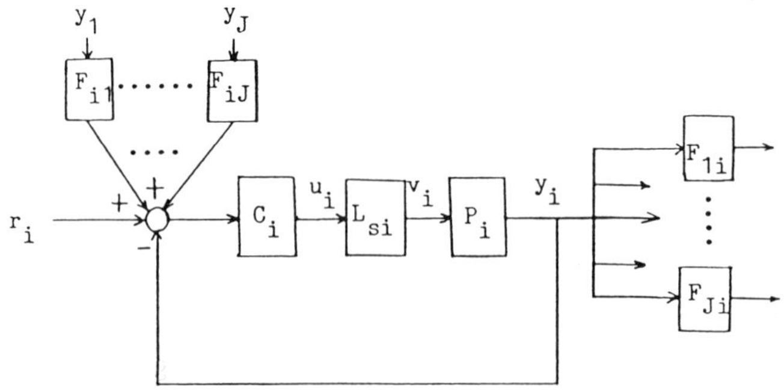

Figure 2. A subsystem of the LSS.

$$\left.\begin{aligned} &u_i=C_i(r_i+\sum_{j\neq i}^{J} F_{ij}y_j-y_i) \\ &v_i=L_{si}u_i\,, \quad i=1,\ 2,\ \ldots,\ J \end{aligned}\right\} \tag{6.3a}$$

$$\left.\begin{aligned} &y_i=p_i v_i \\ &L_{si}u_i=\mathrm{sat}(u_i)=\begin{bmatrix}\mathrm{sat}(u_{i1})\\ \mathrm{sat}(u_{i2})\\ \vdots \\ \mathrm{sat}(u_{im})\end{bmatrix} \\ &\mathrm{sat}(u_{ij})=\begin{cases} \bar{u}_{ij} & \text{if } u_{ij}>\bar{u}_{ij} \\ u_{ij} & \text{if } \underline{u}_{ij}\leq u_{ij}\leq \bar{u}_{ij} \\ \underline{u}_{ij} & \text{if } u_{ij}<\underline{u}_{ij} \end{cases} \end{aligned}\right\} \tag{6.3b}$$

where j=1, 2, ..., m, $\underline{u}_{ij}\leq 0\leq \bar{u}_{ij}$. Furthermore, the ith subsystem–plant P_i: $L_{2e}^{m_i}\to L_{2e}^{n_i}$, the controller C_i: $L_{2e}^{n_i}\to L_{2e}^{m_i}$, and interconnection F_{ij}: $L_2^{n_i}\to L_2^{n_i}$, are all linear time invariant causal maps as well as L_{si} is a saturating actuator in the ith subsystem. The main purpose is to derive the sufficient conditions under which the LSS is L_2–bounded. Before considering the main problem, we need more definitions. Let us define

$$\begin{aligned} &U^T=[\|u_1\|,\ \ldots,\ \|u_J\|], && Y^T=[\|y_1\|,\ \ldots,\ \|y_J\|] \\ &E^T=[\|e_1\|,\ \ldots,\ \|e_J\|], && R^T=[\|r_1\|,\ \ldots,\ \|r_J\|] \end{aligned} \tag{6.4}$$

Definition 6–3 [7]: An LSS specified as (6.3) is said L_2–bounded (I/O –stable) if and only if there are nonnegative constants k_s, t_s, h_s, $\hat{k}_s$, $\hat{t}_s$ and $\hat{h}_s$ such that $\|U\|_2^2\leq k_s\|R\|_2^2+\hat{k}_s$, $\|E\|_2^2\leq t_s\|R\|_2^2+\hat{t}_s$, and $\|Y\|_2^2\leq h_s\|R\|_2^2+\hat{h}_s$ for any $r_i\in L_2^{n_i}$, i=1, ..., J.

§6–3. Main Derivations and Results

First, we consider an isolated (without interconnections F_{ij}) linearized subsystem as shown in Fig.3 and define

$$S_i(s)=(I+0.5P_i(s)C_i(s))^{-1}$$

to be a sensitivity matrix of the ith linearized subsystem. $S_i(s)$ is said to be internally stable (or realizable) if some choice of controller $C_i(s)$ makes the closed–loop of the isolated linearized subsystem internally stable, i.e. there is free of unstable hidden mode between controller and plant [21]. Let the plant be factorized as follows

$$0.5P_i(s)=A_i^{-1}(s)B_i(s)=\hat{B}_i(s)\hat{A}_i^{-1}(s) \tag{6.5}$$

where the pairs $(A_i(s), B_i(s))$ and $(\hat{A}_i(s), \hat{B}_i(s))$ respectively constitute any left and right coprime stable rational decompositions of $0.5P_i(s)$.

Lemma 6.4 [21]: The closed loop of the ith isolated linearized subsystem of Fig.3 is internally stable, if and only if

$$S_i(s)=X_i(s)A_i(s), \tag{6.6a}$$

and

$$I-S_i(s)=\hat{B}_i(s)Y_i(s) \tag{6.6b}$$

where both $X_i(s)$ and $Y_i(s)$ are stable rational matrices. #

Lemma 6.5: If the ith isolated linearized subsystem is internally stable then $S_i(s)P_i(s)$, $C_i(s)S_i(s)$ and $C_i(s)S_i(s)P_i(s)$ are all stable. #
Proof: The proof is given in Appendix A.

Remark 1: If $S_i(s)$ is internally stable then the controller can be obtained by

$$C_i(s)=2P_i^{-1}(s)(I-S_i(s))S_i^{-1}(s) \tag{6.7}$$

without worrying about any unstable hidden mode between $C_i(s)$ and $P_i(s)$.

Lemma 6.6: Consider the isolated linearized subsystem shown in Fig.3. Suppose $\lim_{s\to\infty} S_i(s)$ is a nonsingular constant matrix and $P_i(s)$ is also nonsingular then controller $C_i(s)$ is guaranteed to be proper if

$$O(I-S_i(s))+O(P_i^{-1}(s))\leq 0 \tag{6.8}$$

#

Proof: The proof is given in Appendix B.

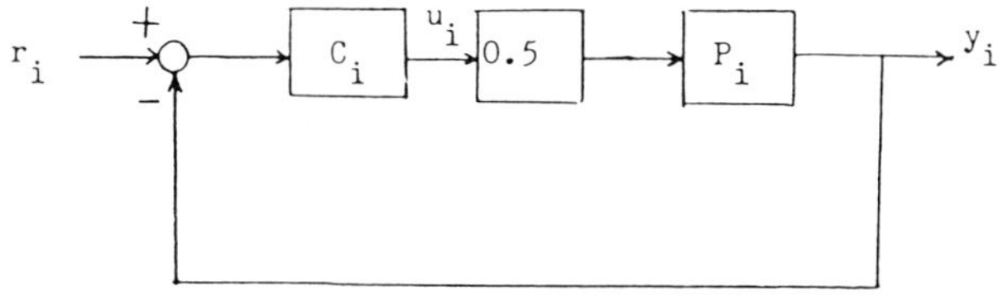

Figure 3. Isolated linearized subsystem.

In the LSS described by (6.3), the control signal of the ith subsystem is given by

$$u_i=C_i(r_i-P_iL_{si}u_i+\sum_{\substack{i=1\\ j\neq i}}^{J} F_{ij}y_j) \tag{6.9a}$$

$$=C_i(r_i-P_iL_{si}u_i+0.5P_iu_i-0.5P_iu_i+\sum_{\substack{i=1\\ j\neq i}}^{J}F_{ij}y_j)$$

Since $S_i=(I+0.5P_iC_i)^{-1}$ then

$$u_i=C_iS_ir_i-C_iS_iP_i(L_{si}-0.5)u_i+C_iS_i\sum_{\substack{i=1\\ j\neq i}}^{J}F_{ij}y_j \tag{6.9b}$$

The next theorem given the stability criterion of the LSS.

Theorem 6.7: An LSS described as (6.3a, b) is considered. Then LSS will be L_2–bounded if the following conditions are satisfied $\forall\ r_i\in L_2^{n_i}$, $i=1, 2, ..., J$.

(i) $\|F_{ij}(jw)\|_2\leq f_{ij}$ is sufficiently small $\forall\ j\neq i$ (6.10a)

(ii) $0.5\|C_i(s)S_i(s)P_i(s)\|_\infty<1$, for $s=jw$, $w\in[0,\infty)$ (6.10b)

where $S_i(s)=(I+0.5P_i(s)C_i(s))^{-1}$ is realizable, and $C_i(s)$ is proper.

(iii) Q is an M–matrix (6.10c)

where

$$Q=[q_{ij}],\ q_{ij}=\begin{cases}-d_{ii}f_{ij} & \text{for } i\neq j\\ 1 & \text{for } i=j\end{cases} \tag{6.11}$$

and

$$d_{ii}=\|I-S_i(s)\|_\infty+0.5z_i\|S_i(s)P_i(s)\|_\infty\ \|C_i(s)S_i(s)\|_\infty \tag{6.12}$$

$$z_i=(1-0.5\|C_i(s)S_i(s)P_i(s)\|_\infty)^{-1};\ s=jw \text{ and } i, j=1, ..., J \tag{6.13}$$

Proof: The proof is given in Appendix C. #

Remark 2: (a) Since $S_i(s)$ is realizable, from Lemma 6.4, $S_i(s)P_i(s)$, $I-S_i(s)$, $C_i(s)S_i(s)$ and $C_i(s)S_i(s)P_i(s)$ of (6.11–6.13) are all free of poles in the closed right half plane, i.e., their values of norm are all bounded.

(b) If the interconnections F_{ij} are nonlinear such that

$$\|F_{ij}(y_i)\| \le \hat{f}_{ij}\|y_j\|$$

and $\hat{f}_{ij}$ is sufficiently small, the above theorem is still true as long as f_{ij} is replaced by $\hat{f}_{ij}$, $\forall$ i, j in (6.11).

(c) In the above analysis, the nonlinearies saturation L_{si} is considered inside the sector (0, 1); we will end up with very conservative results in this case. For the SISO subsystems with stable plants, $0.5\|C_i(s)S_i(s)P_i(s)\|_\infty = \|I-S_i(s)\|_\infty$ which can be arbitrarily small for some choices of controllers $C_i(s)$ [29]; hence, L_{si} can be treated as inside the sector (0,1) and (6.10b) is always solvable. However, in this section some subsystems may be MIMO or/and have unstable plants, then according to "Nivanlinna–Pick theorem" [20], (6.10b) is not guaranteed to be solvable. In order to let (6.10b) be more solvable and to relax the conservativeness of the results, we assume that only finite part of the nonlinearity is used in actual system operation, i.e., the operation range of L_{si} is assumed to be inside the sector $(a_i, 1)$ (see Fig.4) instead of (0, 1). In this case, we can obtain

$$\|\text{sat } u_i - 0.5(1+a_i)u_i\| \le 0.5(1-a_i)\|u_i\| \quad \text{for } 0 \le a_i \le 1 \tag{6.14}$$

Hence, Theorem 6.7 can be modified as follows:

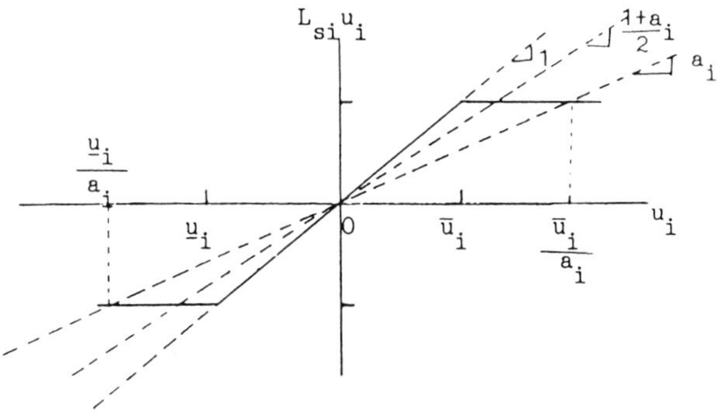

Figure 4. Saturating non-linearity only operating inside the sector $(a_i, 1)$.

Theorem 6.8 (Modification of Theorem 6.7):
The same LSS of Theorem 6.7 will be L_2–bounded if the following conditions are satisfied $\forall\ r_i \in L_2$ for $i=1, \ldots, J$.
(i) $\|F_{ij}(jw)\|_2 \leq f_{ij}$ is sufficiently small $\forall\ j \neq i$.

(ii) $0.5(1-a_i)\|C_i(s)S_i(s)P_i(s)\|_\infty < 1$ for $s=jw$, $w>0$. (6.15)

where $S_i(s)=(I+0.5(1+a_i)P_i(s)C_i(s))^{-1}$ is realizable and $C_i(s)$ is proper.
(iii) Q is an M–matrix where

$$Q=[q_{ij}],\ q_{ij}=\begin{cases} -d_{ii}f_{ij} & \text{for } i \neq j \\ 1 & \text{for } i=j \end{cases} \tag{6.16}$$

and

$$d_{ii}=\|I-S_i(s)\|_\infty+0.5(1-a_i)z_i\|S_i(s)P_i(s)\|_\infty\|C_i(s)S_i(s)\|_\infty \tag{6.17}$$

$$z_i=(1-0.5(1-a_i)\|C_i(s)S_i(s)P_i(s)\|_\infty)^{-1} \quad \text{for } s=jw \tag{6.18}$$

Proof: Similar to the proof of Theorem 6.7. #
It is clearly that Theorem 6.8 is more flexible than Theorem 6.7.

Remark 3: Under the assumption that the operation range of actuator L_{si} is always inside the sector $(a_i, 1)$, i.e., if control signal $u_i(t)$ always satisfies

$$\frac{\underline{u}_i}{a_i} \leq u_i \leq \frac{\bar{u}_i}{a_i}$$

(see Fig.4), for all i, the above theorem is applicable to the LSS. However, there is no prior knowledge, as to whether or not the control signal is inside the interval $[\underline{u}_i/a_i,\ \bar{u}_i/a_i]$. It is possible that the designed LSS may be unstable if $u_i(t)$ moves outside the interval, i.e., the assumed sector (a_i, 1) is not suitable. Consequently, after system design the system simulation is necessary to ensure the system stabilization. The detail can be seen in the paper of Wang and Chen [13].

§6–4. Controller Synthesis Algorithm and Examples

The controller synthesis algorithm is summarized as follows.

Step 1: Check the upper bounds f_{ij} of $F_{ij}(s)$ for i, j=1, ..., J, i≠j.

Step 2: Factorizing $0.5(1+a_i)P_i(s)$ as (6.5) for all i.

Step 3: Choose the suitable form of $I-S_i(s)$ satisfying (6.6b) and (6.8) for each subsystem with desired closed–loop poles in which partial zeros of $I-S_i(s)$ are yet unknown.

Step 4: Determine the remaining unknown zeros of $I-S_i(s)$ by the requirements of internal stability (6.6a) and (6.15)–(6.18).

Step 5: Find the corresponding controllers $C_i(s)$, i=1, ..., J from (6.7).

Step 6: Give the system simulation to assure the stabilization of LSS.

Remark 4 : In Step 3, there must be enough degrees of freedom (i.e., enough unknown coefficients) of the numerator of $I-S_i(s)$ to satisfy (6.7a) and (6.15)–(6.18).

Remark 5: We do not guarantee, however, that Step 4 is solvable. Indeed, the values of a_i, f_{ij} and z_i all can affect the results of Step 4. In other words, it may be possible that a_i is too small or/and f_{ij} is too large to find the stabilizing controllers. We now illustrate this algorithm in more detail by an example.

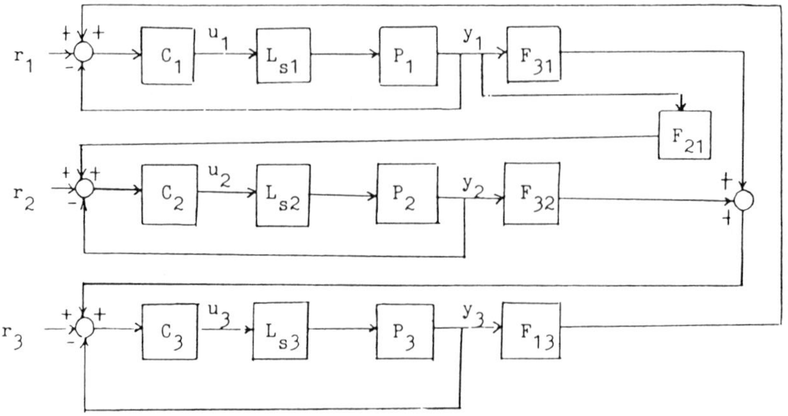

Figure 5. Interconnected system LSS

Example [13]: Consider an LSS (shown in Fig.5). Each subsystem contains a saturating actuator L_{si} and assume the operation range of L_{si} is inside the sector(a_i, 1), where $a_1=a_2=0.6$ and $a_3=1/3$. All the plants of the LSS are as follows:

$$\left.\begin{aligned} P_1(s)&=\frac{(s+2)}{(s-1)(s+3)}, \quad P_2(s)=\frac{(s-13)}{(s-1)(s+9)} \\ P_3(s)&=\begin{bmatrix} \frac{s+5}{s(s+1)} & \frac{-s-5}{s(s+1)} \\ 0 & \frac{s+5}{s(s+1)} \end{bmatrix} \end{aligned}\right\} \qquad (6.19)$$

Three saturating actuators are as described as (6.3), where

$(\bar{u}_1=3,\ \underline{u}_1=-3)$, $(\bar{u}_2=2,\ \underline{u}_2=-2)$, $(\bar{u}_{31}=2,\ \underline{u}_{31}=-2)$ and $(\bar{u}_{32}=1,\ \underline{u}_{32}=-1)$.

Moreover, the interconnections are as follows:

$$\left.\begin{array}{ll} F_{12}(s)=0, & F_{13}(s)=\left[\dfrac{1}{\sqrt{2}(s+6)} \quad \dfrac{1}{\sqrt{2}(s+6)}\right] \\ F_{21}(s)=\dfrac{1}{s+8} & F_{23}(s)=0 \\ F_{31}(s)=\left[\dfrac{1}{5(s+10)} \quad 0\right]^T, & F_{32}(s)=\left[0 \quad \dfrac{1}{5(s+10)}\right]^T, \end{array}\right\} \quad (6.20)$$

Following the "controller synthesis algorithm", we have

Step 1: $f_{12}=f_{23}=0,\ f_{13}=1/6,$ and $f_{21}=1/8,\ f_{31}=f_{32}=1/50$ (6.21)

Step 2: $A_1(s)=\hat{A}_1(s)=\dfrac{5(s-1)(s-3)}{4(s+1)(s+4)},\ B_1(s)=\hat{B}_1(s)=\dfrac{(s+2)}{(s+1)(s+4)}$

$$A_2(s)=\hat{A}_2(s)=\frac{5(s-1)(s+9)}{4(s+1)(s+4)},\ B_2(s)=\hat{B}_2(s)=\frac{(s-13)}{(s+1)(s+4)}$$

$$A_3(s)=\hat{A}_3(s)=\frac{3s}{2(s+1)}\begin{bmatrix}1 & 0\\ 0 & 1\end{bmatrix},\ B_3(s)=\hat{B}_3(s)=\frac{(s+5)}{(s+1)^2}\begin{bmatrix}1 & -1\\ 0 & 1\end{bmatrix}$$

Step 3: We choose $I-S_i(s)$, $i=1, 2, 3$ with desired poles as follows:

$$\left.\begin{array}{l} I-S_1(s)=\dfrac{c_1(s+t_1)}{(s+1)(s+2)} \\ I-S_2(s)=\dfrac{C_2(s+t_2)(s-13)}{(s+1)^2(s+13)} \\ I-S_3(s)=\dfrac{c_3(s+t_3)}{(s+4)(s+6)}I_2 \end{array}\right\} \quad (6.22)$$

where c_i, t_i for $i=1, 2, 3$ are unknown constants. Note that each $I-S_i(s)$ of (6.22) satisfies the requirements of Lemma 6.6 and (6.6b). In order to satisfy the internal stability (6.6a), t_1, t_2, and t_3 are obtained as

$$t_1=\frac{6}{c_1}-1,\ t_2=\frac{-14}{3c_2}-1 \text{ and } t_3=\frac{24}{c_3} \tag{6.23}$$

In order to satisfy (6.15) with unknown $C_i(s)$ (which is going to be found), $0.5(1-a_i)C_i(s)S_i(s)P_i(s)$ must be replaced by $\frac{1-a_i}{1+a_i}(I-S_i(s))$, i=1, 2 (SISO subsystem) and by $\frac{1-a_3}{1+a_3}P_3^{-1}(s)(I-S_3(s))P_3(s)$ (MIMO subsystem). By computer calculation, we have the following inequalities:

$$\begin{aligned}
&4.472>c_1>2 \quad \text{satisfies} \quad 0.2\|C_1(s)S_1(s)P_1(s)\|<0.5,\\
&-1.667>c_2>-4.243 \quad \text{satisfies} \quad 0.2\,\|C_2(s)S_2(s)P_2(s)\|<0.75,\\
&7.21>c_3>-7.21 \quad \text{satisfies} \quad 0.5\|C_3(s)S_3(s)P_3(s)|<0.5
\end{aligned} \tag{6.24}$$

where s=jw for all of the above inequalities. Hence, the requirement (6.15) is satisfied. From (6.22)–(6.24), we have

$$z_1<2, \quad z_2<4, \quad z_3<2 \tag{6.25}$$

consequently, from (6.17) and with the aid of (6.24), (6.25), we get

$$\begin{aligned}
&d_{11}<2+0.4\,\|S_1(s)P_1(s)\|_\infty\,\|C_1(s)S_1(s)\|_\infty,\\
&d_{22}<3+0.8\,\|S_2(s)P_2(s)\|_\infty\,\|C_2(s)S_2(s)\|_\infty,\\
&d_{33}<1+\frac{2}{3}\|S_3(s)P_3(s)\|_\infty\,\|C_3(s)S_3(s)\|_\infty,
\end{aligned} \tag{6.26}$$

From Lemma 2.6, (6.21) and (6.26), Q will be an M–matrix, if the following inequalities are satisfied,

$$\left.\begin{aligned} &\|S_1(s)P_1(s)\|_\infty \ \|C_1(s)S_1(s)\|_\infty < 10 \\ &\|S_2(s)P_2(s)\|_\infty \ \|C_2(s)S_2(s)\|_\infty < 6.25 \\ &\|S_3(s)P_3(s)\|_\infty \ \|C_3(s)S_3(s)\|_\infty < 36 \end{aligned}\right\} \tag{6.27}$$

By Lemma 2.11 and (6.24), we conclude that when

$$4.472 > c_1 > 2, \ -1.9 > c_2 > -4.24 \text{ and } 7.21 > c_3 > -4.31 \tag{6.28}$$

(6.27) is satisfied. (Note that c_3 can not be zero, otherwise $I-S_3(s)$ in (6.22) is meaningless.) The corresponding controllers are obtained by (6.7c)

$$\left.\begin{aligned} &C_1(s) = \frac{5(s+3)(c_1 s+6-c_1)}{4(s+2)(s+4-c_1)} \\ &C_2(s) = \frac{5(s+9)(c_2 s-14/3-c_2)}{4(s\ \ +(16-c_2)s+143/3+13c_2} \\ &C_3(s) = \frac{3(s+1)(c_3 s+24)}{2(s+5)(s+10-c_3)} \begin{bmatrix} 1 & 1 \\ 0 & 1 \end{bmatrix} \end{aligned}\right\} \tag{6.29}$$

where c_1, c_2, and c_3 are under the constraints of (6.28), i.e., when the undetermined coefficients of above equations satisfy (6.28) respectively, then controllers in (6.29) can stabilize the whole LLS under the assumption of that the operation range of L_{si} is inside the sector $(a_i, 1)$ for all i where a_i was specified at the begining of this example. In order to illustrate the simulation results, we choose $c_1=4$, $c_2=-2$, and $c_3=3$ then (6.29) becomes

$$\left.\begin{aligned} C_1(s) &= \frac{5(s+3)(2s+1)}{2s(s+2)} \\ C_2(s) &= \frac{5(s+9)(-2s-(8/3))}{4(s^2+18s+65/3)} \\ C_3(s) &= \frac{9(s+1)(s+8)}{2(s+5)(s+7)}\begin{bmatrix} 1 & 1 \\ 0 & 1 \end{bmatrix} \end{aligned}\right\} \qquad (6.30)$$

respectively.

Step 6: The reference inputs of LSS in this example are $r_1=U$, $r_2=0$ and $r_3=[0.5U, 0.5U]$. where U is a unit step. Fig.6a, 6b, 6c and 6d are the simulation results of this example, where Fig.6a shows the outputs y_1 and y_2 with initial conditions $y_1(0)=0$ and $y_2(0)=1$, Fig.6b shows the control signals u_1 and u_2 of subsystem–1 and subsystem–2 respectively; Fig.6c shows the output y_{31} and control signal u_{31} of subsystem–3 with $y_{31}(0)=0$, and Fig.6d shows the output y_{32} and control signal u_{32} with $y_{32}(0)=0$ of subsystem–3. It is seen that the designed controllers can stabilize the whole LSS. It is noted that, the stabilizing controllers obtained in Step 5 must be checked by the simulation in Step 6. That is to say, after completing Step 5 of the algorithm, we cannot yet assure that the designed controllers will stabilize the systems. [13] has given further two examples to illustrate this problem in detail.

§6–5. Discussion and Conclusion

The main results of this section are Theorem 6.7 and Theorem 6.8, which provide the sufficient conditions for the stability of a large–scale system with saturating actuators. These theorems are applicable to a subsystem with minimum or nonminimum phase (including unstable) plant. An algorithm is also introduced to synthesize the stabilizing controller for each subsystem so that the whole LSS is stabilized. This section also stresses that system simulation is very important and necessary to assure the system stabilization.

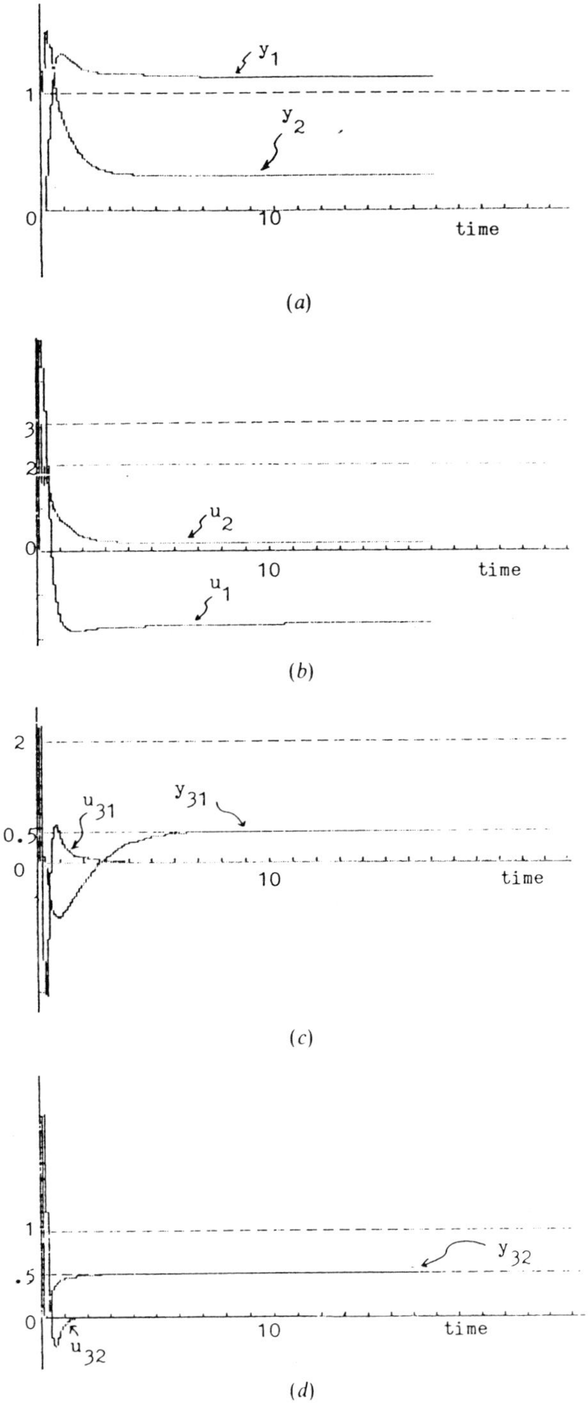

Figure 6. Control signals of subsystem 1 and subsystem 2.

VII. ROBUST STABILIZATION OF NONLINEARLY PERTURBED LARGE–SCALE SYSTEMS BY DECENTRALIZED OBSERVER–CONTROLLER COMPENSATORS

§7–1. Introduction

This section is concerned with the problem of the robust stabilization for nonlinearly perturbed large–scale systems via decentralized observer–controller compensators. The large–scale is composed of several interconnected perturbed subsystems and each isolated subsystem contains a nonlinearly perturbed plant and an observer–controller compensator. The main features of this section are as follows: (i) the nominal plant of each subsystem is not constrained to be stable and/or minimum phase, (ii) the perturbations of the plants and/or interconnections are considered, and (iii) two degree observer–controller compensating scheme is employed to treat the perturbed large–scale systems.

§7–2. System Description and Problem Formulation

Let us consider a large–scale system LS which consists of J nonlinearly perturbed subsystems with nonlinearly perturbed interconnections. One of the subsystems is shown in Fig.1, in which nominal plant $P_i: L_{2e}^{n_i} \to L_{2e}^{m_i}$, observer–controller compensator parameters Q_i, R_i: $L_2^{n_i} \to L_2^{n_i}$ and T_i: $L_2^{m_i} \to L_2^{n_i}$, and nominal interconnection F_{ij}: $L_2^{n_i} \to L_2^{m_i}$. Furthermore, N_i and N_{ij} are the nonlinear perturbation of P_i and F_{ij}, which are inside the sectors $[-b_i, b_i]$ and $[-b_{ij}, b_{ij}]$, respectively. From Fig.1, the LS can be described as follows:

$$u_i = r_i - Q_i^{-1}(R_i + T_i(P_i + N_i))u_i - Q_i^{-1}T_i \sum_{j \neq i}^{J} (F_{ij} + N_{ij})u_j \tag{7.1}$$

$$y_i = (P_i + N_i)u_i + \sum_{j \neq i}^{J} (F_{ij} + N_{ij})u_j, \tag{7.2}$$

for i=1, ..., J. The main purpose is to derive a sufficient condition under which the observer–controller compensators can robustly stabilize the overall perturbed large–scale system LS and propose a synthesis algorithm for robust decentralized observer–controller compensators.

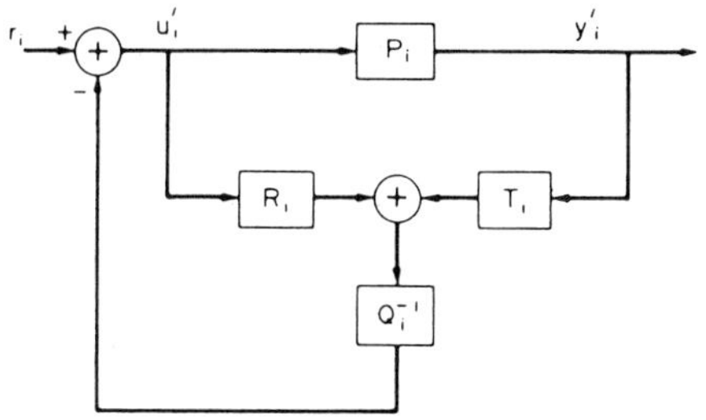

FIG. 1. The *i*th subsystem of the LS.

§7–3. Stability of An Isolated Nominal Subsystem

Before further solving the main problem, an isolated nominal subsystem shown in Fig.2 will be considered to establish some basic concepts at first. For convenience, let M[R], M[S] and M[U] denote the sets of proper rational matrices, stable and proper rational matrices, and stable proper rational unimodular matrices, respectively, in the following subsections. Firstly, let the i–th nominal plant of Fig.2 can be factorized as follows [24]

$$P_i(s)=A_i^{-1}(s)B_i(s)=\hat{B}_i(s)\hat{A}_i^{-1}(s) \tag{7.3}$$

where the pairs $A_i(s)$, $B_i(s) \in M[S]$ and $\hat{A}_i(s)$, $\hat{B}_i(s) \in M[S]$ constitute any left and right coprime decompositions (LCF and RCF) of $P_i(s)$ respectively. From Fig.2 and (7.1) we have

$$\begin{bmatrix} \hat{u}_i(s) \\ \hat{y}_i(s) \end{bmatrix} = G(P,C)r_i(s) = \begin{bmatrix} \hat{A}_i(s)K_i^{-1}(s)Q_i(s) \\ \hat{B}_i(s)K_i^{-1}(s)Q_i(s) \end{bmatrix} r_i(s) \tag{7.4}$$

and

$$K_i(s)=(Q_i(s)+R_i(s))\hat{A}_i(s)+T_i(s)\hat{B}_i(s) \tag{7.5}$$

Here, we use $\hat{u}_i$ and $\hat{y}_i$ to distinguish the signals u_i and y_i of the large–scale subsystem. It is easy to verify that if determinant $|K_i(s)|\neq 0$ for $Re(s)\geq 0$, (7.4) is well–posed [24]. Thus, it is obvious that the subsystem shown in Fig.2 is internally stable if and only if $K_i^{-1}(s)$ exists and is stable, i.e., $G(P,C)\in M[S]$ if and only if $K_i(s)\in M[U]$.

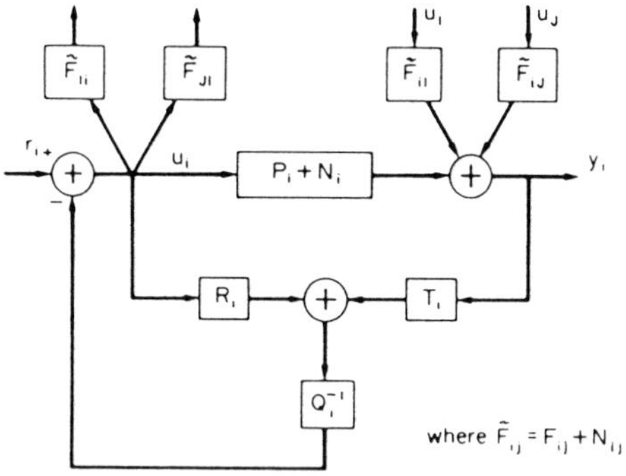

FIG. 2. The *i*th isolated nominal subsystem.

Lemma 7.1: For a given plant $P_i(s)\in M[R]$ with (7.3), the solution of (7.5) with a given $K_i(s)\in M[U]$ is given by

$$Q_i(s)+R_i(s)=K_i(s)\hat{X}_i(s)+H_i(s)B_i(s) \tag{7.6a}$$

$$T_i(s)= K_i(s)\hat{Y}_i(s)-H_i(s)A_i(s) \tag{7.6b}$$

for any $H_i(s) \in M[S]$, where $\hat{X}_i(s)$ and $\hat{Y}_i(s) \in M[S]$ and satisfy

$$\hat{X}_i(s)\hat{A}_i(s)+\hat{Y}_i(s)\hat{B}_i(s) = I \tag{7.7}$$

#

Proof: Given in Appendix D.

In view of this lemma, the stabilizing observer–controller compensator of the isolated nominal subsystem can be obtained as follows:

Lemma 7.2: For the isolated nominal subsystem shown in Fig.2, assume $P_i(s) \in M(R)$ with (7.3), then the observer–controller compensator, which stabilizes the subsystem, can be gotten from (7.6) for an arbitrary $K_i(s) \in M[U]$ and $H_i(s) \in M[S]$. #

Remark 1: (i) Without loss of generality and for simplification, we can let $K_i(s)=I$ (an unity matrix) in Lemma 7.2.

(ii) From (7.6a), we only get the sum of $Q_i(s)$ and $R_i(s)$. The choice of $Q_i(s)$ and $R_i(s)$ from the sum is almost free so long as $Q_i^{-1}(s)T_i(s)$ and $Q_i^{-1}(s)R_i(s)$ are proper simultaneously. Here, for example, we will show that $Q_i(s)$ can be determined by achieving reference signal tracked with zero steady state error for isolated nominal subsystems. The tracking error of the isolated nominal subsystem is given by

$$e_i(s)=r_i(s)-\hat{y}_i(s)= [I-\hat{B}_i(s)K_i^{-1}(s)Q_i(s)]r_i(s) \tag{7.8}$$

In order to track $r_i(s)$ with zero steady state error in this nominal subsystem, $Q_i(s)$ must be chosen such that (7.8) is a vector with no pole in $Re(s) \geq 0$. Up to now, the solution is concerned with isolated nominal subsystem only.

§7–4. Robust Stability of Large Scale System

According to the system shown in Fig.1 we shall state below a theorem regarding the main problem.

Theorem 7.3: In the large–scale system LS of Fig.1, $P_i(s) \in M[R]$, and $Q_i(s)$, $R_i(s)$ and $T_i(s) \in M[S]$ $\forall$ i. Suppose (a) nonlinear perturbations N_i and N_{ij} are inside the sector $[-b_i, b_i]$ and $[-b_{ij}, b_{ij}]$ respectively; and (b) each nominal interconnection is bounded as $\|F_{ij}(jw)\|_\infty \leq f_{ij}$. If

(i) $b_i\|\hat{A}_i(s)K_i^{-1}(s)T_i(s)\|_\infty < 1$ for $s=j\omega$, (7.9)

where $K_i(s) \in M[U]$ is defined as (7.5).

(ii) W is an M–matrix,

then the large–scale system LS is L_2–stable robustly,

where

$$W=[w_{ij}],\ w_{ij}=\begin{cases}1, & i=j\\ -z_i\|L_i(s)\|_\infty(f_{ij}+b_{ij}), & i\neq j\end{cases} \tag{7.10}$$

and

$$z_i= (1-b_i\|L_i(s)\|_\infty)^{-1} \tag{7.11a}$$

$$L_i(s)=\hat{A}_i(s)K_i^{-1}(s)T_i(s),\ \text{for } s=j\omega \text{ and } i, j=1, 2, \ldots, J. \tag{7.11b}$$

#

Proof: The proof is given in Appendix E.

Remark 2: (1) The i–th isolated perturbed subsystem is L_2–stable if assumption (a) for N_i and (7.9) hold.

(2) From (7.11a), for unstable plant $P_i(s)$, if z_i and f_{ij} are smaller then it is easy to find the stabilizing compensators. In other words, it is also possible that some f_{ij} are too large or some N_i or/and N_{ij} are too "nonlinear" (i.e.,the values of b_i or/and b_{ij} are too large) for us to

find the compensators. However if $P_i(s)$ are stable for all i, we are always able to find the stabilizing compensators to treat the perturbed LS, (which will be discussed in detail in Remark 3–(2)).
(3) If N_i and N_{ij} are linear perturbation, we have a similar result as Theorem 7.3 (see Corollary 1 of [19]).

§7–5. Design Algorithm of Observer–Controller Compensators

The task of this subsection is that: under assumptions (a) and (b) of Theorem 7.3, how to choose $Q_i(s)$, $R_i(s)$ and $T_i(s)\in M[S]$ to satisfy conditions (i) and (ii) so that the perturbed LS is robustly stabilized ? For convenience of design, conditions (i) and (ii) are modified as follows. Substituting (7.6b) into (7.11b), we have

$$L_i(s)=\hat{A}_i(s)(\hat{Y}(s)-V_i(s)A_i(s)) \qquad (7.12)$$

where

$$V_i(s)=K_i^{-1}(s)H_i(s) \in M[S] \qquad (7.13)$$

and $K_i(s) \in M[U]$. From (7.9) and Lemma 2.6 [2, 37], it is seen that if we can choose a suitable $V_i(s)\in M[S]$ such that

$$(1)\ \|b_i\hat{A}_i(s)(Y_i(s)-V_i(s)A_i(s))\|_\infty<1 \quad \text{for } i=1, \ldots, J \qquad (7.14)$$

and

$$(2)\ z_i\|\hat{A}_i(s)(\hat{Y}_i(s)-V_i(s)A_i(s))\|_\infty \sum_{j\neq i}^{J} (f_{ij}+b_{ij})<1 \text{ for } i=1, \ldots, J. \qquad (7.15)$$

then the LS will be L_2–stable. Hence by (7.11a) and (7.12), if we choose a suitable $V_i(s)\in M[S]$ such that

$$\|\hat{A}_i(s)(\hat{Y}_i(s)-V_i(s)A_i(s))\|_\infty < \frac{1}{b_i + \sum_{j\neq i} f_{ij} + b_{ij}}, \quad i=1, \ldots, J \qquad (7.16)$$

Then not only (7.9) holds but also W is an M–matrix.

The above analysis is summarized to the following theorem.

Theorem 7.4: For the same LS described as Theorem 7.3, if (7.16) holds then LS is L_2–stable robustly. #

Remark 3: (1) The choice of $V_i(s)$ for satisfying (7.16) is not unique. Once we have chosen one of the suitable $V_i(s)$ then factorize $V_i(s)$ as $K_i^{-1}(s)H_i(s)$ where $K_i(s) \in M[U]$ and $H_i(s) \in M[S]$ and substitute $K_i(s)$ and $H_i(s)$ into (7.6) to obtain compensator parameters $Q_i(s)+R_i(s)$ and $T_i(s)$.

(2) Suppose the i–th subsystem in the LS has stable and exact proper nominal plants, in this case $A_i^{-1}(s)$ and $\hat{A}_i^{-1}(s) \in M[S]$, if we choose

$$V_i(s)=(\hat{Y}_i(s)-\rho_i\hat{A}_i^{-1}(s))A_i^{-1}(s) \qquad (7.17)$$

for the i–th subsystem, where ρ_i is an arbitrarily small scalar with $|\rho_i| << 1$, then $\|\hat{A}_i(s)(\hat{Y}_i(s)-V_i(s)A_i(s))\|_\infty = |\rho_i|$ can be as small as possible, thus (7.9) always holds for the i–th subsystem with stable nominal plant, i.e., it is always possible to synthesize a robustly stabilizing observer–controller compensator to override the nonlinear perturbations in the i–th subsystem. This concept can be extended to the LS. Suppose all the nominal plants of the subsystems in the LS are stable, the equation (7.17) with sufficiently small ρ_i $\forall$ i always implies (7.16), i.e., the perturbed LS is always stabilized. For example: we can choose $V_i(s)=\hat{Y}_i(s)A_i^{-1}(s)$ to yield $\rho_i=0$, then by (7.6a) and (7.6b), we have $T_i(s)=0$ and $Q_i(s)+R_i(s)=\hat{A}_i^{-1}(s)$, where $K_i(s)=I$ and $H_i(s)=\hat{Y}_i(s)A_i^{-1}(s)$. Consequently, when $T_i(s)=0$, $Q_i(s)=\hat{A}_i^{-1}(s)$ and $R_i(s)=0$ are chosen, the feedback path of each subsystem disappears. That means when each nominal subsystem is stable, the whole LS is stable without any feedback control for each subsystem.

(3) However, if m subsystems, where $m \leq J$, have unstable nominal plants, (7.16) is not guaranteed to be solvable for the m subsystems, i.e., the corresponding robust observer–controller compensator may not exist. From the view point of robust optimization, one may solve (7.16) by H^{∞} norm minimization techniques (see [21], [24], [34]) to get an optimal $V_i(s)$. In other words, in the case of unstable nominal plants, it is also possible that some b_i, b_{ij} and/or f_{ij} are not so sufficiently small that we may not find a suitable $V_i(s)$ to satisfy (7.16) (see Nivanlinna–Pick theorem [20] and [39]).

(4) No matter in Theorem 7.3 or Theorem 7.5, the zeros (or $B_i(s)$ and $\hat{B}_i(s)$) of nominal plant $P_i(s)$ are not important. Since in our derivation process, we are not concerned whether the nominal plant of each subsystem is minimum–phase or not.

Let us now summarize the procedure for designing the observer–controller compensator in the nonlinearly perturbed LS.

Step 1: Evaluate the upper–bounds f_{ij}, b_{ij} and b_i of nominal interconnections F_{ij}, perturbations N_{ij} and N_i, respectively.

Step 2: Factor nominal plant $P_i(s)$ as (7.3) to get $(A_i(s),B_i(s))$ and $(\hat{A}_i(s),\hat{B}_i(s))$ as well as solve (7.7) to obtain $(\hat{X}_i(s),\hat{Y}_i(s))$.

Step 3: Choose an adequate $V_i(s) \in M[S]$ to satisfy inequality (7.16).

Step 4: Arbitrarily factor $V_i(s)=K_i^{-1}(s)H_i(s)$ as long as such that $K_i(s) \in M[U]$ and $H_i \in M[S]$. Without loss of generality, we can let

$K_i(s)=I$, i.e., let $H_i(s)=V_i(s)$.

Step 5: Substituting $K_i(s)$ and $H_i(s)$ into (7.6a, b) to get $(Q_i(s)+R_i(s))$ and $T_i(s)$ respectively.

Step 6: Choose $Q_i(s)$ to satisfy the requirement of reference tracking in (7.8) and $R_i(s)$ is obtained simultaneously under the constraint of which $Q_i^{-1}(s)R_i(s)$ and $Q_i^{-1}(s)T_i(s)$ are ensured proper.

§7–6. An Example [19]

Consider a large–scale system shown in Fig.3

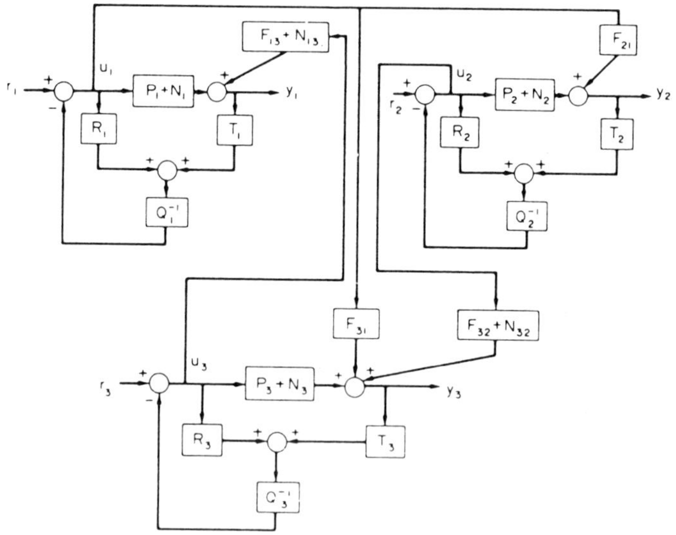

FIG. 3. The large scale system in the example.

in which

$$P_1(s)=\frac{9}{(s-3)(s+3)}, \qquad P_2(s)=\frac{s-3}{(s-1)(s+1)}$$

and (7.18)

$$P_3(s)=\begin{bmatrix} \frac{1}{s-2} & 0 \\ \frac{3}{(s+1)(s-2)} & \frac{1}{s+1} \end{bmatrix}$$

Each nonlinear perturbation is described as follows:

$$N_1(u_1)=0.1u_1\sin(0.1u_1), \quad N_2(u_2)=0.15u_2\sin(0.15u_2)$$

and

$$N_3=\mathrm{diag}[\hat{N}_{31}, \hat{N}_{32}],$$

where

$$\hat{N}_{31}(u_{31})=0.1u_{31}\sin(0.1u_{31}); \ \hat{N}_{32}(u_{32})=0.1u_{32}\sin(0.1u_{32}). \tag{7.19}$$

For simplification, suppose $F_{12}(s)=F_{23}(s)=0$ and

$$F_{21}(s)=\frac{1}{2(s+10)}, \qquad F_{31}(s)=\left[\frac{1}{5(s+10)}, \ 0\right]^T$$

$$F_{32}(s)=\left[0, \frac{1}{5(s+10)}\right]^T \qquad F_{13}(s)=\left[\frac{1}{\sqrt{2}(s+25)}, \frac{1}{\sqrt{2}(s+25)}\right]$$

Moreover there are some interconnections perturbations as follows:

$$N_{13}(u_3)=[0.05u_{31}\sin(0.05u_{31}), \ 0]; \ N_{32}(u_2)^T=[0, \ 0.02u_2\sin(0.02u_2)]$$

According to the design algorithm, we will synthesize three decentralized observer–controller compensators to stabilize the whole large–scale system.

Step 1: It is easy to obtain $f_{12}=f_{23}=0$, $f_{21}=0.05$, $f_{31}=0.02$, $f_{32}=0.02$ and $f_{13}=0.04$; and $b_1=0.1$, $b_2=0.15$, $b_3=0.1$, $b_{13}=0.05$ and $b_{32}=0.02$.

Step 2: Factorizing $P_i(s)$, we get

$$A_1(s)=\hat{A}_1(s)=\frac{(s-3)(s+3)}{(s+1)^2},\quad B_1(s)=B'_1(s)=\frac{9}{(s+1)^2}$$

$$A_2(s)=\hat{A}_2(s)=\frac{(s+1)(s-1)}{(s+3)(s+4)},\quad B_2(s)=\hat{B}_2(s)=\frac{s-3}{(s+3)(s+4)};$$

$$A_3(s)=\begin{bmatrix} 1 & -1 \\ \frac{(s-2)}{3(s+1)} & 0 \end{bmatrix},\quad B_3(s)=\begin{bmatrix} \frac{1}{s+1} & \frac{-1}{s+1} \\ \frac{1}{3(s+1)} & 0 \end{bmatrix}$$

$$\hat{A}_3(s)=\begin{bmatrix} 0 & \frac{(s-2)}{3(s+1)} \\ 1 & 0 \end{bmatrix},\quad \hat{B}_3(s)=\begin{bmatrix} 0 & \frac{1}{3(s+1)} \\ \frac{1}{s+1} & \frac{1}{(s+1)^2} \end{bmatrix}.$$

Solving (7.7), we have

$$\hat{X}_1(s)=\frac{s+3}{s+1},\quad \hat{Y}_1(s)=\frac{12s+28}{9(s+1)};\quad \hat{X}_2(s)=\frac{s+18}{s+1},\quad \hat{Y}_2(s)=-10$$

and $\hat{X}_3(s)=\begin{bmatrix} 0 & 1 \\ 3 & 0 \end{bmatrix}$, $\hat{Y}_3(s)=\begin{bmatrix} 0 & 0 \\ 9 & 0 \end{bmatrix}$.

Step 3: According to H^∞–norm minimization technique [21], [24], and [33], we choose $V_1(s)$ and $V_2(s)$ such that $L_1(s)$ and $L_2(s)$ are of all–pass forms , e.g.,

$$V_1(s)=\frac{-8(s+1)(3s^2+10s+9)}{9(s+3)^3}, \text{ and } V_2(s)=\frac{-2(4s+7)(s+3)(s+4)}{(s+1)^3}$$

will yield

$$L_1(s)=\frac{4(s-3)}{(s+3)}, \text{ and } L_2(s)=\frac{-2(s-1)}{(s+1)}$$

Hence min $\|L_1(j\omega)\|_\infty=4$ and min $\|L_2(j\omega)\|=2$ which satisfy (7.16). For the MIMO subsystem, we may choose

$$V_3(s)=\frac{1}{s+2}\begin{bmatrix} k(2-s) & 3k(s+1) \\ 0 & -9(s+1) \end{bmatrix}$$

to get

$$L_3(s)=\begin{bmatrix} \frac{4(s-2)}{s+2} & 0 \\ 0 & \frac{-(s-2)k}{s+2} \end{bmatrix},$$

when $-4\leq k\leq 4$ implies $\|L_3(j\omega)\|_\infty=4$, which also satisfies (7.16).

Step 4: Let $K_i(s)=1$ (or I), then $H_i(s)=V_i(s)$, $\forall$ i.

Step 5 and Step 6: Substituting $H_i(s)$ and $K_i(s)$ into (7.6a, b) and choosing $Q_i(s)$ to satisfy the reference tracking requirement (7.8), for example $r_1(s)=1/s$, $r_2(s)=-1/s$ and $r_3(s)=[2/s \;\; 1/s]^T$, then we have

$$Q_1(s)=\frac{(s+1)^2}{(s+3)^2},\ R_1(s)=\frac{6(s+1)^3}{(s+3)^3},\ T_1(s)=\frac{4(s+1)^2}{(s+3)^2} \tag{7.20}$$

and

$$Q_2(s)=\frac{s-4}{s+1},\ R_2(s)=\frac{14s^2+54s+64}{(s+1)^3},\ T_2(s)=\frac{-2(s+4)(s+3)}{(s+1)^2} \tag{7.21}$$

For simplicity, in the third subsystem we only try to track the first step input signal $r_{31}(s)=2/s$, then we choose

$$Q_3(s)=\begin{bmatrix} \frac{s-2}{s+1} & \frac{3s+2}{s+2} \\ 3 & 0 \end{bmatrix},\ T_3(s)=\begin{bmatrix} 0 & -\frac{k_1(s-2)}{s+2} \\ \frac{12(s+1)}{s+2} & 0 \end{bmatrix}$$

and

$$R_3(s)=\begin{bmatrix} \frac{-s^2+3k+4}{(s+1)(s+2)} & \frac{-2s^2+(k-2)s-2k}{(s+1)(s+2)} \\ \frac{-3}{s+2} & 0 \end{bmatrix}, \tag{7.22}$$

The results of computer simulation are shown in Fig.4 with $k=1$ and specified input: $r_1(t)=U$, $r_2(t)=-U$ and $r_3(t)=[2U,\ U]^T$ where U is a unit step function and with zero initial conditions. Note that the choice of $Q_i(s)$ by (7.8) is only for reference tracking for the isolated nominal subsystems. In Fig.3, the large–scale system of the example contains many nonlinearities and interconnections, hence the output of each interconnected subsystem may not actually track its reference signal respectively with zero steady state error.

§7–7. Discussion and Conclusion

The main results of this section are Theorem 7.3 and Theorem 7.4. The main features and contributions are that (i) we take the structure of the LS into account and let the so–called "margins of boundedness" and "gain factors" of the subsystems [7] be in terms of the norm measures of system's operators without plotting Nyquist diagram. (ii) the two–degree observer–controller compensating scheme is employed to the robust stability problem of large–scale systems; (iii) the main results are still applicable to the following cases (1) the LS contains nonlinear and/or time–varying perturbations in the plants and interconnections, (2) plants of the subsystems are unstable and/or nonminimum–phase. In order to be more easy for the compensators design, Theorem 7.4 proposes a simple but useful inequality to establish the decentralized compensator synthesis algorithm. It should be noted that since Theorem 7.3 and Theorem 7.4 are only sufficient conditions for the robust stability and stabilization of the LS respectively, if (7.16) is not solvable which does not mean that LS is not stabilizable. To derive the necessary and sufficient conditions of the robust stability for the same structural perturbed large–scale systems is an open problem.

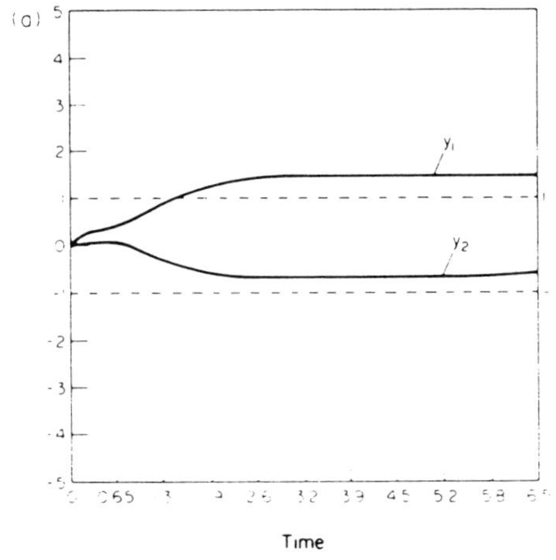

FIG. 4a. The output signals of the first and the second subsystem.

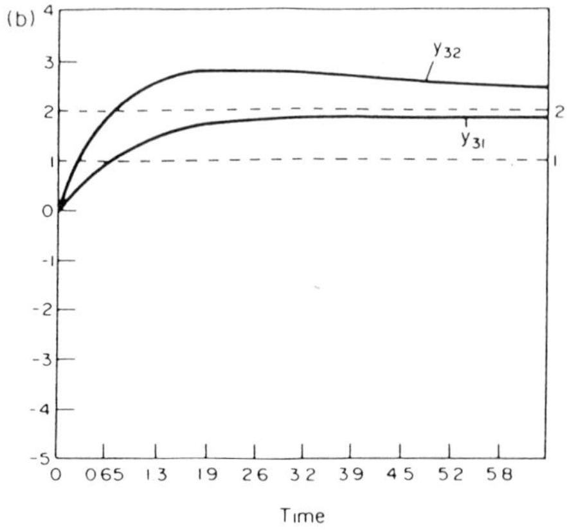

FIG. 4b. The output signals of the third system.

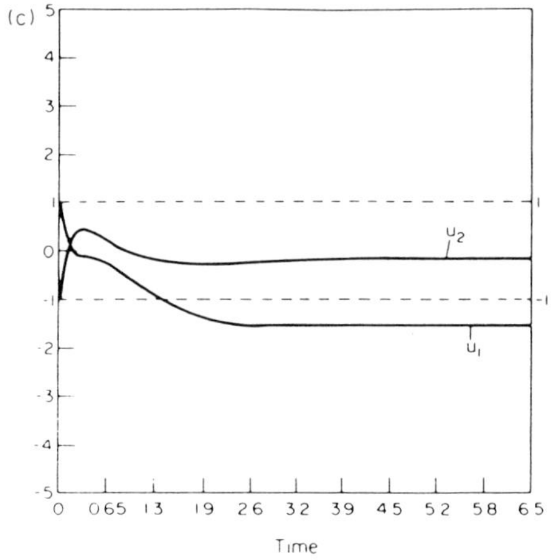

FIG. 4c. The control signals of the first and second subsystem.

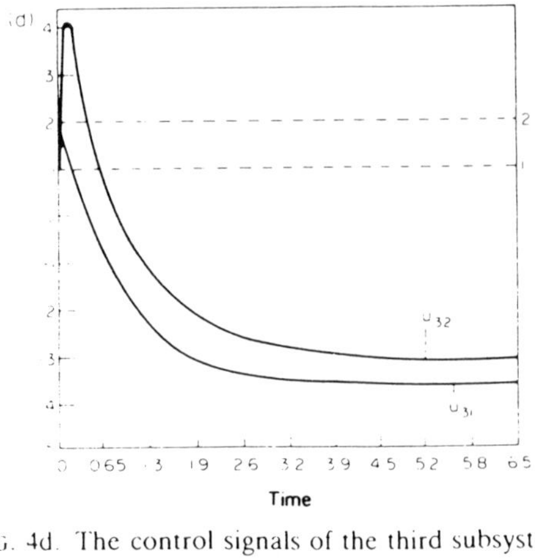

FIG. 4d. The control signals of the third subsystem.

Appendix A (Proof of Lemma 6.5)

From (6.5) and (6.6), it is clear that $S_i(s)P_i(s)=2X_i(s)B_i(s)$ is free of poles in $Re(s)\geq 0$. Moreover, $I-S_i(s)=0.5P_i(s)C_i(s)S_i(s)$, i.e., $C_i(s)S_i(s)=2P_i^{-1}(s)(I-S_i(s))$. By (6.5) and (6.6), we have

$$C_i(s)S_i(s)=\hat{A}_i(s)Y_i(s)$$

to be stable. Let $C_i(s)= C_{ni}(s)C_{di}^{-1}(s)$ be any right–coprime stable rational decompositions of $C_i(s)$ then

$$S_i(s)=C_{di}(s)[A_i(s)C_{di}(s)+B_i(s)C_{ni}(s)]^{-1}A_i(s)$$

Since $S_i(s)$ is stable $\forall$ i, $\det(A_i(s)C_{di}(s)+B_i(s)C_{ni}(s))$ is analytic in $Re(s)\geq 0$. Hence

$$C_i(s)S_i(s)P_i(s)=C_{ni}(s)[A_i(s)C_{di}(s)+B_i(s)C_{ni}(s)]^{-1}B_i(s)$$

is stable too. □.

Appendix B (Proof of Lemma 6.6)

In the system of Fig.3 of Section VI, the closed–loop transfer function matrix is

$$I-S_i(s)=0.5P_i(s)C_i(s)S_i(s) \tag{B.1}$$

then $O(I-S_i(s))+O(P_i^{-1}(s))>O(C_i(s))$, (because $O(S_i(s))=0$). Hence, $O(I-S_i(s))+O(P_i^{-1}(s))<0$ implies that $C_i(s)$ is proper. □

Appendix C (Proof of Theorem 6.7)

Consider the subsystem shown in Fig.2 of Section VI. Since P_i, and C_i, $\forall$ i, are all causal operations, evaluating the norm $\|(.)\|$ of both sides of (6.9b) yields

$$\|(u_i)_d\| \leq \|(C_iS_ir_i)_d\| + 0.5\|(C_iS_iP_iu_i)_d\| + \left\|(C_iS_i \sum_{j\neq i}^{J} F_{ij}y_j)_d\right\| \quad (C.1)$$

By Lemma 6.5, we have bounded $\|C_iS_i\|$ and $\|C_iS_iP_i\|$; furthermore, L_{si} is inside the sector (0, 1), then rewrite (C.1) as following form

$$\begin{aligned}\|(u_i)_d\| &\leq \|C_iS_i\|\ \|(r_i)_d\| + \|0.5C_iS_iP_i\|\ \|(u_i)_d\| \\ &+ \sum_{j\neq i}^{J} f_{ij}\|C_iS_i\|\ \|(y_j)_d\| = (1-0.5\|C_i(s)S_i(s)P_i(s)\|_\infty)^{-1} \\ &\times\left[\|C_i(s)S_i(s)\|_\infty\|(r_i)_d\| + \sum_{j\neq i}^{J} f_{ij}\|C_i(s)S_i(s)\|_\infty\|(y_i)_d\|\right] \\ &\text{(from Lemma 2.11)}, \forall\ s=jw,\ w\geq 0 \qquad (C.2)\end{aligned}$$

Define $z_i=(1-0.5\|C_i(s)S_i(s)P_i(s)\|_\infty)^{-1}$, hence from (C.2), z_i must be larger than zero. Consequently, we have

$$(u_i)_d\| \leq z_i\left[\|C_i(s)S_i(s)\|_\infty\|(r_i)_d\| + \sum_{j\neq i}^{J} f_{ij}\|C_i(s)S_i(s)\|_\infty\|(y_i)_d\|\right] \quad (C.3)$$

From Fig.2, output y_i can be written as follows

$$y_i=(r_i+\sum_{j\neq i}^{J} F_{ij}y_j)-(I+P_iL_{si}C_i)^{-1}(r_i+\sum_{j\neq i}^{J} F_{ij}y_j)$$

$$=(r_i+\sum_{j\neq i}^{J} F_{ij}y_j)+\Big\{(I+0.5P_iC_i)^{-1}[I+L_{si}P_iC_i-(I+0.5P_iC_i)]$$

$$(I+P_iL_{si}C_i)^{-1}-(I+0.5P_iC_i)^{-1}\Big\}(r_i+\sum_{j\neq i}^{J} F_{ij}y_j)$$

$$=(r_i+\sum_{j\neq i}^{J} F_{ij}y_j)+\Big\{(I+0.5P_iC_i)^{-1}(P_i(L_{si}-0.5)C_i)(I+P_iL_{si}C_i)^{-1}$$

$$-\,(I+0.5P_iC_i)^{-1}\Big\}\ (r_i+\sum_{j\neq i}^{J}F_{ij}y_j)$$

where

$$C_i(I+P_iL_{si}C_i)^{-1}\,(r_i+\sum_{j\neq i}^{J}F_{ij}y_j)=u_i \tag{C.4}$$

then

$$y_i=(I-S_i)(r_i+\sum_{j\neq i}^{J}F_{ij}y_j)+S_iP_i(L_{si}-0.5)u_i \tag{C.5}$$

Performing the $\|.\|$ operation and by (6.2), we get

$$\|(y_i)_d\|\leq\|I-S_i\|(\|(r_i)_d\|+\sum_{j\neq i}^{J}f_{ij}\|(y_j)_d\|)+0.5\|S_iP_i\|\ \|(u_i)_d\| \tag{C.6}$$

Substitute (C.3) into above inequality and obtain

$$\|(y_i)_d\|\leq(\|I-S_i(s)\|_\infty+0.5z_i\|S_i(s)P_i(s)\|_\infty\|C_i(s)S_i(s)\|_\infty)\|(r_i)_d\|$$
$$+(\|I-S_i(s)\|_\infty+0.5z_i\|S_i(s)P_i(s)\|_\infty\|C_i(s)S_i(s)\|_\infty)+\sum_{j\neq i}^{J}f_{ij}\|(y_j)_d\| \tag{C.7}$$

for $i=1,\ldots,J$. Since $r_i\in L_2^n$, then (C.7) also holds as $d\to\infty$. Rewrite (C.7) to be a matrix form:

$$Q\ Y < D\ R \tag{C.8}$$

where Q and $D=\text{diag}[d_{11},\ldots,d_{JJ}]$ are defined as (6.11) and (6.12) respectively. If Q is an M–matrix then by condition (c) of Lemma 2.5, $Q^{-1}=[\hat{q}_{ij}]$, and each $\hat{q}_{ij}$ is non–negative. Equation (C.8) is written as

$$Y = Q^{-1}DR \tag{C.9}$$

then by the properties of norm function, we have

$$\|Y\|_2^2 \leq h_s \|R\|_2^2 \tag{C.10}$$

where $h_s \geq \sigma_M^2(Q^{-1}D)$ (maximum singular value). Since $r_i \in L_2^n$, $\forall i$, then $\|Y\|_2^2 < \infty$. Let $D_u = \mathrm{diag}[d_{u1}, \ldots, d_{uJ}]$ and

$$Q_u = [\hat{q}_{ij}],\ \hat{q}_{ij} = \begin{cases} 0 & i=j \\ z_i f_{ij} \|C_i(s) S_i(s)\|_\infty, & i \neq j \end{cases} \tag{C.11}$$

where $d_{ui} = z_i \|C_i(s) S_i(s)\|$. Then (C.3) can be written as

$$U \leq D_u R + Q_u Y \tag{C.12}$$

Substituting (C.9) into (C.12), we have

$$U < D_u R + Q_u Q^{-1} D R = (D_u + Q_u Q^{-1} D) R \tag{C.13}$$

Let us define $(D_u + Q_u Q^{-1} D) = D_{qu}$ (note that whose elements are all nonnegative), thus

$$U < D_{qu} R \Longrightarrow \|U\|_2^2 \leq k_s \|R\|_2^2 \tag{C.14}$$

where $k_s \geq \sigma_M^2(D_{qu})$. Since $e_i = r_i + \sum_{j \neq i}^{J} f_{ij} y_j - y_j$, then we have

$$\|e_i\| \leq \|r_i\| + \sum_{j \neq i}^{J} f_{ij} \|y_j\| - \|y_j\| \Longrightarrow E < I_J R + F_e Y \tag{C.15}$$

where I_J is a $J \times J$ unit matrix, and

$$F_e=[f_{ij}],\quad f_{ij}=\begin{cases}1, & i=j\\ f_{ij}, & i\neq j\end{cases}$$

Substituting (C.9) into (C.16), similarly, there is

$$E\leq(I_J+F_eQ^{-1}D)R=F_{J_e}R \tag{C.16}$$

It is obvious that there is a constant $t_s\geq\sigma_M^2(F_{J_e}R)$ so that

$$\|E\|_2^2\leq t_s\|R\|_2^2 \tag{C.17}$$

From (C.10), (C.14), (C.17) and Definition 6.3, we can conclude that LSS is L_2–bounded robustly. □

APPENDIX D (Proof of Lemma 7.1)

The solution of (7.6) can be directly separated into two parts: the homogeneous and particular solutions. The homogeneous solution is solved as follows. Let

$$(Q_i(s)+R_i(s))\hat{A}_i(s)+T_i(s)\hat{B}_i(s)=0 \tag{D.1}$$

by (7.3) which is satisfied with

$$Q_i(s)+R_i(s)=H_i(s)B_i(s) \tag{D.2a}$$

$$T_i(s)=-H_i(s)A_i(s) \tag{D.2b}$$

where $H_i(s)\in M[S]$ is an arbitrary stable rational matrix. Here (D.2) is the homogeneous solution of (7.6). Since

$$\hat{X}_i(s)\hat{A}_i(s)+\hat{Y}_i(s)\hat{B}_i(s)=I \tag{D.3}$$

Both sides of the above equation are multiplied by $K_i(s)$ yielding

$$K_i(s)\hat{X}_i(s)\hat{A}_i(s)+K_i(s)\hat{Y}_i(s)\hat{B}_i(s)=K_i(s) \tag{D.4}$$

Comparing (7.6) and (D.4) the particular solution is given by

$$\begin{aligned} Q_i(s)+R_i(s)&=K_i(s)\hat{X}_i(s) \\ T_i(s)&=K_i(s)\hat{Y}_i(s) \end{aligned} \tag{D.5}$$

where $K_i(s)\epsilon M[U]$. Consequently, the general solution of (7.6) can be obtained from (D.2) and (D.5).

APPENDIX E (Proof of Theorem 7.3)

From Fig.1 of Section VII, (7.1) can be rewritten as

$$u_i=(Q_i+R_i+T_iP_i)^{-1}[Q_ir_i-T_i\sum_{j\neq i}(F_{ij}+N_{ij})u_j-T_iN_iu_i] \tag{E.1}$$

Substituting (7.3) into (E.1), we have

$$u_i=\hat{A}_i(Q_i\hat{A}_i+R_i\hat{A}_i+T_i\hat{B}_i)^{-1}$$

$$\times[Q_ir_i-T_i\sum_{j\neq i}(F_{ij}+N_{ij})u_i-T_iN_iu_i] \tag{E.2}$$

Since $\hat{A}_i$, $\hat{B}_i$, Q_i, R_i and T_i are all bounded operators, then

$$\begin{aligned}\|(u_i)_d\| \le & \|\hat{A}_i(Q_i\hat{A}_i+R_i\hat{A}_i+T_i\hat{B}_i)^{-1}Q_i\| \ \|(r_i)_d\| \\ & +\|\hat{A}_i(Q_i\hat{A}_i+R_i\hat{A}_i+T_i\hat{B}_i)^{-1}T_i\| \sum_{j\neq i} \|(F_{ij}+N_{ij})(u_j)_d\| \\ & +b_i\|\hat{A}_i(Q_i\hat{A}_i+R_i\hat{A}_i+T_i\hat{B}_i)^{-1}T_i\| \ \|(u_i)_d\| \end{aligned} \tag{E.3}$$

Hence, the following inequality is true [27]

$$\begin{aligned}\|(u_i)_d\| \le & \|\hat{A}_i(s)K_i^{-1}(s)Q_i(s)\|_\infty \|(r_i)_d\| \\ & + \sum_{j\neq i} \|\hat{A}_i(s)K_i^{-1}(s)T_i(s)\|_\infty (f_{ij}+b_{ij})\|(u_j)_d\| \\ & +b_i\|\hat{A}_i(s)K_i^{-1}(s)T_i(s)\|_\infty \|(u_i)_d\| \\ & \text{for } s=j\omega \text{ and } i=1, \ldots, J. \end{aligned} \tag{E.4}$$

where $K_i(s)=Q_i(s)\hat{A}_i(s)+R_i(s)\hat{A}_i(s)+T_i(s)\hat{B}_i(s)$.

Let $z_i(s)=(1-b_i\|\hat{A}_i(s)K_i^{-1}(s)T_i(s)\|_\infty)^{-1}$

and $L(s)=\hat{A}_i(s)K_i^{-1}(s)T_i(s)$ (E.5)

Consequently, we obtain

$$\begin{aligned}\|(u_i)_d\| \le z_i[& \|\hat{A}_i(s)K_i^{-1}(s)Q_i(s)\|_\infty \|(r_i)_d\| \\ & +\|L_i(s)\|_\infty \sum_{j\neq i} (f_{ij}+b_{ij})\|(u_j)_d\|] \\ & \text{for all } i \text{ and } s=j\omega \end{aligned} \tag{E.6}$$

For $r_i \in L_2^{n_i}$, $i=1, \ldots, J$, then (E.6) can be expressed as

$$W\ U \le \text{diag}[z_i\|\hat{A}_i(s)K_i^{-1}(s)Q_i(s)\|_\infty]R, \quad i=1,\ldots, J. \tag{E.7}$$

where W, and (U, R) are defined as (7.10) and (6.4) respectively. If W is an M–matrix, then W is nonsingular and each entry of W^{-1} is nonnegative. This implies that

$$U \leq W^{-1}\,\mathrm{diag}[z_i \|\hat{A}_i(s)K_i^{-1}(s)Q_i(s)\|_\infty]R \tag{E.8}$$

Since $r_i \in L_2^{n_i}$ then $\|u_i\|<\infty$ for all i, hence it is clear that

$$\sum_i \|u_i\| \leq k_s \sum_i \|r_i\| + \hat{k}_s \text{ for some nonnegative } k_s \text{ and } \hat{k}_s. \tag{E.9}$$

Similarly, from (7.10) and (E.2), we have

$$\begin{aligned} \|(y_i)_d\| \leq & \|\hat{B}_i(s)K_i^{-1}(s)Q_i(s)\|_\infty \|(r_i)_d\| \\ & + \sum_{j\neq i} (1+\|\hat{B}_i(s)K_i^{-1}(s)T_i(s)\|_\infty)(b_{ij}+f_{ij})\|(u_j)\| \\ & + b_i(1+\|\hat{B}_i(s)K_i^{-1}(s)T_i(s)\|_\infty)\|(u_i)\| \end{aligned} \tag{E.10}$$

If W is an M–matrix, from (E.8), $\|u_i\|<\infty$ for all $r_i \in L_2^{n_i}$. It is obvious that, from (E.10), $\|y_i\|<\infty$ for all i. Thus the following inequality holds

$$\sum_i \|y_i\| \leq k_t \sum_i \|r_i\| + \hat{k}_t$$

for some nonnegative constants k_t and $\hat{k}_t$. □

ACKNOWLEDGEMENT

We would like to thank Mr. Jia–Ling Lee very much for his careful editing and typewriting of this manuscript.

REFERENCES

1. R. V. Patel , M. Toda, and B. Sridhar," Robustness of linear quadratic state feedback designs in the presence of system uncertainty," IEEE Trans. Automat. Contr., AC–22, pp. 945–949 (1977).
2. M. Fielder and V. Ptak, "On matrices with non–positive off–diagonal elements and positive principle minors," Czech. Math. J., 12, pp. 382–400 (1962).
3. D. D. Siljak, *Large–Scale Dynamic Systems: Stability and Structure,* Elsevier North – Holland (1978).
4. M. Ikeda and D.D. Siljak, "Decentralized stabilization of linear time–varying systems," IEEE Trans. Automat. Contr., AC–25, pp. 106–107 (1980).
5. W. J. Wang, and C. F. Cheng, "Robustness of perturbed large–scale systems with local constant state feedback," Int. J. Control, 50, pp. 373–384 (1989).
6. W. J. Wang, C. F. Cheng, and T. T. Lee, "Stabilization of large–scale systems with nonlinear perturbations via local feedback," Int. J. System Science, 20, pp. 1003–1010 (1989).
7. A. N. Michel and R. K. Miller, *Qualitative analysis of large–scale system,* New York: Academic Press (1977).
8. Lj. T. Grujic' and D. D. Siljak, "Asymptotic stability and instability of large scale systems," IEEE Trans. Automat. Contr., AC–18. pp. 636–645 (1973).
9. D. D. Siljak, "On stability of large scale systems under structural perturbations," IEEE Trans. Syst. Man and Cyber., SMC–3 (1973).
10. W. E. Thompson, "Exponential stability of interconnected systems," IEEE Trans. Automat. Contr., AC–15, pp. 504–506 (1970).
11. E. J. Davison, "The decentralized stabilization and control of unknown nonlinear time varying systems," Automatica, 10, pp.

309–316 (1974).

12. M. E. Sezer and O. Huseyin, "Stabilization of linear time–invariant interconnected systems using local state feedback," IEEE Trans. Syst., Man, Cybern., SMC–8, pp. 751–756 (1978).
13. W. J. Wang, and B. S. Chen, "Stability of large–scale systems with saturating actuators," Int. J. Control, 47, pp. 827–850 (1988).
14. M. Vidyasagar, *Nonlinear systems analysis,* Prentice–Hall, New Jersey (1978).
15. M. Araki,"Applications of M–Matrices to the stability problems of composite dynamical systems," J. Math. Anal. Application, 52, pp. 309–321 (1975).
16. M. Araki, "Input–output stability of composite feedback system," IEEE Trans. Automat. Contr. AC–21, pp. 254–259 (1976).
17. M. Araki, "Stability of large–scale nonlinear systems –quadratic order theory of composite system method using M–matrices," IEEE Trans. Automat. Contr., AC–23, pp. 129–142 (1978).
18. M. Araki, and B. Kondo, "Stability and transient behavior of composite nonlinear systems," IEEE, Trans. Automat. Contr. AC–17, pp. 537–541 (1972).
19. B. S. Chen and W. J. Wang, "Robust stabilization of nonlinearly perturbed larde–scale systems by decentralized observer–controller compensators," Automatica, 26, pp. 1035–1041 (1990).
20. H. Kimura, "Robust stabilizability for a class of transfer function," IEEE Trans. Automat. Control, AC–29, pp. 788–793 (1984).
21 M. G. Safonov and B. S. Chen, "Multivariable stability–margin optimization with decoupling and output

regulation, "IEE Proc., 129, pt.D, pp. 276–282 (1982).

22. G. Zames, "On the input–output stability of time–varying feedback systems–Part I and II," IEEE Trans. Automat. Control, AC–11, Apr. and July respectively (1966).

23. P. Delsarte, Y. Genin and Y. Kamp, "Schur parametrization of positive definite Block–Toeplitz systems," SIAM J. Appl. Math., 36(1) pp. 34–46 (1979).

24. M. Vidyasagar, *Control system synthesis: A factorization approach,* MIT Press, Massachusetts (1985).

25. D. C. Youla, H. A. Jabr, and J. J. Bongiorno, Jr., "Modern Wiener–Hopf design of optimal controllers–Part II: the multivariable case," IEEE Trans. Auto. Control, AC–21, pp. 319–338 (1976).

26 K. J. Astrom and B. Wittenmark, Computer Controlled Systems: Theory and Design, Prentice–Hall (1984).

27. I. W. Sandberg, "On the L_2–boundedness of solutions of nonlinear functional equations," Bell Syst. Tech. J., 43, pp. 1581–1599 (1964).

28. F. N. Bailey, "The application of Lyapunov's second method to interconnected system," SIAM J. Control, 3, pp. 443–462, (1966).

29. B. S. Chen, "Controller Synthesis of optimal sensitivity: multivariable case" IEE Proc., Pt.D, 131, pp. 47 (1984).

30. F. M. Callier. W. S. Chan and C. A. Desoer, "Input–output stability of interconnected systems using decompositions: an improved formulation," IEEE Trans. Auto. Control, AC–23, pp. 150–163 (1978).

31. C. T. Chen, *Linear system theory and design,* N.Y. (1980).

32. M. J. Chen, and C. A. Desoer, "Algebraic theory for robust stability of interconnected systems: necessary and sufficient conditions," IEEE Trans. Auto. Control, AC–29, pp. 511–519 (1984).

33. M. Araki, and T. Mori, Trans. Soc. Instrum. Control Engrs., 15, pp. 267 (in Japanese) (1979).

34. B. A. Francis, and G. Zames, "On H^{∞} optimal sensitivity theory for SISO feedback systems," IEEE Trans. Automatic. Control, AC–29, pp. 9–16 (1974).

35. Y. S. Hung, and D. J. N. Limebeer, "Robust stability of additively perturbed interconnected systems," IEEE Trans. Auto. Control, AC–29, pp.1069–1075 (1984).

36. T. Kailath, *Linear systems,* Englewood Cliffs, N.J., Prentice–Hall (1980).

37. E. L. Lasley. and A. N. Michel, "Input–output stability of interconnected systems," IEEE Trans. Auto. Control, AC–21, pp. 84–89 (1976).

38. D. J. N. Limebeer, and Y. S. Hung, "Robust stability of interconnected system," IEEE Trans. Circuit and System, CAS–30, pp. 397–403 (1983).

39. M. Vidyasagar, and H. Kimura, "Robust controllers for uncertain linear multivariable systems," Automatica, 22(1), pp. 85–94 (1986).

40. W. A. Wolovich, *Linear multivariable systems,* Spring–Verlag, New York (1974).

41. A. H. Glattfelder and W. Schaufelberger, "Stability analysis of single loop control systems with saturation and antireset–windup circuits," IEEE Trans. Automat. Contr., AC–28, pp. 1074–1081 (1983).

42. G. F. Franklin, J. D. Powell and A. Emami–Naeini, *Feedback control of dynamic systems,* New York, Addision–Wesley, (1986).

43. T. Mori, N. Fukuma, and M. Kuwahara, "Simple stability criteria for single and composite linear systems with time–delays," Int. J. Control, pp.1175–1184 (1981).

44. R. Bellman, *Introduction to matrix analysis,* New York,

McGraw–Hill (1960).

45. W. S. Chan and C. A. Desoer, " Eigenvalue assignment and stabilization of interconnected systems using local feedbacks," IEEE Trans. Automat. Contr., AC–24, pp. 312–317 (1979).

46. W. J. Wang and T. T. Lee, " Robust stability criteria for single– and large–scale perturbed systems, " Int. J. Systems Science, 19, pp.405–413 (1988).

EXTENSION IN TECHNIQUES FOR STOCHASTIC DYNAMIC SYSTEMS

Ren-Jung Chang

National Cheng Kung University
Department of Mechanical Engineering
Tainan, Taiwan 701
R. O. C.

I. INTRODUCTION

The investigation of the dynamic behavior of a stochastic system has been attracted numerous researchers in the areas of physics, chemistry, biology, mathematics, sociology, and engineering applications [1]. For the control of a real system, an appropriate mathematical model is usually required for the description of the a priori information of input-output relation of the system. In a real system, the internal and external disturbances on the system always cannot be excluded due to the complex interaction between a system and its environment. The disturbance on the system will drive its response to nonlinear behavior with increasing excitation amplitude and the system response may behave in a stable or unstable manner. In addition to the real disturbance, disturbance can be employed for describing unmodeled dynamics to formulate a system under disturbance. Actually, a formulated model always will be under the risk of incomplete information about the complex system and its environment. With minimum amount of a priori information about the environment and its interaction with a system,

CONTROL AND DYNAMIC SYSTEMS, VOL. 55

a nonlinear dynamic equation with parametric and external disturbances plays a key role to model the system of interest by ignoring the dynamic model of its complex environment. By formulating a dynamic system with disturbances in a stochastic manner, the system behavior can be inferred and controlled with minimum amount of risk of modeling uncertainties.

In recent years, the analysis of nonlinear stochastic dynamic systems has reached a high amount of research interest in various engineering applications. Mathematical formulations extended from linear Langevin equation to Ito's nonlinear equation have been widely employed. By formulating a nonlinear stochastic system in Ito's sense, in principle, the dynamic behavior of the system can be investigated and controlled through employing the stochastic model. Actually, for a general nonlinear stochastic system, the associated partial differential equation with appropriate boundary and initial conditions which governing the propagation of probability density function of states always can be formulated. Several techniques have been developed for obtaining the stationary and nonstationary statistical response of a nonlinear dynamic system under stochastic parametric and/or external excitations. However, even for the stationary response, the exact density function can be obtained only for some classes of stochastic nonlinear systems and is almost impossible to be derived for a general nonlinear stochastic system including parametric-noise excited terms. Only partial and limited information such as the second-moment behavior usually can be estimated in developing some approximate techniques. Therefore, it is essential to develop or extend some analysis techniques for robust, efficient, and transparent obtaining more accurate and complete statistical information in investigating the dynamic behavior of a general stochastic system.

In this chapter, seven sections are included for the investigation of the developments of analytical techniques for stochastic dynamic systems. The main interest is to provide some extensions of existing analytical approaches for externally-excited stochastic system to techniques for the system including stochastic parametrically-excited terms. By following the introduction, a brief review of historical development in stochastic analysis starting from Einstein's work will be presented in section II. Section III describes Langevin equation and its various extensions. The analytical technique extended to general linear systems with external and/or parametric excitations is described in section IV. Section V provides various techniques for investigating nonlinear systems under external excitation. Recent techniques for analyzing nonlinear stochastic parametrically and externally excited systems are introduced and presented in section VI. Finally, the present chapter will be concluded and future research is recommended as given in section VII.

II. HISTORICAL DEVELOPMENT

A. BRIEF REVIEW [1-12]

Since Robert Brown (1827) first described the observation of minute particles contained in the pollen of plants suspended in a liquid, the irregular and perpetual motion of these particles has become known as Brownian motion. This concept of random motion has performed impact on almost all fields of scientific research. The physical interpretation of irregular and perpetual motion of Brownian particles was first provided by Einstein in 1905. In his paper, a mathematical analysis was formulated for the Brownian motion of a free particle. The analysis was also independently developed by Smoluchowski (1906). Langevin (1908) provided a hydrodynamic description of Brownian motion first through differential equation formulation. The mathematical analysis of Brownian motion was supported and verified by Perrin (1909) through microscope observations. Later, the physicists, Fokker (1914) and Planck (1917) developed the Brownian motion theory by constructing a differential equation for distribution. The differential equation obtained by Fokker and Planck was later derived by Kolmogorov (1931) with concrete mathematical foundation and the equation has become known as FPK equation.

Earlier work in engineering applications of stochastic analysis was initiated by Raleigh (1919). He first used the words "random vibrations" in describing an acoustical problem. A very useful tool of correlation function was introduced by Taylor (1920) in turbulence analysis. A fluctuation-dissipation theorem for noise and dissipation in electric resistors was established by Nyquist (1928). Application of Brownian motion theory on continuous solid bodies began with the work of Houdijk (1928) on strings.

The first application of FPK equation for investigating the dynamic behavior of stochastic systems was contributed by Andronov et al. (1933).

Earlier mathematical foundations on the stochastic evolution of Brownian systems were provided by Wiener, Levy, and Kolmogorov in late 1920s and early 30s. The foundation of stationary random process in terms of correlation function and spectral density function was laid by Wiener (1930) and later by Khintchine (1934). Wiener's stochastic integral with integrand of deterministic real function was generalized by Ito (1944) to the integral with stochastic function as integrand. Because of the erratic nature of Brownian motion, the integral was defined in the mean square sense. With Ito's integral, the existence and uniqueness conditions of a general stochastic differential equation were established by Ito (1951) and the equation has become known as Ito's equation.

More recent analyses of stochastic dynamic systems in physics, chemistry, biology, and engineering were initiated in 1940. The physical problems of Brownian motion began with the contributions by Kramers (1940), Chandrasekhar (1943), Wang and Uhlenbeck (1945), and Kac (1947). Pioneer work of zero-crossing and maxima distributions of noise current with later important applications to reliability studies was initiated by Rice (1944). Important contribution to information theory was also given by Rice after the theory was developed by Shannon (1948) through entropy function. A new area called Cybernetics for the communication and control in animal and machine was coined by Wiener and Rosenblueth (1947). A numerical scheme called Monte Carlo method for simulating Brownian motion of neutron particle was developed by Metropolis and Ulam (1949). The Wiener model of Brownian motion, called the Wiener process, was later combined with independent work of Feynman (1948) in treating path integral formulation of quantum mechanics. Applications of FPK equation to various engineering problems including electronic, control, and structural systems had been made by Stratonovitch, Barret, Chung, Tikhonov, Ariaratnam, Lyon, Crandall, and Caughey in middle 1950s and early 60s.

Successful applications of stochastic dynamic equations and FPK equation have further stimulated a wide variety of physical and engineering applications in 60s. Investigations of stochastic stability for dynamic systems with stochastic coefficients were mainly presented by Stratonovitch, Ariaratnam, Infante, Khasminskii, Kushner, Samuels, Caughey, and Kozin.

Related researches in the estimator theory with applications were given by Kalman, Bucy, Zakai, and Kushner. In middle 1960s, several nonlinear stochastic dynamic problems in mechanics, electronics, and control were formulated; however, the approximate solutions usually were derived under very restrictive conditions. Analog techniques were employed for experimental investigations of stochastic systems. Digital techniques for experimental spectrum estimation have been developed since the efficient fast Fourier transform (FFT) algorithm was first obtained by Cooley and Turkey (1965). A definition of stochastic integral in mean-square sense for physical Brownian motion was later proposed by Stratonovitch (1966). The study of generalized Langevin equation with generalized fluctuation-dissipation theorem was contributed by Zwanzig (1964) and Kubo (1966).

The applications of stochastic dynamic analysis for further improvement of engineering designs and scientific investigations of natural behavior have been undertaken since 1970. For the improvement of system performance and reliability design under stochastic disturbances, the techniques have been widely employed for various fields of engineering design problems. The stochastic design problems consist of airplane in atmospheric turbulence, offshore structure subject to ocean waves, structure and buildings under earthquake ground motions, vehicles on rough road and track, signal transmission and receiving through noisy channel, servo-system operation under electromagnetic noise, etc. Contributions to the structural problems have been made by Bolotin, Caughey, Crandall, Lin, Roberts, Shinozuka, Spanos, Vanmarcke, etc. For the control problems, contributions were given by Åström, Athans, Hedrick, Leondes, Wonham, etc. The design and control of a system for reliable operation under natural environment has brought impact to various scientific interests of stochastic analyses. Investigation of self-organization in a natural system has coined the development of a new area called Synergetics by Haken (1971).

Recent work in treating quantum fluctuation and field theory has begun with contribution by Nelson (1966). Some exact solutions of path integral for quantum oscillators were obtained by Duru and Kleinert (1979). Mathematical theory of functional integration are strongly connected with modern statistical mechanics and quantum physics. Semi-martingales theory finally established in 1974-1979 provided the mathematical foundation of stochastic calculus by Doleans-Dade and Meyer (1970). The study of relativistic stochastic mechanics on Minkowski space without using stochastic differential equations was presented by Dohrn, Guerra, and Ruggiero (1979).

Recent engineering researches on a stochastic system interacted with its environment were further investigated in 80s. Techniques for analyzing stochastic response of a parametrically-excited nonlinear system were extensively investigated by Ibrahim, To, Lin, and Chang. Applications of stochastic dynamic analysis have been extended to robotic systems for analyzing operational precision of end effector by Chang in late 80s.

B. CHRONOLOGICAL EVOLUTION

A historical review of evolution history starting from Einstein's analysis to modern relativistic stochastic mechanics is given above. A chronological order of historical development with important contributions will be also summarized as follows.

1905 A. Einstein	First give theoretical investigation of Brownian motion
	First obtain a fluctuation-dissipation relation in statistical physics
1906 M. v. Smoluchowski	Give independent analysis of Brownian motion of a free particle
1908 P. Langevin	First give differential equation formulation of Brownian motion
1909 J. B. Perrin	Perform experimental verification of theory of Brownian motion
1914 A. D. Fokker 1917 M. Planck	First derive a differential equation for the distribution function describing Brownian motion
1919 Lord Raleigh	First use the words "random vibrations" in an acoustical problem

1920 G. I. Taylor	First introduce correlation function
1923 N. Wiener	First to analyze Brownian motion as a continuous process with probability one Introduce a differential space, later called Wiener space, to study Brownian motion
1928 A. Houdijk	First work on Brownian motion of continuous solid bodies
1928 H. Nyquist	First obtain fluctuation-dissipation theorem for electric noise
1930 N. Wiener	Develop generalized harmonic analysis
1931 A. N. Kolmogorov	Give mathematical foundation of Fokker-Planck equation
1933 A. A. Andronov A. A. Witt L. S. Pontryagin	First apply FPK equation for studying one-dimensional dynamic systems
1934 A. Ya. Khintchine	Lay the foundation for theory of stationary random processes
1940 H. A. Kramers	Use FPK equation for studying chemical reaction rate
1944 K. Ito	Define stochastic integral
1944 S. O. Rice	Investigate stochastic properties of level-crossing and maxima distributions
1947 N. Wiener A. Rosenblueth	Coin the word "Cybernetics" for control and communication in animal and machine
1948 R. P. Feynman	Introduce path integral formulation for quantum oscillators
1948 C. E. Shannon	Characterize information transmission through entropy function

1949 N. Metropolis, S. Ulam	Coin the words "Monte Carlo method"
1951 K. Ito	Give existence and uniqueness conditions of stochastic differential equation
1953 R. L. Stratonovitch	Apply FPK equation to electronic engineering
1958-1959	
J. F. Barrett K. Chung V. I. Tikhonov	Apply FPK equation to nonlinear control system
1960 R. E. Kalman R. S. Bucy	First develop optimal linear estimator in state space formulation
1960-1961	
S. T. Ariaratnam R. H. Lyon T. K. Caughey S. H. Crandall	Apply FPK equation to nonlinear vibration
1964 H. J. Kushner	First obtain a stochastic partial differential equation for conditional density in nonlinear filtering theory
1964 R. S. Bucy	Find a connection of stochastic stability with martingale-convergence theorems
1965 J. Cooley J. Turkey	First develop FFT algorithm
1965 E. Wong, M. Zakai	Establish convergence of ordinary integrals to stochastic integrals
1966 R. L. Stratonovich	Interpret stochastic integral of physical Brownian motion
1966 E. Nelson	Introduce stochastic mechanics formulation of quantum fluctuation

1967 H. J. Kushner	Introduce definition of stable with probability one
1971 H. Haken	Establish a new area called Synergetics
1979 I. H. Duru H. Kleinert	Use Kustaanheimo-Stiefel transformation for exact evaluation of path integral
1980-1992 R. A. Ibrahim C. W. S. To Y. K. Lin R. J. Chang	Analyze stochastic parametrically-excited nonlinear systems

Previous work and development on the dynamic analysis of stochastic systems have been briefly reviewed. The Langevin approach for analyzing motion of Brownian particles under friction force and irregular excitation force has been employed and extended for various applications. Generalization and various extensions of Langevin equation will be introduced in section III.

III. LANGEVIN EQUATION AND ITS EXTENSIONS

Langevin equation provides a phenomenological description of the erratic motion of a heavy particle immersed in a fluid bath of light particles. In this formulation, the Brownian particle, and fluid bath can be described as a system, and its environment, respectively from the viewpoint of system engineering. The interactions of the Brownian particle with its complex environment are attributed by irreversible dissipative force arised from Stokes law and by fluctuating force due to instantaneous random imbalance in the number of impacting particles on Brownian particle. The formulation of Langevin equation for a particle of mass m^b moving with velocity v in a fluid is given as

$$m^b \dot{v} + \beta v = F_d(t) \tag{1}$$

The fluctuating force F_d is described as a stochastic process with zero mean and δ-correlation function. The correlation function is proportional to absolute temperature of fluid bath T and friction coefficient β as

$$E[F_d(t)F_d(s)] = 2R_0T\beta\delta(t-s) \tag{2}$$

where R_0 is a Boltzmann constant. The features of this model for a system with its environmental interaction are: (1) the dissipation force is linear and asymptotically achieves thermal equilibrium with the environment, (2) the fluctuating force is independent of the system states, (3) Brownian particle is not in the field of force, (4) particle is in one-dimensional motion, (5) the Reynold number of the motion of Brownian particle is small, and (6) the correlation time of fluctuation force is much less than the "mechanical" time constant of Brownian particle. Based on the above features in Langevin model of a Brownian particle and its environment, several directions of extension are made possible.

The extension of Langevin equation to a general formulation has been undertaken in various physical and engineering applications [2,3,12,13]. In many physical problems, the extension of the first feature has been widely investigated. A generalized Langevin equation with generalized fluctuation-dissipation theorem, which can be derived on the grounds of statistical mechanics, in thermal equilibrium has been developed. The extension of the equilibrium theory to thermal fluctuation has been further presented as governed by Navier-Stokes Langevin equation. Another direction of extension in physical and engineering applications can be considered as describing Brownian motion through a phenomenological n-dimensional state equation with state-noise multiplicative terms as

$$\dot{\underline{X}}(t) = \underline{F}(\underline{X}(t),t) + G(\underline{X}(t),t)\underline{V}(t) \tag{3}$$

where $\underline{X}$ is an nx1 vector of states process, $\underline{F}$ is an nx1 vector nonlinear function of states, G is an nxm matrix nonlinear function of states, and $\underline{V}$ represents a zero-mean ideal Gaussian white noise process with intensity

$$E[\underline{V}(t)\underline{V}^T(s)] = Q_W\delta(t-s) \tag{4}$$

With the conditions of existence and uniqueness as Lipschitz and uniform growth conditions for the solution process $\underline{X}$ to Eq. (3), the Ito's equation associated with the corresponding heuristic formulation of Eq. (3) can be expressed as

$$d\underline{X}(t) = \underline{F}(\underline{X}(t), t)dt + G(\underline{X}(t), t)d\underline{W}(t) \tag{5}$$

where $\underline{W}$ is the zero-mean Wiener process with intensity $Q_W dt$. If the excitation process is provided as physical white noise, Wong-Zakai correction [5] can be employed to convert the stochastic equation to Ito's equation. Based on the Ito's Eq. (5), analytical techniques extended for various dynamic models of stochastic system have been developed. The extension of analytical technique to general linear systems will be presented in the following section.

IV. EXTENSION TO LINEAR SYSTEMS WITH EXCITATIONS

A. TIME-INVARIANT SYSTEMS UNDER EXTERNAL EXCITATION

Dynamic analysis of this class of system constitutes the main research in classical stochastic dynamic analysis. Since the Gaussian density is invariant under linear transformation, the analysis of the first and second moment response thus provides a full description of output behavior. When only the stationary behavior is investigated, the analysis is usually carried out in the frequency domain through Wiener-Khintchine relation [14]. The formulation is derived originally for single-input single-output (SISO) systems and can be extended to multiple-input multiple-output (MIMO) systems through frequency transfer matrix. If the MIMO system has uncoupled modes of oscillation, the total response of the linear system is easily obtained by the superposition of modal responses. The uncoupled formulation is usually accounted for when the system is with light damping and widely spaced modal frequencies. When different modes are highly coupled, the ignorance of cross-correlation terms between different modes may cause severe error in mean square prediction. If the number of modes involved is large, the computation of modal sum is usually prohibitive. Approximate estimation by asymptotic method or statistical energy method is usually employed [15,16]. In addition to the above analysis problem, a synthesis problem for obtaining shaping filter with specified input white noise and output spectrum can be realized through spectral factorization technique [17]. With the shaping filter to filter the white noise input for generating specified correlated noise, the filter equation can be augmented to a system under the correlated noise excitation for providing an augmented system equation under white noise excitation. Thus, the analysis technique applicable for the white noise model can be readily employed for investigating the system behavior under correlated-noise excitation.

B. LINEAR SYSTEMS WITH PARAMETRIC EXCITATION

In investigating linear time-varying stochastic systems, several important researches have been given to the stochastic parametrically-excited systems [9,18]. For analyzing the dynamic behavior of this class of system, two important problems concerning the stochastic stability and statistical response of stable system are usually investigated. Several stability definitions and theorems have been established by investigators with motivation on designing adaptive control systems and structural dynamic systems. Earlier work in stability analysis was contributed by Rosenbloom et. al. (1954), Tikhonov (1958), Samuels, and Eringen (1959) [18]. The techniques for investigating moment stability (including mean-square stability) and almost sure asymptotic sample stability have been widely investigated [9,19-22]. Extensions of theorems provided by Infante (1968) and by Khasminskii (1962) have been usually employed to determine the almost sure stability of the static equilibrium of linear systems with parametric Gaussian white noise excitation [9,19-22]. Moment stability of arbitrary order can be investigated by formulating a set of closed moment equations and employing Hurwitz stability test. For the analysis of statistical output response of a parametrically and externally Gaussian white-noise excited system, the first and second moment response is closed and easily solvable although the response is usually non-Gaussian. For a second-order stochastic equation with a specific intensity relation between spring and damping noise, exact stationary non-Gaussian density has been obtained by Dimentberg [23]. An important result of response of a linear SISO system under parametric and external Gaussian white-noise excitations is that the output spectral density can be derived from the existing result of spectral response of the corresponding externally-excited only linear system multiplied by an amplification factor which is attributed solely by the excitation of parametric noise [24-28]. The above property can be derived by several different approaches; however, the utilizing of the concept of equivalent external excitation is the simplest and most staightforward approach [26,29].

The extension of dynamic analysis from Langevin equation to general linear stochastic systems has been introduced in this section. However, due to the excitation model as with Gaussian distribution, the output response of a real physical system cannot be in linear proportion to input value when it is ranged from $-\infty$ to ∞. The above mathematical model is improper for predicting response behavior and consequently, the model of Gaussian excitation should be incorporated with system nonlinearities in analyzing the dynamic behavior of a real physical system.

V. EXTENSION TO NONLINEAR SYSTEMS UNDER EXTERNAL EXCITATION

A. TECHNIQUES FOR SINGLE-DOF (SDOF) SYSTEMS

1. Solution of FPK Equation [30]

a. Exact Solution of FPK Equation

The first application of FPK equation for studying nonlinear dynamic system was given by Andronov et al. (1933) [31]. They showed that dynamic systems with a nonlinear spring under external white noise excitation were solvable. The same result was also independently obtained by Kramers (1940) [32]. Exact solution of FPK equation usually can be found only for stationary solution of some classes of systems. The exact stationary density function for first-order SISO systems can be obtained through direct integration [31]. For few specific class of first-order nonlinear system, the transition density function has been derived [6]. For second-order systems, exact stationary solutions have been derived for some classes of oscillators with drift and diffusion coefficients which satisfy some restrictive conditions [33,34]. Satisfaction of detailed balance can be verified and employed for constructing a class of stationary solution. The solvable density functions which are physically interpreted are closely related to the Maxwell-Boltzmann distribution in kinetic theory of gases. The corresponding class of systems then enjoy equipartition of energy among the various degrees of freedom in stationary [4].

b. Approximate Solution of FPK Equation

A variety of solution schemes have been proposed for obtaining approximate solutions of FPK equation associated with wide-band excited nonlinear stochastic systems. All of the approaches investigated appear to require tedious derivation and computation. The available solution techniques include eigenfunction expansion, variational method, iterative solution, reduction to Hermitian problem, reduction by adiabatic elimination, and density expansion by complete sets.

2. Statistical Linearization Method [35]

The method of statistical linearization was independently developed by Booton (1954), Kazakov (1954), and Caughey (1959). The objective of this approach is to replace the nonlinear element in a model by a linear one with linearization coefficient obtained from a specified linearization scheme. In application, minimization of mean-square error is usually selected to approximate a nonlinear element by a linear one. If Gaussian (non-Gaussian) density function is applied to express the linearization coefficients as functions of undetermined Gaussian (non-Gaussian) moment response, the linearization method is called Gaussian (non-Gaussian) linearization scheme. After formulating the statistical linearization systems, a final set of moment balance equations is obtained and solved for the first and second moment response. This technique is useful and practical for obtaining mean-square response; however, the nonliner effects on the response distribution cannot be obtained. This technique can be also employed for hysteretic systems by utilizing and incorporating a first-order differential equation for hysteresis model [36]. For the improvement of approximation, a non-Gaussian density through Gram-Charlier series has been employed. However, density expansion by this series may provide an unacceptable negative distribution.

3. Stochastic Averaging Method [37, 38]

This averaging method was initially proposed by Stratonovitch (1963) based on physical argument for dealing with nonlinear lightly-damped systems under wideband excitation [39, 40]. The mathematical foundation was later given by Khasminskii [41]. The application of original Stratonovitch method will eliminate the effect of nonlinear restoring forces. Therefore, the original method, which is called classical or standard method, has been extended to a generalized method for energy envelope process by Stratonovitch [39] and later by Roberts [37] to recover the difficulties in original method. The rigorous mathematical basis is provided by a theorem due to Khasminskii [42]. The third method is the averaging of coefficients in FPK equation or averaging method of Khasminskii. This method is essentially equivalent to stochastic averaging method of Stratonovitch but with different order in averaging and forming FPK equation. In application of averaging method, one usually involves three steps. First, the stochastic equation is transformed into a standard form. Next, an averaged FPK or Ito equation is derived. Finally, the averaged FPK equation or equivalent moment equation is solved.

4. Closure Method [18, 43]

This approach for obtaining statistical moment or correlation response has been given quite extensive investigations in the literature on turbulence. The scheme is to set up specified truncation procedure by which an infinite hierarchy of moment or correlation equations can be truncated such that lower order one's can be closed and computed. If all higher moments are expressed in terms of the first and second moments in the hierarchy moment equations, the simplest Gaussian closure method can be formulated. This approximate approach is identical to statistical linearization scheme in employing Gaussian properties. In order to improve the moment accuracy over that by the Gaussian closure scheme, a non-Gaussian closure method is usually employed. Several other closure methods have been developed since the non-Gaussian scheme was proposed by Dashevskii and Liptser (1967) [44]. The truncation of moments and semi-moments (cumulants) were most extensively investigated. The truncation of third order and higher cumulants will lead to Gaussian closure method. Truncation of cumulants beyond the fourth order were utilized to improve predicted accuracy of moment. Accurate results can be obtained for some ranges of parameter; however, totally unacceptable results are also obtained for some other ranges of parameter. In addition, this approach will become rather difficult when the higher order cumulants need to be retained. For example, the retaining of the cumulants up to the fourth order used to solve a second-order nonlinear oscillator usually requires the solution of ten simultaneous nonlinear algebraic equations if one cannot eliminate certain variables by trivial substitution. Furthermore, it is formidable to apply this method for nonpolynomial-type nonlinearities. Non-Gaussian closure schemes based on the truncated Gram-Charlier or Edgeworth expansion have been also developed for improving stationary moment accuracy [45-47]. Accurate results can be obtained by this approach; however, the resulting nonlinear algebraic equations for solving stationary moment response might have no real solution if inappropriate constraints or trial density function are employed. Actually, the same drawback as in implementing non-Gaussian linearization by this series expansion is also carried over to this method for stationary and non-stationary analysis [48].

5. Statistical Equivalent Method [35]

The difficulties and usefulness in developing this method for investigating stochastic system depend on the definition of equivalence in statistical sense. The concept of equivalence may be specified in a strong or weak sense ranging from density function to moment response. Thus, the equivalent sense indicates that certain specified statistical properties of all sample solutions are the same even though individual sample solutions can be all different. If the equivalent sense can be extended to the sense of mean-square error between a nonlinear function and linear one, the minimization of mean-square error by statistical linearization method may be considered as a special class of statistical equivalent method. The statistical equivalence between two nonlinear stochastic systems was first employed by Lutes (1970) [49] for obtaining response of a hysteretic stochastic system. Some research work on mean-square equivalent nonlinear models in accompany with numerical computation for analyzing nonlinear stochastic systems are referred to [50].

6. Perturbation Approach

This technique was developed by Crandall (1963) [51] for obtaining approximate solutions of weakly nonlinear stochastic systems. By representing system response as a series involving powers of some small parameter describing the size of nonlinearity, terms of like power in ϵ are collected to formulate a cascade of linear stochastic equations which can be solved sequentially. This technique is versatile; however, the scheme for every sample function of response represented as a convergent series of powers in ϵ has not been proved. In addition, the computational difficulties rapidly increase as the higher order terms included. Hence, usually only first order terms in this method are utilized and evaluated.

7. Other Methods

Apart from the aforementioned analysis techniques, several other methods have been developed. Combination of perturbation and statistical linearization methods was employed for spectral response of Duffing equation by Dienes (1961) [52]. Decomposition method based on stochastic operator theory was developed by Adomian (1961) [53]. Functional series expansion method based on Wiener-Hermite series was employed by Orabi and Ahmadi [54].

B. TECHNIQUES FOR MULTIPLE-DOF (MDOF) SYSTEMS

The techniques for analyzing SDOF stochastic systems can be extended for MDOF systems in principle. However, some techniques may not be practically implemented due to analytical difficulties and/or computational inefficiency. For the exact solution of FPK equation, certain restricted classes of MDOF nonlinear systems have been obtained [4]. The statistical linearization method is easily extended and implemented for MDOF systems [55, 56]. Extension of perturbation method to MDOF systems can be directly implemented [51]. The problem of expansion of response as powers of ϵ is also carried over to MDOF formulation. Application of stochastic averaging method to deal with MDOF systems can be implemented only by standard technique [37, 38]. The restrictions of applicable lightly-damped systems and elimination of nonlinear effects of restoring force also exist in MDOF formulation. In addition to the schemes extended from analyzing SDOF systems, a normal-mode approximate approach can be extended from modal technique for linear systems.

Normal Mode Approximation

For a MDOF nonlinear system which can be expressed as a set of generally coupled nonlinear second-order nonlinear systems, one technique is first to reduce the system equation to a set of modal equations coupling through nonlinear terms. The reduced modal equations then can be solved by incorporating statistical linearization or some other approximate schemes [57-59].

The above nonlinear model provides a disturbance-response description for a system under external stochastic disturbance. When the internal stochastic disturbance cannot be ignored, as usually attributed by the thermal dissipation accompanied with signal transmission, the nonlinear stochastic system can be extended to include parametric stochastic excitation.

VI. EXTENSION TO NONLINEAR SYSTEM UNDER PARAMETRIC EXCITATION

The development of analysis techniques for investigating the response behavior of general nonlinear systems subjected to both stochastic parametric and external excitations is the most important objective in the area of nonlinear dynamic analysis. For investigating the response behavior of a general nonlinear stochastic system, an approximate analysis technique is usually required since the exact solution exists only for very few special oscillators. In the investigation of output response by approximate techniques, two important problems concerning the statistical stability and moment response usually need to be investigated. The stability problem is very important for a stochastic parametrically excited nonlinear system; however, the analytical investigation has not gained good progress for a general stochastic system in the past few years. Even for the SISO second-order nonlinear systems with a parametric noise, the mean-square stability problem has not been completely investigated. Actually, most analytical techniques published are valid and reliable only for the statistical moment response of a stable stochastic parametrically-excited nonlinear system. The applications of approximate techniques in analyzing the stability and bifurcation problems are still not suitable.

A. TECHNIQUES FOR SINGLE-DOF SYSTEMS

The techniques developed for analyzing externally-excited nonlinear system are expected to be extended for system with parametrically-excited terms. Exact solution of FPK equation can be derived for a certain class of first-order systems [31]. For second-order systems, the class of solvable systems has been expanded by extension of the method of detailed balance [60] for solving reduced FPK equation [61, 62]. Other than the existed solvable physical systems, the extended solvable class of system by this approach is applicable only for mathematically stochastic system with solvable reduced FPK equation. For the extension of the solvable mathematical stochastic system to physically realizable systems, the method of Cai and Lin [62] has been further extended to equivalent nonlinear system by incorporating least mean-square criterion and Euler equation [63]. This approach is actually an extension of statistical equivalent method for externally excited system. Other statistical equivalent methods have been implemented through developing the concept of equivalent external excitation [29], or utilizing the scheme of stochastic averaging for lightly-damped systems [64, 65]. Statistical linearization method has been extended for the parametrically excited system [31]. The method of Chang and Young [66] is relatively much more practical and simpler to use than the scheme by Bruckner and Lin [31]. In addition, the mean-square stability condition can be obtained by the approach proposed by Chang and Young [66, 67]. Stochastic averaging method has been extended since the pioneering work of Stratonovitch and Romanovskii [68] for investigating the stability of system by parametric noise. Closure method is directly applicable for the system with parametrically stochastic excitation [9, 18, 69, 70]. The cumulant closure method has been widely employed and investigated for predicting mean-square response; however, it is not suitable for analyzing stability and bifurcation problem. Although the accuracy of mean-square response may be improved by retaining higher-order cumulants, totally unacceptable results can be obtained for some ranges of parameters. When a stochastic system with hard constraints as in a robotic system, a Fourier closure method was developed for investigating the dynamic behavior of a stochastic system [71]. First-order approximation method as perturbation approach has been extended to second-order method by using a successive approximation scheme for obtaing spectral response of weakly excited nonlinear systems [72]. However, the instability due to parametric excitation will be removed by this technique.

B. TECHNIQUES FOR MULTIPLE-DOF SYSTEMS

1. Equivalent External Excitation Method [26, 29]

The first scheme for obtaining approximate solution of stochastic systems subjected to both parametric and external excitations can be employed when the response of the corresponding systems subjected to external excitation can be solved by using the FPK equation. By employing this technique, the effect of parametrically stochastic disturbance can be directly included for the prediction of stable statistical response of the associated nonlinear systems subjected to external stochastic excitation. The implementation of this technique based on the method for SISO system [29] has been developed for the coupled matrix second-order systems by the following procedure. Let the nonlinear stochastic system in the Ito's sense be described heuristically as

$$M\ddot{\underline{q}} + C\dot{\underline{q}} + K\underline{q} + \sum_{i=1}^{n_1}(a_i + \alpha_i(t))\underline{g}_i(\underline{q}, \dot{\underline{q}}) = \underline{w}(t) \tag{6}$$

where $\underline{q}$ is a generalized displacement, $\underline{g}_i(\underline{q},\dot{\underline{q}})$ is a nonlinear function, and $\underline{w}(t), a_i(t)$ are mutually independent zero-mean Gaussian white noises with intensities

$$E[\underline{w}(t)\underline{w}^T(s)] = Q_0\delta(t-s)$$
$$E[\alpha_i(t)\alpha_i(s)] = Q_i\delta(t-s) \tag{7}$$

By assuming that M is a nonsingular matrix, Eq. (6) can be rewritten as

$$\ddot{\underline{q}} + M^{-1}C\dot{\underline{q}} + M^{-1}K\underline{q} + \sum_{i=1}^{n_1}(a_i + \alpha_i(t))M^{-1}\underline{g}_i(\underline{q},\dot{\underline{q}}) = M^{-1}\underline{w}(t) \tag{8}$$

If the Ito's differential rule is applied to Eq. (8), then the moment propagation equation can be derived as

$$\dot{E}[\underline{q}\underline{q}^T] = E[\dot{\underline{q}}\underline{q}^T] + E[\underline{q}\dot{\underline{q}}^T]$$

$$\dot{E}[\underline{q}\dot{\underline{q}}^T] = E[\dot{\underline{q}}\dot{\underline{q}}^T] - E[\underline{q}(M^{-1}C\dot{\underline{q}})^T] - E[\underline{q}(M^{-1}K\underline{q})^T] - \sum_{i=1}^{n_1} a_i E[\underline{q}(M^{-1}\underline{g}_i)^T]$$

$$\begin{aligned}\dot{E}[\dot{\underline{q}}\dot{\underline{q}}^T] = &- E[M^{-1}C\dot{\underline{q}}\dot{\underline{q}}^T] - E[M^{-1}K\underline{q}\dot{\underline{q}}^T] - E[\dot{\underline{q}}(M^{-1}C\dot{\underline{q}})^T] - \\ &E[\dot{\underline{q}}(M^{-1}K\underline{q})^T] - \sum_{i=1}^{n_1} a_i E[(M^{-1}\underline{g}_i)\dot{\underline{q}}^T] - \\ &\sum_{i=1}^{n_1} a_i E[\dot{\underline{q}}(M^{-1}\underline{g}_i)^T] + \sum_{i=1}^{n_1} Q_i E[(M^{-1}\underline{g}_i)(M^{-1}\underline{g}_i)^T] + Q_0\end{aligned} \tag{9}$$

Now, by moving the parametric noise excited terms to the right hand side, Eq. (6) is rewritten as

$$M\ddot{\underline{q}} + C\dot{\underline{q}} + K\underline{q} + \sum_{i=1}^{n_1} a_i\underline{g}_i(\underline{q},\dot{\underline{q}}) = \underline{w}(t) - \sum_{i=1}^{n_1}\alpha_i(t)\underline{g}_i(\underline{q},\dot{\underline{q}}) \tag{10}$$

or

$$M\ddot{\underline{q}} + C\dot{\underline{q}} + K\underline{q} + \sum_{i=1}^{n_1} a_i\underline{g}_i(\underline{q},\dot{\underline{q}}) = \underline{w}_e(t) \tag{11}$$

From Eqs. (9)-(11), it is observed that the exact second-moment propagation equation can be derived from Eq. (11) if the $\underline{w}_e$ is interpreted as an equivalent external Gaussian white noise with the intensity

$$E[\underline{w}_e(t)\underline{w}_e{}^T(s)] = (Q_0 + \sum_{i=1}^{n_1} Q_i U_i)\delta(t-s)$$

$$U_i = E[(M^{-1}\underline{g}_i)(M^{-1}\underline{g}_i)^T] \tag{12}$$

After deriving equivalent external excitation system, the method can be implemented by the following procedure.

(1) Replace the stochastic parametrically and externally excited nonlinear system by an equivalent one under only external excitation as Eq. (11).

(2) Obtain the stationary probability density functions of the states of the equivalent model from Step 1 by using the FPK equation.

(3) Match the expected value in each element of U_i in Eq. (12) by taking the expected value of corresponding each element through the probability density functions of the states obtained from Step 2.

(4) Solve the simultaneous algebraic or integral equations by iterative methods.

It has been seen that the concept and technique of equivalent external excitation can be extended from a SDOF to a MDOF one. However, there are two important assumptions which need to be further discussed behind the concept of equivalent external excitation. First, the technique is developed for a nonlinear stochastic system with no correlation between the parametric and external noise. Second, the uncorrelated property between the response and parametric noise is assumed in Eq. (12) for developing this technique. The first assumption is usually employed for developing analysis technique since the parametric and external excitations are usually generated from different sources. As for the second assumption, the validity is supported by Ito's equation since in $\underline{g}_i(\underline{q}, \dot{\underline{q}})$ the value of $\underline{q}$ and $\dot{\underline{q}}$ should be taken just before the pulse is applied. The validity of these assumptions has been proved mathematically for a linear stochastic parametrically and externally excited system [24, 25]. Thus, these assumptions are acceptable in developing this approximate technique. Implementing this method allows the straightforward extension of existing solutions of response of nonlinear externally excited systems to predict the response of nonlinear systems subjected to both stochastic parametric and external excitations.

2. Gaussian Linearization Method [73]

This linearization method has been extended from Gaussian linearization technique for externally excited system. Three different approaches have been developed [35]; however, the instability due to parametrically excited terms will be removed by the scheme of Kottalam et al. [78 of 35] and computational burden will be encountered by the approach of Brückner and Lin [17 of 35]. The linearization scheme presented can be efficiently employed for predicting stable mean-square performance and stability boundary in mean-square sense. The linearization method has been formulated for the system as

$$d\underline{X} = \underline{F}(\underline{X}(t))dt + G(\underline{X}(t))d\underline{W}(t) \tag{13}$$

where $\underline{X}$ is an nx1 vector of state processes, $\underline{F}$ is an nx1 vector nonlinear function of states, G is an nxm matrix nonlinear function of states, and $\underline{W}$ represents a zero-mean Wiener process with intensity

$$E[d\underline{W}(t)d\underline{W}^T(t)] = Q_W dt \tag{14}$$

For the stochastic system, a Gaussian linearization system will be first derived by the following procedure. First, let the nonlinear functions in Eq. (13) be approximated as

$$\underline{F}(\underline{X}(t)) = \underline{A}(t) + K_e(t)(\underline{X}(t) - \underline{M}_e(t)) \tag{15a}$$

and

$$G(\underline{X}(t)) = B(t) + \sum_{i=1}^{n} L_i(t)(x_i - m_i) \tag{15b}$$

with

$$\underline{M}_e(t) = [m_i] = E[\underline{X}(t)] \tag{16a}$$

and

$$\underline{X} = [x_i] \tag{16b}$$

Then the error between the nonlinear functions and the corresponding linear functions in Eq. (15) can be obtained as

$$\underline{E}_1 = \underline{F}(\underline{X}(t)) - \underline{A}(t) - K_e(t)(\underline{X}(t) - \underline{M}_e(t)) \tag{17a}$$

and

$$E_2 = G(\underline{X}(t)) - B(t) - \sum_{i=1}^{n} L_i(t)(x_i(t) - m_i(t)) \tag{17b}$$

By following the usual statistical linearization technique, the unknown $\underline{A}$, K_e, B, and L_i are derived such that the mean-square error by Eq. (17) will be minimized. From the necessary conditions for the minimization of the mean-square error,

$$\frac{\partial E[\underline{E}_1{}^T\underline{E}_1]}{\partial \underline{A}} = \frac{\partial E[\underline{E}_1{}^T\underline{E}_1]}{\partial K_e} = 0 \tag{18a}$$

and

$$\frac{\partial E[\underline{E}_{2,i}^T\underline{E}_{2,i}]}{\partial B} = \frac{\partial E[\underline{E}_{2,i}^T\underline{E}_{2,i}]}{\partial L_i} = 0 \tag{18b}$$

where $\underline{E}_{2,i}$ is an i-th partitioned column vector of E_2, the parameters of the statistical linearization system can be derived as

$$\underline{A}(t) = E[\underline{F}(\underline{X}(t))] \tag{19a}$$

$$K_e(t) = E[\underline{F}(\underline{X}(t))(\underline{X}(t) - \underline{M}_e(t))^T]P(t)^{-1} \tag{19b}$$

$$B(t) = E[G(\underline{X}(t))] \tag{19c}$$

$$L_p = E[G(\underline{X}(t))(\underline{X}(t) - \underline{M}_e(t))_p^T](P_p(t))^{-1} \tag{19d}$$

where $P(t)$ is a covariance matrix defined as

$$P(t) = E[(\underline{X}(t) - \underline{M}_e(t))(\underline{X}(t) - \underline{M}_e(t))^T] \tag{20a}$$

and the partition matrices are defined by

$$L_p = [L_1|L_2|\ldots|L_n] \tag{20b}$$

$$(\underline{X}(t) - \underline{M}_e(t))_p^T = \big(x_1(t) - m_1(t) | x_2(t) - m_2(t) | \ldots | x_n(t) - m_n(t)\big) \quad (20c)$$

$$P_p = \left(\begin{array}{c|c|c|c} P_{11} & P_{12} & \ldots & P_{1n} \\ \hline P_{21} & P_{22} & \ldots & P_{2n} \\ \hline \vdots & \vdots & \ddots & \vdots \\ \hline P_{n1} & P_{n2} & \ldots & P_{nn} \end{array}\right) \quad (20d)$$

The statistical linearization system then can be expressed by substituting Eq. (15) for the corresponding nonlinear functions in Eq. (13) to obtain

$$\begin{aligned} d\underline{X}(t) =& K_e(t)(\underline{X}(t) - \underline{M}_e(t))dt + \sum_{i=1}^{n} L_i(t)(x_i(t) - m_i(t))d\underline{W}(t) + \\ & \underline{A}(t)dt + B(t)d\underline{W}(t) \end{aligned} \quad (21)$$

From Eq. (21), the mean and covariance equations can be derived to yield

$$\dot{\underline{M}}_e(t) = \underline{A}(t) \quad (22a)$$

$$\begin{aligned} \dot{P}(t) =& K_e(t)P(t) + P(t){K_e}^T(t) + \sum_{i=1}^{n}\sum_{j=1}^{n}[P(t)]_{ij} L_i(t) Q_W L_j(t)^T + \\ & B(t) Q_W B^T(t) \end{aligned} \quad (22b)$$

Hence, the statistical linearization system of Eq. (13), and the associated mean and covariance equations are given by Eq. (21), and Eq. (22), respectively. By employing Gaussian density function for evaluating system parameters in Eqs. (19)-(20) for the system by Eqs. (21)-(22), the mean and covariance resposne of Eq. (13) can be investigated. As for the stability problem, the mean-square stable condition is obtained by requiring that $P(t)$ be a positive definite matrix after setting zero external excitation.

3. Non-Gaussian Linearization Method [73]

This technique, extended from Gaussian linearization method, has been developed for the dynamic response of a highly nonlinear system under both stochastic parametric and external excitations. By constructing a non-Gaussian density as the weighted sum of undetermined Gaussian densities, the Gaussian linearization method is directly extended to a non-Gaussian one by this scheme. By following the same approach as Gaussian linearization technique, the undetermined Gaussian parameters are derived through solving a set of nonlinear algebraic moment relations. The non-Gaussian technique formulated for the stochastic Eq. (13) can be implemented by evaluating Eqs. (19)-(20) based on the weighted sum of Gaussian densities with adjustable means and variances as given by

$$p_k(\underline{X}(t)) = \sum_{i=1}^{k} a_i N_i(\underline{\mu}_i, S_i) \tag{23}$$

where the coefficient a_i satisfy

$$\sum_{i=1}^{k} a_i = 1, \qquad a_i > 0 \tag{24}$$

and N_i is an n-dimensional Gaussian density as

$$N_i(\underline{\mu}_i, S_i) = \frac{1}{(2\pi)^{\frac{n}{2}} |S_i|^{\frac{1}{2}}} \exp\left[-\frac{1}{2}(\underline{X} - \underline{\mu}_i)^T S_i^{-1} (\underline{X} - \underline{\mu}_i)\right] \tag{25}$$

By forming the non-Gaussian density as Eq. (23) and substituting the density into Eqs. (19)-(20), the system parameters can be further expressed explicitly as functions of a_i, $\underline{\mu}_i$, and S_i to give

$$\underline{A}(t) = \sum_{i=1}^{k} a_i E[\underline{F}(\underline{X}(t))]|_{\underline{X}=N_i(\underline{\mu}_i, S_i)} \tag{26a}$$

$$K_e(t) = [\sum_{i=1}^{k} a_i E[\frac{\partial \underline{F}(\underline{X}(t))}{\partial \underline{X}}]|_{\underline{X}=N_i(\underline{\mu}_i, S_i)} S_i + \sum_{i=1}^{k}\sum_{j=1}^{k} a_i(\delta_{ij} - a_j) E[\underline{F}(\underline{X}(t))]|_{\underline{X}=N_i(\underline{\mu}_i, S_i)} \underline{\mu}_j^T] P(a_i, S_i, \underline{\mu}_i)^{-1} \tag{26b}$$

$$B(t) = \sum_{i=1}^{k} a_i E[G(\underline{X}(t))]|_{\underline{X}=N_i(\underline{\mu}_i, S_i)} \tag{26c}$$

L_p is given by Eq. (19d) with

$$E[G(\underline{X}(t))(x_l(t) - m_l(t))] = \sum_{i=1}^{k} a_i E[\frac{\partial G(\underline{X}(t))}{\partial x_l}]|_{\underline{X}=N_i(\underline{\mu}_i, S_i)} S_{ll,i} + \sum_{i=1}^{k}\sum_{j=1}^{k} a_i(\delta_{ij} - a_j) E[G(\underline{X}(t))]|_{\underline{X}=N_i(\underline{\mu}_i, S_i)} \mu_{l,j} \tag{26d}$$

and

$$P(a_i, S_i, \underline{\mu}_i) = \sum_{i=1}^{k} a_i [S_i + \underline{\mu}_i \underline{\mu}_i^T] - \sum_{i=1}^{n}\sum_{j=1}^{n} a_i a_j \underline{\mu}_i \underline{\mu}_j^T \tag{26e}$$

$$S_i = [s_{kl,i}] \tag{26f}$$

$$\underline{\mu}_i = [\mu_{k,i}] \tag{26g}$$

where δ_{ij} is a Kronecker delta function. For the Gaussian-sum density by Eq. (23), the corresponding characteristic function can be expressed as

$$K_{\underline{X}}(\underline{\psi}) = \sum_{i=1}^{k} a_i \exp(j \underline{\mu}_i^T \underline{\psi} - \frac{1}{2} \underline{\psi}^T S_i \underline{\psi}) \tag{27a}$$

where

$$\underline{\psi} = [\psi_i] \tag{27b}$$

By employing the characteristic function, the moments of $\underline{X}$ can be derived through the following equation

$$E[x_1^{l_1} x_2^{l_2} \ldots x_n^{l_n}] = (-1)^{l_1 + l_2 + \ldots l_n} \frac{\partial^{l_1 + l_2 + \ldots l_n}}{\partial \psi_1^{l_1} \partial \psi_2^{l_2} \ldots \partial \psi_n^{l_n}} K_{\underline{X}(\underline{\psi})}|_{\underline{\psi}=0} \tag{28}$$

Hence, all the monents of $\underline{X}$ can be obtained from Eqs. (27)-(28) if the unknown a_i, $\underline{\mu}_i$, and S_i can be derived. For the nonlinear stochastic system by Eq. (13), the moment relations can be derived from

$$\sum_{i=1}^{n} E[\frac{\partial \phi(\underline{X})}{\partial x_i} f_i] + 0.5 \sum_{i=1}^{n}\sum_{j=1}^{n} E[\frac{\partial^2 \phi(\underline{X})}{\partial x_i \partial x_j} (G(\underline{X}) Q_W G^T(\underline{X}))_{ij}] = 0 \tag{29}$$

where

$$\underline{F}(\underline{X}) = [f_i] \tag{30}$$

$$\phi(\underline{X}) = x_1^{l_1} x_2^{l_2} \ldots x_n^{l_n} \tag{31}$$

Clearly, the moment relations derived usually will not be "closed" in the lower-order moments. By employing the Gaussian-sum density, all the third- and higher-order moments generated by employing Eqs. (29)-(31) can be expressed as functions of the first two moments. Thus, the unknown parameters a_i, $\underline{\mu}_i$, and S_i can be derived by formulating a set of algebraic nonlinear moment relations through Eqs. (29)-(31) and utilizing Eqs. (27)-(28). As for the stability analysis, the mean-square stability can be verified by following the same procedure as given in above Gaussian linearization method. The above approach formulates a non-Gaussian closure technique. As a result, the non-Gaussian linearization system for the nonlinear stochastic system is given by Eq. (21) with the system parameters obtained from Eq. (26) and by the moment relations from Eq. (29).

The non-Gaussian linearization method for the nonlinear systems under both stochastic parametric and external excitations has been developed. For the investigations of the performance by this approach, numerical results for the Duffing-type oscillator under external excitation have been obtained to show that within five percent accuracy can be obtained by the two Gaussian-sum scheme. The simulation results of non-stationary covariance response by the Gaussian, non-Gaussian linearization system, and 200-run Monte Carlo simulation also have revealed that very accurate stationary and non-stationary responses can be predicted by this non-Gaussian linearization system. By employing this approach, the techniques in deriving a non-Gaussian linearization system are no more than those in deriving the usual Gaussian linearization system. Although the computational effort will be increased due to the increase in the number of algebraic nonlinear equations for improving the accuracy in predicting the response statistics, the computational problem can be easily taken care by employing one of the standard packages directly, e.g., IMSL. Here, it is noted that the applications of Gaussian-sum approach are quite possible useful in predicting the bifurcation phenomena since the peak and trough in the non-Gaussian density can represent the stable and unstable states, respectively for a bifurcated nonlinear stochastic system. In addition, it is noted that the application of Gaussian-sum density has several advantages: it is a positive function in the state space, it has the property of L_1 convergence, and the computational complexity is under the framework of Gaussian linearization technique. Thus, in solving the statistical response of a general n-dimensional nonlinear stochastic system, a non-Gaussian density is better expressed as a Gaussian-sum density than as a Gram-Charlier series.

4. Maximum Entropy Method [74]

The maximum entropy method has been developed for deriving the stationary probability density function of a stable nonlinear stochastic system. This technique is implemented by employing the density function with undetermined parameters from the entropy method and solving a set of algebraic moment equations from the associated nonlinear stochastic system for the unknown parameters. By employing the entropy method, the existing exact density function of a nonlinear system can be easily extended to include the effect of a perturbed nonlinear function of states and noises. For a wide class of nonlinear systems, an explicit approximate density function can be further derived by employing this technique. This technique can be formulated for the Ito's equation as

$$d\underline{X} = \underline{F}_d(\underline{X})dt + G_d(\underline{X})d\underline{W}_d + \epsilon\underline{F}_p(\underline{X})dt + \epsilon^{\frac{1}{2}}G_p(\underline{X})d\underline{W}_p \tag{32}$$

where $\underline{X}$ is an nx1 vector of states process, $\underline{F}_d$ is a dominant nx1 vector nonlinear function of states, $G_d(\underline{X})d\underline{W}_d$ is a dominant nx1 noise excited part, ϵ is a constant, $\underline{F}_p$ is a perturbed nx1 vector nonlinear function of states, and $G_p(\underline{X})d\underline{W}_p$ is a perturbed nx1 vector noise excited part. The random excitations are assumed to be mutually independent Wiener processes with independent increments and with intensities

$$E[d\underline{W}_d(t)d\underline{W}_d^T(t)] = Q_d dt$$

$$E[d\underline{W}_p(t)d\underline{W}_p^T(t)] = Q_p dt \tag{33}$$

With maximum entropy principle, an exponential density function can be constructed as

$$p^t(\underline{X}) = N \exp[\sum_{l_1=0}^{m_1} \dots \sum_{l_n=0}^{m_n} \lambda_{l_1 \dots l_n} x_1^{l_1} \dots x_n^{l_n}] \tag{34}$$

$$except \;\; all \;\; l_i = 0$$

where N is a normalization constant given by

$$N = 1/(\int_R \exp[\sum_{l_1=0}^{m_1} \dots \sum_{l_n=0}^{m_n} \lambda_{l_1 \dots l_n} x_1^{l_1} \dots x_n^{l_n}] d\underline{X}) \tag{35}$$

$$except \;\; all \;\; l_i = 0$$

The associated moment relation is given by

$$\begin{aligned} &\sum_{i=1}^{n} E[\frac{\partial \phi(\underline{X})}{\partial x_i}(f_d^i + \epsilon f_p^i)] + \\ &\frac{1}{2}\sum_{i=1}^{n}\sum_{j=1}^{n} E[\frac{\partial^2 \phi(\underline{X})}{\partial x_i \partial x_j}(G_d(\underline{X})Q_d G_d^T(\underline{X}) + \epsilon G_p(\underline{X})Q_p G_p^T(\underline{X}))_{ij}] = 0 \end{aligned} \tag{36}$$

The unknown $\lambda_{l_1 l_2 \dots l_n}$ in Eq. (34) can be derived if $(m_1+1)(m_2+1)\dots.(m_n+1)-1$ independent moment relations are formulated by selecting appropriate functions $\phi(\underline{X})$ in Eq. (36). The numerical solution technique may utilize a weighted summation scheme for evaluating the expected value in Eq. (36) with Eq. (34) and solving a set of nonlinear algebraic equations for obtaining the unknown in density function. If the value of ϵ in Eq. (32) is small, this scheme will be further simplified for solving an approximate stationary density function. Hence, a set of linear equations can be derived from Eq. (36) to give

ϵ^0 :

$$\sum_{i=1}^{n} E_d[\frac{\partial \phi(\underline{X})}{\partial x_i} f_d^i] + \frac{1}{2}\sum_{i=1}^{n}\sum_{j=1}^{n} E_d[\frac{\partial^2 \phi(\underline{X})}{\partial x_i \partial x_j}(G_d(\underline{X})Q_d G_d^T(\underline{X}))_{ij}] = 0 \tag{37}$$

$$\begin{aligned}
&\epsilon^1 : \\
&\sum_{l_1=0}^{m_1} \cdots \sum_{l_n=0}^{m_n} \lambda^p_{l_1 \ldots l_n} \{ \sum_{i=1}^{n} E_d[\frac{\partial \phi(\underline{X})}{\partial x_i} f^i_d x_1^{l_1} \ldots x_n^{l_n}] + \\
&\quad except \ \ all \ \ l_i = 0 \\
&\frac{1}{2} \sum_{i=1}^{n} \sum_{j=1}^{n} E_d[\frac{\partial^2 \phi(\underline{X})}{\partial x_i \partial x_j} (G_d(\underline{X}) Q_d G_d^T(\underline{X}))_{ij} x_1^{l_1} \ldots x_n^{l_n}]\} + \\
&\sum_{i=1}^{n} E_d[\frac{\partial \phi(\underline{X})}{\partial x_i} f^i_p] + \\
&\frac{1}{2} \sum_{i=1}^{n} \sum_{j=1}^{n} E_d[\frac{\partial^2 \phi(\underline{X})}{\partial x_i \partial x_j} (G_p(\underline{X}) Q_p G_p^T(\underline{X}))_{ij}] = 0
\end{aligned} \tag{38}$$

where $E_d[.]$ is defined as

$$E_d[.] = \int_R (.) p^t_d(\underline{X}) d\underline{X} \tag{39}$$

The unknown $\lambda^p_{l_1 \ldots l_n}$ can be derived from Eq. (38) if $E_d[.]$ can be evaluated by employing the joint probability density function of the dominant part of Eq. (32) obtained from Eq. (37) or from the associated FPK equation. As a result, the stationary probability density function of Eq. (32) can be expressed by employing Eqs. (34)-(35) to yield

$$p^t(\underline{X}) = N p_d(\underline{X}) \exp(\epsilon \sum_{l_1=0}^{m_1} \ldots \sum_{l_n=0}^{m_n} \lambda^p_{l_1 \ldots l_n} x_1^{l_1} \ldots x_n^{l_n}) \quad except \ \ all \ \ l_i = 0 \tag{40}$$

where N is given by

$$N = 1/(\int_R p_d(\underline{X}) \exp(\epsilon \sum_{l_1=0}^{m_1} \ldots \sum_{l_n=0}^{m_n} \lambda^p_{l_1 \ldots l_n} x_1^{l_1} \ldots x_n^{l_n}) d\underline{X}) \quad except \ \ all \ \ l_i = 0 \tag{41}$$

The development of the maximum entropy method has been formulated. The application of this technique for deriving the stationary probability density function has been illustrated by selecting three examples.

The validity of the approximate stationary probability density function has been supported by the estimated density function through Monte Carlo simulation. By employing this technique, a minimally prejudiced probability density function can be derived by solving a set of nonlinear algebraic equations. For a wide class of nonlinear stochastic system including perturbed nonlinear functions of states and parametric noise excited terms, an explicit approximate probability density function can be derived from extending a given exact probability density function of the corresponding dominant stochastic system by this scheme. From illustrated example, this scheme has provided exact density function for some stochastic systems and accurate expression of the stationary density function can be assured if the order of perturbed part is less than one tenth of the corresponding dominant part in a nonlinear stochastic system.

5. Spectral Analysis Method [26]

This spectral analysis approach has been implemented through the combined methods of equivalent external excitation and Gaussian linearization along with the Wiener-Khintchine relation for a nonlinear system subjected to both parametric and external Gaussian white noise excitations. By employing this method, the spectral response of general stable nonlinear systems under both stochastic parametric and external excitations can be efficiently derived. In implementation, an equivalent linearization system under external Gaussian white-noise excitation with undetermined intensity is first formulated. The intensity is expressed as a function of covariance matrix. Next, the formulation of spectral analysis technique is implemented through the Wiener-Khintchine input/output spectral relation of a linear system with the matching condition for covariance matrix. The formulation of this technique has been presented for the system as given by Eq. (6). By utilizing the Gaussian linearization method given in this section, the linearization of the nonlinear coupling function can be directly obtained as

$$\underline{g}_i = C_i\underline{\dot{q}} + K_i\underline{q} \tag{42}$$

with

$$C_i = E[\frac{\partial \underline{g}_i}{\partial \underline{\dot{q}}}]$$

$$K_i = E[\frac{\partial \underline{g}_i}{\partial \underline{q}}] \tag{43}$$

where the vector derivative of a differentiable vector is defined as

$$\frac{\partial \underline{y}}{\partial \underline{x}} = \begin{pmatrix} \frac{\partial y_1}{\partial x_1} & \cdots & \frac{\partial y_1}{\partial x_n} \\ \vdots & \ddots & \vdots \\ \frac{\partial y_n}{\partial x_1} & \cdots & \frac{\partial y_n}{\partial x_n} \end{pmatrix}$$

$$\underline{y} = (y_1 y_2 \ldots y_n)^T$$

$$\underline{x} = (x_1 x_2 \ldots x_n)^T \tag{44}$$

By substituting Eq. (42) into the equivalent external excitation system as Eq. (11), then Eq. (11) becomes

$$M\underline{\ddot{q}} + (C + \sum_{i=1}^{n_1} a_i C_i)\underline{\dot{q}} + (K + \sum_{i=1}^{n_1} a_i K_i)\underline{q} = \underline{w}_e(t) \tag{45}$$

and the equivalent excitation intensity by Eq. (12) becomes

$$E[\underline{w}_e(t)\underline{w}_e^T(s)] = Q_e(E[\underline{qq^T}], E[\underline{q\dot{q}^T}], E[\underline{\dot{q}\dot{q}^T}])\delta(t-s) \tag{46}$$

with

$$\begin{aligned} Q_e =& Q_0 + \sum_{i=1}^{n_1} Q_i(M^{-1}C_i E[\underline{\dot{q}\dot{q}^T}]C_i^T M^{-1^T} + M^{-1}K_i E[\underline{q\dot{q}^T}]C_i^T M^{-1^T} \\ &+ M^{-1}C_i E[\underline{\dot{q}q^T}]K_i^T M^{-1^T} + \\ &M^{-1}K_i E[\underline{qq^T}]K_i^T M^{-1^T}) \end{aligned} \tag{47}$$

Since Eq. (45) is an equivalent linear time-invariant system subjected to the equivalent external Gaussian white-noise excitation with intensity by Eq. (46), the spectral response for the equivalent system can be obtained by employing the input/output spectral relation of a linear system to obtain

$$\Phi(\omega) = H(\omega)\frac{Q_e(E[\underline{qq^T}], E[\underline{q\dot{q}^T}], E[\underline{\dot{q}\dot{q}^T}])}{2\pi}H^{T^*}(\omega) \tag{48}$$

where ω is in rad/sec, H^{T^*} represents the complex conjugate of H^T , and the frequency transfer function $H(\omega)$ is given by

$$H(\omega) = (-\omega^2 M + j\omega(C + \sum_{i=1}^{n_1} a_i C_i) + K + \sum_{i=1}^{n_1} a_i K_i)^{-1} \tag{49}$$

Since the spectral density function Eq. (49) is a function of unknown stationary covariance matrix, a set of matching conditions for the covariance matrix can be formulated from the Wiener-Khintchine relation to obtain

$$\begin{aligned} E[q_k q_l] &= \int_{-\infty}^{\infty} \phi_{kl}(\omega) d\omega \\ E[q_k \dot{q}_l] &= j \int_{-\infty}^{\infty} \omega \phi_{kl}(\omega) d\omega \\ E[\dot{q}_k \dot{q}_l] &= \int_{-\infty}^{\infty} \omega^2 \phi_{kl}(\omega) d\omega \end{aligned} \tag{50}$$

where ϕ_{kl} is the kl-th element of Φ and q_j is the j-th element of $\underline{q}$. From Eq. (50), the stationary output covariance matrix can be obtained by solving a set of nonlinear integral or algebraic equations.

As a result, the output power spectral density of state q_i can be directly obtained by substituting the derived stationary output covariance matrix into Eq. (48). From the derived spectral response, the correlation function of response also can be directly derived by performing the inverse Fourier transform on Eq. (48). These statistical information, which cannot be obtained by the previous time-domain formulation, thus completely characterize the property of stationary response of the equivalent linearization system. As far as the concern of computational complexity and efficiency in online implementation of the scheme, it is noted that the computational effort for solving Eq. (50) is exactly the same as that for spectral implementation of a Gaussian linearization method on an externally excited nonlinear system. After Eq. (50) has been solved, the computational effort for obtaining the output spectrum from Eq. (48) will depend on computational algorithms selected for implementing fundamental matrix operations. The formulation of a practical technique for deriving the output power spectral density of an n-dimensional stochastic parametrically and externally excited nonlinear system has been developed. Three examples have been selected for illustrating the application and accuracy of this technique. Exact solution can be obtained for a linear stochastic system with parametric noise excitation. The validity of predicted spectrum for nonliner Duffing-type oscillator and hysteretic system has been supported by FFT spectral analysis with stationary data collected from Monte Carlo simulation.

The above technique has been provided for the practical analysis of output power spectral density of a general nonlinear system under both stochastic parametric and external excitations. By employing this approach, the spectral response of a nonlinear system under stochastic external excitation can be directly extended to include the effects of parametric noise excitation. Hence, the effects of parametric noise on the information of energy distribution shared by different frequency ranges can be obtained and employed to facilitate the control of random systems. The effects on the correlation property of response also can be investigated if the "memory" behavior of the response of a stochastic dynamic system needs to be analyzed. For a linear SISO system under both stochastic parametric and external excitations, it has been concluded that both the spectrum and autocorrelation of output response obtained by the practical approach are exactly the same as those derived by tedious time-domain approaches. The concept and condition of spectral stability have been also presented and the stability condition has been shown to be in consistent with that obtained from a time-domain approach. It has been also concluded that the output spectrum of a linear SISO system without parametric noise excitation will be uniformly amplified by the combined effects of solely parametric noise and system parameters when the parametric noise is entering into a system. For the spectrum response of a nonlinear parametrically excited system, the simulated results from a Duffing-type oscillator have revealed that the predicted spectrum can underestimate low-frequency spectral response when the system is with "strong" nonlinearity. As far as the spectrum response of a hysteretic nonlinear stochastic system, it has been observed that very accurate prediction of output power spectral density can be derived when the equivalent intensity of parametric noise is less than that of external noise excitation.

VII. CONCLUSIONS AND RECOMMENDATIONS

Extensions of analytical techniques for investigating the dynamic behavior of a general stochastic system from Langevin approach for Brownian motion have been presented. With mathematical foundation laid by Ito, the general stochastic differential equation formulation, extended from linear Langevin equation, has been successfully employed in various fields to model the disturbance-response information of a real system interacting with its complex environment. Development and extension of analytical techniques for various classes of model described as a linear or nonlinear one with parametric and/or external stochastic excitations are introduced. Practical and versatile techniques of statistical linearization and closure schemes are emphasized for engineering applications. Extensions of techniques from time-invariant stochastic system to stochastic parametrically-excited one are described. The analytical techniques described are mainly focused on time-domain stationary response and second-monent stability of a multiple-degree-of-freedom nonlinear system under stochastic parametric and external excitations. Time-domain implementation of various approaches including equivalent external excitation, Gaussian linearization, non-Gaussian linearization, and maximum entropy techniques is provided. In addition to the time-domain approach, a frequency-domain scheme for practical spectral analysis is also introduced.

Although several analytical approaches have been developed for investigating stochastic behavior of a general stochastic dynamic system, these techniques are still under development in light of capturing the complex and sophisticated nonlinear non-Gaussian nonstationary behavior of a real system. Further developments of investigating techniques including analytical, numerical, and experimental schemes are required to enhance the understanding and analyzing of qualitative behavior, such as stability, and quantitative performance when a real system is interacting with its environment. Successful a priori analysis and prediction of local and/or global response in a short and/or long time operation of a stochastic disturbed system play a key role to ensure a posteriori design of an effective and robust continuous/discrete control scheme to manage input for desired response. Applications of high-speed digital processing techniques in real-time Monte

Carlo simulation and control of stochastic parametrically and externally excited continuous systems are essential issues to challange precision system and control engineer.

In addition to the developments of engineering application of stochastic dynamic techniques, the stochastic analysis has stimulated various scientific investigations from classical system to modern quantum relativistic system. Successful scientific achievements from Einstein's work on Brownian motion in 1905 have brought impact to the later development of Cybernetics in 1947 and recent development of Synergetics in 1970. New area for providing a unify theory in answering the stochastic mechanism of self-organization in our universe should be challanged.

REFERENCES

1. H. Haken, "Synergetics: An Introduction," Springer-Verlag, Berlin, 1977.

2. N. Wax, ed., "Selected Papers on Noise and Stochastic Processes," Dover, New York, 1954.

3. E. Nelson, "Dynamical Theories of Brownian Motion," Princeton University Press, New Jersey, 1967.

4. A. T. Fuller, "Analysis of Nonlinear Stochastic Systems by Means of the Fokker-Planck Equation," Int. J. Cont., 9, 603-655 (1969).

5. A. H., Jazwinski, "Stochastic Processes and Filtering Theory," Academic Press, New York, 1970.

6. T. K. Caughey, "Nonlinear Theory of Random Vibrations," in "Advances in Applied Mechanics," Vol. 11, Academic Press, New York, 1971.

7. H. Hida, "Brownian Motion," Springer-Verlag, Berlin, 1980.

8. S. H. Crandall and W. Q. Zhu, "Random Vibration: A Survey of Recent Developments," ASME J. Appl. Mech., 50, 953-962 (1983).

9. R. A. Ibrahim, "Parametric Random Vibration," Research Studies Press, New York, 1985.

10. E. Nelson, "Quantum Fluctuations," Princeton University Press, New Jersey, 1985.

11. R. J. Chang and T. C. Jiang, "Formulation of Dynamic Equations of Robot Manipulators with Joint Irregularities," to appear in ASME J. Dyn. Sys. Meas. Cont. (1993).

12. R. Kubo, M. Toda and N. Hashitsume, "Statistical Physics I. II," Springer-Verlag, Berlin, 1983.

13. B. J. West and K. Lindenberg, "State-Dependent Fluctuations in Open Systems: Simple Models," in "Simple Models of Equilibrium and Nonequilibrium Phenomena," (J. L. Lebowitz ed.), Elsevier Science Publishers, 1987.

14. Y. K. Lin, "Probabilistic Theory of Structural Dynamics," McGraw Hill, New York, 1976.

15. V. V. Bolotin, "Random Vibrations of Elastic Systems," Martinus Nijhoff Publishers, The Hague, 1984.

16. R. H. Lyon, "Statistical Energy Analysis of Dynamical Systems," The MIT Press, Massachusetts, 1975.

17. P. S. Maybeck, "Stochastic Models, Estimation, and Control," Vol. 1, Academic Press, New York, 1982.

18. T. T. Soong, "Random Differential Equations in Science and Engineering," Academic Press, New York, 1973.

19. F. Kozin, "Some Results on Stability of Stochastic Dynamical Systems," J. Prob. Engrg. Mech., 1, 13-22 (1986).

20. S. T. Ariaratnam and W. C. Xie, "Stochastic Sample Stability of Oscillatory Systems," ASME J. Appl. Mech., 55, 458-460 (1988).

21. S. T. Ariaratnam and Wei-Chau Xie, "Effect of Correlation on the Almost-Sure Asymptotic Stability of Second-order Linear Stochastic Systems," ASME J. Appl. Mech., 56, 685-690 (1989).

22. S. T. Ariaratnam, D. S. F. Tam and Wei-Chau Xie, "Lyapunov Exponents and Stochastic Stability of Coupled Linear Systems Under White Noise Excitation," J. Prob. Engrg. Mech., 6(2), 51-56 (1991).

23. M. F. Dimentberg, "Statistical Dynamics of Nonlinear and Time-Varying Systems," Research Studies Press, New York, 1988.

24. H. Benaroya and M. Rehak, "Response and Stability of a Random Differential Equation: Part I- Moment Equation Method," ASME J. Appl. Mech., 56, 192-195 (1989).

25. H. Benaroya, and M. Rehak, "Response and Stability of a Random Differential Equation: Part II- Expansion Method," ASME J. Appl. Mech., 56, 196-201 (1989).

26. R. J. Chang, "A Practical Technique for Spectral Analysis of Nonlinear Systems Under Stochastic Parametric and External Excitations," ASME J. Vib. Acoust., 113, 516-522 (1991).

27. Z. K. Hou and W. D. Iwan, "Nonstationary Response of Linear Systems Under Uncorrelated Parametric and External Excitations," J. Prob. Engrg. Mech., 6(2), 74-82 (1991).

28. W. V. Wedig and A. Ams, "Parametric Amplification Effects in Acoustical Test Facilities," Intl. J. Nonlin. Mech., 26(6), 975-984 (1991).

29. G. E. Young and R. J. Chang, "Prediction of the Response of Non-Linear Oscillators Under Stochastic Parametric and External Excitations," Intl. J. Nonlin. Mech., 22(2), 151-160 (1987).

30. H. Risken, "The Fokker-Planck Equation: Methods of Solution and Applications," 2nd ed. Springer-Verlag, Berlin, 1989.

31. V. S. Pugachev and I. N. Sinitsyn, "Stochastic Differential Systems: Analysis and Filtering," John Wiley & Sons, New York, 1987.

32. H. A. Kramers, "Brownian Motion in a Field of Force and Diffusion Model of Chemical Reactions," Physica 7, 284-304 (1940).

33. T. K. Caughey and Fai Ma, "The Exact Steady-State Solution of a Class of Non-Linear Stochastic Systems," Intl. J. Nonlin. Mech., 17(3), 137-142 (1982).

34. L. Garrido and J. Masoliver, "On a Class of Exact Solutions to the Fokker- Planck Equations," J. Math. Phys., 23 (6), 1155-1158 (1982).

35. Leslaw Socha and T. T. Soong, "Linearization in Analysis of Nonlinear Stochastic Systems," Appl. Mech. Rev., 44(10), 399-422 (1991).

36. Y. K. Wen, "Methods of Random Vibration for Inelastic Structures," Appl. Mech. Rev., 42(2), 39-52 (1989).

37. J. B. Roberts and P. D. Spanos, "Stochastic Averaging: An Approximate Method of Solving Random Vibration Problems," Intl. J. Nonlin. Mech., 21(2), 111-134 (1986).

38. W. Q. Zhu, "Stochastic Averaging Methods in Random Vibration," Appl. Mech. Rev., 41(5), 189-199 (1988).

39. R. L. Stratonovitch, "Topics in the Theory of Random Noise," Vol. 1, Gordon and Breach, New York, 1963.

40. R. L. Stratonovitch, "Topics in the Theory of Random Noise," Vol. 2, Gordon and Breach, New York, 1967.

41. R. Z. Khasminskii, "A Limit Theorem for the Solutions of Differential Equations with Random Right-Hand Sides," Theor. Prob. Appl., 11, 390- 405 (1966).

42. R. Z. Khasminskii, "On the Averaging Principle for Stochastic Differential Ito Equations," Kibernetika, 4(3), 260-279 (1968).

43. J. O. Hinze, "Turbulence," 2nd edition, McGraw-Hill, New York, 1975.

44. M. L. Dashevskii and R. Sh. Liptser, "Application of Conditional Semi-Invariants in Problems of Non-Linear Filtering of Markov Processes," Avtom. i Telemekh., 28(6), 63-74 (1967).

45. Sh. A. Assaf and L. D. Zirkle, "Approximate Analysis of Non-Linear Stochastic Systems," Int. J. Cont., 23, 477-492 (1976).

46. J. J. Beaman, "Statistical Linearization for the Analysis and Control of Non-Linear Stochastic Systems," Sc.D. Thesis, M.I.T., Massachusetts, 1978.

47. S. H. Crandall, "Non-Gaussian Closure Techniques for Stationary Random Vibration," Intl. J. Nonlin. Mech., 20(1), 1-8 (1985).

48. Q. Liu and H. G. Davies, "The Non-Stationary Response Probability Density Functions of Non-Linearly Damped Oscillators Subjected to White Noise Excitations," J. Sound Vib., 139 (3), 425-435 (1990).

49. L. D. Lutes, "Approximate Technique for Treating Random Vibration of Hysteretic Systems," J. Acoust. Soc. Amer., 48, 299-306 (1970).

50. A. Lin, "A Numerical Evaluation of the Method of Equivalent Nonlinearization," Ph.D. Thesis, California Inst. Tech., 1988.

51. S. H. Crandall, "Perturbation Techniques for Random Vibrations of Nonlinear Systems," J. Acous. Soc. Amer., 35(11), 1700-1705 (1963).

52. J. K. Dienes, "Some Applications of the Theory of Continuous Markoff Processes to Random Oscillation Problems," Ph.D. Thesis, California Inst. Tech., 1961.

53. G. Adomian, "Stochastic Systems," Academic Press, New York, 1983.

54. I. I. Orabi and G. Ahmadi, "Nonstationary Response Analysis of a Duffing Oscillator by the Wiener-Hermite Expansion Method," ASME J. Appl. Mech., 54, 434-440 (1987).

55. I-M. Yang, "Stationary Random Response of Multi-Degree of Freedom Systems," Ph.D. Thesis, California Inst. Tech., 1970.

56. Atalik, T. S., "Stationary Random Response of Nonlinear Multi-Degree of Freedom Systems by a Direct Equivalent Linearization Technique," Ph.D. Thesis, Duke University, 1974.

57. T. Fang and Z. Wang, "A Generalization of Caughey's Normal Mode Approach to Nonlinear Random Vibration Problems," AIAA J., 24 (3), 531-534 (1986).

58. H. J. Pradlwarter and Wenlung Li, "On the Computation of the Stochastic Response of Highly Nonlinear Large MDOF-Systems Modeled by Finite Elements," J. Prob. Engrg. Mech., 6(2), 109-116 (1991).

59. R. Villaverde and M. M. Hanna, "Efficient Mode Superposition Method for Seismic Analysis of Nonlinear Systems," Appl. Mech. Rev., 44(11), 264-272 (1991).

60. Y. Yong and Y. K. Lin, "Exact Stationary Response Solution for Second Order Nonlinear Systems Under Parametric and External White Noise Excitations," ASME J. Appl. Mech., 54, 414-418 (1987).

61. Y. K. Lin and G. Q. Cai, "Exact Stationary Response Solution for Second Order Nonlinear Systems under Parametric and External White Noise Excitations: Part II," ASME J. Appl. Mech., 55, 702-705 (1988).

62. G. Q. Cai and Y. K. Lin, "On Exact Stationary Solutions of Equivalent Nonlinear Stochastic Systems," Intl. J. Nonlin. Mech., 23, 315-325 (1988).

63. C. W. S. To and D. M. Li, "Equivalent Nonlinearization of Nonlinear Systems to Random Excitations," J. Prob. Engrg. Mech., 6(3/4), 184-192 (1991).

64. G. Q. Cai and Y. K. Lin, "A New Approximate Solution Technique for Randomly Excited Nonlinear Oscillators," Intl. J. Nonlin. Mech., 23 (5/6), 409-420 (1988).

65. W. Q. Zhu and J. S. Yu, "The Equivalent Nonlinear System Method," J. Sound Vib., 129(3), 385-395 (1989).

66. R. J. Chang and G. E. Young, "Methods and Gaussian Criterion for Statistical Linearization of Stochastic Parametrically and Externally Excited Nonlinear Systems," ASME J. Appl. Mech., 56, 179-185 (1989).

67. R. J. Chang and G. E. Young, "Prediction of Stationary Response of Robot Manipulators Under Stochastic Base and External Excitations - Statistical Linearization Approach," ASME J. Dyn. Sys., Meas. Cont., 111, 426-432 (1989).

68. R. L. Stratonovich and Yu. A. Romanovskii, "Parametric Effect of a Random Force on Linear and Nonlinear Oscillating Systems," in "Non-Linear Transformations of Stochastic Processes," (P. I. Kuznetsov, R. L. Stratonovich, V. I. Tikhonov, ed.), Pergamo Press, New York, 1965.

69. W. F. Wu and Y. K. Lin, "Cumulant- Neglect Closure for Non-Linear Oscillators Under Random Parametric and External Excitations," Intl. J. Nonlin. Mech., 19(4), 349-362 (1984).

70. Jian-Qiao, Sun and C. S. Hsu, "Cumulant-Neglect Closure Method for Nonlinear Systems Under Random Excitations," ASME J. Appl. Mech., 54, 649-665 (1987).

71. R. J. Chang, "Stationary Response of States-Constrained Non-Linear Systems Under Stochastic Parametric and External Excitations," ASME J. Dyn. Sys. Meas. Cont., 113(4), 575-581 (1991).

72. N. C. Menh, "Responses of Weakly Nonlinear Dynamical Systems Subjected to Random Parametric and External Excitations," J. Sound and Vib., 113, 1-8 (1987).

73. R. J. Chang, "Non-Gaussian Linearization Method for Stochastic Parametrically and Externally Excited Nonlinear Systems," ASME J. Dyn. Meas. Cont., 114(1), 20-26 (1992).

74. R. J. Chang, "Maximum Entropy Approach for Stationary Response of Nonlinear Stochastic Oscillators," ASME J. Appl. Mech., 58(1), 266-271 (1990).

ADAPTIVE CONTROL OF DISCRETE-TIME SYSTEMS: A PERFORMANCE-ORIENTED APPROACH

Romeo Ortega

Sophia University
Department of Mechanical Engineering
1, Kioicho 7-chome
Chiyoda-ku, 102 Tokyo
JAPAN

I. INTRODUCTION

Parameter adaptive control is a technique to address the problem of regulating uncertain plants. Its distinctive feature is that it consists of a parametrized controller and an identifying mechanism (an estimator) whose aim is to provide on-line modification of system behavior in response to current measured performance. The effectiveness of an adaptive design is characterized by its ability to achieve a performance close to a hypotetical nonadaptive controller designed with the same objective but with a reduced level of uncertainty. The most typical case being a tracking objective with uncertainty expressed in terms of plant parameters.

Consistent with this formulation the major thrust of the research in the area has been to develop convergence theories to demonstrate the ability of the adaptive schemes to cope with parametric uncertainty. The outcome of these efforts is a class of important results insuring global convergence and asymptotic optimality. However, most currently existing results, even in this ideal situation when no unstructured uncertainty is present, are qualitative in nature insuring only *asymptotic properties and boundedness* of the signals. It has been shown in several examples that the signals in an adaptively controlled system may exhibit bursting behavior in quiescent periods or practically unacceptable excursions highly dependent upon initial conditions even when tracking rich references (see, e.g., [1] and the references in [2]). We believe this issue to be very important, and unfortunately not always appreciated, since it shows that *global theories are not enough to guarantee that the transient performance will be satisfactory*. The situation is, of course, further complicated with the inclusion of unstructured effects when the global properties are lost in a broad range of practical situations. Assesment of the stability properties of adaptive schemes in more realistic situations, e.g., in the face of nonparametric uncertainties, is a major driving force of the field and is refered to as *robust adaptive control theory*.

There are basically two approaches to address the adaptive control robustness problem, see e.g. [2] for a recent survey. In the *first approach* the main concern is to develop modified schemes for which some global stability properties can be established. To this end, the plant output is typically decomposed into the "ideal" system response and a signal, that accounts for the unmodelled part of the response, which is assumed to be bounded by a scaled normalization signal. It has been shown in [3] that this "nonstandard" characterization of uncertainty captures a broad class of "small" unmodelled effects. To insure boundedness of all signals and smallness in the mean of the tracking error the update law is modified introducing the aforementioned normalization together with leakages or relative dead-zones. These modifications, which are qualitatively equivalent, may be thought of as a mean to insure the continuity of the stability property vis a vis the scaled operator that generates the unmodelled response, that is to insure the existence of some neighborhood of the ideal system for which global boundedness is still preserved. This approach is qualitative in nature and overlooks structural features that might help to reduce the conservatism and improve the design (via, e.g., suitable filtering, signal injection). Furthermore, recent fundamental results [4-6] show that *parameter projection is the only estimator modification needed to insure signal boundedness for this class of uncertainties*. Thus, it is the author's opinion that the new interesting open question is the usefulness of the modifications to enhance performance. At the moment this seems possible only using local theories.

The *second research avenue* is, in principle,

less concerned with global convergence issues but concentrates instead in understanding the mechanisms of the adaptive system that determine its stability. To this end, it exploits the dissipation properties of the adaptive system and seeks to characterize the unmodelled effects and operating modes (signals frequency content) which preserve them. As thoroughly discussed in [7] these dissipation properties (which are given in terms of a positivity condition of a signal dependent operator) are *not just sufficient but also necessary* to insure robustness of adaptive systems. Therefore, we tend to believe this research avenue provides a *performance oriented approach to robust adaptive control*. Our objective in this paper is to present in a tutorial manner the global convergence theory of discrete time adaptive systems developed in [10] which follows this approach. This results complement with the averaging results of [7] to provide a fairly complete understanding of the adaptive systems possibilities and limitations.

Roughly speaking, there are two ways to study

the stability of adaptive systems – the state-space Lyapunov approach and the input-output (I/O) approach. See [8] for a general discussion on both approaches and some recent applications of the latter. For an excellent account of the existing results in adaptive systems theory using the Lyapunov formalism we refer the reader to [9]. The two streams have gradually merged leading to a fruitful crossfertilization whose most visible expression is the inclusion of concepts from both approaches in all standard adaptive systems textbooks. However, it is this author's opinion that the I/O perspective provides the "natural" framework to carry out the analysis of adaptive systems in the performance oriented approach. Our belief stems from the following two key characteristics of the I/O formalism:

- First, it provides a convenient generalization (to the nonlinear time varying case) of the fact that stability of a feedback interconnection depends on the amounts of "gain" and "phase shift" introduced in the loop. Furthermore, and perhaps more importantly, the measures of signal amplification and signal shift are physically motivated properties related with the systems energy dissipation. In the case of adaptive systems this fact is elegantly illustrated in the stability/instability boundary conditions of Riedle and Kokotovic [7] and the excitation *vs* fast adaptation tradeoff of the ℓ_∞ results of [10].

• Second, the I/O viewpoint that complicated systems are best thought of as being interconnections of simpler subsytems yields a *design-oriented methodology* since it allows us to isolate the "free" subsytems. In adaptive systems it helps us to clarify the effect of signal filtering and the choices of the estimator.

The remaining of the paper is organized as follows. In Section II we present the controller structure and derive the error equations. Section III is devoted to the global ℓ_2-stability analysis while Section IV presents the results of ℓ_∞-stability under persistency of excitation assumptions. We wrap up in Section V with some concluding remarks. Some preliminary material in I/O theory is given in the Appendix.

The results reported here heavily rely on the techniques developed in [10]. In that paper both ℓ_2 and ℓ_∞ stability conditions are derived. Unfortunately, several serious typographical errors in [10] and the fact that in the application of the sectoricity theorem the estimator initial conditions were not explicitely taken into account (see point 1 in the discussion of paragraph 4.E below) have led several authors to believe that the results in [10] are incomplete. For this reason, we have prefered to provide detailed proofs of all the statements in [10] in the present paper. Also, we (straightforwardly) extend these results to derive some transient bounds.

II. ADAPTIVE CONTROLLER AND ERROR MODEL

A. The Plant

Consider the single input single output (SISO) LTI discrete-time system described by

$$A(q^{-1})y(t)=B(q^{-1})u(t-d)+\zeta(t) \qquad (1)$$

where y, u, ζ denote the output, input and bounded disturbance sequences respectively, A, $B\in\mathbb{R}[q^{-1}]$ (the set of polynomials in q^{-1} with real coefficients), of degrees $\partial(A)=n_A$ and $\partial(B)=n_B$ respectively, A is monic, and $d\geq 1$ is the system pure delay, which we *assume to be known*. The polynomial coefficients and their degrees are however *unknown*.

B. *The Controller Structure*

We will pursue an all zero cancelling pole placement objective with reference model q^{-d}/C_R, $C_R\in\mathbb{R}[q^{-1}]$. That is, our "ideal" objective is to insure that, with all internal signals bounded,

$$y(t)-\frac{1}{C_R}w(t)\to 0 \text{ as } t\to\infty$$

where w is the reference signal (assumed known d steps ahead). To this end, we consider a controller structure of the form

$$S_* u(t)+R_* y(t)=w(t+d)$$

where $S_*,R_*\in\mathbb{R}[q^{-1}]$ of orders n_S and n_R, respectively. In compact notation the parametrized regulator is given by

$$S_* u(t)+R_* y(t)=\theta_*^T \phi(t)=w(t+d) \tag{2}$$

$$\phi(t) \overset{\Delta}{=} [u(t),\ldots,u(t-n_S),y(t),\ldots,y(t-n_R)]^T$$

$$\theta_* \overset{\Delta}{=} [s_0,s_1,\ldots,s_{n_S},r_0,r_1,\ldots,r_{n_R}]^T$$

We will need in the sequel the following stabilizability assumption:

- Let $\mu \in (0,1)$ be a given scalar. Then, we assume there exists a controller parametrization, i.e., $\theta_* \in \mathbb{R}^n$, $n:=n_S+n_R+2$, such that the closed loop poles are strictly inside a disk centered at the origin with radius μ.

□□□

The adaptive controller is, as usual, obtained by replacing the controller parameters by its current estimate, namely

$$w(t+d)=\hat{\theta}(t)^T \phi(t) \tag{3}$$

C. Error Equations.

Define, as usual, the parameter error as

$$\tilde{\theta}(t) \stackrel{\Delta}{=} \hat{\theta}(t) - \theta_*$$

which replacing in (3) and taking into account (2) yields

$$\tilde{\theta}(t)^T \phi(t) + S_* u(t) + R_* y(t) = w(t+d)$$

We will find convenient to define

$$\psi(t) \stackrel{\Delta}{=} \tilde{\theta}(t-d)^T \phi(t-d)$$

and a filtered tracking error

$$e(t) := C_R y(t) - w(t)$$

With these notations and after some calculations using the plant equation (1) one gets the first error equation

$$e(t) = -\mathcal{H}_2(q^{-1})\psi(t) + e_*(t) \tag{4}$$

$$e_*(t) := (\mathcal{H}_2 - 1)w(t) + C_R C^{-1} S_* \zeta(t)$$

$$\mathcal{H}_2 := C_R C^{-1} B, \quad C := AS_* + q^{-d} BR_*$$

Remark 1. From (4) we see that $\mathcal{H}_2$ is the transfer function $w \to C_R y$ that results when we fix the controller parameters to θ_* and noise is absent. Also, it is important to remark that in the definition of $\mathcal{H}_2$ we have chosen to factor the delay term. Therefore, $\mathcal{H}_2$ is *of relative degree zero*. This feature will be of fundamental importance in the analysis of the implications of the global ℓ_2-stability conditions.

□□□

The second error equation relates the regressor with the signal ψ. To this end, notice from (4) and the definition of ϕ that there exists stable transfer matrices $\mathcal{W}_1$ and $\mathcal{W}_2$ such that

$$\phi(t-d)=-\mathcal{W}_1\psi(t)+\phi_*(t-d) \tag{5}$$

$$\phi_*(t-d):=\mathcal{W}_1 w(t)+\mathcal{W}_2\zeta(t)$$

D. Parameter Estimators

To complete the error model we need to specify the parameter update law. In the next section, where we present the ℓ_2-stability results, we will consider the following gradient algorithm with normalization factor

$$\hat{\theta}(t)=\hat{\theta}(t-d)+\gamma\,\frac{\phi(t-d)e(t)}{\rho^2(t)}, \quad 2>\gamma>0$$

$$\rho^2(t)=\mu^2\rho^2(t-1)+\max\{|\phi(t-d)|^2,(1-\mu^2)\underline{\rho}^2\},$$

$$0<\mu<1,\quad \rho(0)=\underline{\rho}>0 \tag{6}$$

For the ℓ_∞ results of section 4 we will use instead a regularized least squares estimator with normalization

$$\hat{\theta}(t)=\hat{\theta}(t-d)+F(t)\bar{\phi}(t-d)\bar{e}(t)$$

$$F(t)=(1-\frac{\lambda_o}{\lambda_1})F'(t)+\lambda_o I$$

$$F'(t):=F(t-d)-\frac{F(t-d)\bar{\phi}(t-d)\bar{\phi}^T(t-d)F(t-d)}{\lambda+\bar{\phi}^T(t-d)F(t-d)\bar{\phi}(t-d)}$$

$$\bar{\phi}(t-d):=\frac{\phi(t-d)}{\rho(t)},\quad \bar{e}(t):=\frac{e(t)}{\rho(t)},\quad 0\le\lambda_o<\lambda_1,\lambda>0 \tag{7}$$

The parameter estimator (6) (or (7)) defines an operator $\mathcal{H}_1:\ell_{pe}\to\ell_{pe}:e\to\psi$ with ϕ as an external (possibly unbounded) input. The complete error model consists then of (4)-(6) (or (7)) *together with the initial conditions* of the LTI filters $\underset{\sim}{\mathcal{H}}_2$, $\underset{\sim}{\mathcal{W}}_1$ and $\underset{\sim}{\mathcal{W}}_2$, and the estimator $\theta(-1)$, $\theta(-2),\ldots,\theta(-d)$, $\rho(0)$ (and $F(-1),\ldots,F(-d)$).

Remark 2. The "nonzero memory" ($\mu \neq 0$) normalization signal ρ was first introduced by Egardt [13], see also [17]. As will be discussed below it is a key ingredient to establish the I/O property of the estimator required for the stability proofs in the reduced order model case [10]. Also, it has proven essential in all global proofs of "robustified" schemes, see e.g. [2]. It is interesting to note that in the global boundedness results of [4-6] normalization, together with other modifications (leakages, dead-zones, etc.), are obviated.

E. Alternative Error Models.

Several variations of the error model described above have been studied in the literature. We will present in this paragraph two of them.

E.1. A Loop Transformation.

In a series of pioneering papers culminated in [11] Gawthrop established some gain properties of the estimator-defined operator

$$\mathcal{H}_a : \vartheta \to e, \quad \vartheta := \psi - e$$

It is clear from the definition above that

$$\mathcal{H}_a = (\mathcal{H}_1 - 1)^I$$

where $(\cdot)^I$ stands for the inverse operation and $\mathcal{H}_1$ as defined above. This transformation induces, through the feedback loop $\mathcal{H}_1$, $\mathcal{H}_2$, a corresponding transformation on $\mathcal{H}_2$. This, so-called loop transformation procedure, is well established in I/O theory [12] and has extensively been used to reduce the conservatism of the I/O stability analysis.

E.2. A Perturbation Error Equation.

The error equation (4) can be written in an alternative form which is more convenient for "small perturbation" analysis [2] as follows. Assume the plant transfer function to be of the form

$$q^{-d}\frac{B}{A} = q^{-d}\frac{B'}{A'}(1+\Delta); \quad B', A' \in \mathbb{R}[q^{-1}], \quad \Delta \in \mathbb{R}(q^{-1})$$

where $\partial(B')$ and $\partial(A')$ are *known* and Δ represents the unstructured uncertainty. In this case we can choose $\partial(S_*)$ and $\partial(R_*)$ so that the polynomial identity

$$C_R B' = A' S_* + q^{-d} B' R_*$$

is satisfied. Using the last identity and the plant equation we can show that the filtered tracking error satisfies

$$e(t) = -\psi(t) + \eta(t)$$

$$\eta(t) := S_* [\Delta u(t) + \frac{A'}{B'} \zeta(t)]$$

The advantage of this formulation with respect to (4) is that η can be treated as a small perturbation of the "ideal" linear regression form. For further details on this approach see [2] and references therein.

III. GLOBAL ℓ_2-STABILITY ANALYSIS

In this section we carry out the global ℓ_2-stability analysis of the error model. Some preliminary results which will be instrumental for our analysis are presented in the Appendix. These results are well known in the input-output stability theory and the reader is referred to [12,14] for further details.

A. Stability Analysis Procedure.

To carry out the stability analysis of the error model we will follow the input-output approach first introduced in adaptive systems by Landau [15] and later used in [10,11,18] to study the robustness problem. That is, we first concentrate in the interconnection of the estimator-defined operator $\mathcal{H}_1 : e \to \psi$ in closed loop with the LTI operator $\mathcal{H}_2 : \psi \to (e_* - e)$. As discussed in [10] it is not possible to establish the required properties of $\mathcal{H}_1$ for possibly unbounded signals. Therefore, we are compelled to introduce a suitably chosen multiplier $\mathcal{M}$ so that the new operator $\bar{\mathcal{H}}_1 := \mathcal{M}\mathcal{H}_1\mathcal{M}^{-1}$ satisfies the conditions of the passivity theorem. It turns out that the required operator is defined by the normalizing factor as

$$\bar{x}(t) := (\mathcal{M}x)(t) = \frac{1}{\rho(t)}\, x(t)$$

The introduction of this multiplier yields the *normalized error equations*

$$\bar{e}(t)=-\bar{\mathcal{H}}_2\bar{\psi}(t)+\bar{e}_*(t),\quad \bar{\mathcal{H}}_2:=\mathcal{M}^{-1}\mathcal{H}_2\mathcal{M}$$

$$\bar{\psi}(t)=\bar{\mathcal{H}}_1\bar{e}(t) \tag{8}$$

Two new problems arise at this point: First, to conclude stability of the original system from stability of the normalized system we must prove that the assumptions involved in fact A.1, i.e. $\mathcal{M}$ and $\mathcal{M}^{-1}$ have *finite gain*, are satisfied. Second, How to *translate the input-output condition on* $\bar{\mathcal{H}}_2$ *into conditions on the LTI operator* $\mathcal{H}_2$?. Answers to these questions will be provided in the subsequent paragraphs.

B. Stability of the Normalized System.

Proposition 1. *(Normalized estimator property).* Consider the operator $\bar{\mathcal{H}}_1:\bar{e}\to\bar{\psi}$ defined by (6) with initial conditions $\tilde{\theta}(-1)$, $\tilde{\theta}(-2),\ldots,\tilde{\theta}(-d)$, $\rho(0)$. Then, for some constant β

$$<\bar{e}|\bar{\psi}>_T\geq-\frac{\gamma}{2}\ \|\bar{e}\|^2_{2,T}+\beta,\quad \forall\bar{e}(t)\in\ell_{2e},\ \forall T\in\mathbb{Z}_+$$

Proof. Let

$$V(t):=|\tilde{\theta}(t)|^2$$

Direct replacement of (6) yields

$$V(t)=V(t-d)+2\gamma\bar{\psi}(t)\bar{e}(t)+\gamma^2|\bar{\phi}(t-d)|^2\bar{e}^2(t)$$

Taking into account that $|\bar{\phi}|^2\leq 1$ and summing up from 0 to T we get

$$\sum_{t=0}^{T}[V(t)-V(t-d)]\leq 2\gamma<\bar{\psi}\ |\ \bar{e}>_T+\gamma^2\|\bar{e}\|_{2,T}^2$$

The proof is completed setting up

$$\beta:=-\frac{1}{2\gamma}[\sum_{t=1}^{d}V(-t)]$$

□□□

Proposition 2. *(Passivity of normalized LTI operator)*. Consider the normalized operator $\bar{\mathcal{H}}_2=\mathcal{M}^{-1}\mathcal{H}_2\mathcal{M}$. Assume that there exists $\delta,\varepsilon>0$ so that

$$\mathbb{R}_e\{\mathcal{H}_2(\mu e^{j\omega})\}\geq\delta|\mathcal{H}_2(\mu e^{j\omega})|^2+\varepsilon,\ \forall\omega\in[0,2\pi] \quad (9)$$

then, $\forall x\in\ell_{2e}$, $\forall T\in\mathbb{Z}_+$ and for some β we have

$$<x|\bar{\mathcal{H}}_2x>_T\geq\delta\|\bar{\mathcal{H}}_2x\|_{2,T}^2+\varepsilon\|x\|_{2,T}^2+\beta$$

Proof. The proof is divided in two steps. First, we will prove that for any stable LTI

$\mathscr{H}_2$ with input ψ and output $r:=e_*-e$ the following relation holds

$$\mu^{-t}r(t)=\mathscr{H}_2[(\mu q)^{-1}]\mu^{-t}\psi(t)$$

with some suitable initial conditions. To this end, notice that

$$r(t)=\mathscr{H}_2(q^{-1})\psi(t)=\left[\sum_{i=0}^{\infty} h(i)q^{-i}\right]\psi(t)$$

whose convolution summation is given by

$$r(t)=\sum_{\tau=0}^{t} h(t-\tau)\psi(\tau)+\beta(t)$$

$$=\sum_{\tau=0}^{t} h(t-\tau)\mu^{\tau}\mu^{-\tau}\psi(\tau)+\beta(t)=\mu^{t}\mu^{-t}[r(t)]$$

where β is the zero input response of the system. Thus,

$$\mu^{-t}r(t)=\sum_{\tau=0}^{t} h(t-\tau)\mu^{-(t-\tau)}\mu^{-\tau}\psi(\tau)+\mu^{-t}\beta(t)$$

The proof is completed from

$$\mathscr{H}_2[(\mu q)^{-1}]=\sum_{i=0}^{\infty} h(i)\mu^{-i}q^{-i}$$

Second, we will show that the input-output properties of the LTI system $\mathscr{H}_2[(\mu q)^{-1}]$ are inherited by the normalized system $\bar{\mathscr{H}}_2$. To do so, first notice that theorem A.2 insures that, for some negative β, the sequence

$$Z(t) \triangleq r(t)\,\psi(t) + \varepsilon|\psi(t)|^2 + \delta|r(t)|^2 + \beta$$

satisfies the series inequality

$$\sum_{t=0}^{T} \mu^{-2t} Z(t) \le 0 \qquad (10)$$

Now, let us define

$$\bar{Z}(t) := \rho^{-2}(t) Z(t)$$

It is clear that the desired result follows if we can establish

$$\sum_{t=0}^{T} \bar{Z}(t) \le 0$$

Notice that the term due to the initial conditions can be easily handled since $\beta<0$ and ρ is bounded from below by $\underline{\rho}$. On the other hand, one has

$$\sum_{t=0}^{T} \bar{Z}(t) = \sum_{t=0}^{T} \mu^{2t} \rho(t)^{-2}\; \mu^{-2t} Z(t) := \sum_{t=0}^{T} a^2(t) b(t)$$

where we have defined

$$a(t)\triangleq\mu^{t}\rho(t)^{-1},\quad b(t)\triangleq\mu^{-2t}Z(t)$$

The sum can be developed as

$$\sum_{t=0}^{T}\bar{Z}(t)=a^{2}(T+1)\sum_{t=0}^{T}b(t)+$$

$$+\sum_{t=0}^{T}\left\{\left[\sum_{j=0}^{t}b(j)\right]\left[a^{2}(t)-a^{2}(t+1)\right]\right\}$$

Now, notice from (6) that since

$$\mu^{-2t}\rho^{2}(t)=\mu^{-2(t-1)}\rho^{2}(t-1)+$$

$$+\mu^{-2t}\max(\left|\phi(t-d)\right|^{2},(1-\mu^{2})\underline{\rho}^{2}) \qquad (11)$$

the sequence $\mu^{2t}\rho(t)^{-2}$ is nonincreasing, i.e. $a^{2}(t)-a^{2}(t+1)\geq 0$. Therefore, from (10)

$$\sum_{t=0}^{T}\bar{Z}(t)\leq a^{2}(T+1)\sum_{t=0}^{T}b(t)\leq 0$$

which completes the proof.

□□□

We are in position to present the main result of this paragraph whose proof follows inmediately from the results above.

Proposition 3. *(Stability of the normalized system).* Consider the normalized error model. Assume there exists $\delta > \gamma/2$ and $\varepsilon > 0$ so that (9) holds. Under these conditions, the feedback system is ℓ_2-stable, i.e., there exists constants $c > 0$ and β such that

$$\|\bar{e}(t)\|_2, \|\bar{\psi}(t)\|_2 \leq c\|e_*(t)\|_2 + \beta$$

□□□

C. Multiplier Boundedness.

In this paragraph we will show that, under the conditions of proposition 3, the multiplier $\mathcal{M}$ and its inverse are bounded operators. This will allows us to conclude in the next paragraph stability of the original system from stability of the normalized one as discussed in paragraph 3.A.

Proposition 4. *(Multiplier boundedness).* Under the conditions of proposition 3, the multiplier and its inverse have finite ℓ_2-gain for all $e_* \in \ell_2$, that is,

$$\gamma_2\{\mathcal{M}\}, \gamma_2\{\mathcal{M}^{-1}\} < \infty.$$

Proof. The following procedure is used to

establish the proof. First, we use the order relationships between ϕ, ψ and ρ and the fact that $e_{*}\in\ell_2$ implies $\bar{\psi}\to 0$ to obtain a bound on ψ suitable for the application of the Bellman-Gronwall lemma ([12], pg. 254). Using the latter we then establish boundedness of ψ and consequently of all remaining signals.

Taking the sum on (11) from 1 to T and recalling that $\rho(0)=\underline{\rho}$ we get

$$\sum_{t=0}^{T} \mu^{-2t}|\phi(t-d)|^2+\sum_{t=0}^{T} \mu^{-2t}(1-\mu^2)\underline{\rho}^2\geq$$

$$\geq\mu^{-2T}\rho^2(T)-\underline{\rho}^2$$

On the other hand, using the series inequality

$$\sum_{t=0}^{T} \mu^{-2t}=\frac{1-\mu^{-2T}}{1-\mu^{-2}} \leq \frac{\mu^{-2T}}{1-\mu^2}$$

we get

$$\mu^{-2T}\rho^2(T)\leq\|\mu^{-t}\phi(t-d)\|^2_{2,T}+c\mu^{-2T} \qquad (12)$$

where c stands here, and throughout the rest of the proof, for some constant. Replacing (12) in $\bar{\psi}$ gives

$$\bar{\psi}^2(T)=\frac{\mu^{-2T}\psi^2(T)}{\mu^{-2T}\rho^2(T)} \geq \frac{\mu^{-2T}\psi^2(T)}{\|\mu^{-t}\phi(t-d)\|^2_{2,T}+c\mu^{-2T}} \quad (13)$$

Now, one has that $\forall\varepsilon>0$, $\exists T>0$ such that $\bar{\psi}(T)^2\leq\varepsilon$. Using this upperbound in (13), we get

$$\mu^{-2T}\psi^2(T)\leq\varepsilon[\|\mu^{-t}\phi(t-d)\|^2_{2,T}+c\mu^{-2T}] \quad (14)$$

Now, from (5) we can, similarly to the proof of proposition 2, establish

$$\mu^{-t}\phi(t-d)=\mathscr{W}_1[(\mu q)^{-1}]\mu^{-t}\psi(t)+\mu^{-t}\phi_*(t-d)$$

Notice that according to the conditions of the proposition $\mathscr{W}_1[(\mu q)^{-1}]$ is stable and ϕ_* is bounded. Therefore,

$$\|\mu^{-t}\phi(t-d)\|^2_{2,T}\leq\alpha_W\|\mu^{-t}\psi(t)\|^2_{2,T}+c\mu^{-2T}$$

$$\alpha_W:=\gamma_2\{\mathscr{W}_1[(\mu q)^{-1}]\}$$

Replacing this bound in (14) yields

$$\mu^{-2T}\psi(T)^2\leq\varepsilon\alpha_W\|\mu^{-t}\psi(t)\|^2_{2,T}+c\mu^{-2T}$$

Let us choose ε such that

$$1-\varepsilon\alpha_W>\mu^2 \tag{15}$$

and factor the last element of the truncated norm, which yields

$$\mu^{-2T}\psi(T)^2\le \frac{\varepsilon\alpha_W}{1-\varepsilon\alpha_W}\ \|\mu^{-t}\psi(t)\|^2_{2,T-1}+c\mu^{-2T}$$

Now, applying the Bellman-Gronwall lemma gives

$$\mu^{-2T}\psi(T)^2\le c\mu^{-2T}+\sum_{t=0}^{T-1}\left(\frac{1}{1-\varepsilon\alpha_W}\right)^{T-t-1}c\mu^{-2t}$$

which can also be written as

$$\psi(T)^2\le c+c\sum_{t=0}^{T-1}\left(\frac{\mu^2}{1-\varepsilon\alpha_W}\right)^{T-t}=c+c\sum_{t=1}^{T}\left(\frac{\mu^2}{1-\varepsilon\alpha_W}\right)^{t}$$

Since the term inside the brackets is smaller than 1 the series is convergent and then $\psi\in\ell_\infty$. From stability of $\mathcal{W}_1$ this implies that ϕ is bounded and so is ρ. The proof is completed from the definition of $\mathcal{M}$ and the fact that ρ is, by construction, bounded away from zero.

□□□

D. Global Convergence of the Adaptive System. The main result of this section, conditions for global convergence of the adaptive system, is contained in the theorem below. The proof follows directly from the previous developments.

Theorem 1. Consider the *known delay* plant (1) in closed loop with the adaptive controller (3) whose parameters are updated with (6). Assume that, for the given n_S, n_R and μ, there exists a parametrization of the controller such that:

A.1 *The closed loop transfer function* $\mathcal{H}_2$ *satisfies* (9) *for some* $\delta>\gamma/2$.

A.2 *Robust servo-behavior is attainable* (i.e., $e_* \to 0$ as $t\to\infty$) *for the given bounded reference and disturbances* ($w,\zeta\in\ell_\infty$).

Under these conditions, global convergence is insured, that is $e\to 0$ as $t\to\infty$ and $\phi,\hat{\theta}\in\ell_\infty^n$.

□□□

E. Discussion.

1. The conditions of the theorem are always satisfied when there are no unmodelled dynamics nor disturbances, since in that case $\mathcal{H}_2 = 1$ and $e_* \equiv 0$. The sectoricity condition defines the ability to control the actual plant as compared to its nominal objective. Under slow adaptation the condition becomes a strict positive realness requirement.

2. As explained above the normalization "memory" μ must be chosen different from zero to establish the result. The importance of this choice is particularly clear if we view the normalizing factor as a signal bounding (pointwise in time) the unmodelled response of the perturbation error equation of paragraph 2.E.2. See also [2]. In the present context it establishes and *alertness robustness tradeoff* since large μ increases the frequency range where the sectoricity condition is satisfied, but it slows down the adaptation by imposing a small stepsize in the gradient search (via large ρ).

3. An interesting open question is to relate the class of systems defined above with the ones *considered in perturbation model based schemes* (with modified estimators). To this end, we need to establish the existence of a sufficiently small scaling factor for the unmodelled dynamics such that $\mathcal{H}_2$ is "close" to $1/C_R$ for all ω. Recall that, under the assumption of known delay, $\mathcal{H}_2$ is relative degree zero. It is naturally expected that it will hold in low frequencies. Thus, it inherits the local stability properties established in [7]. Furthermore, we believe our analysis reveals some structural properties, otherwise hidden in a scaling factor, and therefore will tend to be useful for practical implementations (see Ch. 6 of [7] and [11]).

IV. STABLITY UNDER PERSISTENCY OF EXCITATION

In this Section we present some ℓ_∞-stability properties of the adaptive controller described above under the assumption that the signals satisfy a persistency of excitation (PE) condition.

A. An Exponentially Weighted Error Model.

A classical procedure in I/O theory to translate ℓ_2-stability results into the more practically interesting case of ℓ_∞-stability is through the use of *exponential weightings* (see [12], Ch. V, and [14], Ch. 9). Roughly speaking, the rationale behind this technique is that ℓ_∞-stability of the original system follows from ℓ_2-stability of its exponentially weighted counterpart. One interesting consequence of this approach is that for LTI operators one obtains practically appealing *frequency domain* criteria for ℓ_∞-stability. This in contrast with the time domain calculations of state-space (total stability) approaches.

As proposed in [10], we carry out this procedure here for the adaptive controller (3) with parameter update (7). To this end, we need to define an *exponentially weighted normalized error equation* as follows

$$\bar{e}^{\alpha}(t) = -\bar{\mathscr{H}}_2^{\alpha}\bar{\psi}^{\alpha}(t) + \bar{e}_*^{\alpha}(t), \quad \bar{\mathscr{H}}_2^{\alpha} := \alpha^{t}\bar{\mathscr{H}}_2\alpha^{-t}$$

$$\bar{\psi}^{\alpha}(t)=\bar{\mathscr{H}}_1^{\alpha}\bar{e}^{\alpha}(t) \quad \bar{\mathscr{H}}_1^{\alpha}:=\alpha^{t}\bar{\mathscr{H}}_1\alpha^{-t} \tag{16}$$

where we use the superscript α to denote

$$x^{\alpha}(t):=\alpha^{t}x(t),\ \alpha>0$$

B. A Signal-dependent I/O Property.

Proposition 5. *(Exponentially weighted normalized estimator property)*. Consider the operator $\bar{\mathscr{H}}_{\tilde{1}}^{\alpha}:\bar{e}^{\alpha}\to\bar{\psi}^{\alpha}$ defined by (7) with initial conditions $\theta(-1),\ldots,\tilde{\theta}(-d)$, $F(-1),\ldots,F(-d)$, $\rho(0)$. Then, for some constant β, $\forall\bar{e}(t)\in\ell_{2e}$ and $\forall T\in\mathbb{Z}_+$ we have

$$<\bar{e}^{\alpha}|\bar{\psi}^{\alpha}>_T\geq-\frac{\sigma}{2}\ \|\bar{e}^{\alpha}\|_{2,T}^{2}-\ \frac{1}{2}\ \|\bar{\psi}^{\alpha}\|_{2,T}^{2}+\beta$$

for α verifying

$$\lambda_{max}\{F^{-1}(t)F'(t)\}\alpha^{2d}\leq 1$$

and all σ satisfying

$$\sigma\geq\frac{\lambda_1}{\lambda+\lambda_1} \tag{16}$$

where $\lambda_{max}\{\cdot\}$ is the maximum eigenvalue.

Proof. See Lemma 3.1 of [10].

□□□

C. Sufficiency of Persistence of Excitation.

From the definitions of F' and the design parameters λ_0, λ_1 in (7) it is clear that $F \geq F'$. Thus, the estimator property defined in the proposition holds for all $\alpha \geq 1$. To be able to infer signal boundedness from ℓ_2 properties of the exponentially weighted signals *we need however the stronger case of* $\alpha > 1$. The proposition below shows that to establish the property for this case it is *sufficient to impose a PE condition.*

Proposition 6. *(PE and Estimator Properties).* If there exists $\varepsilon > 0$ and $T_0 \geq 0$ such that

$$\sum_{t=0}^{T} \Lambda^{T-t} \bar{\phi}(t) \bar{\phi}^T(t) \geq \varepsilon I, \quad \forall T \geq T_0$$

$$\Lambda := \frac{\lambda(\lambda_1 - \lambda_0)}{\lambda(\lambda_1 - \lambda_0) + \lambda_1(\lambda + \lambda_0)}$$

and d=1, then $\bar{\mathcal{H}}_1^{\alpha}$ satisfies the I/O property of proposition 5 for all

$$\alpha \geq \left[\frac{1}{1-\varepsilon\varkappa}\right]^{1/2} > 1$$

$$\varkappa := \frac{\lambda_0 \lambda_1}{\lambda(\lambda_1 - \lambda_0) + \lambda_1(\lambda + \lambda_0)} \in (0,1)$$

Proof. Follows inmediately from the proof in the Appendix of [10] that the PE conditon above insures

$$\lambda_{max}\{F^{-1}(t)F'(t)\} \leq 1 - \varepsilon\varkappa$$

□□□

D. Boundedness and a Performamce Measure.

The proposition above gives conditions for the estimator-defined operator to satisfy the desired I/O properties. Analogously to the previous Section, if the operator $\bar{\mathscr{H}}_2^{\alpha}$ satisfies the required passivity properties, we can establish ℓ_2-stability of the exponentially weighted normalized error model It is clear from the proof of proposition 2 that the latter is insured if the sectoricity condition (9) holds for $\mathscr{H}_2[(\mu q/\alpha)^{-1}]$.

Now, to conclude stability of the original system from stability of the normalized one we need to prove, as in proposition 4, that the multipliers have finite gain. A problem arises at this point since in the present ℓ_∞ context we cannot claim that $\bar{\psi}^{\alpha}$ converges to zero, since $\bar{e}_*^{\alpha} \notin \ell_2$. Therefore, the bound on $|\bar{\psi}^{\alpha}|$, that we get from stability of the feedback interconnection, will depend on the initial conditions of the estimator $\tilde{\theta}$ and the ℓ_∞ norm of e_*. On the other hand, to complete the proof we require some smallness condition on the bound of $|\psi|$, see (15). Fortunately, as will be shown below, we can insure the latter condition *for all initial conditions and tuned tracking errors with a sufficiently slow adaptation gain* as stated in theorem 5.3 of [10]. We summarize this discussion in the theorem below.

Theorem 2. Consider the *known delay* plant (1) in closed loop with the adaptive controller (3) whose parameters are updated with (7). Assume that, for the given n_S, n_R λ, λ_0, λ_1, $\hat{\theta}(-1),\ldots\hat{\theta}(-d)$ and μ:

A.1. *There exists a parametrization of the controller such that the closed loop transfer function* $\mathcal{H}_2$ *satisfies (9) for some* $\varepsilon>1/2$ *and some* δ *verifying*

$$2\delta>\frac{\lambda_1}{\lambda+\lambda_1}$$

A.2. *The signals in the system are such that*

$$\lambda_{max}(F^{-1}(t)F'(t))\leq\mu^d$$

A.3. *The normalization factor is bounded from below by a sufficiently large constant, i.e.* $\varrho\geq\varrho_*>0$, *with* ϱ_* *sufficiently large.*

Under these conditions, all signals in the system are bounded and there exist constants c_1 and c_2 such that the following *performance measure* holds (uniformly in t)

$$|e(t)|\leq c_1\|e_*\|_\infty+c_2$$

Proof. From the discussion above we see that **A.1** and **A.2** insure ℓ_2-stability of the exponentially weighted normalized error model with $\alpha=1/\mu$. That is, the existence of c>0 and β (dependent on the initial parameter error) such that for all $T \in \mathbb{Z}_+$

$$\|\bar{\psi}^{\alpha}\|^2_{2,T} \le c\|\bar{e}^{\alpha}_{*}\|^2_{2,T} + \beta \tag{17}$$

On the other hand, we have the bounds

$$\|\bar{e}^{\alpha}_{*}\|^2_{2,T} \le \frac{\alpha^{2T}}{(1-\alpha^{-2})\underline{\rho}}\|e_{*}\|^2_{\infty}$$

$$\|\bar{\psi}^{\alpha}\|^2_{2,T} \ge [\alpha^T\bar{\psi}(T)]^2$$

which, together with the definition of the normalization, provide us with the uniform upperbound

$$|\bar{\psi}^2(T)| \le \frac{c}{(1-\alpha^{-2})\underline{\rho}}\|e_{*}\|^2_{\infty} + \alpha^{-2T}\beta$$

To proceed with the proof as done in proposition 4 we need this bound to satisfy (15). Notice that (for all β and $\|e_{*}\|_{\infty}$) this is insured by assumption 3 if we take T sufficiently large. Application of the Bellman-Gronwall lemma proves that ψ is bounded and via stability of $\mathcal{H}_2$ and $\mathcal{W}_1$ we can complete the proof.

□□□

E. Discussion.

1. The result presented above is essentially contained in [10] and the idea to provide the performance bounds is borrowed from [1]. However, in [10] we disregarded the effect of the initial conditions of the estimator in the passivity theorem (eq. (4.4) in lemma 4.2 should be replaced by (17) above). As shown above this does not invalidate the claim of theorem 5.3 in [10] but only requires an additional step in the proof.

2. As mentioned above the condition $\alpha>1$ is indispensable for our analysis. In the case of slow adaptation (λ sufficiently large and in a neigbourhood of the stabilizing parametrization set) and when the reference is sufficiently rich, the regresssor will be PE and we can invoke Proposition 6 to insure the strict inequality. However, the condition seems difficult to verify in the global case.

3. As discussed in [1] we expect that the *transient response bounds obtained in our global analysis will tend to be very bad.* We

refer the reader to [1] for tighter but local performance bounds and an interesting interpretation of the condition $\alpha>1$ as a state reachability condition.

V. CONCLUDING REMARKS

We have presented a detailed account of the global convergence theory for discrete adaptive systems of [10]. It is our belief that this theory nicely complements with the averaging analysis of [7] to provide a performance oriented approach to adaptive control. A major strength of this approach is that it is not just analytical but it can lead to ideas for the synthesis of new adaptation algorithms and performance enhancing modifications.

The global theory presented here clearly displays the importance of key issues in adaptive control like the level of signal PE and speed of adaptation. But it also preserves, in the stability conditions, the essence of the "classical" performance measures of allowable signal phase shift and disturbance rejection capabilities.

APPENDIX

Input-Output Stability Theory Preliminaries

We will denote the space of square integrable vector sequences as

$$\ell_2^n \stackrel{\Delta}{=} \{ x(\cdot): \mathbb{Z}_+ \to \mathbb{R}^n \mid \|x\|_2^2 \stackrel{\Delta}{=} \sum_{t=0}^{\infty} |x(t)|^2 < \infty \}$$

and its extension as

$$\ell_{2e}^n \stackrel{\Delta}{=} \{ x(\cdot): \mathbb{Z}_+ \to \mathbb{R}^n \mid \|x\|_{2,T}^2 \stackrel{\Delta}{=} \sum_{t=0}^{T} |x(t)|^2 < \infty, \ \forall T \in Z_+ \}$$

They are equipped with the inner product

$$\langle x | y \rangle_T \stackrel{\Delta}{=} \sum_{t=0}^{T} x(t)^T y(t)$$

We treat here finite state dynamical systems as p-input q-output mappings from ℓ_{2e}^p to ℓ_{2e}^q. They are causal and dependent on the initial state. The effect of the latter appears as a bias term (denoted in the paper β) in the various input-output properties. See, e.g. [16], for a detailed discussion of this issue. ℓ_2 gain of an operator $\mathcal{A}:\ell_{2e}^{p}\to\ell_{2e}^{q}$ is denoted by $\gamma_2\{\mathcal{A}\}$. We will use $X^*(\cdot)$ to denote the complex conjugate transpose of an LTI system frequency response.

Theorem A.1. *(Passivity, [13]).* Consider the following feedback interconnection (Fig. A.1)

$$e_1(t)=z_1(t)-\mathcal{A}e_2(t)$$

$$e_2(t)=z_2(t)+\mathcal{H}e_1(t)$$

where $\mathcal{A},\mathcal{H}:\ell_{2e}\to\ell_{2e}$. Assume that for any z_1 and z_2 in ℓ_2, there are solutions e_1 and e_2 in ℓ_{2e}. Suppose that there are constants δ_1, δ_2, ε_1, ε_2 and β (not the same throughout) such that $\forall x\in\ell_{2e}$, $\forall T\in\mathbb{Z}_+$

$$<x|\mathcal{A}x>_T\geq\varepsilon_1\|x\|_T^2+\delta_1\|\mathcal{A}x\|_T^2+\beta$$

$$<x|\mathcal{H}x>_T\geq\varepsilon_2\|x\|_T^2+\delta_2\|\mathcal{H}x\|_T^2+\beta$$

Under these conditions, if

$$\delta_1+\varepsilon_2>0 \text{ and } \delta_2+\varepsilon_1>0$$

then the closed loop operator

$$\Sigma : \begin{bmatrix} z_1 \\ z_2 \end{bmatrix} \to \begin{bmatrix} e_1 \\ e_2 \end{bmatrix}$$

is ℓ_2 stable, that is there exist constants $\xi>0$ and β such that

$$\left\| \begin{matrix} e_1 \\ e_2 \end{matrix} \right\|_2 \leq \xi \left\| \begin{matrix} z_1 \\ z_2 \end{matrix} \right\|_2 + \beta$$

□□□

Fact A.1. *(Multiplier theory, [12]).* Consider the systems of Fig. A.1, A.2, with the latter defining a map

$$\bar{\Sigma} = \begin{bmatrix} \bar{z}_1 \\ \bar{z}_2 \end{bmatrix} \to \begin{bmatrix} \bar{e}_1 \\ \bar{e}_2 \end{bmatrix}, \quad (\bar{\cdot}) \stackrel{\Delta}{=} \mathcal{M}(\cdot)$$

where $\mathcal{M}:\ell_{2e} \to \ell_{2e}$. Assume $\mathcal{M}$ is linear and $\mathcal{M}$ and its inverse have finite ℓ_2 gain. Under these conditions

$$\bar{\Sigma} \text{ is } \ell_2\text{-stable} \iff \Sigma \text{ is } \ell_2\text{-stable}$$

□□□

Theorem A.2. *(LTI operator passivity, [22]).* Consider the causal discrete-time LTI stable

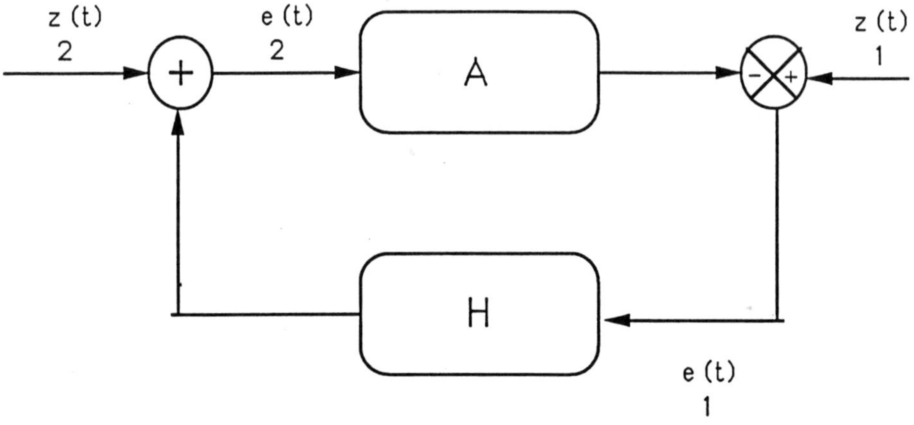

Fig. A.1

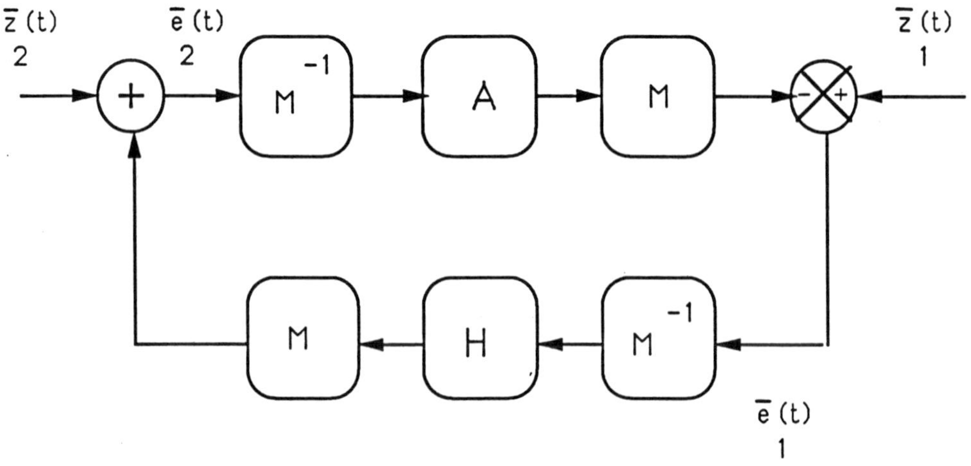

Fig A.2

system

$$y(t)=\mathcal{H}(q^{-1})u(t), \quad \mathcal{H}\in\mathbb{R}^{m\times m}(q^{-1})$$

Then, there exists $\beta<0$ (dependent on the initial conditions) such that

$$<y\,|\,u>_T \geq \delta\,\|y\|^2_{2,T} + \varepsilon\,\|u\|^2_{2,T} + \beta, \quad \forall u\in\ell_{2e}, \quad \forall T\in\mathbb{Z}_+$$

for some real numbers δ and ε, iff $\forall\omega\in[0,2\pi]$

$$\frac{1}{2}\{\mathcal{H}^*(e^{j\omega})+\mathcal{H}(e^{j\omega})\}\geq\delta\mathcal{H}^*(e^{j\omega})\mathcal{H}(e^{j\omega})+\varepsilon I$$

□□□

REFERENCES

[1] Zang, Z. and R. Bitmead, "Transient bounds for adaptive control systems", *Proc. IEEE CDC*, Dec. 5-7, 1990. Hawaii, USA.

[2] Ortega, R. and Y. Tang, "Robustness of adaptive controllers: A survey", *Automatica*, Vol. 25, No. 5, September 1989.

[3] Praly, L., "The almost exact modelling assumption in adaptive linear control", *Int. J. Control*, 1989.

[4] Ydstie, B., "Stability of discrete model reference adaptive control – revisited", *Syst. & Cont. Letters*, Vol. 13, No.5, 1989.

[5] Wen, C. and D. Hill, "Robustness of adaptive control without deadzones, data normalization or persistency of excitation", *Automatica*, Vol. 25, No. 6, Nov. 1989.

[6] Naik, S., P. Kumar and B. Ydstie, "Robust continuous time adaptive control by parameter projection", *Proc. 1990 Grainger Lect. Adaptive Control*, Univ. of Illinois, Sept 28 – Oct. 1, 1990.

[7] Anderson, B. D. O., et al, "*Stability of Adaptive Systems: Passivity and Averaging Analysis*", MIT Press, 1986.

[8] Ortega, R., "Applications of Input-Output techniques to control problems", *Proc. 1st Europ. Cont. Conf.*, Grenoble, France, June 2-5, 1991.

[9] Narendra, K. and A. Annaswamy, *Stable Adaptive Systems*, Prentice-Hall, NJ, 1989.

[10] Ortega, R., L. Praly and I. Landau, "Robustness of discrete-time adaptive controllers", *IEEE Trans. Aut. Cont.*, Vol. AC-30, No. 12, December 1985, pg. 1179-1187.

[11] Gawthrop, P., *"Continuous-Time Self-Tuning Control"*, Research Studies Press, John Wiley & Sons, 1987.

[12] Desoer, C. and M. Vidyasagar, *"Feedback systems: Input-Output properties"*, New-York, Academic Press, 1975.

[13] Egardt, B., *"Stability of Adaptive Systems"*, Springer-Verlag, Berlin, 1979.

[14] Vidyasagar, M., *"Input-Output Analysis of Large Scale Interconnected Systems"*, Springer-Verlag, Berlin, 1981.

[15] Landau, I., *"Adaptive control: The model reference approach"*, Marcel-Dekker, New-York, 1979.

[16] Hill, D. and P. Moylan, "Dissipative dynamical systems: Basic input-output and

state properties", *J. Franklin Inst.*, Vol. 309, No.5, 1980.

[17] Praly, L., "Robustness of indirect adaptive control based on pole placement", *Proc. IFAC Work. on Adapt. Syst. and Signal Proc.*, San Francisco, Cal., USA, June 1983.

[18] Kosut, R. and B. Friedlander, "Robust Adaptive Control: Conditions for Global Stability", *IEEE Trans. Autom. Contr.*, AC-30, 7, pp. 610-624, 1985.

INDEX

M

N

P

R

S

T

W

ISBN 0-12-012755-5

90051

9 780120 127559